Lecture Notes in Computer Science 15495

Founding Editors

Gerhard Goos
Juris Hartmanis

AF148471

The series Lecture Notes in Computer Science (LNCS), including its subseries Lecture Notes in Artificial Intelligence (LNAI) and Lecture Notes in Bioinformatics (LNBI), has established itself as a medium for the publication of new developments in computer science and information technology research, teaching, and education.

LNCS enjoys close cooperation with the computer science R & D community, the series counts many renowned academics among its volume editors and paper authors, and collaborates with prestigious societies. Its mission is to serve this international community by providing an invaluable service, mainly focused on the publication of conference and workshop proceedings and postproceedings. LNCS commenced publication in 1973.

Sourav Mukhopadhyay · Pantelimon Stănică
Editors

Progress in Cryptology – INDOCRYPT 2024

25th International Conference on Cryptology in India
Chennai, India, December 18–21, 2024
Proceedings, Part I

 Springer

Editors
Sourav Mukhopadhyay
Indian Institute of Technology
Kharagpur, India

Pantelimon Stănică
Naval Postgraduate School
Monterey, CA, USA

ISSN 0302-9743   ISSN 1611-3349 (electronic)
Lecture Notes in Computer Science
ISBN 978-3-031-80307-9   ISBN 978-3-031-80308-6 (eBook)
https://doi.org/10.1007/978-3-031-80308-6

This Springer imprint is published by the registered company Springer Nature Switzerland AG
The registered company address is: Gewerbestrasse 11, 6330 Cham, Switzerland

If disposing of this product, please recycle the paper.

A Special Note on Silver Jubilee Edition
by Bimal Kumar Roy

It's a matter of great joy and satisfaction to learn that INDOCRYPT is celebrating its silver jubilee edition. Since its beginning in 2000 in Kolkata, the conference has put the Indian Cryptology community in the international arena. INDOCRYPT is well recognized by IACR and all researchers worldwide.

All the success came as the output of great efforts by Indian researchers and premier institutions who have supported the conference all along. Also, the Cryptology Research Society of India (CRSI) played a significant role all along.

This edition was organized jointly by three premier research institutions in Chennai: SETS, IMSc and CMI. My special thanks to the Directors of these institutions, viz. N. Subramanian, V. Ravindran and Madhavan Mukund, who are also General Co-Chairs for the event. My special thanks to the Program Co-Chairs, Pante Stanica and Sourav Mukhopadhyay, who are my long-time friends too.

I celebrate the grand success of the Silver Jubilee edition and hope to see the Golden Jubilee in 2049 even though I may not be physically present then.

December 2024   Bimal Kumar Roy

Foreword

The Silver Jubilee edition of the Indocrypt series was jointly organised by the Society for Electronic Transactions and Security [SETS], The Institute of Mathematical Sciences [IMSc] and Chennai Mathematical Institute [CMI]. The conference was held in Chennai between Dec. 18 and 21, 2024. We, the General Chairs, are extremely happy about this. The Cryptology Research Society of India [CRSI] under the guidance of Bimal Roy has played a key role in promoting R&D in the subject. We are grateful to the guidance given by Roy and R. Balasubramanian. The conference page is available in the link: https://setsindia.in/indocrypt2024/indocrypt.

The Programme Co-Chairs decided to invite Vincent Rijmen, Lucca Hirschi and Luca De Feo to deliver Invited Talks. They also invited Pantelimon Stanica and Subhamoy Maitra to deliver the tutorials. They also finalized the list of contributed papers. The papers cover various sub-areas in cryptology. We are confident that the programme provided insights into niche areas of cryptography to aspiring researchers.

The Society for Electronic Transactions and Security [SETS] is an R&D lab focussing on cryptology, including hardware, quantum and network security areas. This lab is funded by the office of the Principal Scientific Adviser to the Government of India. The conference venue was the Institute of Mathematical Sciences [IMSc], which is an R&D institution funded by the Department of Atomic Energy [DAE], Government of India, and which is famous for its doctoral programmes. Chennai Mathematical Institute [CMI] is a premier R&D institute located in Chennai which is famous for its undergraduate, postgraduate and doctoral programmes.

We are thankful to all the people who were involved in organising this event. We enjoyed hosting the delegates at Chennai for this conference. We are confident that the deliberations were beneficial to the community and the society.

December 2024

N. Subramanian
V. Ravindran
Madhavan Mukund

Preface

It is our pleasure and honour to invite you to peruse the proceedings of the Silver Jubilee Anniversary of the International Conference on Cryptology in India, Indocrypt 2024. The conference was held from 18–21 December 2024 in Chennai, India. It was held under the aegis of the Cryptology Research Society of India (CRSI), jointly organized by the Society for Electronic Transactions and Security (SETS), the Institute of Mathematical Sciences (IMSc) and Chennai Mathematical Institute (CMI).

We were fortunate to receive the help of 69 brilliant researchers who graciously agreed to be part of the Technical Program Committee (TPC). Our Call for Papers generated 96 full paper submissions. Based upon a rigorous process and discussions among the TPC members, we selected 31 papers, which were finally accepted based on double-blind (at least 3) reviews and extensive discussions among the TPC members. The acceptance rate was 30.09%, lower than the previous two editions of this conference (namely, 31.39%, respectively, 35.23%). The accepted papers were organized under the following themes: Blockchain and Cloud Computing, Cryptanalysis, Cryptographic Constructions, Foundations, Post-Quantum Cryptography, Quantum Cryptography, Symmetric-Key Cryptography.

As part of the technical program, we also solicited tutorial speakers on symmetric-key systems (Pantelimon Stanica of Naval Postgraduate School, Monterey CA, USA), and Quantum Cryptography (Subhamoy Maitra, ISI-Kolkata India) panel discussions and inspiring keynote speeches from Luca de Feo (IBM Research, Switzerland), Lucca Hirschi (Inria - France) and Vincent Rijmen (KU Leuven - Belgium, co-inventor, with Joan Daemen, of the Advanced Encryption Standard). We are very thankful to several people and organizations who played a huge supporting role behind the successful organization of this conference. The following list is our humble attempt to acknowledge their service and support. First and foremost, we would like to thank our sponsors, SETS, IMSc, CMI, Odyssey, Thales and Exclusive Networks, listed in no order of priority. We are very thankful for the service of the entire organization committee for their hard work through the last few months of event organization. Last but not least, we would like to thank our General Chairs, N. Subramanian (SETS, India), V. Ravindran (IMSc, India) and Madhavan Mukund (CMI, India), the Organizing Committee, M. Prem Laxman Das (Senior Scientist, SETS) and Anirban Mukhopadhyay (IMSc Chennai), the Advisory Committee, R. Balasubramanian (President, CRSI & Former Director, the Institute of Mathematical Sciences, Chennai), Bimal Roy, (General Secretary, CRSI & Former Director, Indian Statistical Institute, Kolkata) and Rajeeva L. Karandikar (Former Director, Chennai Mathematical Institute), for guidance and motivation, as well as our students and scholars who helped in many ways to make the conference a success.

We lastly thank the many authors who submitted and presented their research at the conference and the many scholars who attended these talks in person or remotely.

December 2024

Sourav Mukhopadhyay
Pantelimon Stănică

Organization

General Chairs

N. Subramanian   Society for Electronic Transactions and Security, India

V. Ravindran   Institute of Mathematical Sciences, India

Madhavan Mukund   Chennai Mathematical Institute, India

Program Committee Chairs

Sourav Mukhopadhyay   Indian Institute of Technology Kharagpur, India

Pantelimon Stanica   Naval Postgraduate School, USA

Advisory Committee

R. Balasubramanian   Institute of Mathematical Sciences, India

Bimal Kumar Roy   Indian Statistical Institute, India

Rajeeva L. Karandikar   Chennai Mathematical Institute, India

Organizing Chairs

M. Prem Laxman Das   Society for Electronic Transactions and Security, India

Anirban Mukhopadhyay   Institute of Mathematical Sciences, India

Organizing Co-chairs

Natarajan V.   Society for Electronic Transactions and Security, India

Suganya A.   Society for Electronic Transactions and Security, India

Publicity Chair

Nageswar Rao   Society for Electronic Transactions and Security, India

Industry Chairs

Thiruppathi K.   Society for Electronic Transactions and Security, India

Foram P. Shingala   Society for Electronic Transactions and Security, India

Web Co-chairs

Santoshkumar T.   Society for Electronic Transactions and Security, India

Samyuktha M.   Society for Electronic Transactions and Security, India

M. Swathi Mithran   Society for Electronic Transactions and Security, India

Publication Co-chairs

Jayashree Dey   IIT Madras, India
Manas Jana   IIT Kharagpur, India
Pratima Jana   IIT Kharagpur, India
Reshmi T. R.   Society for Electronic Transactions and Security, India

Shashikant Pandey   Society for Electronic Transactions and Security, India

Easwaridevi N.   Society for Electronic Transactions and Security, India

Program Committee

Abhay Kumar Singh   IIT Dhanbad, India
Alfred Menezes   University of Waterloo, Canada
Ana Salagean   Loughborough University, UK

Andre Esser	Technology Innovation Institute, UAE
Anubhab Baksi	Nanyang Technological University, Singapore
Anupam Chattopadhyay	Nanyang Technological University, Singapore
Arka Rai Choudhuri	NTT Research, USA
Arpita Maitra	IAI, TCG CREST, Kolkata, India
Avijit Dutta	IAI, TCG CREST, Kolkata, India
Avishek Adhakari	Presidency University, India
Bhupendra Singh	CAIR, DRDO, India
Captain Ritesh Wahi	Indian Navy, India
Chaoyun Li	University of Surrey, UK
Chester Rebeiro	IIT Madras, India
Christina Boura	University of Versailles, France
Colonel Sai Shankar P.	Indian Army, India
Debdeep Mukhopadhyay	IIT Kharagpur, India
Deepak Kumar Dalai	NISER, Bhubaneswar, India
Dipanwita Roy Chowdhury	IIT Kharagpur, India
Dirmanto Jap	Nanyang Technological University, Singapore
Francesco Regazzoni	University of Amsterdam, The Netherlands and Università della Svizzera italiana, Switzerland
Gopalan Raghvan	Defense Institute of Advanced Technology, India
Goutam Paul	Indian Statistical Institute, India
Haoyang Wang	Shanghai Jiao Tong University, China
Indivar Gupta	SAG DRDO, India
Jason LeGrow	Virginia Polytechnic Institute and State University, USA
Jayaprakash Kar	LNM Institute of Information Technology, India
Jun Xu	Institute of Information Engineering, Chinese Academy of Sciences, China
Katsuyuki Takashima	Waseda University, Japan
Kazuhiko Minematsu	NEC, Japan
Lilya Budaghyan	University of Bergen, Norway
Luca De Feo	IBM Research, Switzerland
Mahabir Prasad Jhanwar	Ashoka University, India
Marine Minier	LORIA, France
Meltem Turan	National Institute of Standards and Technology, USA
Mridul Nandi	Indian Statistical Institute, India
Mustafa Khairallah	Lund University, Sweden
Nicolas Sendrier	Inria, France
Nilanjan Datta	IAI, TCG CREST, Kolkata, India
Pandu Rangan C.	IIT Madras, India
Prasanna Ravi	Nanyang Technological University, Singapore

Pratish Datta	NTT Research, USA
Prem Laxman Das	SETS, India
Ratna Dutta	IIT Kharagpur, India
Rei Ueno	Kyoto University, Japan
Reshmi T. R.	SETS, India
Saibal Pal	SAG, DRDO, Delhi, India
Sanjit Chatterjee	Indian Institute of Science, India
Santanu Sarkar	IIT Madras, India
Satrajit Ghosh	IIT Kharagpur, India
Sherman Chow	Chinese University of Hong Kong, China
Sihem Mesnager	Paris 8 University, France
Somitra Sanadhya	IIT Jodhpur, India
Souradyuti Paul	IIT Bhilai, India
Sourav Sen Gupta	IMEC, Belgium
Stjepan Picek	Radboud University, Netherlands
Subhamoy Maitra	Indian Statistical Institute, India
Subidh Ali	IIT Bhilai, India
Sugata Gangopadhyay	IIT Roorkee, India
Sumanta Sarkar	University of Warwick, UK
Sushmita Ruj	University of New South Wales, Australia
Suvradip Chakraborty	Visa Research, USA
Svetla Nicova	KU Leuven, Belgium
Takanori Isobe	University of Hyogo, Japan
Tapas Pal	Karlsruhe Institute of Technology, Germany
Venkata Koppula	IIT Delhi, India
Yixin Shen	Inria, Univ. Rennes, IRISA, France.

Additional Reviewers

Abishanka Saha	Floyd Zweydinger
Aditya Singh Rawat	George Lu
Akiko Inoue	Gireesh Pandey
Ali Raya	Haomeng Xu
Amaury Pouly	Harry W. H. Wong
Amit Jana	Harshal Bhadreshkumar Shah
Anand Ravi	Harshdeep Singh
André Schrottenloher	Jack P. K. Ma
Animesh Singh	Jiafan Wang
Anuja Modi	Jianing Zhang
Anum Sajjad	Jun Song
Arindam Mandal	Lejla Batina
Dilara Toprakhisar	Minxin Du

Mohit Pal
Monosij Maitra
Mostafizar Rahman
Nabanita Chakraborty
Nikhil Vanjani
Nimish Mishra
Nupur Gupta
Oleksandr Kholosha
Paolo Santini
Pierrick Dartois
Prashant Agrawal
Pratima Jana
Sanajit Patra
Sanjay Kumar
Sarah Arpin

Shubhi Shukla
Sriram Vasudevan
Subhranil Dutta
Suman Ghosh
Suparna Kundu
Suprava Roy
Tapas Pandit
Thales Paiva
Tianning Wang
Travis Morrison
Valerio Cini
Vikas Kumar
Yanli Zou
Ying-Yu Pan
Zhipeng Wang

Invited Talks

SQIsign: Past, Present and Future

Luca De Feo

IBM Research Europe

Abstract. 5 years ago, while everyone was raving about the compactness of SIKE's public keys and ciphertexts, the isogeny community was facing a conundrum: why isn't there any compact isogeny-based signature?

The solution came from a technique that was originally devised to attack SIKE: thanks to the KLPT algorithm, it became possible for the first time to "rerandomize" isogeny walks, provided knowledge of endomorphism rings, which could then be used as cryptographic trapdoors.

As our understanding of the connections between elliptic curves and their endomorphism rings expanded, new cryptographic discoveries kept coming. Among them, the celebrated equivalence between the supersingular isogeny path problem and the endomorphism ring problem.

Some feared the devastating attacks on SIKE would spell the end of isogeny-based cryptography. Instead, the technique of higher-dimensional embeddings they introduced has generated a remarkable stream of research, using the new tools to construct and improve all sorts of schemes.

As we celebrate SQIsign advancing to the second round of the on-ramp NIST call for signatures today, higher-dimensional embeddings have entered SQIsign in various declinations (SQIsignHD, SQIsign2D, etc.), revolutionizing performance and proof techniques for isogeny-based signatures and heralding a new era for SQIsign.

Logic-Based Verification and Testing for Cryptographic Protocol Design and Implementation

Lucca Hirschi

Inria, France

Abstract. Today's information society crucially relies on *cryptographic protocols*. These protocols leverage cryptographic primitives to ensure confidentiality, integrity, or other security goals. Any attack on these protocols can have dramatic consequences, amplified by their ubiquity and our dependence on them for example in finance, business, and communication. And yet, critical and widely used cryptographic protocols have repeatedly been found to contain flaws in both their design and their implementation, opening the way for serious attacks (such as on TLS in web browsing, EMV in credit card transactions, 5G in mobile networks, WPA2 in WiFi, and e-voting systems). A widespread class of these vulnerabilities is *logical attacks*, which exploit flawed protocol logic.

This keynote will examine two logic-based methods to preclude logical attacks from protocol designs and implementations. First, we will discuss automated formal verification methods based on *Dolev-Yao (DY) models*, which formally define and excel at finding such flaws on abstract specifications. This approach has evolved over 40 years and yielded significant results, which we will highlight. Nevertheless, these methods alone cannot secure protocol implementations, as bugs may introduce implementation-level logical attacks. We will present a recent research avenue that aims to integrate formal DY models with *fuzz* testing techniques to capture logical attacks on cryptographic protocol implementations.

Challenges in Symmetric-Key Cryptography

Vincent Rijmen[1,2]

[1] KU Leuven, Belgium
[2] University of Bergen, Norway

Abstract. Almost 25 years ago, the block cipher Rijndael was selected to become the Advanced Encryption Standard (AES). This concluded two research lines that were of great importance in the 1990s: the security evaluation of the DES and the design of a replacement mainstream block cipher. Since then the community has identified a number of new challenges. Symmetric-key primitives are used in a wide range of new applications: zero-knowledge proofs, fully homomorphic encryption schemes, quantum-secure signature schemes, etc.

In this talk we will give an overview of challenges in symmetric-key cryptography and the progress that has been made on them. We will also present our own selection of the most important research topics in symmetric-key cryptography for the future.

Contents – Part I

Cryptographic Constructions

Quantum Cryptography

Contents – Part II

Blockchain and Cloud Computing

Foundations

An Efficient Noncommutative NTRU from Semidirect Product

Vikas Kumar[1](✉) , Ali Raya[2] , Aditi Kar Gangopadhyay[1],
Sugata Gangopadhyay[2], and Md Tarique Hussain[3]

[1] Department of Mathematics, Indian Institute of Technology Roorkee,
Roorkee 247667, India
{v_kumar,aditi.gangopadhyay}@ma.iitr.ac.in
[2] Department of Computer Science and Engineering, Indian Institute of Technology
Roorkee, Roorkee 247667, India
{ali_r,sugata.gangopadhyay}@cs.iitr.ac.in
[3] Department of Information Technology, Indian Institute of Engineering Science and
Technology Shibpur, Howrah 711103, India

Abstract. NTRU is one of the most extensively studied lattice-based
schemes. Its flexible design has inspired different proposals constructed
over different rings, with some aiming to enhance security and others
focusing on improving performance. The literature has introduced a line
of noncommutative NTRU-like designs that claim to offer greater resistance to existing attacks. However, most of these proposals are either
theoretical or fall short in terms of time and memory requirements when
compared to standard NTRU. To our knowledge, DiTRU (Africacrypt
2024) is the first noncommutative analog of NTRU provided as a complete package. Although DiTRU is practical, it operates at two times
slower than NTRU with no decryption failure. Additionally, key generation, encryption, and decryption are 1.2, 1.7, and 1.7 times slower,
respectively, with negligible decryption failure. In this work, we introduce
a noncommutative version of NTRU that offers comparable performance
and key sizes to NTRU while improving upon DiTRU. Our cryptosystem is based on the GR-NTRU framework, utilizing the group ring of
a semidirect product of cyclic groups over the ring of Eisenstein integers. This design allows for an efficient construction with key generation
speeds approximately two (three) times faster than NTRU (DiTRU).
Further, the proposed scheme provides roughly a speed-up by a factor
of 1.2 (2) while encrypting/decrypting messages of the same length over
NTRU (DiTRU). We provide a reference implementation in C for the
proposed cryptosystem to prove our claims.

Keywords: NTRU · GR-NTRU · Semidirect product · Group rings ·
Eisenstein integers

1 Introduction

NTRU [19], first introduced by Hoffstein, Pipher, and Silverman, is one of the
most prominent and efficient lattice-based postquantum cryptosystems. The long

© The Author(s), under exclusive license to Springer Nature Switzerland AG 2025
S. Mukhopadhyay and P. Stănică (Eds.): INDOCRYPT 2024, LNCS 15495, pp. 3–27, 2025.
https://doi.org/10.1007/978-3-031-80308-6_1

cryptanalytic history and absence of effective attacks against well-defined parameter sets of NTRU place it at a strong position in the pyramid of postquantum schemes. The trust in NTRU is further deepened as three NTRU-style schemes [7,9,25] reached the third round of NIST's postquantum standardization process. The flexibility in the design of NTRU has resulted in many variants aimed at improving the efficiency and security of NTRU. We will refer to the NTRU version presented in [18] as *standard* NTRU.

Crypanalytic Landscape. Broadly, the security of NTRU is based on the NTRU hard assumption, formulated as:

Given a public key h, computed as $h = f^{-1} * g$ (mod q), where f, g are short private polynomials and q is a modulus, find f', g' with small coefficients such that $f' * h = g'$ (mod q).

The most straightforward way to attack the problem is to search for small elements from the underlying ring satisfying the NTRU key equation. One can optimize the search process by incorporating approaches like Meet in the middle attack [20]. The NTRU problem can also be solved by finding short vectors in lattices of particular structures [13] using lattice reduction algorithms [12,32]. In this sense, NTRU is classified as a lattice-based cryptosystem. The two previous approaches can be further combined, resulting in a hybrid attack [21]. Other attacks against NTRU exploit the selected parameters, like decryption failure attacks [22] and subfield attacks [14]. Hence, to propose a set of parameters that target a certain level of security, one needs to consider the cost of all previous attacks. However, in some other scenarios, the attacker may have access to extra information about the cryptosystem that enables different cryptanalysis tools. For example, the NTRU learning problem, phrased as:

Given NTRU public keys $h_i = f^{-1} * g_i$ (mod q), for a fixed f and a number of independently sampled g_i, find f,

was believed to be hard until recently, Kim and Lee [27] introduced a polynomial-time attack that can break it if the attacker has access to n different samples of h_i (where n refers to the extension degree of the NTRU ring $\mathbb{Z}[x]/(x^n - 1)$). A simple analysis of the Kim and Lee attack shows that their method works when the underlying ring is commutative since building the system of equations that leads to attacking the NTRU learning problem is possible only if the attacker can reformulate the equations using commutativity. We refer the reader to the original work [27] for the attack details. Therefore, employing noncommutative algebras to generalize NTRU appears to be a promising research direction. Furthermore, Coppersmith and Shamir [13], in the initial work of lattice attack on NTRU, also hinted that noncommutative structure might prevent their attack and other possible attacks that take benefit of the commutative structure.

Related Works. Although several proposals exist for noncommutative NTRU-like cryptosystems, many of them do not maintain the hard assumption of NTRU. The first noncommutative variant of NTRU by Hoffstein and Silverman is an example under this category, where the scheme was vulnerable to attack, which does not apply to standard NTRU. For details of this attack, we refer the readers to [40]. Other proposals uphold the general assumption of NTRU but fall behind in terms of the efficiency and compactness of the parameter sets compared to standard NTRU. For example, QTRU [34], SQTRU [39], OTRU [33] based on quaternion, split quaternion, and octonion algebras, respectively, are 4, 4, and 16 times slower than NTRU for the same level of security. BQTRU's [5] security analysis raises concerns as the authors discuss their parameter selection, conjecturing that Gentry's attack [17] does not challenge the security of their scheme without a rigorous analysis. Further, none of the above constructions provided a full implementation, keeping it unclear how efficiently one can address some of the design aspects, like inverting elements in the new setting of the noncommutative ring.

To our knowledge, DiTRU [35] is the only noncommutative NTRU-like design provided with a full-package implementation. DiTRU is structured as a group ring NTRU over the dihedral group of order $2N$. The hard assumption of NTRU is maintained as the key recovery attack is equal to finding 'short' elements from the underlying noncommutative ring. However, according to the authors, the associated lattice with DiTRU is susceptible to a one-layer Gentry attack, which can reduce the dimension of lattice attacks from $4N$ to $2N$. Consequently, the parameters chosen for DiTRU are twice as large as those used for NTRU to achieve equivalent levels of security without allowing decryption failure. This ratio can be scaled down slightly when a negligible decryption failure is deemed acceptable. In summary, while DiTRU offers a practical noncommutative analog to NTRU, it fails to maintain NTRU performance for equivalent parameter sets.

Our Contribution. We design a noncommutative NTRU variant in the GR-NTRU framework [42]. Although GR-NTRU is usually designed over the group rings $\mathbb{Z}G$. To achieve faster multiplication, we make minor modifications and build it over the group ring RG where R is the ring of Eisenstein integers as in ETRU [26]. The group $G = C_N \rtimes C_3$ is the noncommutative semidirect product of cyclic groups C_N and C_3 of order N and 3, respectively. For our construction, we clear all the implementation details and consider the following points:

- **Inversion algorithm:** We provide an inversion algorithm (Algorithm 2) to find invertible elements in the underlying group ring. This algorithm constitutes an essential part of the key generation process. The proposed algorithm introduces a way to check/find invertible elements by mapping the units over the proposed ring $R(C_N \rtimes C_3)$ to the ring RC_N where R is the ring of Eisenstein integers. We provide the constant-time implementation for our algorithm following the Bernstein-Yang algorithm [8]. Our findings demonstrate that the proposed key generation process is faster than the key generation processes for NTRU and DiTRU by a factor of 2 and 3, respectively.
- **Analysis of lattice security:** We give a detailed cryptanalysis of the security of the associated lattices with our construction and analyze the hardness of retrieving the decryption key using the lattice reduction algorithms.

- **Concrete parameter selections:** We model the decryption failure with respect to the chosen design, and accordingly, we provide two sets of parameters: one with zero decryption failure rate, and the other allows a negligible decryption failure. These parameters have been selected considering the best combinatorial and lattice-based attacks against our construction.
- **Reference implementation:** We provide a C reference implementation to prove the claimed results on the performance and compactness of our proposal. Table 4 compares the performance of our construction vs. NTRU and DiTRU while encrypting/decrypting messages of the same length. Our cryptosystem demonstrates improvement over NTRU and DiTRU by a factor of 1.2 and 2, respectively. The implementation is available and can be accessed at https://github.com/The-Isogeniest/Ei_TRU.

1.1 Paper Layout

Section 2 contains the required notations and preliminaries. The proposed cryptosystem is given in Sect. 3. Section 4 gives an analysis of different attacks on the new design. Finally, the cryptosystem's parameters, its performance analysis, and comparison with NTRU and DiTRU are provided in Sect. 5.

2 Notation and Preliminaries

$\mathbb{C}, \mathbb{R}$, and $\mathbb{Z}$ denote the set of complex numbers, real numbers and integers, respectively. Symbol $*$, wherever it occurs, denotes the multiplication of two elements with respect to the underlying algebraic structure, which should be clear from the context. For a positive integer n, $\mathbb{Z}/n\mathbb{Z}$ is the ring of integers modulo n. R denotes a commutative ring with unity and R^n is cartesian product of n copies of R. The norm of a vector $u = (u_1, u_2, \ldots, u_n) \in \mathbb{R}^n$ is defined as $||u|| = \sqrt{\sum_{i=1}^{n} u_i^2}$. The length/norm of a complex number $\xi = a + \iota b$ is $|\xi| = \sqrt{a^2 + b^2}$, where $\iota = \sqrt{-1} \in \mathbb{C}$ is the imaginary root of unity. Let Re and Im denote the real and imaginary parts of a complex number, respectively. We denote the primitive cube root of unity by ω, i.e., $\omega = e^{\frac{2\pi i}{3}}$, $\omega^3 = 1$ and $\omega \neq 1$. U_n denote the set of nth roots of unity. $U_3 = \{1, \omega, \omega^2 = -1 - \omega\}$ and $U_6 = \{\pm 1, \pm \omega, \pm \omega^2\}$. $M_n(R)$ denotes the ring of $n \times n$ matrices with entries from the ring R. Sampling an element s uniformly at random from a set S is denoted by $s \xleftarrow{\$} S$. We may define more notations in the course of the paper, wherever required.

2.1 Lattices

Definition 1 (Lattice). *Let $B \in \mathbb{R}^{n \times m}$ with linearly independent rows b_i, for $i = 1, 2, \ldots, n$. A lattice L_B generated by the matrix B is the set of integer linear combination of rows of B, i.e.,*

$$L_B = \left\{ \sum_{i=1}^{n} \gamma_i b_i : \gamma_i \in \mathbb{Z} \right\}. \tag{1}$$

The matrix B is called a basis matrix of the lattice L_B. The determinant of the lattice L_B is given by $\sqrt{det(B^T B)}$ and is independent of the choice of basis. If all the rows of B have integer entries, we say that the lattice is an integral lattice. This paper deals with only full-rank integral lattices, i.e., $n = m$. A full rank lattice $L_B \subset \mathbb{Z}^n$ is called q-ary lattice for some $q > 0$, if $q\mathbb{Z}^n \subset L_B \subset \mathbb{Z}^n$.

Definition 2 (SVP). *The Shortest Vector Problem (SVP) is to find a non-zero vector $u \in L_B$ such that*

$$||u|| = \min_{w \in L_B - \{0\}} ||w||.$$

We denote the length of the shortest vector in lattice L_B by $\lambda_1(L_B)$.

Definition 3 (CVP). *Closest Vector Problem (CVP) is to find a vector $v \in L_B$ closest to the given target vector $t \in \mathbb{R}^d$, i.e., $||v - t|| \leq ||w - t||$ for all $w \in L_B$.*

Definition 4 (Gaussian heuristic). *Suppose $L_B \subset \mathbb{R}^n$ is a lattice generated by matrix $B \in \mathbb{R}^{n \times n}$. Gaussian heuristic estimates the length of the shortest vector in the lattice L_B to be*

$$\sigma(L_B) = \sqrt{n/2\pi e} \cdot \ det(B)^{1/n}. \tag{2}$$

2.2   Semidirect Product of Cyclic Groups

Definition 5. *[16, Definition 2.2] Given two groups G and H and a group homomorphism $\phi : H \to Aut(G)$ (the automorphism group of G), the Semidirect Product of G and H with respect to ϕ, denoted $G \rtimes_\phi H$ (or, simply, $G \rtimes H$) is a new group with set $G \times H$ and multiplication operation $(g_1, h_1)(g_2, h_2) = (g_1\phi(h_1)(g_2), h_1 h_2)$.*

The fact that $Aut(C_N) \cong \mathbb{Z}_{N-1}$ gives the following result:

Theorem 1. *[16, Proposition 2.1] Let $C_N \cong \frac{\mathbb{Z}}{N\mathbb{Z}}$ and $C_M \cong \frac{\mathbb{Z}}{M\mathbb{Z}}$ be two cyclic groups of order N and M, respectively. A semidirect product $C_N \rtimes_k C_M$ corresponds to a choice of integer k such that $k^M \equiv 1 \mod N$. The semidirect product group is given by*

$$C_N \rtimes_k C_M = \langle x, y \mid x^N = y^M = 1, yxy^{-1} = x^k \rangle. \tag{3}$$

When there is no confusion of k, we denote the semidirect product by $C_N \rtimes C_M$.

In our work, we consider the case when N is prime, $M = 3$, and $3|(N-1)$ so that we have a noncommutative semidirect product $C_N \rtimes_k C_3$ for some $k \not\equiv 1 \mod N$ such that $k^3 \equiv 1 \mod N$. Let us fix such a k and order the elements of the group $C_N \rtimes_k C_3$ as follows:

$$C_N \rtimes_k C_3 = \{1, x, \ldots, x^{N-1}, y, yx, \ldots, yx^{N-1}, y^2, y^2 x, \ldots, y^2 x^{N-1}\}.$$

Theorem 2. *[15, Section 5.5] Let H be a finite cyclic group and N be an arbitrary group. Suppose $\phi_1, \phi_2 : H \to Aut(N)$ ($Aut(N)$ is the group of automorphisms on N) are homomorphisms such that $Im(\phi_1)$ and $Im(\phi_2)$ are conjugate subgroups of $Aut(N)$. Then $N \rtimes_{\phi_1} H \cong N \rtimes_{\phi_2} H$.*

Corollary 1. *Let N be a prime number such that $3|(N-1)$, then there exists only one noncommutative semidirect product $C_N \rtimes_k C_3$ unique up to isomorphism.*

Proof. Since N is prime, therefore $Aut(C_N) \cong C_{N-1}$ (cyclic group of order $N-1$). Hence, there is one and only one subgroup of order 3 of $Aut(C_N)$ because $3|(N-1)$. Consequently, for any two non-trivial homomorphisms $\phi_1, \phi_2 : C_3 \to Aut(C_N)$, we have $Im(\phi_1) = Im(\phi_2)$. Thus, $C_N \rtimes_{\phi_1} C_3 \cong C_N \rtimes_{\phi_2} C_3$. □

2.3   Ring of Eisenstein Integers

We briefly discuss the essential properties of Eisenstein integers that are given in [26] in detail. The ring of Eisenstein integers is defined as

$$\mathbb{Z}[\omega] = \{a + b\omega : a, b \in \mathbb{Z}\} = \left\{a - \frac{b}{2} + i\frac{b\sqrt{3}}{2} : a, b \in \mathbb{Z}\right\}. \tag{4}$$

The length of an Eisenstein integer $z = a + b\omega$ is $|z| = \sqrt{a^2 + b^2 - ab}$. The product of two Eisenstein integers is given by

$$(a + b\omega) * (c + d\omega) = ac - bd + (ac + (a - b)(d - c))\omega. \tag{5}$$

Therefore, one product in $\mathbb{Z}[\omega]$ requires 3 multiplications and 4 additions over $\mathbb{Z}$. The map $\langle \cdot \rangle : \mathbb{Z}[\omega] \to M_2(\mathbb{Z})$ given by

$$\langle z \rangle = \begin{pmatrix} a & b \\ -b & a-b \end{pmatrix} \tag{6}$$

is a ring homomorphism and the map $a + b\omega \to (a, b) \in \mathbb{Z}^2$ is an isomorphism. The multiplication $(a + b\omega) * (c + d\omega) \in \mathbb{Z}[\omega]$ can be realized as $(a, b) \cdot \langle c + d\omega \rangle \in \mathbb{Z}^2$. The Voronoi cell V_q of an element $q \in \mathbb{Z}[\omega]$ is the region bounded by a certain regular hexagon inscribed between circles of radius $|q|/2$ and $|q|/\sqrt{3}$ as shown in Fig. 1.

Theorem 3. *[26, Theorem 1] The set U_6 consists of exactly all units (invertible elements) of $\mathbb{Z}[\omega]$. The primes of $\mathbb{Z}[\omega]$ are (up to multiplication by a unit): $1 - \omega$; rational primes $p \in \mathbb{Z}$ satisfying $p \equiv 2 \pmod 3$; and those $q \in \mathbb{Z}[\omega]$ for which $|q|^2 = p$ is a rational prime satisfying $p \equiv 1 \pmod 3$.*

Division in $\mathbb{Z}[\omega]$. For any α and a nonzero q in $\mathbb{Z}[\omega]$, we say that $\beta \in \mathbb{Z}[\omega]$ is residue or reduced element modulo q corresponding to α, i.e., $\alpha \pmod q = \beta$, if we can write $\alpha = rq + \beta$ where $r \in \mathbb{Z}[\omega]$ is the closest element to $q^{-1}\alpha \in \mathbb{C}$, or equivalently $rq \in \mathbb{Z}[\omega]$ is the nearest multiple of q to α. The set of residues/reduced elements modulo q is denoted as D_q. It should be observed that $\mathbb{Z}[\omega]$ is a regular hexagonal lattice in $\mathbb{C} \cong \mathbb{R}^2$ with basis $\{1, \omega\}$ over $\mathbb{Z}$, and the ideal $\langle q \rangle$ is again a lattice with basis $\{q, q\omega\}$. Therefore, finding $r \in \mathbb{Z}[\omega]$ closest to $q^{-1}\alpha \in \mathbb{C}$ is equivalent to solving Closest Vector Problem (CVP) in the lattice $\mathbb{Z}[\omega]$. A division algorithm over $\mathbb{Z}[\omega]$ is discussed in [26, Algorithm 1] that costs 27 integer multiplications and 32 integer additions, which is significantly costlier than computing an integer modulus.

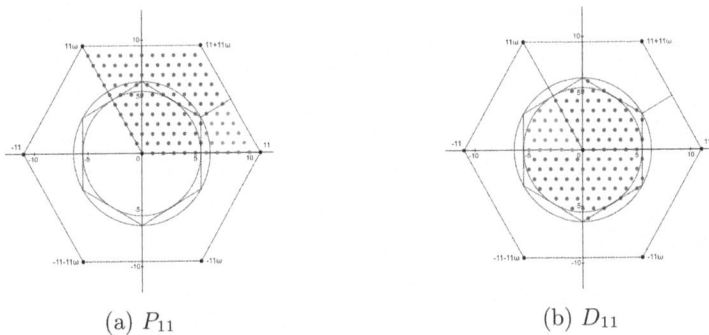

(a) P_{11} (b) D_{11}

Fig. 1. Division in $\mathbb{Z}[\omega]$ by $q = 11$. The different colors in P_{11} represent regions close to different multiples of $q = 11$, which are possibly $0, 11, 11\omega$, or $11 + 11\omega$. Each colored region in D_{11} represents the residues of elements in its corresponding part in P_{11} translated according to Algorithm refalg:division depending on their closeness to multiples of 11.

In this work, we propose a more efficient division algorithm (Algorithm 1) that works for division by elements of the form $q + 0\omega$, with a cost of 4 integer multiplications and 4 integer additions. We briefly explain the working of Algorithm refalg:division. Let $a, b \in \mathbb{Z}$ then there exist unique integers r, s and $0 \le x, y < q$ such that $a = rq + x$, $b = sq + y$. Therefore, $a + b\omega = q(r + s\omega) + (x + y\omega) \equiv x + y\omega \pmod{q}$. Let $P_q = \{x + y\omega : 0 \le x, y < q\}$, then it is enough to find residues of elements in P_q modulo q. For an element $x + y\omega \in P_q$, Algorithm refalg:division returns the residue modulo q by locating the nearest multiple of q in $\mathbb{Z}[\omega]$ as follows: If the nearest multiple of q is $x_1 + y_1\omega$, where $x_1, y_1 \in \{0, q\}$, then the residue is $(x - x_1) + (y - y_1)\omega$. When a point is equidistant from two multiples of q, then the algorithm chooses the one on the left. We have shown the regions in P_q closer to different multiples of q in Fig. 1b, and the corresponding residues D_q in Fig. 1a, for $q = 11$.

Algorithm 1: Division by integers in $\mathbb{Z}[\omega]$

    Input: $\alpha = a + b\omega \in \mathbb{Z}[\omega]$, and an element $q = q + 0\omega \in \mathbb{Z}[\omega]$.
    Output: $\beta \in \mathbb{Z}[\omega]$ such that $\alpha = rq + \beta$ where $r \in \mathbb{Z}[\omega]$ is nearest to $q^{-1}\alpha$.
1   $x = a \pmod{q}$, $y = b \pmod{q}$, $X = 2x$, $Y = 2y$
2   **if** $x + y > q$, $X > y$, $Y \ge x$ **then return** $\beta = (x - q) + (y - q)\omega$ **if**
    $X - y > q$, $Y < x$ **then return** $\beta = (x - q) + y\omega$ **if** $Y - x \ge q$, $X \le y$ **then**
    return $\beta = x + (y - q)\omega$ **else return** $\beta = x + y\omega$

Remark 1. Algorithm   refalg:division returns the set of residues D_q modulo q that is almost symmetrically distributed around 0, which is needed to decrease the decryption failure. However, $q = 2$ is a special case where we get $D_2 = \{0, 1, -\omega, -\omega^2\}$ which is not distributed around 0. Observing that $\omega, \omega^2 \equiv -\omega, -\omega^2 \pmod 2$ and $|\pm\omega| = |\pm\omega^2| = 1$. We redefine D_2 as $D_2 = \{0, 1, \omega, \omega^2\}$ by mapping $-\omega \to \omega$ and $-\omega^2 \to \omega^2$. Conclusively, $a + b\omega \pmod 2 = a \pmod 2 + b \pmod 2\,\omega$, and if $a \pmod 2 = b \pmod 2 = 1$ then $a + b\omega \pmod 2 = -1 - \omega = -\omega^2$.

Lemma 1. *For a rational prime $q \in \mathbb{Z}[\omega]$, inverse of a nonzero element $z = a + b\omega$ modulo q is given by $|z|^{-2}((a - b) - b\omega)$, where $|z|^{-1}$ is computed modulo integer q in $\mathbb{Z}_q$.*

Proof. Consider $(a + b\omega) * ((a - b) - b\omega) = a^2 - ab + b^2 = |z|^2$, and $|z|^{-1} = |z|^{q-2} \pmod q$ (Fermat's theorem) exists since $\mathbb{Z}_q$ is a field as q is prime integer.

2.4   Group Rings

Definition 6 (Group rings). *The group ring of a group $G = \{g_i : i = 1, 2, \ldots, n\}$ over a ring R is the set of formal sums*

$$RG = \left\{ a = \sum_{i=1}^{n} \alpha_i g_i : \alpha_i \in R \quad for \quad i = 1, 2, \ldots, n \right\} \tag{7}$$

that forms a ring under the following operations. Suppose $a = \sum_{i=1}^{n} \alpha_i g_i$ and $b = \sum_{i=1}^{n} \beta_i g_i$ in RG.

1. *The sum of a and b is given by $a + b = \sum_{i=1}^{n}(\alpha_i + \beta_i)g_i$.*
2. *The product of a and b is given by $a * b = \sum_{i=1}^{n} \left(\sum_{g_h g_k = g_i} \alpha_h \beta_k \right) g_i$.*

Definition 7 (Coefficient vector). *Every element $a = \sum_{i=1}^{n} \alpha_i g_i$ can be mapped uniquely to its coefficient vector $(\alpha_1, \alpha_2, \ldots, \alpha_n) \in R^n$. We freely use the same notation 'a' to denote the elements of the group ring and their corresponding coefficient vectors depending on the context.*

Definition 8 (RG-matrix). *[24] For every element $a = (\alpha_{g_1}, \alpha_{g_2}, \ldots, \alpha_{g_n}) \in RG$, we construct the RG-matrix of a as follows:*

$$M_{RG}(a) = \begin{pmatrix} \alpha_{g_1^{-1} g_1} & \alpha_{g_1^{-1} g_2} & \cdots\cdots & \alpha_{g_1^{-1} g_n} \\ \alpha_{g_2^{-1} g_1} & \alpha_{g_2^{-1} g_2} & \cdots\cdots & \alpha_{g_2^{-1} g_n} \\ \vdots & \vdots & \ddots & \vdots \\ \alpha_{g_n^{-1} g_1} & \alpha_{g_n^{-1} g_2} & \cdots\cdots & \alpha_{g_n^{-1} g_n} \end{pmatrix}. \tag{8}$$

The set $M_{RG} = \{M_{RG}(a) : a \in RG\}$ is the subring of $M_n(R)$. We say a matrix $A \in M_n(R)$ is an RG-matrix if there is an $a \in RG$ such that $A = M_{RG}(a)$.

Theorem 4. *[24, Thereom 1] The mapping $\tau : RG \to M_{RG} \subset M_n(R)$ defined as $\tau(a) = M_{RG}(a)$ is a bijective ring homomorphism.*

Theorem 5. *[24, Thereom 2] An element $a \in RG$ is a unit if and only if $M_{RG}(a)$ is invertible in $M_n(R)$. In that case, inverse of $M_{RG}(a)$ is also an RG-matrix.*

2.5 Group ring $R(C_N \rtimes C_3)$

In this section, we derive some results on the group ring $R(C_N \rtimes_k C_3)$. Particularly, we give the matrix representation of elements in $R(C_N \rtimes_k C_3)$ and then derive an inversion algorithm to check the invertibility and find the inverses of elements in this group ring.

The group ring $R(C_N \rtimes_k C_3)$ can be defined as

$$R(C_N \rtimes_k C_3) = \{\alpha(x) + y\beta(x) + y^2\gamma(x) : \alpha(x), \beta(x), \gamma(x) \in RC_N\}, \tag{9}$$

where $yxy^{-1} = x^k$, $k^3 \equiv 1 \bmod N$. Consider

$$(yxy^{-1})^{k^2} = (x^k)^{k^2}$$

$$yx^{k^2}y^{-1} = x \qquad \text{since} \quad k^3 \equiv 1 \bmod N \quad \text{and} \quad x^N = 1$$

$$yx^t = xy \qquad \text{where} \quad t \equiv k^2 \bmod N.$$

Therefore,

$$C_N \rtimes_t C_3 = \langle x, y \mid x^N = y^3 = 1, xy = yx^t \rangle \tag{10}$$

where $3|(N-1), t^3 \equiv 1 \bmod N$, and $t \not\equiv 1 \bmod N$. As a result, $\alpha(x)y = y\alpha(x^t)$ for every $\alpha(x) \in RC_N$. Consequently, the product of two elements $z = u(x) + yv(x) + y^2w(x)$, $a = \alpha(x) + y\beta(x) + y^2\gamma(x) \in R(C_N \rtimes_k C_3)$ is given by

$$z * a = u(x)\alpha(x) + w(x^t)\beta(x) + v(x^{t^2})\gamma(x) + y\left(v(x)\alpha(x) + u(x^t)\beta(x) + w(x^{t^2})\gamma(x)\right)$$

$$+ y^2\left(w(x)\alpha(x) + v(x^t)\beta(x) + u(x^{t^2})\gamma(x)\right). \tag{11}$$

Lemma 2 (Matrix representation). *Let $G = C_N \rtimes_k C_3$ then RG-matrix of an element $z \in R(C_N \rtimes_k C_3)$ is of the form*

$$M_{RG}(z) = \begin{pmatrix} M_0 & M_1 & M_2 \\ M_2 & M_0 & M_1 \\ M_1 & M_2 & M_0 \end{pmatrix} \in R^{3N \times 3N}, \tag{12}$$

i.e., $M_{RG}(z)$ is a block circulant matrix of order $3N$ where each submatrix M_i is an order N matrix.

Proof. We divide the matrix $M_{RG}(z)$ into blocks as

$$M_{RG}(z) = \begin{pmatrix} A_{00} & A_{01} & A_{02} \\ A_{10} & A_{11} & A_{12} \\ A_{20} & A_{21} & A_{22} \end{pmatrix},$$

where A_{rs} is an $N \times N$ matrix over R, for $r, s \in \{0, 1, 2\}$. From the Definition 8, for every $0 \leq i, j \leq N - 1$, we have

$$(A_{rs})_{i,j} = \text{ coefficient of } \quad (y^r x^i)^{-1}(y^s x^j) \quad \text{in} \quad z.$$

Use $xy = yx^t$, where $t \equiv k^2 \pmod N$, to get $x^i y^j = y^j x^{it^j}$. Further, using $x^N = y^3 = 1$ and $t^3 \equiv 1 \bmod N$, we get

$$(y^r x^i)^{-1}(y^s x^j) = x^{N-i} y^{(s-r) \bmod 3} x^j = y^{(s-r) \bmod 3} x^{(j-it^{(s-r) \bmod 3}) \bmod N}.$$

Therefore, $A_{00} = A_{11} = A_{22}$, $A_{01} = A_{12} = A_{20}$, and $A_{02} = A_{10} = A_{21}$. $\qquad\square$

Theorem 6 (Units). *Let $z = u(x) + yv(x) + y^2 w(x) \in R(C_N \rtimes_k C_3)$, and $t \equiv k^2 \pmod N$. Then, z is a unit in $R(C_N \rtimes_k C_3)$ if and only if the element*

$$det(u,v,w) = u(x)u(x^t)u(x^{t^2}) + v(x)v(x^t)v(x^{t^2}) + w(x)w(x^t)w(x^{t^2})$$
$$- u(x)v(x^t)w(x^{t^2}) - v(x)w(x^t)u(x^{t^2}) - w(x)u(x^t)v(x^{t^2}) \qquad (13)$$

is a unit in RC_N. In this case, the inverse of z is given by

$$det(u,v,w)^{-1} * \begin{pmatrix} u(x)u(x^t) - v(x^t)w(x^{t^2}) + y(w(x)w(x^{t^2}) - v(x)u(x^{t^2})) \\ + y^2(v(x)v(x^t) - w(x)u(x^t)) \end{pmatrix}. \qquad (14)$$

Proof. Element z is a unit if and only if there exists a unique $a = \alpha(x) + y\beta(x) + y^2\gamma(x) \in R(C_N \rtimes_k C_3)$ such that $z * a = a * z = 1$. From (11), we have

$$z * a = u(x)\alpha(x) + w(x^t)\beta(x) + v(x^{t^2})\gamma(x) + y\left(v(x)\alpha(x) + u(x^t)\beta(x) + w(x^{t^2})\gamma(x)\right)$$
$$+ y^2\left(w(x)a(x) + v(x^t)b(x) + u(x^{t^2})\gamma(x)\right) = 1 \qquad (15)$$

Rewriting Eq. (15) as

$$\begin{pmatrix} u(x) & w(x^t) & v(x^{t^2}) \\ v(x) & u(x^t) & w(x^{t^2}) \\ w(x) & v(x^t) & u(x^{t^2}) \end{pmatrix} \begin{pmatrix} \alpha(x) \\ \beta(x) \\ \gamma(x) \end{pmatrix} = \begin{pmatrix} 1 \\ 0 \\ 0 \end{pmatrix}. \qquad (16)$$

By uniqueness of inverse, such an a exists if and only if the matrix in Eq. (16) is invertible over RC_N. Consequently, the determinant of this matrix, given precisely by $det(u,v,w)$, is a unit in RC_N. Furthermore, a is obtained as defined in Eq. (14). $\qquad\square$

3   GR-NTRU over the Group Ring $\mathbb{Z}[\omega](C_N \rtimes C_3)$

The Group ring NTRU or GR-NTRU [42] provides a general framework to design NTRU-like cryptosystems by employing different group rings. The standard NTRU operates over the truncated ring of polynomials $\mathbb{Z}[x]/\langle x^N - 1 \rangle$. If we let $C_N = \langle x : x^N = 1 \rangle$ to be the cyclic group of order N, then $\mathbb{Z}[x]/\langle x^N - 1 \rangle$ can be viewed as a group ring of C_N over $\mathbb{Z}$, i.e., $\mathbb{Z}[x]/\langle x^N - 1 \rangle \approx \mathbb{Z}C_N$.

Definition 9 (GR-NTRU). *The GR-NTRU generalizes NTRU by replacing the cyclic group ring $\mathbb{Z}C_N$ in NTRU with any group ring $\mathbb{Z}G$ of a finite group G and keeping all other procedures the same with a little modification depending on the requirements.*

3.1 $\mathbb{Z}[\omega](C_N \rtimes C_3)$-NTRU

Let N be a prime number, and $p, q \in \mathbb{Z}[\omega]$ be two primes chosen using Theorem 3 such that $gcd(p, q) = 1$ and $|p| \ll |q|$. We fix $p = 2$ for this work. Our scheme operates over the following rings:

$$R^\omega = \mathbb{Z}[\omega](C_N \rtimes C_3) \quad \text{and} \quad R_\alpha^\omega = \frac{\mathbb{Z}[\omega]}{\langle \alpha \rangle}(C_N \rtimes C_3), \tag{17}$$

where $\alpha \in \{p, q\}$ and R_α^ω is the set of elements in R^ω whose coefficients are reduced modulo α. Let $r = 2/3$, t be the nearest integer to $r(3N) = 2N$ and s be multiple of 3 nearest to $2N$. The set $L_f \subset R^\omega$ consists of elements with exactly t nonzero coefficients from U_6, and other coefficients are 0. Sets $L_g, L_\phi \subset R^\omega$ consists of elements with $s/3$ triples of coefficients each from sets either U_3 or $-U_3$ in a random order, and other coefficients are 0. The message space is $L_m = R_p^\omega$. In other words, a message is an element of the group ring R^ω whose coefficients belong to the set $D_2 = U_3 \cup \{0\}$. The basic framework of the scheme is similar to NTRU [18] and is sketched as follows:

Key Generation	Encryption	Decryption
1. Sample $F \xleftarrow{\$} L_f$ until $f = 1 + pF$ is invertible in R_q^ω. 2. $f_q \leftarrow$ inverse of f in R_q^ω. 3. Sample $g \xleftarrow{\$} L_g$. 4. **Public key** $h = f_q * g$ (mod q). 5. **Private key** F.	1. Sample $\phi \xleftarrow{\$} L_\phi$. 2. For messgae $m \in L_m$, compute $e = ph * \phi + m$ (mod q). 3. **return** e.	1. Compute $a = f * e$ (mod q). 2. **return** $m = a$ (mod p).

Correctness of Decryption. We have $a = p(g * \phi + F * m) + m$ (mod q). If the absolute value of the largest coefficient of $p(g * \phi + F * m) + m$ is less than $|q|/2$, then $a = p(g * \phi + F * m) + m$ without modulo q. Since g, ϕ, and F have maximum $3rN = 2N$ nonzero coefficients and every coefficient has norm 1, also the coefficients of m belong to $U_3 \cup \{0\}$ thus have norm 1. Therefore, the absolute value of the largest possible coefficient of $p(g * \phi + F * m) + m$ is bounded by $4N|p| + 1$. So, if we choose q such that $|q| > 8N|p| + 2$, then we can eliminate decryption failure entirely. In particular, for $p = 2$, choose q such that $|q| > 16N + 2$.

Inversion. For generating the keys, we need an efficient way to find the inverses of elements in the group ring R_q^ω, where $q \in \mathbb{Z}[\omega]$ is a prime. There exist algorithms [8,37,41] to check the invertibility and find inverses of elements in the ring $\mathbb{Z}_q C_N$ where q is a prime or prime power. These algorithms can easily be modified to work for the ring $\mathbb{Z}[\omega]/\langle q \rangle C_N$. We use the constant-time modular inversion by Bernstein and Yang [8] in our implementation to compute inverses in $\mathbb{Z}[\omega]/\langle q \rangle C_N$ with some modifications as it requires to find inverses in $\mathbb{Z}[\omega]/\langle q \rangle$. That can be done in constant time using the Square-and-Multiply algorithm [38, Page 200] in Lemma 1. Finally, combining the inversion in $\mathbb{Z}[\omega]/\langle q \rangle C_N$ with Theorem 6, one can find invertible elements in the ring R_q^ω as shown in Algorithm 2.

The complexity of the inversion algorithm for our scheme and its efficiency over NTRU is discussed in Sect. 5.

Algorithm 2: Inversion in R_q^ω

Input: $z = u(x) + yv(x) + y^2 w(x) \in R_q^\omega$
Output: $z^{-1} = \alpha(x) + y\beta(x) + y^2\gamma(x) \in R_q^\omega$ as inverse of f, or a failure
1 $d(x) \leftarrow det(u, v, w)$ /* as in Eq.(13) */
2 $inv(x), found \leftarrow$ `find-inverse-of-d(x)-in-`$\frac{\mathbb{Z}[\omega]}{<q>}C_N$
3 **if** *not found* **then return** *failure*
4 $\alpha(x) \leftarrow inv(x) * \left(u(x^t)u(x^{t^2}) - v(x^t)w(x^{t^2})\right)$ /* product in $\frac{\mathbb{Z}[\omega]}{<q>}C_N$ */
5 $\beta(x) \leftarrow inv(x) * \left(w(x)w(x^{t^2}) - v(x)u(x^{t^2})\right)$ /* product in $\frac{\mathbb{Z}[\omega]}{<q>}C_N$ */
6 $\gamma(x) \leftarrow inv(x) * \left(v(x)v(x^t) - w(x)u(x^t)\right)$ /* product in $\frac{\mathbb{Z}[\omega]}{<q>}C_N$ */
7 **return** $z^{-1} = \alpha(x) + y\beta(x) + y^2\gamma(x)$

Probability of Decryption Failure. Allowing negligible decryption failure in accordance with NIST guidelines can help reduce the key sizes. To model the probability of decryption failure, we follow a similar approach as [26] and make the following assumptions regarding the distribution of coefficients of F, g, ϕ, m.

Assumption 1. *We assume that $r(3N) = 2N$ is evenly divisible by 6 so that the number of nonzero coefficients in g and ϕ is $2N$. Further, assume that all the $2N$ nonzero coefficients of $F, g,$ and ϕ are equi-probable and uniformly distributed over U_6. Similarly, assume all the coefficients of m are uniformly distributed over $U_3 \cup \{0\}$.*

Let $a' = p(g * \phi + F * m) + m$, then the ith coefficient of a' is given by

$$a_i' = p\left(\sum_{j+k\equiv i} g_j\phi_k + \sum_{j+k\equiv i} F_j m_k\right) + m_i$$

for each $0 \leq i \leq 3N$. For a fixed pair (j, k), the terms $g_j\phi_k$ and $F_j m_k$ take the values from the set $U_6 = \{\pm 1, -\frac{1}{2} \pm \frac{\sqrt{3}}{2}\iota, \frac{1}{2} \pm \frac{\sqrt{3}}{2}\iota\}$ each with probabilities $r^2/6 = 2/27$ and $r/8 = 1/12$, respectively. Therefore, the expected mean values of the real and imaginary parts of $g_j\phi_k$ and $F_j m_k$ are zero, i.e., $E(Re(g_j\phi_k)) = E(Im(g_j\phi_k)) = 0$ and $E(Re(F_j m_k)) = E(Im(F_j m_k)) = 0$. Further, their variances are given by

$$Var(Re(g_j\phi_k)) = \frac{r^2}{2} = \frac{2}{9}, \; Var(Re(F_j m_k)) = \frac{3r}{8} = \frac{1}{4}.$$

Similarly, $Var(Im(g_j\phi_k)) = 2/9$ and $Var(Im(F_j m_k)) = 1/4$. By the central limit theorem for large N, the real and imaginary parts of a_i' can be modeled as a bivariate normal distribution $(\mathcal{R}, \mathcal{I})$. Then, the means of $\mathcal{R}$ and $\mathcal{I}$ are $\mu_{\mathcal{R}} = \mu_{\mathcal{I}} = 0$ and their variances are

$$\sigma^2 = \sigma_{\mathcal{R}}^2 = \sigma_{\mathcal{I}}^2 = 3Np^2\left(\frac{r^2}{2} + \frac{3r}{8}\right) + \frac{3}{8} = \frac{17N}{3} + \frac{3}{8},$$

since, $p = 2$, $E(Re(m_i)) = E(Im(m_i)) = 0$, and $Var(Re(m_i)) = Var(Im(m_i))$ $= 3/8$, for each i. The probability distribution function for the random variable $(\mathcal{R}, \mathcal{I})$ at each $x + \iota y$ is $P(x, y) = \frac{1}{2\pi\sigma_{\mathcal{R}}\sigma_{\mathcal{I}}} \exp\left(-\frac{x^2 + y^2}{2\sigma_{\mathcal{R}}\sigma_{\mathcal{I}}}\right)$. For successful decryption, we need all the coefficients of a' to be reduced modulo q. Therefore, the probability of successful decryption is given by

$$P_{success}(N, q) = \left(\iint_{V_q} P(x, y)dxdy\right)^{3N}. \tag{18}$$

We underestimate the probability of successful decryption to get a closed form of the expression (18):

$$\tilde{P}_{success}(N, q) = \left(\iint_{C} P(x, y)dxdy\right)^{3N} = \left(1 - \exp\left(-\frac{|q|^2}{8\sigma^2}\right)\right)^{3N}, \tag{19}$$

where C is a closed disk of radius $\frac{|q|}{2}$ inscribed inside the voronoi cell V_q. We experimentally confirmed the validity of our model in Fig. 2.[1]

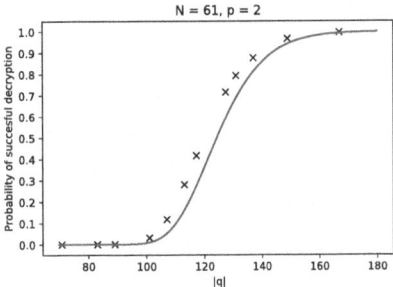

Fig. 2. The probability of successful decryption as a function of $|q|$ for $N = 61, p = 2$. The curve represents $\tilde{P}_{success}(N, q)$, and the crosses represent the ratio of the successful decryption out of $10,000$ randomly generated messages for each prime q.

4   Security Analysis

4.1   Combinatorial Search Attack

Given the public key h and other public parameters, the adversary can try brute force search for some element $f' \in L_f$ such that $f' * h \in L_g$. Therefore, the size of the search spaces is

---

[1] The curve in Fig. 2 lies slightly below the experimental observations since $\tilde{P}_{success}(N, q)$ (19) gives the underestimated value of the probability of successful decryption while the actual value of our model is given by $P_{success}(N, q)$ (18).

$$\frac{|L_f|}{3N} = \frac{1}{3N}\binom{3N}{2N}6^{2N}. \tag{20}$$

We have divided by $3N$ to account for all the $3N$ rotations associated with f'. Further, the meet-in-the-middle attack on private key f proposed by Odlyzkoa presented in [20] decreases the size of search space to $\sqrt{|L_f|/3N}$. Table 2 gives the cost (log base 2) of combinatorial or MITM attacks, denoted by **Comb**, against the parameters recommended in Sect. 5.

4.2 Lattice Attacks

Lattice reduction attacks are the most prominent against NTRU-like schemes. With the knowledge of the public information, the adversary constructs a lattice containing the private key as a short vector that can be recovered by solving SVP or its approximation. First, we discuss the state-of-art cost of lattice reduction algorithms, particularly BKZ, that depends on an important parameter called blocksize β that dominates the runtime. The greater the value of β, the more the runtime and the better the quality of the reduced basis. We call BKZ with blocksize β to be BKZ-β. BKZ has many advancements like [3,12]. In the literature, many estimators estimate the value of β in higher dimensions. NTRU fatigue estimator [14] is the most accurate one, which is itself based on *2016-estimator* [2]. According to 2016-estimator, for a basis matrix $B = [b_1, b_2, \ldots, b_n]$ of the lattice L_B, BKZ-β detects a unique short vector u if

$$||\pi_{n-\beta}(u)|| < ||b^*_{n-\beta}||, \tag{21}$$

where b^*_i for $i = 1, 2, \ldots, n$ denote the Gram-Schmidt orgthogolization vectors of rows of B, and π_i is the orthogonal projection over $(b_1, b_2, \ldots, b_{i-1})^\perp$. The projected norm is expected to be $\sqrt{\beta/n}||u||$. 2016-estimator adopts the GSA(*Geometric Series Assumption*) [36] that says, for a BKZ reduced lattice with blocksize β, the Gram-Schmidt orthogonalized vectors follow $||b^*_i|| = \delta_\beta^{n-2i-1}det(B)^{\frac{1}{n}}$, where δ_β is called the root Hermite factor of BKZ-β. For $\beta \geq 50$, Chen [11] estimated that

$$\delta_\beta \approx \left(\frac{\beta}{2\pi e}(\pi\beta)^{\frac{1}{\beta}}\right)^{\frac{1}{2(\beta-1)}}. \tag{22}$$

Ducas et al. [14] introduced alternative heuristics, called Z-GSA, for the lengths of Gram-Schmidt vectors of a BKZ-β reduced basis of a q-ary lattice as follows:

Definition 10 (Z-GSA). *[14, Heuristic 2.8] Let B be a basis of a $2n$-dimensional q-ary lattice L_B with n q-vectors. After BKZ-β reduction, the lengths of Gram-Schmidt vectors have the following shape: $m = \frac{1}{2} + \frac{ln(q)}{ln(\delta_\beta)}$ and*

$$||b^*_i|| = \begin{cases} q, & if\, i \leq n - m \\ \sqrt{q} \cdot \delta_\beta^{2n-1-2i}, & if\, \, n - m < i < n + m \\ 1, & if\, \, i \geq n + m \end{cases} \tag{23}$$

Since we deal with the lattices of the same nature as described in Z-GSA. Therefore, we employ Z-GSA in 2016-estimation instead of GSA. However, both the models coincide for the successful blocksize when B is a basis of a $2n$-dimensional q-ary lattice with $det(B) = q^n$, as for the lattices in our case.

BKZ uses two approaches to solve SVP: Sieving and Enumeration. Empirical results [29] show that sieving outperforms enumeration starting from a dimension greater than or equal to 65. Therefore, we use the BKZ with Sieving model, denoted as **BKZ(S)**, to compute the cost of lattice attacks. The cost of **BKZ(S)** is modeled as $2^{0.292\beta+o(\beta)}$ (classically) [6] and $2^{0.265\beta+o(\beta)}$ (quantumly) [31].

Primal Attack. Gentry [17] introduced a dimension reduction attack on an NTRU variant by factoring the ring $\mathbb{Z}[x]/\langle x^n - 1\rangle$, where n is composite, using the Chinese remainder theorem (CRT). This technique has a possible extension to different algebraic structures, as shown in [30]. Therefore, for any new NTRU-like proposal, it is essential to discuss the possibility of Gentry's attack for a fair security estimate. The underlying algebra in our construction can also be subjected to one layer of Gentry's dimension reduction attack. We discuss the possible homomorphisms that can help the adversary reduce the dimension of the lattice attacks and show that recovering the private key is equivalent to solving SVP in $8N$ dimensional lattices rather than $12N$. However, it is important to point out that the lattices to be attacked in our cryptosystem are difficult to reduce by lattice reduction algorithms in practice.

Notations: We represent every element $a = (\alpha_1 + \omega\beta_1, \ldots, \alpha_N + \omega\beta_N) \in \mathbb{Z}[\omega]C_N$ by its integral coefficient vector as $\mathbf{a} = (\alpha_1, \beta_1, \ldots, \alpha_N, \beta_N) \in \mathbb{Z}^{2N}$. Similarly, we represent every element $f = (f_0, f_1, f_2) \in \mathbb{Z}[\omega](C_N \rtimes C_3)$ where $f_i \in \mathbb{Z}[\omega]C_N$ by its integral coefficient vector $\mathbf{f} = (\mathbf{f}_0, \mathbf{f}_1, \mathbf{f}_2) \in \mathbb{Z}^{6N}$. For a matrix $A \in M_n(\mathbb{Z}[\omega])$, we define a $2n \times 2n$ integral matrix $\mathbf{A}$ by replacing every entry A_{ij} with a 2×2 integral matrix $\langle A_{ij}\rangle$ as in (6).

The public key equation can be expressed as

$$\mathbf{f} * \mathbf{H} = \mathbf{g}\,(\mathrm{mod}\,q), \tag{24}$$

where $\mathbf{H} \in \mathbb{Z}^{6N \times 6N}$ is the corresponding integer matrix of R^ω-matrix of the public key h given by $M_{R^\omega}(h) \in M_{3N}(\mathbb{Z}[\omega])$ (Theorem 2). Similar to standard NTRU, the private key $(\mathbf{f}, \mathbf{g})$ can be recovered in a naive way by solving SVP in a 12-dimensional lattice $L_{\mathbf{H}}$ generated by the matrix

$$\mathbf{M_H} = \begin{pmatrix} \mathbf{I}_{6N} & \mathbf{H} \\ \mathbf{0}_{6N} & q\mathbf{I}_{6N} \end{pmatrix}. \tag{25}$$

As discussed in Theorem 2, the matrix $M_{R^\omega}(h)$ and consequently the matrix $\mathbf{H}$ has a special structure

$$\mathbf{H} = \begin{pmatrix} \mathbf{H}_0 & \mathbf{H}_1 & \mathbf{H}_2 \\ \mathbf{H}_2 & \mathbf{H}_0 & \mathbf{H}_1 \\ \mathbf{H}_1 & \mathbf{H}_2 & \mathbf{H}_0 \end{pmatrix} \in M_{6N}(\mathbb{Z}) \tag{26}$$

where each $\mathbf{H}_i \in \mathbb{Z}^{2N \times 2N}$. We discuss the scenarios of how an adversary can take advantage of this structure in the context of dimension-reduction attacks. Generally, the goal is to *homomorphically* reduce the size of the public matrix and recover information about the private key that can be lifted back to the original key. Since the value of N is selected to be prime, it rules out the possibility of reducing the size of the matrices $\mathbf{H}_i$ (see [17] for details). The other ring homomorphisms that preserve the information about the private key and prevent the norms of the target vector from growing too large are of the form

$$\mathbf{H} \rightarrow \alpha \mathbf{H}_0 + \beta \mathbf{H}_1 + \gamma \mathbf{H}_2 \qquad (27)$$

where α, β, γ are small constants. Consequently, it reduces the public key equation to

$$(\alpha \mathbf{f}_0 + \beta \mathbf{f}_1 + \gamma \mathbf{f}_2) * (\alpha \mathbf{H}_0 + \beta \mathbf{H}_1 + \gamma \mathbf{H}_2) = \alpha \mathbf{g}_0 + \beta \mathbf{g}_1 + \gamma \mathbf{g}_2 \,(\mathrm{mod}\, q). \qquad (28)$$

It can easily be checked that map 27 is a ring homomorphism, i.e., preserve the matrix addition and multiplication, if and only if $(\alpha, \beta, \gamma) \in \{(0,0,0), (1,1,1), (1,\omega,\omega^2), (1,\omega^2,\omega)\}$. The case $(\alpha, \beta, \gamma) = (0,0,0)$ is of no use, therefore, we consider the others only. This way, one is able to reduce the size of the public matrix but end up in matrices with complex entries, apart from when $(\alpha, \beta, \gamma) = (1,1,1)$. In practice, for applying lattice reduction algorithms, such matrices are mapped to the real matrices, which leads to an increase in the dimension. In our case, the matrices

$$\mathbf{H}_{01} = \mathbf{H}_0 + \omega \mathbf{H}_1 + \omega^2 \mathbf{H}_2 = (\mathbf{H}_0 - \mathbf{H}_2) + \omega(\mathbf{H}_1 - \mathbf{H}_2),$$
$$\mathbf{H}_{02} = \mathbf{H}_0 + \omega^2 \mathbf{H}_1 + \omega \mathbf{H}_2 = (\mathbf{H}_0 - \mathbf{H}_1) + \omega(\mathbf{H}_2 - \mathbf{H}_1)$$

belonging to $M_{2N}(\mathbb{Z}[\omega])$ can be mapped to $4N \times 4N$ integer matrices $\mathcal{H}_{01}$ and $\mathcal{H}_{02}$, respectively, as done before. Suppose $L_{\mathcal{H}_{01}}$ and $L_{\mathcal{H}_{02}}$ are the $8N$-dimensional lattices generated by the matrices

$$\mathbf{M}_{\mathcal{H}_{01}} = \begin{pmatrix} \mathbf{I}_{4N} & \mathcal{H}_{01} \\ \mathbf{0} & q\mathbf{I}_{4N} \end{pmatrix} \quad \text{and} \quad \mathbf{M}_{\mathcal{H}_{02}} = \begin{pmatrix} \mathbf{I}_{4N} & \mathcal{H}_{02} \\ \mathbf{0} & q\mathbf{I}_{4N} \end{pmatrix}. \qquad (29)$$

Let $\mathbf{f}_{01}, \mathbf{f}_{02} \in \mathbb{Z}^{4N}$ be the integer vectors corresponding to the element $(\mathbf{f}_0 - \mathbf{f}_2) + \omega(\mathbf{f}_1 - \mathbf{f}_2)$, $(\mathbf{f}_0 - \mathbf{f}_1) + \omega(\mathbf{f}_2 - \mathbf{f}_1) \in \mathbb{Z}[\omega]^{2N}$, respectively. Similarly are defined the vectors $\mathbf{g}_{01}, \mathbf{g}_{02} \in \mathbb{Z}^{4N}$. Then, the vectors $(\mathbf{f}_{01}, \mathbf{g}_{01})$, $(\mathbf{f}_{02}, \mathbf{g}_{02})$ belong to the lattices $L_{\mathcal{H}_{01}}$ and $L_{\mathcal{H}_{02}}$, respectively. According to Assumption 1, for the secret vector $(\mathbf{f}, \mathbf{g}) = (1 + p\mathbf{F}, \mathbf{g})$, we have

$$\|\mathbf{f}_i\| \approx |p| \cdot \|\mathbf{F}_i\| \approx 2\sqrt{\frac{8N}{9}}, \ \|\mathbf{g}_i\| \approx \sqrt{\frac{8N}{9}}, \qquad (30)$$

and the length of the private vector $(\mathbf{f}, \mathbf{g})$ is approximately $\sqrt{40N/3}$. Therefore,

$$\|(\mathbf{f}_{01}, \mathbf{g}_{01})\| \approx \|(\mathbf{f}_{02}, \mathbf{g}_{02})\| \lesssim \sqrt{\frac{320N}{9}} \approx \sqrt{\frac{8}{3}} \|(\mathbf{f}, \mathbf{g})\|. \qquad (31)$$

While the Gaussian heuristic predicts the length of the shortest vectors in the lattices $L_{\mathcal{H}_{01}}$ and $L_{\mathcal{H}_{02}}$ to be

$$\sigma(L_{\mathcal{H}_{01}}) = \sigma(L_{\mathcal{H}_{02}}) = \sqrt{\frac{4N|q|}{\pi e}} \approx 2.738N. \tag{32}$$

Therefore, the vectors $(\mathbf{f}_{01}, \mathbf{g}_{01})$ and $(\mathbf{f}_{02}, \mathbf{g}_{02})$ are $O(\frac{1}{\sqrt{N}})$ times shorter than the Gaussian expected length. Hence, for large values of N, they are the shortest vectors in the corresponding lattices with a high probability. Thus, the problem of recovering the key is equivalent to solving SVP in $8N$-dimensional lattices that is equal to lattice dimension in the case of NTRU over $\mathbb{Z}C_{N'\approx 4N}$. Conclusively, in our design, the dimension reduction attack reduces the dimension of the lattice by a factor of 1.5, i.e., from $12N$ to $8N$, while DiTRU suffers a dimension loss by a factor of 2. This shows the benefit of working with the semidirect product $C_N \rtimes C_3$.

Hardness of Lattice Reduction. It is a known fact that the hardness of solving the SVP in a lattice increases with the ratio of the length of the shortest vector to the Gaussian heuristic called *lattice gap* [14]. For NTRU over $\mathbb{Z}C_{N'\approx 4N}$, this ratio[2] is $0.731/\sqrt{N}$, while for our scheme, the lattice gap is $1.54/\sqrt{N}$. Therefore, the lattices associated with our cryptosystem are practically more resistant to lattice attacks compared to the standard NTRU in equal dimensions. To further investigate, corresponding to every parameter set (N', q', p') for NTRU HPS in [35, Table 3], we choose a prime $N \approx N'/4$ and the smallest rational prime $q \in \mathbb{Z}[\omega]$ such that $q > 16N + 2$. Then, according to 2016-estimation, we estimate the blocksize β required for recovering the short vectors in lattices $L_{\mathcal{H}_{01}}, L_{\mathcal{H}_{02}}$, and compare with β' required for NTRU in Table 1. It suggests that one can select smaller values of N such that $4N < N'$ for our scheme and still achieve the same security as NTRU over $\mathbb{Z}C_{N'}$.

Table 1. Blocksize estimation for NTRU vs. our scheme for approximately the same dimensions.

	NTRU HPS	Our scheme	
(N', q', p')	β'	(N, q, p)	β
$(587, 2048, 3)$	456	$(139, 2237, 2)$	506
$(863, 2048, 3)$	701	$(211, 3389, 2)$	777
$(1109, 4096, 3)$	893	$(277, 4451, 2)$	1025

The value of the required blocksize is higher for the new proposal compared to NTRU HPS. It confirms that the lattices associated with our design offer more resistance against lattice reduction techniques, thus resulting in smaller values of N in Table 2.

---

[2] For NTRU, the length of the key is assumed to be $\sqrt{4N'/3 + 1}$ and the value of q' that achieves no decryption failure is $q' \geq 16N'/3$. For our scheme, although the norm of the target vector has upper bound $\sqrt{320N/9}$. However, it is empirically observed that the norm of the target vector is approximately $\sqrt{160N/9}$. Therefore, for a conservative estimation of the lattice gap and the blocksize, we consider the latter value of the norm.

Hybrid Attack. As the name suggests, hybrid attack [21] combines two attacks, the lattice and the combinatorial search attacks. It involves searching for some coefficients of the key in its tail region and reducing a part of the lattice to recover the full secret using the nearest neighborhood algorithm [4]. The parameters of the recent NTRU proposals [9,25], whose keys are ternary and sparse, are evaluated based on hybrid attacks. However, it is observed that the primal attack outperforms the hybrid attack when the secret key is not ternary, which increases the search cost, as in the case of DiTRU [35]. In our design also, the partial information of the key stored in lower dimensional lattices consists of coefficients from the set $\{0, \pm 1, \pm 2, \pm 3\}$. Expectedly, the overall cost of the hybrid attack exceeds the cost of the primal attack. Therefore, we have selected parameters by considering only the primal attack.

4.3   Overstreched NTRU Attack

An NTRU variant with a very large modulus is referred to as overstretched. The attacks exploiting the presence of specific algebraic structures in overstretched NTRU lattices are presented in [1,28]. Later, Ducas and Woerdon [14] narrowed down the estimation on modulus q that separates the overstretched regime from the standard regime. They call this *fatigue point* and showed that for an NTRU lattice of dimension $2n$ with modulus q, the fatigue point is $q \approx 0.004n^{2.484}$. One can verify that the suggested parameter sets in Table 2 for GR-NTRU over $\mathbb{Z}[\omega](C_N \rtimes C_3)$ satisfy $|q| \ll 0.004(4N)^{2.484}$. Therefore, our cryptosystem does not fall under the category of overstretched NTRU and is safe against these kinds of attacks.

5   Parameters and Performance Analysis

For our scheme, we are proposing two categories of parameters targeting 128-bit (Level I), 192-bit (Level III), and 256 (Level V) according to NIST definition. Table 2 provides the memory and time requirements for the two selected parameter sets, where the first set provides no decryption failure while the other allows a negligible decryption rate.

Memory Requirements. According to [26, Theorem 3], any element $a + b\omega$ reduced modulo q satisfies $a, b \in [-2|q|/3, 2|q|/3]$. Therefore, the size of the public key $h = f^{-1}g \in \mathbb{Z}[\omega]/\langle q \rangle(C_N \rtimes C_3)$ is $(6N/8) \cdot \lceil \log_2(4|q|/3) \rceil$ bytes. The private key $F \in \mathbb{Z}[\omega](C_N \rtimes C_3)$ is such that its coefficients are of the form $a + b\omega$ where $a, b \in \{0, \pm 1\}$. Since, $-1 \equiv 2 \mod 3$, and $\sum_{i=0}^{4} 2 \cdot 3^i \leq \sum_{i=0}^{7} 2^i$. Therefore, every five coefficients of F can be stored in 8 bits or 1 byte. Thus, the size of the private key is $\lceil 6N/5 \rceil$ bytes (Table 3).

Table 2. Parameters for $\mathbb{Z}[\omega](C_N \rtimes C_3)$-NTRU with no decryption failure and negligible decryption failure.

	No decryption failure			Neglible decryption failure		
Security level	I	III	V	I	III	V
(N, q, p)	$(127, 2039, 2)$	$(181, 2903, 2)$	$(241, 3863, 2)$	$(109, 701, 2)$	$(157, 1013, 2)$	$(211, 1361, 2)$
sk (bytes)	153	218	290	131	189	254
pk (bytes)	1143	1629	2350	818	1296	1741
β	461	664	890	464	663	886
BKZ(S) [classical]	134	193	259	135	193	258
BKZ(S) [quantum]	122	175	235	122	175	234
Comb	505	719	957	433	624	838
Dec failure	–	–	–	2^{-135}	2^{-199}	2^{-269}
CPU cycles $\times 10^3$						
KeyGen	38 163	72 545	131 162	27 498	58 308	103 094
Enc	6 692	11 442	20 452	4 907	9 878	16 313
Dec	12 125	21 308	38 147	8 712	18 109	30 619

Table 3. Memory requirements of the considered NTRU variants.

	NTRU HPS		DiTRU	
Level	sk	pk	sk	pk
I	118	808	217	1488
III	173	1187	319	2391
V	221	1664	416	3116

This demonstrates the memory benefits of the proposed scheme as the size of the private (**sk**) and public key (**pk**) (in bytes) of parameters allowing negligible decryption failure for our design are less than DiTRU, while are approximately equal to NTRU HPS.

Performace Analysis. In order to analyze the performance of the proposed scheme, we provide a full reference implementation in C. All the provided measurements are evaluated on a single core of 12th Gen Intel(R) Core(TM) i7-1255U with 32 GB RAM and running Linux (Ubuntu 22.04.3 LTS) with TurboBoost and hyper-threading disabled. We compile the code using GCC version 11.4.0-1ubuntu1 22.04 with no optimization flags enabled. Table 2 presents the average CPU cycles required to generate a key, encrypt, and decrypt a message over 10,000 runs. In Table 4, we compare the performance of our work with other prominent NTRU variants in the literature by comparing the CPU cycles needed for key generation and for encrypting/decrypting messages of the same length, **not** only a single message (see Sect. 5.1). The design rationale of the IND-CCA2 PKE in our work, as well as DiTRU and NTRU HPS, is similar to the one used in the NTRUEncrypt submission [10] (see Appendix A).

5.1 Discussion

The computational cost of key generation, encryption, and decryption is mainly determined by the 'polynomial' multiplications over the underlying ring. For simplicity, we will discuss the results using the conventional polynomial multi-

Table 4. Performance benchmark (*CPU cycles* $\times 10^3$) of this work vs. NTRU and DiTRU for *Key generation*, *Encryption*, and *Decryption* for messages of equal lengths.

	NTRU HPS ($N,q,p=3$)			This work ($N,q,p=2$)			Ratio[a]
	$(587,2048)$	$(863,2048)$	$(1109,4096)$	$(109,701)$	$(157,1013)$	$(211,1361)$	(r_1,r_2,r_3)
Gen:	62 311	146 706	224 363	27 498	58 308	103 094	$(2.27,2.52,2.18)$
Enc:	3 132 799	9 105 932	19 790 178	2 772 310	7 569 493	16 294 397	$(1.13,1.20,1.21)$
Dec:	5 800 643	17 201 618	37 829 256	4 988 320	13 965 567	30 569 442	$(1.16,1.23,1.24)$
	DiTRU ($N,q,p=3$)						
	$(541,2048)$	$(797,4096)$	$(1039,4096)$	$(109,701)$	$(157,1013)$	$(211,1361)$	(r_1,r_2,r_3)
Gen:	84 756	189 770	308 543	27 498	58 308	103 094	$(3.08,3.05,2.99)$
Enc:	9 777 811	29 658 528	66 558 364	5 092 057	14 373 555	30 551 756	$(1.92,2.06,2.19)$
Dec:	18 682 243	57 329 287	129 664 570	9 180 125	26 540 407	57 287 299	$(2.04,2.16,2.26)$

The ratio is provided as a tuple (r_1, r_2, r_3), where r_1 represents the ratio of CPU cycles needed for key generation, encryption, and decryption by NTRU (DiTRU) to the cycles required by our work in the first level of security. Similarly, r_2 and r_3 represent the ratios measured for the third and the fifth levels of security.

plications. The cost of a polynomial multiplication is $27N^2$ scalar multiplications for this work versus $16N^2$ and $32N^2$ for NTRU and DiTRU, respectively.

Analysis of Key Generation. We discuss the performance of the key generation algorithm when implemented in constant time using the Bernstein-Yang algorithm [8]. For NTRU HPS, the inversion algorithm performs 8 polynomial multiplication in $\mathbb{Z}C_{N'\approx 4N}$ (of cost $\approx 8 \times 16N^2$). Similarly, for DiTRU, the inversion algorithm [35, Algorithm 1] over $\mathbb{Z}D_{N'}$ finds an inverse in $\mathbb{Z}C_{N'}$ plus does 4 extra multiplications over $\mathbb{Z}C_{N'}$, that costs approximately $12 \times 16N^2$. On the other hand, inversion for this work (Algorithm 2) requires 15 multiplications over $\mathbb{Z}[\omega]C_N$ costing $3 \times 15N^2$ scalar multiplications plus an inversion over $\mathbb{Z}[\omega]C_N$. The cost of finding the inverse in $\mathbb{Z}[\omega]C_N$ is upper bounded by $57N^2$ scalar multiplications as detailed in Appendix B. Hence, the cost of constant time implementation of our key generation process is dominated by $102N^2$ scalar multiplications, which is roughly 1.3 and 1.9 faster than NTRU HPS and DiTRU, respectively, when $N' \approx 4N$. In practice, the decryption failure model and a higher lattice gap of this work allow smaller values of $N(< N'/4)$ for equivalent levels of security. As a result, the key generation is roughly two times (three times) faster than the one used in NTRU (DiTRU) in practice. See Table 4.

Analysis of Encryption/Decryption. As in the key generation, the cost of encryption/decryption is dominated by the polynomial multiplications cost. The length of a message encrypted using $\mathbb{Z}C_{N'}$ is $N' \approx 4N$, using $\mathbb{Z}D_{N'}$ is $2N' \approx 8N$, whereas the length of a message encrypted using $\mathbb{Z}[\omega](C_N \rtimes C_3)$ is $6N$ (integer coefficients). Therefore, for a fair comparison of efficiency, we compare the cost of encrypting/decrypting messages of the same length, that is 3 message processings by $\mathbb{Z}C_{N'}$ and 3 message processings by $\mathbb{Z}D_{N'}$ with 2 and 4 message processings, respectively, by $\mathbb{Z}[\omega](C_N \rtimes C_3)$. Therefore, in general, for $N' = 4N$, our cryptosystem is approximately 1.125 times slower than the standard NTRU, while it is approximately 1.7 times faster than DiTRU. However, this is not the case in practice where the parameters selection (considering the smaller value of modulus q and the hardness of the Core-SVP) leads to values of N smaller than $N'/4$. As a result, our cryptosystem is faster than NTRU and DiTRU by approximately a factor of 1.2 and 2, respectively, while encrypting/decrypting messages of the same length. Refer to Table 4.

Acknowledgment. The authors want to express their gratitude to the reviewers whose valuable suggestions have greatly helped improve the editorial quality of the paper. Vikas Kumar would like to thank CSIR for supporting his research through grant no. 09/143(1038)/2020-EMR-I.

A   Sketched Design Rationale

We follow the same design framework as adopted in NTRUEncrypt [10] to construct a Probabilistic Public Key Encryption (PPKE) scheme. The proposal derives its CPA security from the NTRU assumption, which is transformed into CCA2 secure by employing the NAEP padding mechanism [23]. All the steps in Fig. 3 are almost identical to the design rationale used in NTRUEncrypt submission [10], except that the operations are now performed in the noncommutative structure $R^\omega(C_N \rtimes C_3)$ modulo q or p.

KEYGEN(seed)	ENCRYPT(h, m)	DECRYPT(f, c)
1. $g \leftarrow$ Sampler($seed, L_g$) 2. $F \leftarrow$ Sampler($seed, L_f$) 3. $f \leftarrow 1 + pF$ 4. if(f invertible mod q) $\quad f_q \leftarrow$ inversemodq(f) $\quad h \leftarrow pf_q * g \pmod q$ $\quad$ return (h, f) 5. else go to step 2	1. $coins \leftarrow$ Hash(h, m) 2. $\phi \leftarrow$ Sampler($coins, L_\phi$) 3. $s \leftarrow \phi * h \pmod q$ 4. $t \leftarrow$ Sampler(Hash(s), L_m) 5. $m' = m - t \pmod p$ 6. $c = s + m' \pmod q$ 7. return c	1. $a \leftarrow c * f \pmod q$ 2. $m' \leftarrow a \pmod p$ 3. $s \leftarrow c - m' \pmod q$ 4. $t \leftarrow$ Sampler(Hash(s), L_m) 5. $m \leftarrow m' + t \pmod p$ 6. if(Encrypt(h, m) $\neq c$) $\quad$ return $\perp$ 7. else return m

Fig. 3. Sketch of the CCA2 secure PPKE for our proposal. The function Sampler randomly samples an element unique to the seed from the input space. The spaces L_f, L_g, L_ϕ, and L_m are defined in the Sect. 3.

B   Constant Time Inversion Algorithm for $\mathbb{Z}[\omega]/\langle q \rangle C_N$

Algorithm 3 is a direct adaptation of the Bernstein-Yang algorithm [8] with the required modifications to our new ring $\mathbb{Z}[\omega]/\langle q \rangle C_N$.

– Multiplication of two Eisenstein integers requires 3 integer multiplications.

Algorithm 3: Constant time inversion in $\mathbb{Z}\omega/\langle q \rangle C_N$

Input: $d(x) \in \mathbb{Z}\omega/\langle q \rangle C_N$
Output: $delta = 0, inv(x) = d(x)^{-1} \in \mathbb{Z}\omega/\langle q \rangle C_N$, if $d(x)$ is invertible, else $delta = -1$

1 $g(x) \leftarrow d(x), f(x) \leftarrow x^N - 1, v(x) \leftarrow 0, r(x) \leftarrow 1$
2 $delta \leftarrow 1$
3 **for** $i = 0$ *to* $2N - 2$ **do**
4 $\quad$ $v(x) \leftarrow x * v(x)$
5 $\quad$ $swap = (-delta < 0)$ & $(g_0 \neq 0)$
6 $\quad$ $delta^\wedge = swap$ & $(delta^\wedge - delta)$
7 $\quad$ $delta = delta + 1$
8 $\quad$ constSwap$(f(x), g(x), swap)$ $\qquad$ /* swap $f(x)$ and $g(x)$ if $swap$ is 1 */
9 $\quad$ constSwap$(v(x), r(x), swap)$ $\qquad$ /* swap $v(x)$ and $r(x)$ if $swap$ is 1 */
10 $\quad$ $g(x) \leftarrow f_0 g(x) - g_0 f(x) (\text{mod } q)$
11 $\quad$ $r(x) \leftarrow f_0 r(x) - g_0 v(x) (\text{mod } q)$
12 $\quad$ $g(x) \leftarrow g(x)/x$

13 $k \leftarrow$ inverse-mod q-in-$\mathbb{Z}[\omega](f_0)$ $\qquad$ /* inverse of f_0 in $\mathbb{Z}[\omega]$ modulo q */
14 $inv(x) \leftarrow k*$reverse$(v(x))(\text{mod } q)$ $\qquad$ /* reverse coefficients of $v(x)$ */
15 **return** $delta, inv(x)$

- Modulo q in $\mathbb{Z}$ (for a prime q) requires 4 scalar multiplications in constant-time, therefore modulo q in $\mathbb{Z}[\omega]$ (for a prime $q + 0\omega$) requires 8 scalar multiplications (Algorithm refalg:division).
- Inversion of an element in $\mathbb{Z}[\omega]$ modulo q is upper bounded by $17 + 10 \log_2(q - 2)$ scalar multiplications as in Lemma 1.

Therefore, lines 10 and 11 contribute to $14(N + 1)$ scalar multiplications each. Line 13 contributes to $17 + 10 \log_2(q - 2)$ multiplications, and line 14 contributes to $11N$ scalar multiplications.

References

1. Albrecht, M., Bai, S., Ducas, L.: A subfield lattice attack on overstretched NTRU assumptions. In: Advances in Cryptology – CRYPTO 2016, pp. 153–178. Springer Berlin Heidelberg (2016). https://doi.org/10.1007/978-3-662-53018-4_6
2. Alkim, E., Ducas, L., Pöppelmann, T., Schwabe, P.: Post-quantum key {Exchange-A} new hope. In: 25th USENIX Security Symposium (USENIX Security 16), pp. 327–343 (2016). https://www.usenix.org/system/files/conference/usenixsecurity16/sec16_paper_alkim.pdf
3. Aono, Y., Wang, Y., Hayashi, T., Takagi, T.: Improved progressive BKZ algorithms and their precise cost estimation by sharp simulator. In: Annual International Conference on the Theory and Applications of Cryptographic Techniques, pp. 789–819. Springer (2016).https://doi.org/10.1007/978-3-662-49890-3_30
4. Babai, L.: On lovász' lattice reduction and the nearest lattice point problem. Combinatorica **6**, 1–13 (1986). https://doi.org/10.1007/BF02579403

5. Bagheri, K., Sadeghi, M.R., Panario, D.: A non-commutative cryptosystem based on quaternion algebras. Des. Codes Crypt. **86**, 2345–2377 (2018). https://doi.org/10.1007/s10623-017-0451-4

6. Becker, A., Ducas, L., Gama, N., Laarhoven, T.: New directions in nearest neighbor searching with applications to lattice sieving. In: Proceedings of the Twenty-Seventh Annual ACM-SIAM Symposium on Discrete Algorithms, pp. 10–24. SIAM (2016). https://doi.org/10.1137/1.9781611974331.ch2

7. Bernstein, D.J., Chuengsatiansup, C., Lange, T., van Vredendaal, C.: NTRU Prime: reducing attack surface at low cost. In: Selected Areas in Cryptography – SAC 2017, pp. 235–260. Springer International Publishing (2018). https://doi.org/10.1007/978-3-319-72565-9_12

8. Bernstein, D.J., Yang, B.Y.: Fast constant-time GCD computation and modular inversion. IACR Trans. Crypt. Hardw. Embed. Syst. **2019**(3), 340–398 (2019).https://doi.org/10.13154/tches.v2019.i3.340-398

9. Chen, C., et al.: PQC round-3 candidate: NTRU. technical report. Tech. rep., NTRU Cryptosystems Technical Report No.11, Version 2, March 2001. Report (2019). https://ntru.org/f/ntru-20190330.pdf

10. Chen, C., Hoffstein, J., Whyte, W., Zhang, Z.: NIST PQ submission: NTRU-Encrypt a lattice based encryption algorithm. NIST (2017). https://csrc.nist.gov/Projects/post-quantum-cryptography/post-quantum-cryptography-standardization/round-1-submissions

11. Chen, Y.: Réduction de réseau et sécurité concrète du chiffrement complètement homomorphe. Ph. D. thesis, l'Université Paris Diderot (2013). http://www.theses.fr/2013PA077242

12. Chen, Y., Nguyen, P.Q.: BKZ 2.0: better lattice security estimates. In: International Conference on the Theory and Application of Cryptology and Information Security, pp. 1–20. Springer (2011). https://doi.org/10.1007/978-3-642-25385-0_1

13. Coppersmith, D., Shamir, A.: Lattice attacks on NTRU. In: Advances in Cryptology — EUROCRYPT '97, pp. 52–61. Springer Berlin Heidelberg, Berlin, Heidelberg (1997). https://doi.org/10.1007/3-540-69053-0_5

14. Ducas, L., van Woerden, W.: NTRU fatigue: how stretched is overstretched? In: Advances in Cryptology – ASIACRYPT 2021, pp. 3–32. Springer International Publishing (2021). https://doi.org/10.1007/978-3-030-92068-5_1

15. Dummit, D.S., Foote, R.M.: Abstract Algebra, 3 edn. Wiley, Inc. (2003). https://www.wiley.com/en-in/Abstract+Algebra%2C+3rd+Edition-p-9780471433347

16. Fox, N.: Spectra of semidirect products of cyclic groups. Rose-Hulman Undergraduate Math. J. **11** (2010). https://scholar.rose-hulman.edu/rhumj/vol11/iss2/7

17. Gentry, C.: Key recovery and message attacks on NTRU-composite. In: Pfitzmann, B. (ed.) Advances in Cryptology — EUROCRYPT 2001, pp. 182–194. Springer Berlin Heidelberg, Berlin, Heidelberg (2001). https://doi.org/10.1007/3-540-44987-6_12

18. Hoffstein, J., Pipher, J., Silverman, J.: An Introduction to Mathematical Cryptography, 1st edn. Springer Publishing Company, Incorporated, NY (2008)

19. Hoffstein, J., Pipher, J., Silverman, J.H.: NTRU: a ring-based public key cryptosystem. In: International Algorithmic Number Theory Symposium, pp. 267–288. Springer, Berlin, Heidelberg (1998). https://doi.org/10.1007/BFb0054868

20. Hoffstein, J., Silverman, J.H., Whyte, W.: Meet-in-the-middle attack on an NTRU private key. Tech. rep., Technical report, NTRU Cryptosystems, July 2006. Report (2006). https://ntru.org/f/tr/tr004v2.pdf

21. Howgrave-Graham, N.: A hybrid lattice-reduction and meet-in-the-middle attack against NTRU. In: Advances in Cryptology - CRYPTO 2007, pp. 150–169. Springer Berlin Heidelberg (2007). https://doi.org/10.1007/978-3-540-74143-5_9

22. Howgrave-Graham, N., et al.: The impact of decryption failures on the security of NTRU encryption. In: Boneh, D. (ed.) Advances in Cryptology - CRYPTO 2003, pp. 226–246. Springer Berlin Heidelberg, Berlin, Heidelberg (2003). https://doi.org/10.1007/978-3-540-45146-4_14

23. Howgrave-Graham, N., Silverman, J.H., Whyte, W.: Choosing parameter sets for NTRUEncrypt with NAEP and SVES-3. In: Menezes, A. (ed.) Topics in Cryptology – CT-RSA 2005, pp. 118–135 (2005). https://doi.org/10.1007/978-3-540-30574-3_10

24. Hurley, T.: Group rings and rings of matrices. Int. J. Pure Appl. Math. **31**, 319–335 (2006). https://www.researchgate.net/publication/228928727_Group_rings_and_rings_of_matrices

25. Hülsing, A., Rijneveld, J., Schanck, J., Schwabe, P.: High-speed key encapsulation from NTRU. In: International Conference on Cryptographic Hardware and Embedded Systems, CHES 2017, pp. 232–252 (2017). https://doi.org/10.1007/978-3-319-66787-4_12

26. Jarvis, K., Nevins, M.: ETRU: NTRU over the eisenstein integers. Des. Codes Cryptogr. **74**, 219–242 (2015). https://doi.org/10.1007/s10623-013-9850-3

27. Kim, J., Lee, C.: A polynomial time algorithm for breaking NTRU encryption with multiple keys. Des. Codes Crypt. **91**, 2779–2789 (2023). https://doi.org/10.1007/s10623-023-01233-5

28. Kirchner, P., Fouque, P.A.: Revisiting lattice attacks on overstretched NTRU parameters. In: Advances in Cryptology – EUROCRYPT 2017, pp. 3–26. Springer International Publishing (2017). https://doi.org/10.1007/978-3-319-56620-7_1

29. Kirshanova, E., May, A., Nowakowski, J.: New NTRU records with improved lattice bases. In: Post-Quantum Cryptography, pp. 167–195 (2023). https://doi.org/10.1007/978-3-031-40003-2_7

30. Kumar, V., Raya, A., Gangopadhyay, S., Gangopadhyay, A.K.: Lattice attack on group ring NTRU: the case of the dihedral group. arXiv:2309.08304 (2023)

31. Laarhoven, T.: Search problems in cryptography: from fingerprinting to lattice sieving. Ph. D. thesis, Eindhoven University of Technology (2015). https://research.tue.nl/en/publications/search-problems-in-cryptography-from-fingerprinting-to-lattice-si

32. Lenstra, A.K., Lenstra, H.W., Lovász, L.: Factoring polynomials with rational coefficients. Mathematische annalen **261**(ARTICLE), 515–534 (1982). https://doi.org/10.1007/BF01457454

33. Malekian, E., Zakerolhosseini, A.: OTRU: a non-associative and high speed public key cryptosystem. In: 2010 15th CSI International Symposium on Computer Architecture and Digital Systems, pp. 83–90 (2010). https://doi.org/10.1109/CADS.2010.5623536

34. Malekian, E., Zakerolhosseini, A., Mashatan, A.: QTRU : a lattice attack resistant version of NTRU PKCS based on quaternion algebra. IACR Cryptol. ePrint Archive **386** (2009). https://eprint.iacr.org/2009/386

35. Raya, A., Kumar, V., Gangopadhyay, S.: DiTRU: a resurrection of NTRU over dihedral group. In: Vaudenay, S., Petit, C. (eds.) Progress in Cryptology - AFRICACRYPT 2024, pp. 349–375. Springer Nature Switzerland, Cham (2024). https://doi.org/10.1007/978-3-031-64381-1_16

36. Schnorr, C.: A hierarchy of polynomial time lattice basis reduction algorithms. Theoret. Comput. Sci. **53**(2), 201–224 (1987). https://doi.org/10.1016/0304-3975(87)90064-8
37. Silverman, J.H.: Almost inverses and fast NTRU key creation. NTRU Cryptosystems Technical Report #14 (1999). https://ntru.org/f/tr/tr014v1.pdf
38. Stinson, D., Paterson, M.: Cryptography: Theory and Practice, 4 edn. CRC Press, Chapman and Hall Book, Taylor & Francis (2017). https://doi.org/10.1201/9781315282497
39. Thakur, K.: A variant of NTRU with split quaternions algebra. Palestine J. Math. **6**(2), 598–610 (2017). https://pjm.ppu.edu/sites/default/files/papers/PJM_April_2017_28.pdf
40. Truman, K.R.: Analysis and extension of non-commutative NTRU. Ph. D. dissertation, University of Maryland (2007). https://drum.lib.umd.edu/handle/1903/7344
41. Venier, D., Cheung, R.C.: A highly parallel constant-time almost-inverse algorithm. In: 2020 IEEE International Conference on Signal Processing, Communications and Computing (ICSPCC), pp. 1–6 (2020). https://doi.org/10.1109/ICSPCC50002.2020.9259505
42. Yasuda, T., Dahan, X., Sakurai, K.: Characterizing NTRU-variants using group ring and evaluating their lattice security. IACR Cryptol. ePrint Arch. 1170 (2015). http://eprint.iacr.org/2015/1170

Improving Tightness Gap of GGM Construction and Its Applications

Mridul Nandi[✉]

Indian Statistical Institute, Kolkata, India
mridul@isical.ac.in

Abstract. The prefix-free PRF (pseudorandom function) security of a cascade function based on a compression function f against a q-query distinguisher is reduced to a q-query PRF security of f with a tightness gap ℓq where ℓ represents the length of the longest query among all q queries. In this paper, we have shown a new reduction which is also applicable to multiuser setup and improves the tightness gap for both adaptive and non-adaptive distinguishers. As an immediate application of our result, we have shown multiuser security of NMAC, HMAC and many other known MACs in the standard model. Moreover, the tightness gap is improved in comparison with known single-user analysis. We also have shown a similar tightness gap for the single-keyed version of NMAC. As a result, the constants ipad and opad used in HMAC and relying upon the PRB (pseudorandom bit) assumption on the underlying compression function become redundant.

Keywords: PRF · HMAC · NMAC · cascade · non-adaptive security

1 Introduction

Verifying the integrity and authenticity of data is a fundamental necessity in computer systems and networks. Two parties communicating over an insecure channel can use a message authentication code (MAC) or a stronger construct called a pseudorandom function (PRF), which enables the receiver to validate that the data was indeed sent by the sender. MACs and PRFs are often constructed from block ciphers, such as CBC-MAC [5,6], PMAC [8], or the NIST-recommended CMAC [10,16].

Historically, hash functions were faster than block ciphers in software implementations. Additionally, hash functions were not subject to export restrictions enforced by the USA and other countries, which led to increased interest in constructing MACs from cryptographic hash functions. However, hash functions were not originally designed to serve as MACs or PRFs and did not naturally accommodate a secret key. One of the earliest techniques for converting a hash function (specifically, the Merkle-Damgård hash [9,25]) into a MAC involved prepending the message with a secret key. It was soon discovered, however, that such constructions are vulnerable to the *length extension attack* (see Example

9.64 in [24]). Fortunately, it has been shown [4] that the construction is secure for prefix-free message spaces, where no two messages are prefixes of each other.

The Envelope MAC, also known as the Sandwich MAC [17, 27, 30], a variant of keyed MD, was demonstrated to be insecure [28]. The vulnerabilities were due to poor formatting of the key block processed at the end. Later, Yasuda [32] and Koblitz and Menezes [20] established the PRF security of Envelope MAC when appropriate message formatting was applied (this variant is also considered in this paper). In CRYPTO 1996, the authors of [3] proposed NMAC and HMAC, and proved their security under certain assumptions on the underlying compression function.

NOTATIONS. In this paper, we fix two positive integers b and c, and denote $\{0,1\}^b$ and $\{0,1\}^c$ by B (referred to as the set of blocks) and C, respectively. Let λ represent the empty string, and let B* (or B$^+$) denote the set of all block tuples (or block tuples with at least one block, respectively). We write $\mathsf{B}^+_{mu} = \mathcal{I} \times \mathsf{B}^+$ and $\mathsf{B}^*_{mu} = \mathcal{I} \times \mathsf{B}^*$ for some set $\mathcal{I}$ (representing user index space). For $m := (m[1], \ldots, m[r]) \in \mathsf{B}^+$ and $1 \le i \le j \le r$, we write

1. the number of blocks $\|m\| = r$,
2. sub-tuple $m[i..j] := (m[i], \ldots, m[j])$,
3. suffix $m[i..] = m[i..r]$ and
4. prefix $m[..j] = m[1..j]$.

The same convention is followed when the index of m starts at 0. For $a \le b$, we call $m[..a]$ a *prefix* of $m[..b]$, and denote it by $m[..a] \preceq m[..b]$. We write x^q (resp. (x^q, y^q)) to represent a q-tuple $(x_1, \ldots, x_q)$ (resp. $((x_1, y_1), \ldots, (x_q, y_q))$). When x is chosen uniformly from a set S and is independent of all random variables defined so far, we denote it as $x \leftarrow_\$ S$.

CASCADE FUNCTION. For $f : \mathsf{C} \times \mathsf{B} \to \mathsf{C}$, we define the *cascade function* recursively as $f^*(h, \lambda) = h$ for all $h \in \mathsf{C}$ and

$$f_h^*(m[..i]) := f^*(h, m[..i]) = f^*\big(f(h, m[..i-1]), m[i]\big), \ \forall m[..i] \in \mathsf{B}^+.$$

One can further extend the domain of f^* to the set of all arbitrary bit strings by applying an appropriate injective padding rule as a preprocessor of the above

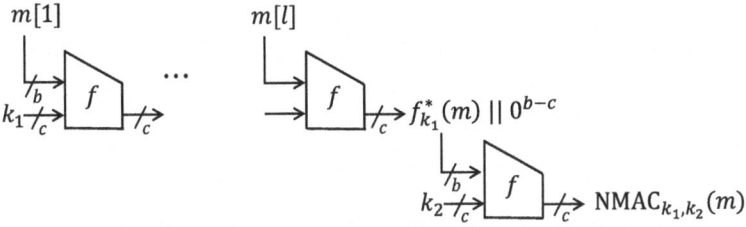

Fig. 1. NMAC$_{k_1, k_2}(m)$: The top layer represents the cascade output and the bottom layer represents the finalization process applied to the output of the cascade.

cascade function. As there is no loss in security, *we assume message space as* B* (Fig. 1).

NMAC AND HMAC. For keys $k, k_1, k_2 \in \{0,1\}^c$, b-bit constants ipad, opad specified in [3], a c-bit initial value IV and message $m \in B^*$

$$\mathsf{NMAC}_{k_1,k_2}(m) = f_{k_2}(f^*_{k_1}(m)\|0^{b-c}), \qquad \mathsf{HMAC}_k(m) = \mathsf{NMAC}_{\mathsf{KDF}(k)}(m),$$

where $\mathsf{KDF}(k) = \big(k_1 := f(IV, k^{\oplus \mathsf{ipad}}), k_2 := f(IV, k^{\oplus \mathsf{opad}})\big)$ and $k^{\oplus \alpha} = (k\|0^{b-c}) \oplus \alpha$. Here, we must assume that $c \le b$, which used to hold for the earlier compression functions.[1]

ENVELOPE MAC. Finally, we define another old cascade-based MAC construction, called the Envelope MAC or EvMAC. Let pad map a k-bit string to B. For example, if $k \le b$, we consider $\mathsf{pad}(K) = K\|0^{b-k}$. We define a dual keyed function (interchanging the position of the input and key)

$$f^{\downarrow}_K(x) := f^{\downarrow}(K, x) := f(x, \mathsf{pad}(K)).$$

For any $m \in B^+$, $K \in \{0,1\}^k$, we define $\mathsf{EvMAC}(K, m) = f^*(IV, (K', m, K'))$ where $K' = \mathsf{pad}(K)$ and $IV \in \{0,1\}^c$ is a fixed constant specified by the MD hash function based on f. Using the dual function notation, we can equivalently write, for $K' = f^{\downarrow}_K(IV)$, $\quad \mathsf{EvMAC}(K, m) = f^{\downarrow}_K\big(f^*_{K'}(m)\big)$ (Fig. 2).

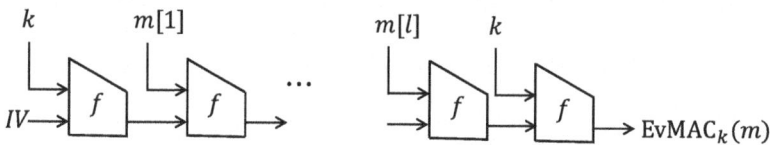

Fig. 2. $\mathsf{EvMAC}_k(m)$: Envelope MAC or Sandwiched MAC.

1.1   Known Results of Hash-Based MAC

For a keyed function F, we denote the maximum PRF advantage against adaptive (resp. non-adaptive) distinguishers as $\mathbf{Adv}^{\mathrm{prf}}_F(q, \ell, \sigma, T)$ (resp. $\mathbf{Adv}^{\mathrm{nprf}}_F(q, \ell, \sigma, T)$), where the maximum is taken over all q-query distinguishers D running in time T, such that the total number of blocks and the maximum size of the query are at most σ and ℓ, respectively. We use the superscripts (i) nprf, (ii) pf_prf, and (iii) pf_nprf when restricting to distinguishers that are (i) non-adaptive (all queries are made before observing any responses),

---

[1] Later in RFC 2014 [21] and the FIPS PUB 198-1 [11] by NIST, the MD hash was replaced by any recommended hash function H while defining HMAC. Similar to the original definition of HMAC, the new definition assumed the hash size to be less than the block size.

(ii) prefix-free (query tuples are prefix-free), and (iii) prefix-free non-adaptive, respectively. The multi-user advantage for at most u users is similarly denoted as $\mathbf{Adv}_F^{\text{mu-prf}}(u, q, q_{\max}, \ell, \sigma, T)$, where $q_{\max}$ represents the maximum number of queries across all users. We use mu to denote a multi-user distinguisher. When $F = f$ (so $\ell = 1$, $\sigma = q$), we ignore the parameters ℓ and σ. Let $\theta := (u, q, q_{\max}, \ell, \sigma)$.

PRF Analysis on Cascade. The security of a fixed-length cascade construction (a special case of a cascade with a prefix-free domain, known as the GGM construction) was first implicitly shown in 1984 [13] and later published in 1986 [14]. The authors proved asymptotically that a c-bit to $2c$-bit PRBG (pseudorandom bit generator) can be extended to a fixed-length PRF. Such a PRBG is equivalent to a PRF with a one-bit domain (i.e., $b = 1$). In 1996, Bellare et al. [4] extended the GGM results to a general value of b and arbitrary prefix-free domains, showing the following (Gazi et al. [12] proved non-adaptive PRF security):

$$\mathbf{Adv}_{f^*}^{\text{pf-prf}}(q, \ell, \sigma, T) \leq \ell q \cdot \mathbf{Adv}_f^{\text{prf}}(q, T'), \quad \mathbf{Adv}_{f^*}^{\text{pf-nprf}}(q, \ell, \sigma, T) \leq \ell q \cdot \mathbf{Adv}_f^{\text{nprf}}(q, T'),$$

where $T' := T + O(\sigma)$ (throughout this paper, we use this notation). The GGM construction is essentially the fixed-length domain cascade construction when viewing the PRBG as a one-bit PRF.

PRF Analysis on NMAC and HMAC. Bellare [1] proved that

$$\mathbf{Adv}_{\text{NMAC}^f}^{\text{prf}}(q, \ell, \sigma, T) \leq \ell q^2 \cdot \mathbf{Adv}_f^{\text{prf}}(2, O(\ell)) + \mathbf{Adv}_f^{\text{prf}}(q, T').$$

Bellare assumed that a good compression function f satisfies $\mathbf{Adv}_f^{\text{prf}}(2, \ell) \approx \ell/2^c$, presuming key-guessing is the best strategy for distinguishing f from a random function. Thus, the security of NMAC is dominated by the bound $\ell^2 q^2/2^c$. Koblitz and Menezes (KM) observed that the reduction used in the preceding proof is non-constructive or existential (see [19,29] for details about different types of reductions). KM, and later Bernstein and Lange [7], showed that for almost all functions $f' : \{0,1\}^c \times B \to \{0,1\}$, there exists a 1-query distinguisher $\mathcal{A}$ running in $O(1)$ time such that $\mathbf{Adv}_{f'}^{\text{prf}}(\mathcal{A}) \geq \frac{1}{2^{c/2}}$. This implies that Bellare's result cannot guarantee security better than $\ell q^2/2^{c/2}$, violating Bellare's tightness claim (see [18] for a detailed discussion). Later in [2], Bellare withdrew and revised the tightness claim. In 2013, Koblitz-Menezes [18] also provided a constructive reduction, proving the following result (ignoring a dominated term):

$$\mathbf{Adv}_{\text{NMAC}^f}^{\text{prf}}(q, \ell, \sigma, T) \leq \ell q \cdot \mathbf{Adv}_f^{\text{prf}}(q, T').$$

In 2014, Gazi et al. [12] proved the following NMAC security through a constructive reduction:

$$\mathbf{Adv}_{\text{NMAC}^f}^{\text{prf}}(q, \ell, \sigma, T) \leq \ell q \cdot \mathbf{Adv}_f^{\text{nprf}}(q, T') + \mathbf{Adv}_f^{\text{prf}}(q, T') + \frac{q^2}{2^c}. \tag{1}$$

From the definition of HMAC, one can easily see that the PRF security of HMAC can be reduced to the PRF security of NMAC and the PRBG security of KDF.

1.2   Our Contributions

1. Multiuser PRF Security of Cascade. In this paper, we provide two reductions for the multiuser PRF security of the cascade construction:

$$\mathbf{Adv}_{f^*}^{\text{mu-pf-prf}}(\theta, T) \leq \sigma \cdot \mathbf{Adv}_f^{\text{prf}}(q_{\max}, T') \tag{2}$$

$$\mathbf{Adv}_{f^*}^{\text{mu-pf-nprf}}(\theta, T) \leq \sigma \cdot \mathbf{Adv}_f^{\text{nprf}}(1, T') + u \cdot \mathbf{Adv}_f^{\text{nprf}}(q_{\max}, T'). \tag{3}$$

The first reduction improves the tightness gap from ℓq to σ. The second reduction further improves the query complexity by bringing down it to 1.

2. Non-adaptive PRF security under weak f. Due to the key guessing attack, the above bounds cannot guarantee a security better than $\sigma T/2^c$ for any function f. Suppose f is a keyed function with a higher non-adaptive PRF advantage, such as $\mathbf{Adv}_f^{\text{nprf}}(D, T) \approx DT/2^c + 2^{-c/2}$. We still prove a similar advantage (up to a logarithmic factor) against a non-adaptive distinguisher:

$$\mathbf{Adv}_{f^*}^{\text{mu-pf\_nprf}}(\theta, T) \leq \frac{(\sigma T + \sigma^2) \cdot \log_2 q_{\max}}{2^c}$$

APPLICATIONS TO HMAC NMAC AND OTHERS. It can be shown that the generic reduction from NMAC to cascade (see Eq. 1) and HMAC to NMAC can be extended for a multiuser set-up. Hence, our results for non-adaptive PRF security of cascade can be directly applied for multiuser security of HMAC and NMAC. Following similar approach, simpler and improved analysis for boosted MD (Asiacrypt 2007 [31]) and MDP (JoC 2007 [15]) are given.

3. Security of single keyed NMAC, constant-free HMAC and EvMAC. We prove the security of the single-keyed NMAC construction $1k\_NMAC_K = NMAC_{K,K}$. This helps not only to eliminate the two constants used in HMAC but also to weaken the PRBG assumption on f. To prove this, we first establish a reduction for single-keyed composition. This single-keyed composition also helps us to prove the multiuser PRF security of Envelope MAC. Our result on Envelope MAC does not require any related-key type assumption appearing in [20].[2] In particular, we show the following three results ($\theta' = (q', q', q_{\max}, \ell+1, \sigma', \sigma_{\max})$):

$$\mathbf{Adv}_{1k\_NMAC^f}^{\text{mu-prf}}(\theta, T) \leq \mathbf{Adv}_{f^*}^{\text{mu-pf\_nprf}}(\theta', T') + u' \cdot \mathbf{Adv}_f^{\text{prf}}(2q_{\max}, T') + \frac{2q'^2}{2^c},$$

$$\mathbf{Adv}_{\text{EvMAC}}^{\text{mu-prf}}(\theta') \leq \mathbf{Adv}_{f^*}^{\text{mu-pf\_nprf}}(\theta', T') + u' \cdot \mathbf{Adv}_f^{\text{prf}}(q_{\max} + 1, T') + \frac{q'^2}{2^c} \text{ and}$$

$$\mathbf{Adv}_{\text{HMAC'}}^{\text{mu-prf}}(\theta') \leq \mathbf{Adv}_{1k\_NMAC}^{\text{mu-prf}}(\theta'),$$

---

[2] Yasuda proved PRF security of Envelope MAC (also called "Sandwich MAC," see [32]), along the lines of Bellare's NMAC security proof in [2]. Thus, the issues for NMAC are also present in his analysis. Koblitz and Menezes [20] proved the constructive reduction, but relies on some related-key security.

where HMAC$'$ is the same as HMAC when the two constants opad and ipad are replaced by the zero-bit string. Moreover, the security of the modified HMAC does not require the PRBG property (it only needs the regular property as KDF does not expand the output size in the modified definition, assuming key size to be as large as the chain size c). We note that all of our non-adaptive multiuser PRF securities for the cascade construction are applicable to these variants. We finally note that all reductions in our analysis are constructive and so the bounds apply to a uniform setting when we naturally extend the result in an asymptotic set-up.

2   Preliminaries

JOINT QUERY SPACE. Let $\mathcal{Q}(u, q, q_{\max}, \ell, \sigma)$, called joint query space, be the set of all q tuples m^q of at most ℓ blocks with altogether at most σ blocks such that

- the number of distinct elements present in $m_1[0], \ldots, m_q[0]$ is at most u and
- for all $\gamma \in \mathcal{I}$, the set $\mathcal{Q}_\gamma := \{i : m_i[0] = \gamma\}$ has at most $q_{\max}$ elements,

When $\ell = 1$ or $u = 1$ (single user) we simply write the set as $\mathcal{Q}^{\mathrm{mu}}(u, q, q_{\max})$ or $\mathcal{Q}^{\mathrm{su}}(q, \ell, \sigma)$ respectively. When $u = 1$, we simply skip the user index space and consider m^q with $m_i \in \mathsf{B}^*$. For any joint query space $\mathcal{Q}$, we write $\mathcal{Q}_{\mathrm{pf}} = \mathcal{Q} \cap \mathcal{P}$ where $\mathcal{P}$ is the set of tuples of prefix-free messages. We now associate the tuple $m^q \in \mathcal{Q}_{\mathrm{pf}}(\theta)$ with a tree $\mathcal{T}_{m^q}$, called *prefix tree*, over the vertex set $V \cup \{\lambda\}$ where

$$V = (m^q) = \{x \in \mathsf{B}^*_{\mathrm{mu}} : x \preceq m_i, \ i \in [q]\}.$$

It consists of all directed edges of the form $\mathsf{chop}(y) \to y$ for $y \in V$ where $\mathsf{chop}(y)$ represents the tuple after removing the last block from y. It is a rooted tree with λ as the root (it is the only vertex with in-degree zero). For every $v \in V \cup \{\lambda\}$, we define the set of outwards nodes, or children nodes, as $\mathsf{ch}(v) = \{u : v \to u\}$. For a set $\mathsf{ch}(S) = \cup_{v \in S}\mathsf{ch}(v)$. Let L denote the set of leaf nodes (having zero out-degree) which is same as the set $\{m_1, \ldots, m_q\}$ (as m^q is a prefix-free). Let $'(m^q) := V \setminus L$ be the set of all intermediate nodes. We denote $d := |V \setminus L| - 1$ and so $d \leq \sigma' := \sigma - q$.

Definition 1 (leave-cut). *A subset $U \subseteq V$ is called leave-cut if it is prefix-free and for every leaf m_i in the prefix tree $\mathcal{T}_{m^q}$ there exists exactly one node $u \in U$ such that $u \preceq m_i$.*

It is easy to see that whenever U is a leave-cut and $S \subseteq U \setminus L$ then $U' = (U \setminus S) \cup \mathsf{ch}(S)$ is also a leave-cut as we cut all leaf nodes m_i with a member of S as a prefix by one of its children nodes.

ORACLE AND KEYED FUNCTION. An $(\mathcal{X}, \mathcal{Y})$-oracle $\mathcal{O}$ is an interactive probabilistic algorithm that takes inputs from the set $\mathcal{X}$ and returns elements from the set $\mathcal{Y}$. A *random function* $\mathsf{RF}_{\mathcal{X} \to \mathcal{Y}}$ is a $(\mathcal{X}, \mathcal{Y})$-oracle which returns $\mathsf{RF}(x)$ on an input $x \in \mathcal{X}$, where $\mathsf{RF} \leftarrow_\$ \mathsf{Func}(\mathcal{X}, \mathcal{Y})$, the set of all functions from $\mathcal{X}$ to $\mathcal{Y}$. We write $\mathsf{RF}_{* \to \mathcal{Y}}$ to denote a random function from some domain $\mathcal{X}$ to $\mathcal{Y}$. A *keyed function* $F : \mathcal{K} \times \mathcal{X} \to \mathcal{Y}$ can be viewed as an $(\mathcal{X}, \mathcal{Y})$-oracle where the key $K \leftarrow_\$ \mathcal{K}$ (key space), and then for every query x it returns $F_K(x) := F(K, x)$. Note that the key is sampled once and is used for every query. We also call it $(\mathcal{X}, \mathcal{Y})$ keyed function (with a key space $\mathcal{K}$). A random function $\mathsf{RF}_{\mathcal{X} \to \mathcal{Y}}$ is an example of $(\mathcal{X}, \mathcal{Y})$ keyed function.

Definition 2 (multiuser keyed function). *Let* $F : \mathcal{K} \times \mathcal{X} \to \mathcal{Y}$ *be a keyed function. An* $\mathcal{I}$-*folded multiuser* $F^{\otimes \mathcal{I}}$ *(or simply* $F^\otimes$*) is an* $(\mathcal{I} \times \mathcal{X}, \mathcal{Y})$ *keyed function which, on an input* (γ, x)*, returns*

$$F^\otimes(\gamma \; ; \; x) := F(\mathsf{RF}_{\mathcal{I} \to \mathcal{K}}(\gamma), x).$$

In other words, a multiuser keyed function samples keys independently for all user index γ from the user index space $\mathcal{I}$ (i.e. $K_\gamma \leftarrow_\$ \mathcal{K}$ for all $\gamma \in \mathcal{I}$) and then it behaves as the original keyed function F_{K_γ} for any query with the user input γ.

A q-query (Without loss of generality, we assume it makes exactly q queries) t-time $(\mathcal{X}, \mathcal{Y})$-oracle algorithm $\mathcal{A}$ is an interactive algorithm that can interact with any $(\mathcal{X}, \mathcal{Y})$-oracle $\mathcal{O}$ (called a compatible oracle with $\mathcal{A}$), that makes at most q queries to its oracle, runs for time t, and finally returns some output z, denoted as $\mathcal{A}^\mathcal{O} \to z$, (if $z \in \{0, 1\}$, we also call it a *distinguisher*). When $\mathcal{O}$ is a multiuser oracle, $\mathcal{A}$ is called a u-user oracle algorithm if the number of distinct user indices queried is at most u. The transcript of interaction between $\mathcal{A}$ and $\mathcal{O}$ is denoted as

$$\tau(\mathcal{A}^\mathcal{O}) := (\tau_{\mathrm{query}}(\mathcal{A}^\mathcal{O}) := x^{q'}, \; \tau_{\mathrm{resp}}(\mathcal{A}^\mathcal{O}) := y^{q'})$$

where x_i denotes the ith query (which includes the user index in case of multiuser oracle algorithm) and y_i denotes the response of the query, $1 \leq i \leq q'$.

Definition 3. *A distinguisher* D *is called a* (θ, t)-*complexity distinguisher (or prefix-free distinguisher) if* D *runs for time* t *and for all compatible oracles* $\mathcal{O}$, *the transcript* $\tau_{\mathrm{query}}(\mathsf{D}^\mathcal{O}) \in \mathcal{Q}(\theta)$ *(or* $\tau_{\mathrm{query}}(\mathsf{D}^\mathcal{O}) \in \mathcal{Q}_{\mathrm{pf}}(\theta)$ *respectively).*

We sometimes ignore the time parameter t. Sometimes, we adjoin a post-processing oracle $\mathcal{O}_{\mathrm{pp}}$ (may be an internal state shared oracle with $\mathcal{O}$) which returns an additional response S after all queries have been made. In this case, we write the response transcript as $\tau_{\mathrm{resp}}(\mathcal{A}^{\mathcal{O}, \mathcal{O}_{\mathrm{PP}}}) := (y^{q'}, S)$.

Definition 4 (distinguishing advantage). *Let* F *and* G *be* $(\mathcal{X}, \mathcal{Y})$-*oracles. We define the* distinguishing advantage *of a distinguisher* D *as* $\Delta_\mathsf{D}(F \; ; \; G) := |\Delta_\mathsf{D}^*(F \; ; \; G)|$ *where* $\Delta_\mathsf{D}^*(F \; ; \; G) = \Pr(\mathsf{D}^F = 1) - \Pr(\mathsf{D}^G = 1)$ *is called signed distinguishing advantage.*

Definition 5. *Two* $(\mathcal{X}, \mathcal{Y})$-*oracles* $\mathcal{O}$ *and* $\mathcal{O}'$ *are called* equivalent *on* $\mathcal{T} \subseteq \mathcal{X}^q \times \mathcal{Y}^q$, *denoted as* $\mathcal{O} \cong_\mathcal{T} \mathcal{O}'$, *if for all* $(x^q, y^q) \in \mathcal{T}$

$$\Pr(\tau_{\mathrm{resp}}(\mathsf{D}^\mathcal{O}_{x^q}) = y^q) = \Pr(\tau_{\mathrm{resp}}(\mathsf{D}^{\mathcal{O}'}_{x^q}) = y^q)$$

When $\mathcal{T} = \mathcal{Q} \times \mathcal{Y}^q$ we simply write $\mathcal{O} \cong_{\mathcal{Q}} \mathcal{O}'$ and we call $\mathcal{O}$ and $\mathcal{O}'$ are equivalent on a joint query space $\mathcal{Q}$.

For $\mathcal{T} \subseteq \mathcal{X}^q \times \mathcal{Y}^q \times \mathcal{S}$, we say that $(\mathcal{O}, \mathcal{O}_{pp}) \cong_{\mathcal{T}} (\mathcal{O}', \mathcal{O}'_{pp})$ if for all $(x^q, y^q, S) \in \mathcal{T}$

$$\Pr(\tau_{\mathrm{resp}}(\mathsf{D}_{x^q}^{\mathcal{O}, \mathcal{O}_{pp}}) = (y^q, S)) = \Pr(\tau_{\mathrm{resp}}(\mathsf{D}_{x^q}^{\mathcal{O}', \mathcal{O}'_{pp}}) = (y^q, S))$$

Observation. When the user index is also considered as a part of the input, the *two random functions* $\mathsf{RF}_{\mathcal{X} \to \mathcal{Y}}^{\otimes \mathcal{I}}$ *and* $\mathsf{RF}_{\mathcal{I} \times \mathcal{X} \to \mathcal{Y}}$ *are equivalent and so we use the notations interchangeably in the paper.*

A similar statement of the following lemma can be found in [22] in the language of a random system. This says that under certain assumptions probability of realizing a collection of transcripts

Lemma 1 (adaptive to non-adaptive). *Suppose $\mathcal{O}$ and $\mathcal{O}_{pp}$ use independent random coins. Let $\mathcal{T} = \mathcal{T}' \times \mathcal{Y}^q$ for some $\mathcal{T}' \subseteq \mathcal{X}^q \times \mathcal{S}$. Then, for every adaptive distinguisher D there is a non-adaptive distinguisher D_0 such that*

$$\Pr(\tau(\mathsf{D}_0^{\mathcal{O}, \mathcal{O}_{pp}}) \notin \mathcal{T}) = \Pr(\tau(\mathsf{D}_0^{\mathcal{O}, \mathcal{O}_{pp}}) \notin \mathcal{T}).$$

2.1   Formalizing Hybrid Reduction Proof

A reduction algorithm is an essential object in every reduction proof. We now formalize the notion of reduction algorithm and hybrid reduction algorithm.

Definition 6. *A **reduction algorithm** or **simulator** Sim is an interactive algorithm such that*

- *for any compatible oracle $\mathcal{O}$, $\mathsf{Sim}^{\mathcal{O}}$ behaves as an oracle and*
- *for any oracle algorithm $\mathcal{A}$ (so that $\mathsf{Sim}^{\mathcal{O}}$ is a compatible oracle of $\mathcal{A}$), $\mathcal{A}^{\mathsf{Sim}}$ behaves as an oracle algorithm.*

The joint interaction among all three is denoted as

$$\mathcal{A}^{\mathsf{Sim}^{\mathcal{O}}}.$$

If for every θ-complexity algorithm $\mathcal{A}$, $\mathcal{A}^{\mathsf{Sim}}$ is an θ'-complexity algorithm, we write $\theta^{\mathsf{Sim}} = \theta'$. Note that the run time for $\mathcal{A}^{\mathsf{Sim}}$ is $t_{\mathsf{Sim}} + t_{\mathcal{A}}$ where t_X represents the time for algorithm X.

Note that a reduction algorithm is neither an oracle nor an oracle algorithm. However, it can be placed in between an oracle algorithm $\mathcal{A}$ and an oracle $\mathcal{O}$. Moreover, $\mathcal{A}^{\mathsf{Sim}}$ is an oracle algorithm and $\mathsf{Sim}^{\mathcal{O}}$ is an oracle.

A set I is called hybrid-index space of Sim if the random coin $\mathcal{C} = I \times \mathcal{C}'$ for some $\mathcal{C}'$. So for every $h \in I$, $\mathsf{Sim}(h)$ represents a reduction algorithm with coin space $\mathcal{C}'$ which behaves like Sim conditioned on the hybrid index h. Let $I \subseteq J$ be a set and $\perp$ represents an arbitrary but a fixed deterministic oracle. Given a reduction algorithm Sim with coin space $I \times \mathcal{C}$, we consider the following reduction algorithm (abusing notation, we also write Sim to denote it) with the coin space $J \times \mathcal{C}'$:

– $h \leftarrow_\$ J$. If $h \notin I$ then it behaves as $\bot$, Else, runs $\mathsf{Sim}(h)$.

2.2   Hybrid Reduction Algorithm

In many proofs, we simply substitute one oracle with another (e.g., a PRF by a random function and a pseudorandom permutation or PRP by a random permutation). This substitution reduction can be described formally as follows.

Definition 7 (substitution reduction). *Let* F, G, F', G' *be four oracles. A simulator* Sim *is called an 1-step reduction (or a substitution reduction) from the pair of oracles* (F', G') *to the pair* (F, G) *on a joint query space* $\mathcal{Q}$*, if* $\mathsf{Sim}^F \cong_\mathcal{Q} F'$ *and* $\mathsf{Sim}^G \cong_\mathcal{Q} G'$*. We simply denote the above 1-step reduction as*

$$F' \xrightarrow{\mathsf{Sim}^{F/G}} G'.$$

We now extend the notion of substitution reduction or 1-step reduction to a d-step reduction in the following way.

Definition 8 (d-step reduction). *A simulator* Sim *with a hybrid index space* $[d]$ *is called a d-step substitution reduction from a pair of oracles* (F', G') *to a pair of oracles* (F, G) *on a joint query space* $\mathcal{Q}$ *if there are* $d - 1$ *intermediate oracles* $\mathcal{O}_1, \ldots, \mathcal{O}_{d-1}$ *such that*

$$\mathcal{O}_0 := F' \xrightarrow{\mathsf{Sim}(1)^{F/G}} \mathcal{O}_1 \xrightarrow{\mathsf{Sim}(2)^{F/G}} \mathcal{O}_2 \cdots \xrightarrow{\mathsf{Sim}(d-1)^{F/G}} \mathcal{O}_{d-1} \xrightarrow{\mathsf{Sim}(d)^{F/G}} G' := \mathcal{O}_d$$

In other words, $\mathsf{Sim}(j)$ *is a substitution reduction from* $(\mathcal{O}_{j-1}, \mathcal{O}_j)$ *to* (F, G) *on a joint query space* $\mathcal{Q}$*. When the simulator* Sim *and the oracles* F, G *are understood, we simply ignore the notation. When* G *and* G' *are random functions, we call* Sim *a d-step PRF-reduction from* F' *to* F*.*

The above definition can be easily extended for any arbitrary hybrid index space I (not necessarily of the form $[d]$). Note that in the above definition, it is not required to define the intermediate oracles explicitly. It is sufficient to show the following two conditions :

– (boundary condition): $\mathsf{Sim}^F(1) \cong_\mathcal{Q} F'$, $\mathsf{Sim}^G(d) \cong_\mathcal{Q} G'$ and
– (transition equivalence): $\mathsf{Sim}^G(j) \cong_\mathcal{Q} \mathsf{Sim}^F(j+1)$ for all $j \in [d-1]$.

The reduction algorithm Sim is a d-step reduction as we set $\mathcal{O}_j = \mathsf{Sim}^F(j+1)$ for all $1 \leq j \leq d-1$. However, sometimes it is easier to first describe the intermediate oracles $\mathcal{O}_j$ in a stand-alone way and then we show the equivalence between oracles. The proof of the following lemma is similar to standard reduction proof and can be found in details in the full version of the paper [26]

Lemma 2 (hybrid reduction).   *Let* Sim *be a d-step substitution reduction from* (F', G') *to* (F, G)*. Then for any* (θ, t)*-complexity distinguisher* D*, we have a* $(\theta^{\mathsf{Sim}}, t + t_{\mathsf{Sim}})$*-complexity distinguisher* $\mathsf{D}' := \mathsf{D}^{\mathsf{Sim}}$ *such that*

$$\Delta_{\mathsf{D}'}(F \; ; \; G) = \frac{1}{d} \cdot \Delta_{\mathsf{D}}(F' \; ; \; G').$$

So, if Sim *is a d-step PRF-reduction from F' to F, we have*

$$\mathbf{Adv}_{F'}^{\mathrm{prf}}(\theta, t) \le d \cdot \mathbf{Adv}_F^{\mathrm{prf}}(\theta', t + t_{\mathsf{Sim}}).$$

3   A Generalized Adaptive Reduction for Cascade

In this section, we provide a general method of reduction proof for the multiuser cascade against both adaptive and non-adaptive distinguishers. Let $\mathcal{Q} := \mathcal{Q}_{\mathrm{pf}}(\theta)$ then it is a prefix-closed (i.e., for all $m^q \in \mathcal{Q}$, and $i \in [q]$, $m^i \in \mathcal{Q}$). Suppose an adversary makes queries $m_1, m_2, \ldots, m_q$ adaptively. On ith query, we represent the state as m^i which captures all queries till the ith query (including the ith query). We write $m_i = m_i[0..\ell_i] \in \mathcal{I} \times \mathsf{B}^{\ell_i}$.

Definition 9 (structure). *A structure for a joint query space $\mathcal{Q}_{\mathrm{pf}}(\theta)$ is a pair $\tau := (R, \rho)$ of functions $(R, \rho) : \mathcal{Q} \to \mathsf{B}_{\mathrm{mu}}^* \times \{\mathsf{orc}, \mathsf{sim}\}$ satisfying the following conditions:*

1. *For all i, $R(m^i) \preceq m_i$ and the set $\{R(m^i) : i \in [q]\}$ is a leave-cut for $\mathcal{T}_{m^q}$,*
2. *$R(m^i) = R(m^j) \Rightarrow \rho(m^i) = \rho(m^j)$ and*
3. *$\rho(m^i) = \mathsf{orc} \Rightarrow R(m^i) \ne m_i$.*

Let $\mathcal{O}_R$ respond $z_1, \ldots, z_q$ on queries $m_1, \ldots, m_q$ respectively where

$$z_i = f^*_{\mathsf{RF}(R(m^i))}(m_i \setminus R(m^i)) \quad \forall i \in [q].$$

The function R puts an independent key at the node $R(m^i)$ on ith query m_i and follows the definition of f^* starting from the key on the node applied to the rest of the message blocks. We can imagine this as a traversal of the path from the root till the leaf node m_i. Two extreme and trivial examples are $R_{\mathrm{root}}(m^i) = m_i[0]$ and $R_{\mathrm{full}}(m^i) = m_i$ for all m^i. It is easy to see that $\mathcal{O}_{R_{\mathrm{root}}} \cong f^{*\otimes}$ and $\mathcal{O}_{R_{\mathrm{full}}} = \mathsf{RF}$ on the query space $\mathcal{Q}$. Any other R induces an immediate or hybrid oracle in between these two extreme examples. The function ρ suggests where to make oracle query and where it would be simulated by the simulator itself. More precisely, we define a simulator Sim_τ below.

Definition 10 (Simulator associated with a structure τ). *Let $\mathcal{C} = \{0,1\}^c$. For every structure $\tau = (R, \rho)$ on a joint query space $\mathcal{Q}$, we associate a simulator Sim_τ defined as follows. Given any multiuser $(\mathsf{B}, \mathcal{C})$-oracle $\mathcal{O}^\otimes$ with user index space $\mathcal{I} \times \mathsf{B}^*$, Sim_τ returns z^q on q adaptive query m^q (from an oracle algorithm) where*

$$z_i = \begin{cases} f^*_{\mathsf{RF}_{* \to \mathcal{C}}(m_i[..s])}(m_i[s+1..]) & \text{if } \rho(m^i) = \mathsf{sim}, \\ f^*_{\mathcal{O}^\otimes(m_i[..s]\,;\,m_i[s+1])}(m_i[s+2..]) & \text{if } \rho(m^i) = \mathsf{orc}, \end{cases}$$

where $R(m^i) = m_i[..s]$. Here, the outputs of $\mathsf{RF}_{* \to \mathcal{C}}(m_i[..s])$ is simulated by the simulator itself (as a part of the simulator's random coin). We write $\mathcal{Q}^\tau = \mathcal{Q}(u, q, q_{\max})$ if for all $m^q \in \mathcal{Q}$,

1. *$Q := \{R(m^i) : \rho(m^i) = \mathsf{orc}\}$ has at most u elements and*

2. $\sum_{x \in Q} |\mathsf{ch}(x)| \leq q$, $\max_{x \in Q} |\mathsf{ch}(x)| \leq q_{\max}$.

In this case we have $\theta^{\mathsf{Sim}_\tau} = (u, q, q_{\max})$.

By definition, whenever $\rho(m^i) = \mathsf{orc}$, $s < \ell_i$ and so $m_i[s+1]$ is defined. Moreover, both $\mathsf{RF}_{* \rightarrow \mathcal{C}}(m_i[..s])(m_i[s+1..])$ and $\mathcal{O}^\otimes(m_i[..s] \, ; \, m_i[s+1])(m_i[s+2..])$ are members of $\mathcal{C}$ and hence f^* outputs are defined. For $\tau = (R, \rho)$,

$$\mathrm{next}(\tau)(m^i) = \begin{cases} m_i[..s] & \text{if } \rho(m^i) = \mathsf{sim} \\ m_i[..s+1] & \text{if } \rho(m^i) = \mathsf{orc}, \end{cases}$$

where $R(m^i) = m_i[..s]$. Whenever $\rho(m^i) = \mathsf{orc}$, $s < \ell_i$ and so $m_i[s+1]$ is defined. Let $U = \{R(m^i) : i \in [q]\}$ and $S = \{R(m^i) : \rho(m^i) = \mathsf{orc}\}$. Then, $\{\mathrm{next}(\tau)(m^i) : i \in [q]\} = (U \setminus S) \cup \mathsf{ch}(S)$ is a leave-cut set for all m^q.

Proposition 1 (generalized reduction for multiuser cascade). *Let θ be some complexity parameter. Suppose $\tau_i := (R_i, \rho_i)$, $0 \leq i \leq d$, are hybrid structures over $\mathcal{Q}_{\mathrm{pf}}(\theta)$ such that*

1. $R_0 = R_{\mathrm{root}}$ and $R_d = R_{\mathrm{full}}$ *(boundary condition)*
2. $\mathrm{next}(\tau_{i-1}) = R_i$ *for all $i \in [d]$ (transition equivalence)*
3. $\mathcal{Q}_{\mathrm{pf}}(\theta)^{\tau_i} = \mathcal{Q}(u, q, q_{\max})$ *for all $i \in [d]$ (complexity reduction)*

Then,

$$\mathbf{Adv}_{f^*}^{\mathrm{mu\text{-}pf\text{-}prf}}(\theta) \leq d \cdot \mathbf{Adv}_f^{\mathrm{mu\text{-}prf}}(u, q, q_{\max}). \tag{4}$$

Proof. Let Sim be a reduction which samples $h \leftarrow_\$ [d]$ and then run $\mathsf{Sim}_{\tau_{h-1}}$. Note that for any prefix-free θ-complexity distinguisher $\mathcal{A}$, $\mathcal{A}^{\mathsf{Sim}}$ is a $(u, q, q_{\max})$-complexity distinguisher. Now we show the following d-step reduction:

$$f^{*\otimes} \xrightarrow{\mathsf{Sim}(1)^{f^\otimes/\mathsf{RF}}} \mathcal{O}_{R_1} \xrightarrow{\mathsf{Sim}(2)^{f^\otimes/\mathsf{RF}}} \mathcal{O}_{R_2} \cdots \xrightarrow{\mathsf{Sim}(d-1)^{f^\otimes/\mathsf{RF}}} \mathcal{O}_{R_{d-1}} \xrightarrow{\mathsf{Sim}(d)^{f^\otimes/\mathsf{RF}}} \mathsf{RF}'$$

From the definition of Sim_τ and oracle $\mathcal{O}_R$, it is straightforward that $\mathsf{Sim}_\tau^{f^\otimes} \cong \mathcal{O}_R$ and $\mathsf{Sim}_\tau^{\mathsf{RF}^\otimes} \cong \mathcal{O}_{\mathrm{next}(\tau)}$. However, we have seen that $\mathcal{O}_{R_0} \cong f^{*\otimes}$ and $\mathcal{O}_{R_d} \cong \mathsf{RF}$. The above d-step reduction follows from the given condition that $\mathrm{next}(\tau_{i-1}) = R_i$. The result follows from the hybrid reduction lemma. $\qquad \square$

3.1   Applications: A Depth-First Reduction

For every $m^q \in \mathcal{Q}$, we associate a prefix-tree $\mathcal{T}_{m^q}$. We now define a bijective function : $\mathsf{DF} : V' \cup \{\lambda\} \rightarrow [0..d]$ mapping λ to 0 where $d = |V'| - 1$. So we can write the elements of $V' \cup \{\lambda\}$ in a sequence $v_0 = \lambda, v_1, \ldots, v_d$ where $\mathsf{DF}(v_i) = i$.

Recursive Definition of DF

---

1. Initialize $ctr = 1$, $\mathsf{DF}(\lambda) = 0$, $v_0 = \lambda$.
2. for $i = 1$ to q and for $j = 1$ to $\ell_i - 1$
       if $\mathsf{DF}(m_i[..j])$ is not defined then
           $\mathsf{DF}(m_i[..j]) = ctr$, $v_{ctr} = m_i[..j]$ and $ctr \leftarrow ctr + 1$.

Note that $d \leq \sigma' = \sigma - q$. When d is smaller than σ', we define $v_{d+1} = \cdots = v_{\sigma'} = v_d$. Note, we order the vertices following the depth first principle. For any i, exactly any one of the three conditions will hold:

type-1 $\mathsf{DF}(m_i[\ell_i - 1]) < h$.
type-2 $\mathsf{DF}(m_i[..j]) = h$.
type-3 $\mathsf{DF}(m_i[..j - 1]) < h$ and $\mathsf{DF}(m_i[..j]) > h$.

Now we define a structure, called depth-first structure.

$$\tau_h(m^i) := (R_h(m^i), \rho_h(m^i)) = \begin{cases} (m_i, \mathsf{sim}) & \text{if type-1} \\ (m_i[..j], \mathsf{orc}) & \text{if type-2} \\ (m_i[..j], \mathsf{sim}) & \text{if type-3} \end{cases}$$

where j is defined in type-2 and type-3 as before.

Lemma 3. τ_h, defined as above, is a structure.

So, following our generalized reduction lemma, the following theorem for the multiuser cascade construction is established.

Theorem 1. Let $\theta = (u, q, q_{max}, \ell, \sigma)$. Every (θ, T)-distinguisher D can be reduced to a $(q_{max}, T' := T + O(\sigma))$-distinguisher $\mathsf{D}' := \mathsf{D}^{\mathsf{Sim}}$ (non-adaptive whenever D is non-adaptive), where Sim is defined as above, so that

$$\mathbf{Adv}_{f^*}^{\text{mu-pf-prf}}(\mathsf{D}) \leq (\sigma - q) \cdot \mathbf{Adv}_f^{\text{prf}}(\mathsf{D}'). \tag{5}$$

Hence, (1) $\mathbf{Adv}_{f^*}^{\text{mu-pf-(n)prf}}(\theta, T) \leq (\sigma - q) \cdot \mathbf{Adv}_f^{\text{(n)prf}}(q_{max}, T')$. In case of single user, (2) $\mathbf{Adv}_{f^*}^{\text{pf-(n)prf}}(q, \ell, \sigma, T) \leq (\sigma - q) \cdot \mathbf{Adv}_f^{\text{(n)prf}}(q, T')$.

Remark 1. We apply padding whenever the message space is not prefix-free, e.g. $\{0,1\}^*$. Let $\mathsf{pad} : \{0,1\}^* \to \mathsf{B}^+$ be defined as follows. Given $x \in \{0,1\}^*$ we find smallest non-negative integer d such that $|x| + 1 + d$ is a multiple of $b - 1$. Let $x \| 10^d = (x_1, \ldots, x_\ell) \in (\{0,1\}^{b-1})^\ell$. Finally, we define $\mathsf{pad}(x) = (x_1\|0, \ldots, x_{\ell-1}\|0, x_\ell\|1)$. Clearly, for any $x \neq x'$, $\mathsf{pad}(x)$ is not a prefix of $\mathsf{pad}(x')$. So, $f_K^* \circ \mathsf{pad}$ is PRF with same security bound as shown before for arbitrary message space.

3.2 Application: Classical Reduction for Cascade

As a first application, we establish the classical reduction proof for cascade. Let $R_h(m^i) = m_i[..h]$ and

$$\rho(m^i) = \begin{cases} \text{orc} & \text{if } h < \ell_i \\ \text{sim} & \text{if } h \geq \ell_i. \end{cases}$$

Then, it is easy to see that for all $h \in [\ell]$, $\text{next}(R_{i-1}, \rho_{i-1}) = R_i$ for all $i \in [\ell]$. Moreover, $R_0 = R_{\text{root}}$ and $R_\ell = R_{\text{full}}$. So, by Proposition 1.

$$\mathbf{Adv}_{f^*}^{\text{mu-pf-prf}}(u, q, q_{\max}, \ell, \sigma) \leq \ell \cdot \mathbf{Adv}_{f_\otimes}^{\text{mu-prf}}(u, q, q_{\max}). \tag{6}$$

Now by using standard multiuser to single user reduction we have

$$\mathbf{Adv}_{f^*}^{\text{mu-pf-prf}}(u, q, q_{\max}, \ell, \sigma) \leq \ell q_{\max} \cdot \mathbf{Adv}_f^{\text{mu-prf}}(q_{\max}). \tag{7}$$

The depth-first reduction has improved tightness gap as $\sigma \leq \ell q_{\max}$.

3.3 Applications: Simple Proofs for MDP and Boosted MD

Let $\alpha : \mathsf{B}' \times \mathcal{C} \to \mathsf{B} \times \mathcal{C}$ be a function. So, $g := f \circ \alpha : \mathsf{B}' \times \mathcal{C} \to \mathcal{C}$. Let pad be a prefix-free padding rule mapping to B'^+. Then, $g^* \circ$ pad is PRF secure whenever g_K is PRF (with the bound mentioned in Theorem 1). Now we discuss two examples of α which has been used to design some constructions (Fig. 3).

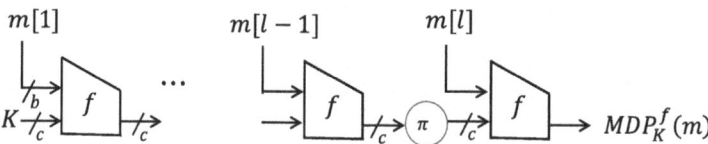

Fig. 3. MDP Keyed PRF.

MDP. Let π be a permutation on $\mathcal{C}$ and $\mathsf{B}' = \mathsf{B} \times \{0,1\}$. Now, we define $\alpha((m,0),K) = (m,K)$ and $\alpha((m,1),K) = (m,\pi(K))$. In JoC 2012 [15] authors defined a special case of related key security of f which is essentially same as the PRF security of $f \circ \alpha$. Authors of [15] provided PRF security of $g^* \circ$ pad based on related key PRF security of f (equivalently PRF security of g) with tightness gap $\ell \cdot q$. Clearly this claim is an immediate corollary of existing prefix-free PRF security of cascade. Moreover, using our reduction of this section, we have σ as tightness gap of the MDP construction.

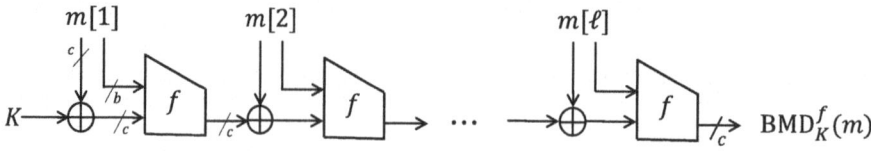

Fig. 4. BMD or Boosted-MD keyed function.

BOOSTED MD OR BMD. In Asiacrypt 2007 [31], author proposed Boosted-MD to provide much faster absorption of message in cascade construction. In particular, we additionally xor c-bit message with chaining values starting from the key (see Fig. 4). Let $\mathsf{B}' = \mathsf{B} \times \{0,1\}^c$ and $\alpha(m_1, m_2, K) = (m_1, K \oplus m_2)$. In [31], author considered another variant of related key security of f which is once again same as the PRF security of $g := f \circ \alpha$. Moreover, we have σ as tightness gap of the MDP construction in contrast to original tightness gap ℓq^2.

4   Improved Bound for Non-adaptive Distinguisher

In the last subsection, we have shown a multiplicative gap of σ for the cascade, which is definitely an improvement over the previously known bound of ℓq. Now we show that we can improve the query complexity for the simulator in case of the non-adaptive bound. In Theorem 1, we reduce to a $q_{\max}$-query algorithm. However, it is easy to see that the number of queries can be much less. More precisely, except about q choices, all other hybrid reduction indices reduce to a single-query algorithm. To conclude this, we need the following simple result on a rooted tree.

Lemma 4. *Let $V(T)$ and $L(T)$ denote the set of nodes and the set of leaf nodes of a rooted tree T respectively and $V' = V \setminus L$. Then, $\sum_{v \in V'}(|\mathsf{ch}(v)| - 1) = |L(T)| - 1$. Hence, we have*

(1) $|\{v : |\mathsf{ch}(v)| > 1\}| \leq |L(T)| - 1$,   (2) $|\{v : |\mathsf{ch}(v)| > i\}| < |L(T)|/i$.

Let $\theta = (u, q, q_{\max}, \ell, \sigma)$. In the previous section, we have proved that Sim is a σ-step hybrid reduction from $(f^{*\otimes}, \mathsf{RF}^{\otimes})$ to (f, RF) with hybrid index space $[\sigma]$. Now, for any $m^q \in \mathcal{Q}(\theta)$, we define d as the number of intermediate nodes of the prefix tree and so $d \leq \sigma - q$. Now, for every $h \in [d]$, we associate a unique node v_h. Moreover, $\mathsf{Sim}(h)$ makes $|\mathsf{ch}(v_h)|$ (the number of children) many queries, whenever v_h is defined, otherwise it does not make any query. Note that the number of children (and hence queries) depends on m^q. Let $\mathcal{N}_i$ denote the set of h such that the number of children of v_h is i. As $\mathsf{Sim}(h)$ is a hybrid reduction, we have

$$\Pr(\mathsf{D}^{f^{*\otimes}} = 1) - \Pr(\mathsf{D}^{\mathsf{RF}^{\otimes}} = 1) = \sum_{h \in [\sigma]} (\Pr(\mathsf{D}^{\mathsf{Sim}^f(h)} = 1) - \Pr(\mathsf{D}^{\mathsf{Sim}^{\mathsf{RF}}(h)} = 1))$$

$$= \sum_{i \geq 1} \sum_{h \in \mathcal{N}_i} \Pr(\mathsf{D}^{\mathsf{Sim}^f(h)} = 1) - \Pr(\mathsf{D}^{\mathsf{Sim}^{\mathsf{RF}}(h)} = 1) \tag{8}$$

Note that the inner sum $\sum_{h \in \mathcal{N}_i} \Pr(\mathsf{D}^{\mathsf{Sim}^f(h)} = 1) - \Pr(\mathsf{D}^{\mathsf{Sim}^{\mathsf{RF}}(h)} = 1)$ the signed advantage for the simulator which makes exactly i queries. Note that the size of $\mathcal{N}_i$ is a random variable depending on m^q. However, we know that for all $i \geq 1$, $\sum_{j > i} |\mathcal{N}_j| \leq q/i$ (due to the above lemma on the tree). Using this bound we can prove our improved reduction for non-adaptive distinguisher.

Theorem 2 (First Improved Non-adaptive Reduction).

$$\mathbf{Adv}^{\mathrm{mu\_pf\_nprf}}_{f^*\otimes}(\theta, T) \leq \sigma \cdot \mathbf{Adv}^{\mathrm{nprf}}_{f}(1, T') + q \cdot \mathbf{Adv}^{\mathrm{nprf}}_{f}(q_{\max}, T') \qquad (9)$$

where $T' = T + O(\sigma)$.

Theorem 3 (Second Improved Non-adaptive Reduction). *For any $\theta :=$ $(u, q, q_{\max}, \ell, \sigma, T)$-non-adaptive distinguisher D, we have 1-query non-adaptive distinguisher D'_1 and $(2^{i-1} + 1)$-query non-adaptive distinguisher D_i for $1 \leq i \leq \lceil \log_2 q_{\max} \rceil$ with run time $T' = T + O(\sigma)$ such that*

$$\mathbf{Adv}^{\mathrm{mu\_pf\_nprf}}_{f^*}(\mathsf{D}) \leq \sigma \cdot \mathbf{Adv}^{\mathrm{nprf}}_{f}(\mathsf{D}'_1) + \sum_{i=1}^{\lceil \log_2 q_{\max} \rceil} \frac{q}{2^{i-1}} \cdot \mathbf{Adv}^{\mathrm{nprf}}_{f}(\mathsf{D}_i).$$

Hence,

$$\mathbf{Adv}^{\mathrm{mu\_pf\_nprf}}_{f^*}(\theta) \leq \sigma \cdot \mathbf{Adv}^{\mathrm{nprf}}_{f}(1, T') + \sum_{i=1}^{\lceil \log_2 q_{\max} \rceil} \frac{q}{2^{i-1}} \cdot \mathbf{Adv}^{\mathrm{nprf}}_{f}(2^i, T'). \quad (10)$$

APPLICATIONS TO HMAC AND NMAC. The generic reduction from NMAC to cascade (see Eq. 1) and HMAC to NMAC can be easily extended for the multiuser set-up in the following way:

$$\mathbf{Adv}^{\mathrm{mu\_prf}}_{\mathsf{NMAC}^f}(\theta, T) \leq \mathbf{Adv}^{\mathrm{mu\_pf\_nprf}}_{f^*}(\theta, T') + \mathbf{Adv}^{\mathrm{prf}}_{f}(q, T') + \frac{q^2}{2^c}, \qquad (11)$$

$$\mathbf{Adv}^{\mathrm{mu\_prf}}_{\mathsf{HMAC}^f}(\theta, T) \leq \mathbf{Adv}^{\mathrm{mu\_prf}}_{\mathsf{NMAC}^f}(\theta, T) + \mathbf{Adv}^{\mathrm{prbg}}_{\mathsf{KDF}}(T'). \qquad (12)$$

Hence our results for non-adaptive PRF security of cascade can be directly applied to the multiuser security of HMAC and NMAC.

4.1 Significance of Improvement in the Standard Model

The known tightness gap ℓq becomes worse when the queries can be both very short as well as large. In that case we can limit $\ell, q, \sigma \leq D$ where D represents the maximum data complexity.[3] With this limit, the known bound for cascade turns out to be

$$\mathbf{Adv}^{\mathrm{prf}}_{f^*}(q, \ell, \sigma, T) \leq \ell q \cdot \mathbf{Adv}^{\mathrm{prf}}_{f}(q, T')$$

$$\leq D^2 \cdot \mathbf{Adv}^{\mathrm{prf}}_{f}(q, T')$$

where $T' = T + O(\sigma)$. In [7,18], authors showed that for almost all designs we can have $\mathbf{Adv}^{\mathrm{prf}}_{f}(q, T') \geq 2^{-c/2}$. Moreover, by key-guessing attack, we also have

---

[3] NIST actually considered this in the call for the standardization process of lightweight cipher [23].

$\mathbf{Adv}_f^{\mathrm{prf}}(q, T') \geq T'/2^c$ (as f^* uses c-bit key). So the known bound can only ensure security as long as

$$D \leq 2^{c/4}, \quad D^2 T \leq 2^c.$$

Using our new bound we have $\mathbf{Adv}_{f^*}^{\mathrm{prf}}(q, \ell, \sigma, T) \leq D \cdot \mathbf{Adv}_f^{\mathrm{prf}}(q, T')$ and so our bound can ensure security as long as

$$D \leq 2^{c/2}, \quad DT \leq 2^c.$$

Thus, it improves the data-time trade-off for the cascade construction. A similar improvement works for multiuser setup.

4.2   Significance of Improvement in the Ideal Model

We have defined PRF security in the ideal model. Now we extend the definition in the ideal model. Let Γ be a random function and F_K be a construction which uses Γ as an oracle. A distinguisher D is an oracle algorithm which has access of two oracles. We define PRF distinguishing advantage of D against F as

$$\mathbf{Adv}_F^{\mathrm{prf}}(D) := |\Pr(D^{F_K, \Gamma} \to 1) - \Pr(D^{\mathsf{RF}, \Gamma} \to 1|)|$$

where RF is a compatible random function independent with Γ. Complexity parameter of D can be written as (θ, η) where θ and η represent the complexity parameter for all construction queries (i.e., the first oracle which is either F_K or RF) and primitive queries (which is always Γ) respectively. η mostly represents the number of queries to Γ.

Convention. Note that we assume that $\eta = O(T)$ where T is the run time of D (since otherwise, D can make additional queries which increases run time by $O(T)$).

A simple example is the cascade based on a random function (or idealized compression function) $\Gamma : \{0,1\}^{b+c} \to \{0,1\}^c$ as $\Gamma_K(x) := \Gamma(K, x)$ where $K \in \{0,1\}^c$ and $\Gamma_K : \{0,1\}^b \to \{0,1\}^c$ be a keyed function based on an ideal random function Γ. The reduction proved in the paper and in the previous papers for the cascade construction Γ_K^* can be translated to the following relations for the complexity parameters $\theta = (u, q, q_{\max}, \ell, \sigma)$, $\theta_1 = (q, \ell, \sigma)$ and primitive query complexity η:

$$\mathbf{Adv}_{\Gamma^*}^{\mathrm{prf}}(\theta_1, \eta) \leq \ell q \cdot \mathbf{Adv}_{\Gamma}^{\mathrm{prf}}(q, \eta + \sigma).$$

Note that reduction algorithm needs to call at most $\eta + \sigma$ many primitive queries to simulate all queries of the distinguisher. It is easy to show that for any (q, η')-query distinguisher D, $\mathbf{Adv}_{\Gamma}^{\mathrm{prf}}(D) \leq \eta'/2^c$ (until we cannot guess the key in the primitive query, all construction oracle queries behave like an independent random function) and the bound is tight. So plugging the bound above we can ensure $\mathbf{Adv}_{\Gamma^*}^{\mathrm{prf}}(\theta_1, \eta) \leq \ell q \cdot (\eta + \sigma)/2^c$. The same bounds hold for non-adaptive

PRF advantage. Similar bound applies when Γ is replaced by Davis-Meyer compression function based on an c-bit ideal cipher E with key space $\{0,1\}^b$. In particular we define $\mathsf{DM}(K,x) := E_x(K) \oplus K$.

Our adaptive PRF advantage shows that

$$\mathbf{Adv}_{\Gamma_*}^{mu\text{-}prf}(\theta,\eta) \leq \sigma \cdot \mathbf{Adv}_{\Gamma}^{mu\text{-}prf}(q_{max},\eta+\sigma)$$

and hence $\mathbf{Adv}_{\Gamma_*}^{mu\text{-}prf}(\theta,\eta) \leq \sigma(\eta+\sigma)/2^c$. The same relation holds for non-adaptive PRF advantage and also for the Davis-Meyer compression function based cascade construction.

Example 1 (Boosted MD in the ideal model). We define $\Gamma_K^{\oplus}(x_1,x_2) = \Gamma(K \oplus x_1, x_2)$, $(x_1,x_2) \in \{0,1\}^{c+b}$. If a distinguisher cannot make guesses of $K \oplus x_1$ (of construction queries) in any primitive queries, then it cannot distinguish $\Gamma^{\oplus}$ from an independent random function RF. Hence, $\mathbf{Adv}_{\Gamma^{\oplus}}^{prf}(q,\eta) \leq q\eta/2^c$. In fact, one can construct a distinguisher making q construction queries and η primitive queries with PRF advantage about $\eta q/2^c$ in an ideal random function model. The cascade based on $\Gamma^{\oplus}$ is called boosted MD which has been studied in Sect. 3 in the standard model. A straightforward application of the PRF advantage of $\Gamma^{\oplus}$ in the ideal model to the cascade construction gives $\mathbf{Adv}_{\Gamma^{\oplus}_*}^{prf}(\theta_1,\eta) \leq \ell q^2(\eta+\sigma)/2^c$. However, our bound provides an improved bound of the form:

$$\mathbf{Adv}_{\Gamma^{\oplus}_*}^{mu\text{-}nprf}(\theta,\eta) \leq (\sigma + q\lceil \log_2 q_{max} \rceil)(\eta+\sigma)/2^c.$$

Hence we have shown birthday bound for boosted cascade function in a modular way instead of cubic bound derived from the existing reduction.

Let us now consider some popular examples of compression functions based on a c-bit ideal cipher E with keyspace $\{0,1\}^b$.

Example 2. MMO (Matyas-Meyer-Oseas) compression function based on an ideal cipher E can be defined as $\mathsf{MMO}(K,x) = E_{K\|0^{b-c}}(x) \oplus x$. Note that it can be distinguished from random function with an advantage about $q^2/2^c + \eta/2^c$ where q and η denote the number of construction and primitive queries respectively. The birthday terms arise due to the following reason: In the real construction there cannot be any collision among the values $z_i \oplus x_i$ where z_i's are the outputs of the queries x_i. However, a collision is observed for a random function RF with probability about $q^2/2^{c+1}$ Similarly, we also have key guessing attack for this compression function. Hence, we have $\mathbf{Adv}_{\mathsf{MMO}}^{prf}(q,\eta) \geq \max\{q^2/2^{c+1}, \eta/2^c\}$. Now if we plug in the existing bound we have a cubic bound:

$$\mathbf{Adv}_{\mathsf{MMO}_*}^{nprf}(\theta,\eta) \leq \frac{\ell q^3}{2^c} + \frac{\ell q\eta}{2^c}.$$

However, if we use our second improved non-adaptive reduction and simplify the sum we have quadratic bound (up to a log factor)

$$\mathbf{Adv}_{\mathsf{MMO}_*}^{nprf}(\theta,\eta) \leq \lceil \log_2 q_{max} \rceil \sigma(\eta+\sigma)/2^c + 2qq_{max}/2^c.$$

Example 3. The similar result like MMO compression function is also applicable for Miyaguchi-Preneel compression function: $\mathsf{MP}(K, x) = E_{K\|0^{b-c}}(x) \oplus x \oplus K$. Once again, we have $\mathbf{Adv}_{\mathsf{f}\oplus}^{\mathrm{prf}}(q, \eta) \geq \max\{q^2/2^{c+1}, \eta/2^c\}$ the same bound as MMO compression function.

5 Single Keyed **NMAC**, Constant-Free **HMAC** and **EvMAC**

5.1 PRF Security of Single Keyed Composition

For every key K in a key space, let $g_K : \mathsf{B} \to \mathcal{K}$ and $F : \mathcal{K} \times \mathcal{X} \to \mathsf{B}$. We now define a keyed function $G_K : \mathsf{B} \times \mathcal{X} \to \mathcal{K}$ by combining g and F as follows:

$$G_K(a, x) := g_K\big(F(g_K(a), x)\big).$$

Here, we recall the notation g_K to denote the function $g(K, \cdot)$. Let $\mathcal{I}$ be a user index space for G. For every $\gamma \in \mathcal{I}$, we sample $K_\gamma \leftarrow_\$ \mathcal{K}$. So, $G^\otimes(\gamma, (a, x)) := G_{K_\gamma}(a, x)$. We denote this oracle $\mathcal{G}_0$ (see Fig. 5). Now we define an intermediate oracle $\mathcal{G}_1$. We obtain the oracle $\mathcal{G}_1$ replacing g by a (multiuser) random function RF. More precisely, on a user index γ and an input (a, x), it returns

$$\mathcal{G}_1(\gamma, (a, x)) := \mathsf{RF}_\gamma\big(F(\mathsf{RF}_\gamma(a), x)\big)$$

Now, for every θ-complexity distinguisher D,

$$\Delta_\mathsf{D}(\mathcal{G}_0, \mathcal{G}_1) = \Delta_{\mathsf{D}_0}(g^{\otimes\mathcal{I}}, \mathsf{RF}^{\otimes\mathcal{I}})$$

by using the substitution reduction where D_0 is a $(u, 2q, 2q_{\max})$-complexity distinguisher (it simply simulates the construction G using its oracle replacing the function g in G).

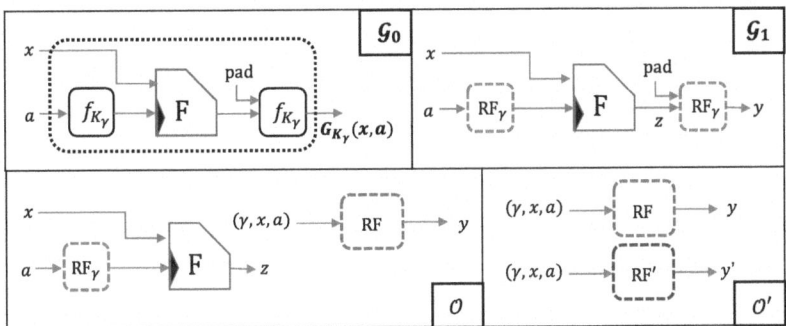

Fig. 5. Games $\mathcal{G}_0$ and $\mathcal{G}_4$ represent the real and ideal world respectively. The games $\mathcal{G}_1, \mathcal{G}_2, \mathcal{G}_3$ are intermediate oracles $\mathcal{G}_1$ is obtained by replacing f by a random function. Game $\mathcal{G}_2$ makes two executions of random functions independent. $\mathcal{G}_3$ replaces F by a random function.

Now we bound the distance between the oracle $\mathcal{G}_1$ and a random function $\mathcal{G}_2$. Let $\mathcal{O}^{\mathrm{PP}}(\gamma, a, x) = F(\mathsf{RF}(\gamma, a), x)$ be a post-processing oracle for both $\mathcal{G}_1$

and $\mathcal{G}_2$. In case of $\mathcal{G}_2$, the random function RF is independent with $\mathcal{G}_2$. Let D be any (u, q, q_{max})-complexity adaptive distinguisher interacting with either $\mathcal{G}_1$ or $\mathcal{G}_2$ and a post-processing oracle which returns responses of all queries made by D after the query-response phase is over. Let $(\gamma_1, (a_1, x_1)), \ldots, (\gamma_q, (a_q, x_q)) \in \mathcal{I} \times B \times \mathcal{X}$ be all q queries and $y_1, \ldots, y_q \in \mathcal{K}$ be the corresponding responses. Let $z_1, \ldots, z_q \in B$ be the responses of post-processing oracle, i.e. $F(\mathsf{RF}(\gamma_i, a_i), x_i) = z_i$. We say that bad holds if

1. either $z_i = z_j$ for some $i \neq j$ or
2. $(\gamma_i, z_i) = (\gamma_j, a_j)$ for some i, j.

So, the event bad satisfies conditions of Lemma 1. Hence there is a non-adaptive distinguisher D_1 with same complexity parameter as D_1 such that

$$\Pr(\text{bad } holds \text{ in } D^{\mathcal{G}_1, \mathcal{O}_{\mathrm{pp}}}) \leq \Pr(\text{bad } holds \text{ in } D_1^{\mathcal{G}_2, \mathcal{O}_{\mathrm{pp}}}).$$

Now for any good transcript $(\gamma^q, (a^q, x^q), y^q, z^q)$, it is easy to see that the both worlds realize the transcript with probability

$$\frac{\Pr(F(\mathsf{RF}(\gamma_i, a_i), x_i) = z_i \forall i)}{|\mathcal{K}'|^q}$$

(as the z_i values are all distinct and different from the other inputs of RF). Now by identical until bad lemma, $\Delta_D(\mathcal{G}_1; \mathcal{G}_2) \leq \Pr(\tau(D^{\mathcal{G}_2}) \text{ is bad})$. So, we have proved

$$\Delta_D(\mathcal{G}_0; \mathcal{G}_2) \leq \Delta_D(\mathcal{G}_0; \mathcal{G}_1) + \Delta_D(\mathcal{G}_1; \mathcal{G}_2)$$
$$\leq \Delta_{D_0}(g^{\otimes \mathcal{I}}, \mathsf{RF}^{\otimes \mathcal{I}}) + \Pr(\tau(D_1^{\mathcal{G}_2, \mathcal{O}_{\mathrm{pp}}}) \text{ is bad})$$

Note that the bad does not depend on the response of the oracle $\mathcal{G}_2$. So we can define a bad event $\mathsf{bad}(D'^{\mathcal{O}})$ for a non-adaptive interaction between a non-adaptive distinguisher D_1 and a $\mathcal{I}$-folded $(B \times \mathcal{X}, B)$-oracle $\mathcal{O}$ whenever

1. either $z_i = z_j$ for some $i \neq j$ or
2. $(\gamma_i, z_i) = (\gamma_j, a_j)$ for some i, j.

hold true where $(\gamma_1, (a_1, x_1)), \ldots, (\gamma_q, (a_q, x_q)) \in \mathcal{I} \times B \times \mathcal{X}$ be all q non-adaptive queries and $z_1, \ldots, z_q \in B$ be the corresponding responses. Thus, we have our single-keyed composition theorem.

Theorem 4 (single-keyed composition). *Let g, F and G as defined above. Then, for any $(u, q, q_{max}, \ell, \sigma)$-complexity distinguisher D,*
(i) there is a $(u, 2q, 2q_{max})$-complexity distinguisher D_0, and
(ii) $(q, q, q_{max}, \ell, \sigma)$-complexity non-adaptive adversary D_1 such that

$$\Delta_D(G^{\otimes \mathcal{I}}, \mathsf{RF}''^{\otimes \mathcal{I}}) \leq \Delta_{D_0}(g^{\otimes \mathcal{I}}, \mathsf{RF}^{\otimes \mathcal{I}}) + \Pr(\mathsf{bad}(D_1^{\mathcal{O}})) \tag{13}$$

where $\mathcal{O}(\gamma, a, x) = F(\mathsf{RF}(\gamma, a), x)$.

5.2   Security of Single-Keyed **NMAC** and Enveloped **MAC**

Now we show that the single keyed $1k\_NMAC_K := NMAC_{K,K}$ based on $f : \{0,1\}^c \times B \to \{0,1\}^c$ has almost the same security as $NMAC_{K_1,K_2}$. Note that in Theorem 4 we can consider $g = f$ and $F(K,m) = f_K^*(m)\|0^{b-c}$. Then, the function G is same as $1k\_NMAC_K$. By our single keyed composition we need to bound the probability of bad event for $D_1^{F^\otimes}$.

   We note that the queries may not be prefix-free and so we cannot replace f^* by a random function. However, we consider another bad event bad' as follows. Let x be a block such that (γ_i, m_i, x)'s are prefix-free. Clearly such a block x exists. Now, we say that bad' holds if

1. either $f(z_i, x) = f(z_j, x)$ for some $i \neq j$ or
2. $(\gamma_i, f(z_i, x)) = (\gamma_j, f(a_j, x))$ for some i, j.

So, bad $\Rightarrow$ bad' and hence $\Pr(\text{bad}) \leq \Pr(\text{bad}')$. Now, bad' is actually bad event for queries (γ_i, m_i, x) which are prefix-free. Thus,

$$\Pr(\text{bad}'(D_1^{F^\otimes})) \leq \Pr(\text{bad}'(D_1^{RF^\otimes})) + \mathbf{Adv}_F^{mu\text{-}nprf}(q, q, q_{max}, \ell+1, \sigma+q)$$
$$\leq 3q^2/2^{c+1} + \mathbf{Adv}_F(u, q, q_{max}, \ell+1, \sigma+q)$$

By using randomness of RF, bad' holds with probability at most $3q^2/2^{c+1}$. So, we have proved our PRF analysis for single keyed NMAC.

Theorem 5 (single-keyed NMAC). *For $T' = T + O(\sigma)$,*

$$\mathbf{Adv}_{1k\_NMAC^f}^{mu\text{-}prf}(\theta) \leq \mathbf{Adv}_{f^*}^{mu\text{-}pf\_nprf}(q, q, q_{max}, \ell+1, \sigma+q, T')+$$
$$+ u\mathbf{Adv}_f^{prf}(2q_{max}, T') + 1.5q^2/2^c.$$

   The previous result can be plugged into the above expression to get the security of the constant-free variant of HMAC, denoted as HMAC', where

$$HMAC'(K, m) := 1k\_NMAC(KDF(K), m)$$

and $KDF(K) = f(IV, K\|0^*)$. If $f(IV, \cdot, 0^*)$ is (almost) regular then $KDF(K)$ is uniformly distributed and hence the security of the variant of HMAC' is reduced to $1k\_NMAC$. So the bound for the single-keyed NMAC can be directly applied to the constant-free variant of HMAC.

   We now similarly prove an improved analysis for Enveloped MAC (we get a better tightness reduction as well as eliminate the related key advantage in the existing analysis).

Theorem 6 (Envelope MAC).

$$\mathbf{Adv}_{EvMAC}^{mu\text{-}prf}(\theta) \leq \mathbf{Adv}_{f^*}^{mu\text{-}pf\_nprf}(q, q, q_{max}, \ell, \sigma, \sigma_{max}, T')+$$
$$+ u' \cdot \mathbf{Adv}_f^{prf}(q_{max}+1, T') + \frac{q'^2}{2^c}.$$

6   Open Problems

We finally conclude our paper by stating the following important open problems those can be studied in the future.

1. Similar to the reduction for non-adaptive distinguishers, can we have an improved trade-off for adaptive reductions? This problem seems to be challenging as the simulator needs to guess a bound on the number of children of a node adaptively.
2. Having an improved reduction for adaptive PRF distinguisher of AMAC is not yet solved. A different approach to handle prefix queries is needed.
3. All known bounds for cascade construction can be at the best $DT/2^c$. However, there is no such generic matching attack (matching attacks are applicable for some pathological examples). Understanding the right PRF security bound when f behaves like a random function is not yet known.

References

1. Bellare, M.: New proofs for NMAC and HMAC: security without collision-resistance. In: Dwork, C. (ed.), Advances in Cryptology - CRYPTO 2006, 26th Annual International Cryptology Conference, Santa Barbara, California, USA, August 20–24, 2006, Proceedings, volume 4117 of Lecture Notes in Computer Science, pp. 602–619. Springer (2006)
2. Bellare, M.: New proofs for NMAC and HMAC: security without collision resistance. J. Cryptol. **28**(4), 844–878 (2015)
3. Bellare, M., Canetti, R., Krawczyk, H.: Keying hash functions for message authentication. In: Koblitz, N. (ed.), Advances in Cryptology - CRYPTO '96, 16th Annual International Cryptology Conference, Santa Barbara, California, USA, August 18–22, 1996, Proceedings, volume 1109 of Lecture Notes in Computer Science, pp. 1–15. Springer (1996). Extended version available at http://cseweb.ucsd.edu/~mihir/papers/kmd5.pdf
4. Bellare, M., Canetti, R., Krawczyk, H.: Pseudorandom functions revisited: the cascade construction and its concrete security. In: 37th Annual Symposium on Foundations of Computer Science, FOCS '96, Burlington, Vermont, USA, 14–16 October, 1996, pp. 514–523. IEEE Computer Society (1996). Extended version available at http://cseweb.ucsd.edu/~mihir/papers/cascade.pdf
5. Bellare, M., Kilian, J., Rogaway, P.: The security of cipher block chaining. In: Desmedt, Y. (ed.), CRYPTO, volume 839 of Lecture Notes in Computer Science, pp. 341–358. Springer (1994)
6. Bellare, M., Kilian, J., Rogaway, P.: The security of the cipher block chaining message authentication code. J. Comput. Syst. Sci. **61**(3), 362–399 (2000)
7. Bernstein, D.J., Lange, T.: Non-uniform cracks in the concrete: the power of free precomputation. In: International Conference on the Theory and Application of Cryptology and Information Security, pp. 321–340. Springer (2013)
8. Black, J., Rogaway, P.: A block-cipher mode of operations for parallelizable message authentication. In: Advances in Cryptology – Eurocrypt 2002, number 2332 in Lecture Notes in Computer Science, pp. 384–397. Springer, Berlin (2002)

9. Damgård, I.B.: A design principle for hash functions. In: Conference on the Theory and Application of Cryptology, pp. 416–427. Springer (1989)
10. Dworkin, M.: Recommendation for block cipher modes of operation: the CMAC mode for authentication. https://nvlpubs.nist.gov/nistpubs/specialpublications/nist.sp.800-38b.pdf
11. PUB FIPS. 198-1: The keyed-hash message authentication code (HMAC) (2008)
12. Gaži, P., Pietrzak, K., Rybár, M.: The exact PRF-security of NMAC and HMAC. In: Annual Cryptology Conference, pp. 113–130. Springer (2014)
13. Goldreich, O., Goldwasser, S., Micali, S.: How to construct random functions (extended abstract). In: 25th Annual Symposium on Foundations of Computer Science, West Palm Beach, Florida, USA, 24–26 October 1984, pp. 464–479. IEEE Computer Society (1984)
14. Goldreich, O., Goldwasser, S., Micali, S.: How to construct random functions. J. Assoc. Comput. Mach. $33(4)$, 792–807 (1986)
15. Hirose, S., Park, J.H., Yun, A.: A simple variant of the merkle-damgård scheme with a permutation. In: International Conference on the Theory and Application of Cryptology and Information Security, pp. 113–129. Springer (2007)
16. Iwata, T., Kurosawa, K.: OMAC: one-key CBC MAC. In: Fast Software Encryption, 10th International Workshop – FSE 2003, number 2887 in Lecture Notes in Computer Science, pp. 129–153. Springer, Berlin (2003)
17. Kaliski, B., Robshaw, M.: Message authentication with MD. CryptoBytes $1(1)$, 5–8 (1995)
18. Koblitz, N., Menezes, A.: Another look at HMAC. J. Math. Cryptol. $7(3)$, 225–251 (2013)
19. Koblitz, N., Menezes, A.: Another look at non-uniformity. Groups Complex. Cryptol. $5(2)$, 117–139 (2013)
20. Koblitz, N., Menezes, A.: Another look at security theorems for 1-key nested MACs. In: Open Problems in Mathematics and Computational Science, pp. 69–89. Springer (2014)
21. Krawczyk, H., Bellare, M., Canetti, R.: HMAC: keyed-hashing for message authentication. RFC 2104, 1–11 (1997)
22. Maurer, U.: Indistinguishability of random systems. In: International Conference on the Theory and Applications of Cryptographic Techniques, pp. 110–132. Springer (2002)
23. McKay, K., Bassham, L., Turan, M.S., Mouha, N.: Report on lightweight cryptography. Technical report, National Institute of Standards and Technology (2016)
24. Menezes, A., van Oorschot, P.C., Vanstone, S.A.: Handbook of Applied Cryptography. CRC Press (1996)
25. Merkle, R.C.: One way hash functions and DES. In: Brassard, G. (ed.), Advances in Cryptology - CRYPTO '89, 9th Annual International Cryptology Conference, Santa Barbara, California, USA, August 20–24, 1989, Proceedings, volume 435 of Lecture Notes in Computer Science, pp. 428–446. Springer (1989)
26. Nandi, M.: A new and improved reduction proof of cascade PRF. Cryptology ePrint Archive, Paper 2021/097 (2021)
27. Piermont, P., Simpson, W.: IP authentication using keyed MD5. IETF RFC 1828 (1995)
28. Preneel, B., Van Oorschot, P.C.: On the security of iterated message authentication codes. IEEE Trans. Inf. Theory $45(1)$, 188–199 (1999)
29. Rogaway, P.: Formalizing human ignorance. In: Nguyen, P.Q. (ed.), Progressin Cryptology - VIETCRYPT 2006, First International Conference on Cryptology

in Vietnam, Hanoi, Vietnam, September 25–28, 2006, Revised Selected Papers, volume 4341 of Lecture Notes in Computer Science, pp. 211–228. Springer (2006)

30. Tsudik, G.: Message authentication with one-way hash functions. ACM SIGCOMM Comput. Commun. Rev. **22**(5), 29–38 (1992)

31. Yasuda, K.: Boosting merkle-damgård hashing for message authentication. In: International Conference on the Theory and Application of Cryptology and Information Security, pp. 216–231. Springer (2007)

32. Yasuda, K.: "sandwich" is indeed secure: how to authenticate a message with just one hashing. In: Pieprzyk, J., Ghodosi, H., Dawson, E. (eds.), Information Security and Privacy, 12th Australasian Conference, ACISP 2007, Townsville, Australia, July 2–4, 2007, Proceedings, volume 4586 of Lecture Notes in Computer Science, pp. 355–369. Springer (2007)

Secure and Efficient Outsourced Matrix Multiplication with Homomorphic Encryption

Aikata Aikata[(✉)] and Sujoy Sinha Roy

Graz University of Technology, Graz, Austria
{aikata,sujoy.sinharoy}@iaik.tugraz.at

Abstract. Fully Homomorphic Encryption (FHE) is a promising privacy-enhancing technique that enables secure and private data processing on untrusted servers, such as privacy-preserving neural network (NN) evaluations. However, its practical application presents significant challenges. Limitations in how data is stored within homomorphic ciphertexts and restrictions on the types of operations that can be performed create computational bottlenecks. As a result, a growing body of research focuses on optimizing existing evaluation techniques for efficient execution in the homomorphic domain.

One key operation in this space is matrix multiplication, which forms the foundation of most neural networks. Several studies have even proposed new FHE schemes specifically to accelerate this operation. The optimization of matrix multiplication is also the primary goal of our work. We leverage the Single Instruction Multiple Data (SIMD) capabilities of FHE to increase data packing and significantly reduce the KeySwitch operation count-an expensive low-level routine in homomorphic encryption. By minimizing KeySwitching, we surpass current state-of-the-art solutions, requiring only a minimal multiplicative depth of two.

The best-known complexity for matrix multiplication at this depth is $\mathcal{O}(d)$ for matrices of size $d \times d$. Remarkably, even the leading techniques that require a multiplicative depth of three still incur a KeySwitch complexity of $\mathcal{O}(d)$. In contrast, our method reduces this complexity to $\mathcal{O}(\log d)$ while maintaining the same level of data packing. Our solution broadly applies to all FHE schemes supporting Single Instruction Multiple Data (SIMD) operations. We further generalize the technique in two directions: allowing arbitrary packing availability and extending it to rectangular matrices. This versatile approach offers significant improvements in matrix multiplication performance and enables faster evaluation of privacy-preserving neural network applications.

Keywords: Fully Homomorphic Encryption · Secure Outsourced Matrix Multiplication · Arbitrary Packing · Privacy-enhancing Techniques

© The Author(s), under exclusive license to Springer Nature Switzerland AG 2025
S. Mukhopadhyay and P. Stănică (Eds.): INDOCRYPT 2024, LNCS 15495, pp. 51–74, 2025.
https://doi.org/10.1007/978-3-031-80308-6_3

1 Introduction

The ability to process encrypted data without decryption has positioned Fully Homomorphic Encryption (FHE) as the "Holy Grail" of privacy-preserving data storage and computation [27]. However, this promising technique faces significant challenges that hinder its widespread adoption, including substantial data expansion and high computational requirements. These issues have sparked numerous research directions aimed at addressing the computing limitations associated with FHE, such as hardware acceleration approaches [1,4,16,26] that seek to enhance server performance.

Most of the efficient and high-performing FHE schemes [7,11,14] are lattice-based. A key limitation of these schemes is their linear slot-wise ciphertext encoding, which can be conceptualized as a one-dimensional array where each plaintext element occupies a single index. This encoding restricts operations requiring permutation, as elements cannot be easily extracted from the array, unlike in plaintext operations. Consequently, performing permutations on homomorphic ciphertext necessitates costly multiplications with masks and rotations. While this is manageable for operations like approximate function evaluation that operate slot-wise, it poses a significant challenge for matrix multiplication.

Matrix multiplication is fundamental in advanced mathematics and is especially critical for secure data analysis and machine learning (ML), particularly within neural networks (NN) [19–21,29,32]. Network components, such as fully connected layers and filters/kernels, depend heavily on efficient matrix multiplication. Although efficient algorithms like Strassen's algorithm exist for plaintext operations, conducting matrix multiplication in the encrypted domain remains an emerging area of research. This field has garnered attention for its potential to facilitate encrypted ML training and inference using FHE schemes like CKKS [10], which support approximate arithmetic.

1.1 Prior Works

The authors in [17] introduced the first technique for multiplying encrypted matrices and vectors, which was later extended to matrix-matrix multiplication by [24,30]. However, these methods require a substantial number of homomorphic multiplications and rotation operations. The literature [19–21,24,29,30,32] on encrypted matrix multiplication (referred to as matrix-matrix multiplication from here on) can be broadly categorized into three types, as summarized in Table 1.

The first type necessitates a multiplicative depth of two and employs a simple row-wise encoding for the initial input data. The work in [29] exemplifies this approach, presenting a technique for d^3 packing availability in the ciphertext. For a square matrix of dimensions $d \times d$, this method requires $2 \cdot d + 3 \cdot \log_2(d) - 2$ rotations and one ciphertext-ciphertext (ct-ct) multiplication. Importantly, these operations are costly, as a KeySwitch operation is required after each to maintain that the ciphertext is decryptable with the same secret key. Thus, the total KeySwitch complexity amounts to $2 \cdot d + 3 \cdot \log_2(d) - 1$. A significant limitation

Table 1. Comparison with secure d-dimensional matrix multiplication techniques. The division of works is done based on the three types of works discussed in Sect. 1.1. # Key-Switches = # ct-ct Mult + # Rotations.

Methodology	Packing	# ct-ct Mult	# Rotations	Required Depth[†]
Naive	1	$\mathcal{O}(d^3)$	-	2
[24,30]	d	$\mathcal{O}(d^2)$	$\mathcal{O}(d^2 \log_2 d)$	2
This Work	$\mathbf{d^2}$	$\mathcal{O}(\mathbf{d})$	$\mathcal{O}(\mathbf{d} \log_2 \mathbf{d})$	**2**
[29]	d^3	$\mathcal{O}(1)$	$\mathcal{O}(d)$	2
This Work	$\mathbf{d^3}$	$\mathcal{O}(\mathbf{1})$	$\mathcal{O}(\log_2 \mathbf{d})$	**2**
[21]	$d^2, d^{3\ddagger}$	$\mathcal{O}(d)$	$\mathcal{O}(d)$	3
[12,19,20,32]	d^2	$\mathcal{O}(d)$	$\mathcal{O}(d \log_2 d)$	2

† This includes (Plaintext-Ciphertext) pt-ct and ct-ct multiplications, which consume the same depth in Libraries like OpenFHE [2].
‡ With $d\times$ more packing, the number of rotations reduce from $3 \cdot d + 5 \cdot \sqrt{d}$ to $d + 2 \cdot \sqrt{d}$.

of this approach is its requirement for d^3 slots of packing availability, which limits scalability for large matrices. Additionally, it does not generalize well for lower packing availability.

The second category, as described in [21], diverges slightly from previous methods by utilizing diagonal-packing for matrix multiplication. In this approach, matrices are packed diagonally rather than row- or column-wise. While this technique is highly complex for a multiplicative depth of two, it allows for some pre-processing at a higher multiplicative depth of three, reducing the complexity to $3d + 5\sqrt{d}$ rotations and d ct-ct multiplications. If d^3 slot packing is available, it can be further optimized to require $d + 2\sqrt{d}$ rotation computations. However, a major drawback is the necessity for three multiplicative depths. Multiplicative depth is the currency in FHE schemes; the less spent, the better, as more depth remains for remaining computations. Although the algorithms proposed in [21] can be adjusted to operate within a multiplicative depth of two, this significantly increases the number of rotations and ct-ct multiplications.

The third and final category of works, as discussed in [19,20,32], diverges significantly from the previous two types. These studies leverage a multivariate variant of the CKKS scheme (m-RLWE) [12], which enables the encoding of matrices into a hypercube structure (tensor packing) instead of a linear array-like structure typical of CKKS. This approach allows for more efficient rotations, making row-wise or column-wise transformations cheaper. Matrix multiplication using this scheme requires only a multiplicative depth of two, with the cost of transformations reduced to $2 \cdot d + 4\sqrt{d}$ rotations and d ct-ct multiplications. However, the multivariate CKKS [12] is incompatible with the original CKKS [10]. Furthermore, the parameters for multivariate CKKS are not standardized, and its initial proposal [28] was found to be insecure [5], limiting its adaptability for existing implementations.

This category also includes recent works [22,31] that propose altering the initial encoding of ciphertexts to facilitate faster multiplication on the server. Such modifications require changing the specifications of the FHE scheme or necessitating client support for different encodings. When client data is already encrypted and stored on the server, these techniques become impractical, as the server would need to adjust the encoding to the desired form. For new computations, the FHE encoding that employs a Discrete Fourier Transform (DFT) is already resource-intensive for the FHE client. Consequently, we opted not to pursue this direction, as it merely shifts the computational burden from the server to the client, which has even less computational capacity.

1.2   Contributions

In this work, we restrict our solution to a multiplicative depth of two and build on the first type of technique discussed above. We observe that the best-known technique in this direction does not fully utilize the SIMD processing capabilities and leaves significant scope for optimization. We bridge this gap and propose a technique that improves with higher packing availability. For d^3 packing, our technique requires only $5 \cdot \log_2 d$ rotation operations and one ct-ct multiplication. Thus, this work contributes an efficient homomorphic matrix-multiplication framework for privacy-preserving applications. Its features are as follows.

– The proposed framework fully exploits the SIMD processing capabilities provided by FHE schemes and their routines. It is generalized for various packing availabilities, ranging from d^2 to d^3 for square matrices of dimension d, with benefits increasing with increasing packing availability in the homomorphic ciphertext.
– The KeySwitch operation in FHE is the most resource-intensive low-level routine and serves as a benchmark for assessing the performance overhead of our proposed techniques. We demonstrate that for d^3 packing, our technique achieves the lowest KeySwitch complexity ($\mathcal{O}(\log_2 d)$) compared to all prior works in the literature. This is also illustrated in Table 1.
– While our initial proposal focuses on square matrices, we further generalize it to accommodate rectangular matrices. To this end, we introduce two techniques based on padding and divide-and-conquer strategies, enhancing the versatility of our framework for various neural network applications and layers, such as filter layers.
– Alongside our proposal, we provide validation artefacts for our technique, which can be accessed at[1]. Our approach leverages the open-source FHE library OpenFHE [2], allowing researchers and practitioners to easily integrate our matrix multiplication framework into their own projects.

---

[1] https://anonymous.4open.science/r/MatMul-0568.

1.3   Roadmap

The paper is organized as follows: In Sect. 2, we provide an overview of the FHE routines and describe prior state-of-the-art matrix multiplication techniques. Section 3 details our proposed technique, explaining how it achieves improved runtime complexity. This section also discusses the generalized approach for arbitrary packing availability within the range of d^2 to d^3. In Sect. 4, we extend this technique to accommodate arbitrary rectangular matrix multiplication. The experimental evaluation is presented in Sect. 5, where we assess the performance of our approach. Section 6 explores scenarios involving simultaneous matrix multiplications, and Sect. 7 concludes the paper.

2   Background

Notations. Let $\mathbb{Z}_Q$ represent the ring of integers in the $[0, Q-1]$ range. $\mathcal{R}_{Q,N} = \mathbb{Z}_Q[x]/(x^N + 1)$ refers to polynomial ring containing polynomials of degree at most $N - 1$ and coefficients in $\mathbb{Z}_Q$. In the Residue Number System (RNS) [15] representation, Q is a composite modulus comprising co-prime moduli, $Q = \prod_{i=0}^{L-1} q_i$. The RNS representation divides a big computation modulo Q into much smaller computations modulo q_i such that the small computations can be carried out in parallel. With the application of RNS, a polynomial $a \in \mathcal{R}_{Q,N}$ becomes a vector, say $\boldsymbol{a}$, of residue polynomials. Let the i-th residue polynomial within $\boldsymbol{a}$ be denoted as $a^i \in \mathcal{R}_{q_i,N}$. $\langle, \rangle$ denotes the dot-product between two ring elements. Matrices are denoted using capital letters. For simplicity, we assume throughout that the values in fractions are divisible.

2.1   FHE Schemes and Routines

Several Fully Homomorphic Encryption (FHE) schemes are documented in the literature, including BFV [14], BGV [7], CGGI [13], CKKS [10,11]. While BGV and BFV encrypt integers, CKKS is designed for fixed-point numbers, making it particularly well-suited for machine learning applications [18,23]. Consequently, this work focuses on the RNS (Residue Number System) variant of CKKS [10]. Below, we briefly outline the main procedures within the RNS CKKS framework [10], where ciphertexts operate at level l (indicating a multiplicative depth of $l - 1$), with $l < L$. A CKKS ciphertext is represented as $c = (c_0, c_1)$, where c_0 and c_1 are polynomial vectors.

Two important terms- Depth and Packing, are used throughout the paper to assess the importance of this work. Multiplicative depth refers to the complexity of the computations that the FHE scheme can support. More specifically, it denotes the maximum number of operations (like multiplications) that can be performed on encrypted data before noise in the ciphertext grows too large and prevents decryption. Understanding depth is crucial for assessing the practicality of FHE in computational tasks. Every ciphertext initially starts with full-depth L. After multiple computations, the noise growth is significant, and

Table 2. CKKS Parameters

Parameter	Definition
$N, n \ (\leq \frac{N}{2})$	Polynomial size, maximum slots packed
Q, q_i	Coefficient modulus, RNS bases $Q = \prod_{i=0}^{L} q_i$
L, l	Multiplicative depth (#RNS bases - 1) $l < L$
P, p_i	Special modulus and its RNS base
L_{boot}, L_{eff}	Multiplicative depth of/after bootstrapping

scaling is done to reduce the noise, which also reduces the computational depth. Thus, the lower the computation depth of a function the more computation can be performed on the data, before the depth is refreshed via the expensive Bootstrapping operation.

Packing, also known as batching, is a technique that significantly improves the efficiency of FHE schemes, particularly those based on RLWE (Ring Learning with Error). Instead of encrypting a single plaintext value into a single ciphertext, packing allows multiple plaintext values to be encoded into a single ciphertext. This is especially valuable for applications requiring parallel computations (SIMD), such as matrix operations, machine learning, or data analytics. In RLWE-based schemes, packing leverages the structure of the underlying ring. Typically, plaintexts are elements of a polynomial ring, and packing encodes several plaintext slots into a single polynomial. Each slot can then store an individual message, enabling the system to simultaneously perform parallel homomorphic operations (like addition and multiplication) across all packed slots. This reduces the number of ciphertexts needed and increases throughput. The CKKS parameters are summarized in Table 2.

1. CKKS.KeyGen( ): This routine generates secret key $\mathsf{sk} = (1, s)$, public key $\mathsf{pk} = (-a \cdot s + e, a) \in \mathcal{R}^2_{Q_L, N}$, and several key-switching keys $\mathsf{ksk}_i = (-a \cdot s + e + P \cdot s', a) \in \mathcal{R}^2_{PQ_L, N}$ for $i \in [0, L)$, where a is uniformly random and s' is a secret polynomial square or permutation, depending on the type of key.
2. CKKS.Enc(m, pk): It encrypts message m , and returns ciphertext $\mathsf{c} = (c_0, c_1) = v \cdot \mathsf{pk} + (m + e, e) \in \mathcal{R}^2_{Q_L, N}$, where e is refreshed after every computation.
3. CKKS.Dec(c, sk): The ciphertext c is decrypted using the secret key sk to return message $m' = \langle \mathsf{c}, \mathsf{sk} \rangle$.
4. CKKS.Add(c, c'): It takes two input ciphertexts c and c' and adds them to compute $\mathsf{c}_{\mathrm{add}} = (d_0, d_1) = (c_0 + c'_0, c_1 + c'_1)$.
5. CKKS.Mult(c, c'): It multiplies the two input ciphertexts $(\mathsf{c}, \mathsf{c}')$, and computes the non-linear ciphertext $\mathsf{d} = (d_0, d_1, d_2) = (c_0 \cdot c'_0, c_0 \cdot c'_1 + c_1 \cdot c'_0, c_1 \cdot c'_1)$. Subsequently, CKKS.KeySwitch is employed to transform d into a linear ciphertext. It is the most expensive routine.
6. CKKS.KeySwitch(d, ksk): It uses a KeySwitch or 'evaluation key ksk to homomorphically transform a ciphertext decryptable under one key into a new ciphertext decryptable under another key. It computes c'' where

$c_0'' = \sum_{i=0}^{l-1} d_2^i \cdot ksk_0^i \in \mathcal{R}_{PQ_l,N}$ and $c_1'' = \sum_{i=0}^{l-1} d_2^i \cdot ksk_1^i \in \mathcal{R}_{PQ_l,N}$. This is followed by $c = ((d_0, d_1) + (CKKS.ModDown(c''))) \in \mathcal{R}_{Q_l,N}^2$. CKKS.ModDown() scales down the modulus (PQ_l to Q_l).

7. CKKS.Rotate(c, rot, ksk_{rot}): It rotates the plaintext slots within c by rot. First, a permutation ρ is applied to the ciphertext polynomial coefficients. This permutation is called automorphism and is determined by the Galoi element $gle = 5^{rot} \mod 2N$. Finally, the permuted ciphertext is processed by CKKS.Keyswitch using the rotation key ksk_{rot}.

8. CKKS.Bootstrap: It refreshes a noisy ciphertext [6,8,9] by producing a new ciphertext with a higher depth or lower noise. As bootstrapping itself consumes a certain number of depths, the depth of a bootstrapped ciphertext, say L_{eff}, is smaller than the initial depth L after fresh encryption.

The fundamental FHE operations include addition, multiplication, and rotation. Among the low-level routines, ModDown and KeySwitch, the KeySwitch operation is the most resource-intensive and is required after every rotation and ct-ct multiplication. Consequently, the frequency of these routines significantly influences the complexity of matrix multiplication. In contrast, bootstrapping is a high-level routine that employs all basic and low-level routines.

2.2 Matrix Multiplication Technique

In this section, we will introduce the state-of-the-art technique for matrix multiplication at multiplicative depth two, presented in [29]. This technique leverages the available packing capability of d^3. To simplify this, let us take an example of a processing system which only operates on the array. The proposal in the work utilizes arrays which can store d^3 elements, where each matrix to be multiplied is $d \times d$. In this approach, the first step is to decide how to effectively pack a matrix in the array so that it facilitates multiplication. Suppose we have two matrices, A and B, of dimension 4×4 ($d = 4$) for multiplication. This means our array must accommodate 64 elements.

$$A = \begin{bmatrix} a_0 & a_1 & a_2 & a_3 \\ a_4 & a_5 & a_6 & a_7 \\ a_8 & a_9 & a_{10} & a_{11} \\ a_{12} & a_{13} & a_{14} & a_{15} \end{bmatrix} \quad B = \begin{bmatrix} b_0 & b_1 & b_2 & b_3 \\ b_4 & b_5 & b_6 & b_7 \\ b_8 & b_9 & b_{10} & b_{11} \\ b_{12} & b_{13} & b_{14} & b_{15} \end{bmatrix}$$

Packing Strategy. To facilitate efficient multiplication, the authors needed to adopt a packing strategy that organizes the elements of matrices A and B into the array. One common approach is to fill the array in a row-major or column-major order, depending on the operations we intend to perform. They chose a row-major order, where the elements of matrix A would be packed into the first 16 slots of the array. Similarly, matrix B would be packed in the first 16 slots of another array, as shown below.

$$A = [a_0 a_1 \quad a_2 a_3 \quad a_4 a_5 \quad a_6 a_7 \quad a_8 a_9 \quad a_{10} a_{11} \quad a_{12} a_{13} \quad a_{14} a_{15}] \quad (1)$$
$$B = [b_0 b_1 \quad b_2 b_3 \quad b_4 b_5 \quad b_6 b_7 \quad b_8 b_9 \quad b_{10} b_{11} \quad b_{12} b_{13} \quad b_{14} b_{15}] \quad (2)$$

Matrix Multiplication Strategy. The authors in [29] employ an efficient multiplication technique that enhances the performance of matrix multiplication in the homomorphic setting. In this approach, the elements of matrix A are duplicated column-wise, while the elements of matrix B are duplicated row-wise. For example, the first column of A and the first row of B are as follows.

$$A_1 = \begin{bmatrix} a_0 \\ a_4 \\ a_8 \\ a_{12} \end{bmatrix} \quad B_1 = \begin{bmatrix} b_0 & b_1 & b_2 & b_3 \end{bmatrix}$$

Once the rows and columns are duplicated, the proposed technique gives the multiplied result A·B as follows.

$$
A \cdot B = \begin{bmatrix} a_0 & a_0 & a_0 & a_0 \\ a_4 & a_4 & a_4 & a_4 \\ a_8 & a_8 & a_8 & a_8 \\ a_{12} & a_{12} & a_{12} & a_{12} \end{bmatrix} \odot \begin{bmatrix} b_0 & b_1 & b_2 & b_3 \\ b_0 & b_1 & b_2 & b_3 \\ b_0 & b_1 & b_2 & b_3 \\ b_0 & b_1 & b_2 & b_3 \end{bmatrix} + \begin{bmatrix} a_1 & a_1 & a_1 & a_1 \\ a_5 & a_5 & a_5 & a_5 \\ a_9 & a_9 & a_9 & a_9 \\ a_{13} & a_{13} & a_{13} & a_{13} \end{bmatrix} \odot \begin{bmatrix} b_4 & b_5 & b_6 & b_7 \\ b_4 & b_5 & b_6 & b_7 \\ b_4 & b_5 & b_6 & b_7 \\ b_4 & b_5 & b_6 & b_7 \end{bmatrix}
$$
$$
+ \begin{bmatrix} a_2 & a_2 & a_2 & a_2 \\ a_6 & a_6 & a_6 & a_6 \\ a_{10} & a_{10} & a_{10} & a_{10} \\ a_{14} & a_{14} & a_{14} & a_{14} \end{bmatrix} \odot \begin{bmatrix} b_8 & b_9 & b_{10} & b_{11} \\ b_8 & b_9 & b_{10} & b_{11} \\ b_8 & b_9 & b_{10} & b_{11} \\ b_8 & b_9 & b_{10} & b_{11} \end{bmatrix} + \begin{bmatrix} a_3 & a_3 & a_3 & a_3 \\ a_7 & a_7 & a_7 & a_7 \\ a_{11} & a_{11} & a_{11} & a_{11} \\ a_{15} & a_{15} & a_{15} & a_{15} \end{bmatrix} \odot \begin{bmatrix} b_{12} & b_{13} & b_{14} & b_{15} \\ b_{12} & b_{13} & b_{14} & b_{15} \\ b_{12} & b_{13} & b_{14} & b_{15} \\ b_{12} & b_{13} & b_{14} & b_{15} \end{bmatrix}
$$

This technique requires d column-wise and row-wise duplications of the matrices. With $d^3(= 64)$ packing available, the matrices are stored as follows (their matrix form visualization is provided afterwards in Eq. 4). Notably, if we assume that matrices A and B can be packed into two ciphertexts (one for A and one for B), as illustrated in Eq. 3, then only one ct-ct multiplication is needed.

$$A \cdot B = [a_0 \ a_0 \ a_0 \ a_0 \ a_4 \ a_4 \ \cdots \ a_{15} \ a_{15}] \odot [b_0 \ b_1 \ b_2 \ b_3 \ b_0 \ b_1 \ \cdots \ b_{14} \ b_{15}] \quad (3)$$

There are two challenges associated with the packing format assumption. The first is transforming the input data into the required form. In prior works, all inputs (matrices A and B) are encoded in a row-wise format (row_enc), as shown in Eqs. 1 and 2. Therefore, it is necessary to convert this row-wise packing into the desired encoding for our approach. The second challenge is how to accumulate the resulting multiplication results. The methods for addressing these two challenges distinguish the work presented in [29] from our proposal, which is discussed in the next section. The technique proposed in [29] is detailed in Algorithm 1.

$$
A \cdot B =
\begin{bmatrix}
a_0 & a_0 & a_0 & a_0 \\
a_4 & a_4 & a_4 & a_4 \\
a_8 & a_8 & a_8 & a_8 \\
a_{12} & a_{12} & a_{12} & a_{12} \\
\hline
a_1 & a_1 & a_1 & a_1 \\
a_5 & a_5 & a_5 & a_5 \\
a_9 & a_9 & a_9 & a_9 \\
a_{13} & a_{13} & a_{13} & a_{13} \\
\hline
a_2 & a_2 & a_2 & a_2 \\
a_6 & a_6 & a_6 & a_6 \\
a_{10} & a_{10} & a_{10} & a_{10} \\
a_{14} & a_{14} & a_{14} & a_{14} \\
\hline
a_3 & a_3 & a_3 & a_3 \\
a_7 & a_7 & a_7 & a_7 \\
a_{11} & a_{11} & a_{11} & a_{11} \\
a_{15} & a_{15} & a_{15} & a_{15}
\end{bmatrix}
\odot
\begin{bmatrix}
b_0 & b_1 & b_2 & b_3 \\
b_0 & b_1 & b_2 & b_3 \\
b_0 & b_1 & b_2 & b_3 \\
b_0 & b_1 & b_2 & b_3 \\
\hline
b_4 & b_5 & b_6 & b_7 \\
b_4 & b_5 & b_6 & b_7 \\
b_4 & b_5 & b_6 & b_7 \\
b_4 & b_5 & b_6 & b_7 \\
\hline
b_8 & b_9 & b_{10} & b_{11} \\
b_8 & b_9 & b_{10} & b_{11} \\
b_8 & b_9 & b_{10} & b_{11} \\
b_8 & b_9 & b_{10} & b_{11} \\
\hline
b_{12} & b_{13} & b_{14} & b_{15} \\
b_{12} & b_{13} & b_{14} & b_{15} \\
b_{12} & b_{13} & b_{14} & b_{15} \\
b_{12} & b_{13} & b_{14} & b_{15}
\end{bmatrix}
\tag{4}
$$

Plaintext Masks. The authors define two plaintext masks π_i and ψ_i, such that multiplication with π_i causes all expected elements of column i to become zero, while multiplication with ψ_i results in all expected elements of row i becoming zero, as illustrated above. This is achieved by placing '1' and '0' in the desired positions within another array of size d^2. Although this multiplication is plaintext-to-ciphertext (pt-ct), it requires an additional multiplicative depth. Consequently, the total multiplicative depth for the overall technique is two.

cMult(A, π_0)	=	$[a_0 0$	00	$a_4 0$	00	$a_8 0$	00	$a_{12} 0$	$00]$
cMult(A, π_1)	=	$[0 a_1$	00	$0 a_5$	00	$0 a_9$	00	$0 a_{13}$	$00]$
cMult(A, π_2)	=	$[00$	$a_2 0$	00	$a_6 0$	00	$a_{10} 0$	00	$a_{14} 0]$
cMult(A, π_3)	=	$[00$	$0 a_3$	00	$0 a_7$	00	$0 a_{11}$	00	$0 a_{15}]$

cMult(B, ψ_0)	=	$[b_0 b_1$	$b_2 b_3$	00	00	00	00	00	$00]$
cMult(B, ψ_1)	=	$[00$	00	$b_4 b_5$	$b_6 b_7$	00	00	00	$00]$
cMult(B, ψ_2)	=	$[00$	00	00	00	$b_8 b_9$	$b_{10} b_{11}$	00	$00]$
cMult(B, ψ_3)	=	$[00$	00	00	00	00	00	$b_{12} b_{13}$	$b_{14} b_{15}]$

After obtaining these results, the values are first right-aligned using the rotations specified in Step 3 and Step 10 for matrices A and B. Once aligned, these matrices are appended next to each other, which requires $d-1$ rotations for both A and B (Steps 4-5 and 11-12). A duplication step is then performed to replicate these values and fill in the zeros in the ciphertexts, as shown in Steps 6-7 and 13-14. After multiplication, the results are accumulated again using $\log_2 d$ rotations,

---

Algorithm 1 `Matrix.Mult` [29]

Require: $A, B \leftarrow \text{row\_enc}(\mathbf{A}_{d\times d}, \mathbf{B}_{d\times d})$
Out: $C = \text{row\_enc}(\mathbf{A}_{d\times d} \times \mathbf{B}_{d\times d})$

    // *Preprocess A*
1: **for** $j = 0$ to $d - 1$ **do**
2:     $\tilde{A}[j] \leftarrow \text{cMult}(A, \pi_j)$ $\triangleright$ Splitting A cols
3:     $\tilde{A}[j] \leftarrow \text{Rot}(\tilde{A}[j], -j)$ $\triangleright$ Right align all cols
4: **for** $i = 1$ to $d - 1$ **do** $\triangleright$ Add the cols
5:     $\tilde{A}[0] + = \text{Rot}(\tilde{A}[j], -i(d^2 - i))$
6: **for** $i = 0$ to $\log_2 d - 1$ **do** $\triangleright$ Replicate cols
7:     $\tilde{A}[0] + = \text{Rot}(\tilde{A}[0], -2^i)$

    // *Preprocess B*
8: **for** $j = 0$ to $d - 1$ **do**
9:     $\tilde{B}[j] \leftarrow \text{cMult}(B, \psi_j)$ $\triangleright$ Splitting B rows
10:     $\tilde{B}[j] \leftarrow \text{Rot}(\tilde{B}[j], -j \cdot d)$ $\triangleright$ Top align all rows
11: **for** $i = 1$ to $d - 1$ **do** $\triangleright$ Add the rows
12:     $\tilde{B}[0] + = \text{Rot}(\tilde{B}[j], -(d^2 - d))$
13: **for** $i = 1$ to $d - 1$ **do** $\triangleright$ Replicate the rows
14:     $\tilde{B}[0] + = \text{Rot}(\tilde{B}[0], -d \cdot 2^i)$

    // *Compute C*
15: $C = \text{Mult}(\tilde{A}[0], \tilde{B}[0])$
16: **for** $j = 0$ to $\log_2 d$ **do** $\triangleright$ Result Accumulation
17:     $C = C + \text{Rot}(C, d^3/2^i)$

---

detailed in Steps 16-17. Overall, this technique requires $2 \cdot d$ pt-ct multiplications, one ct-ct multiplication, and $2 \cdot d + 3 \log_2 d - 2$ rotations, while consuming a multiplicative depth of two. The authors limit this proposal to square matrices and do not extend it to rectangular matrices or arbitrary packing.

In contrast, other techniques in the field [21] that utilize diagonal-based packing for matrix multiplication incur significantly higher rotation costs due to the required transformations. These techniques perform poorly at a multiplicative depth of two. However, by employing some pre-generation at the expense of additional multiplicative depth, they can reduce the complexity to $\mathcal{O}(d)$ rotations for lower packing availability $\mathcal{O}(d^2)$; unfortunately, this advantage does not translate effectively to higher packing availabilities.

3   Proposed Matrix Multiplication Technique

In this work, we optimize the technique outlined in [29]. We still perform ct-ct multiplication using Eq. 3 (as discussed in Sect. 2). However, what sets our approach apart is how we achieve the packing format required for Eq. 3. Before presenting our technique for d^3 packing, we note a gap in the literature: no existing technique addresses matrix multiplication complexity using Eq. 3 for d^2

packing. This requirement highlighted in [29] poses a limitation for medium- to large-sized matrices.

Furthermore, prior techniques are constrained to specific packing availabilities and do not generalize to accommodate arbitrary slot availability. This limitation restricts their applicability, particularly when the available packing varies with the matrix size across different layers. This flexibility is essential for effectively handling diverse matrix dimensions in practical applications.

For instance, in a CKKS ciphertext, the available packing ranges from 2^{13} – 2^{15} depending on the polynomial degree. Consequently, the technique from [29] can only be applied to matrices of dimensions upto 2^5. This limitation makes it unsuitable for processing larger datasets, such as high-resolution images of size 64×64 [3,25] or 100×100 [25]. In contrast, a technique that utilizes d^2 packing could facilitate fast matrix multiplications for data sizes up to 2^7. Therefore, we begin with a technique designed for d^2 packing and demonstrate how it can be extended to accommodate higher packing availability up to d^3. While packing more than d^3 is possible, it does not enhance the complexity of a single matrix multiplication.

3.1   Technique for d^2 Packing

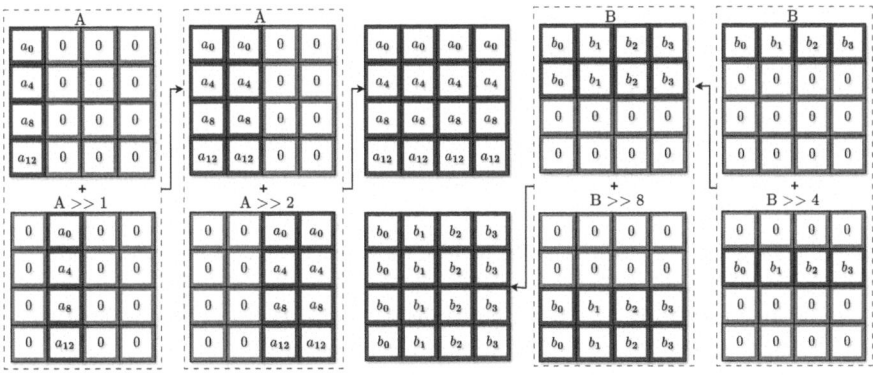

Fig. 1. Matrix Column and Row duplication for A and B respectively

Each matrix consists of d^2 elements, which means that one ciphertext which has d^2 packing capability, can only pack a single matrix entirely. To ensure compatibility with previous works, we assume these ciphertexts are initially row-wise encoded and that d is a power of two.

Our proposed algorithm is outlined in Algorithm 2. We initiate the process of multiplying by splitting matrix A column-wise in Step 2, which allows for more efficient handling of its elements. This is followed by immediate left-alignment in Step 3 to ensure that the data is properly structured for subsequent operations. Similarly, matrix B is split row-wise in Step 7 and left-aligned in Step 8, maintaining consistency in data organization. The duplication of these matrices necessitates $\log_2 d$ steps, as depicted in Fig. 1. This duplication process occurs in Steps 4-5 for matrix A and Steps 9-10 for matrix B, allowing us to leverage the SIMD capabilities of the encryption scheme. Since we create a distinct ciphertext for each row/column split, we can perform multiplication and addition separately, as shown in Steps 11-12.

Algorithm 2 Matrix.Mult_d^2

Require: $A, B \leftarrow$ row_enc($A_{d \times d}, B_{d \times d}$)
Out: $C =$ row_enc($A_{d \times d} \times B_{d \times d}$)

    // *Preprocess A*
1: **for** $j = 0$ to $d - 1$ **do**
2:      $\tilde{A}[j] \leftarrow$ cMult(A, π_j)   ▷ Splitting A cols
3:      $\tilde{A}[j] \leftarrow$ Rot($\tilde{A}[j], j$)   ▷ Right align all cols
4:      **for** $i = 0$ to $\log_2(d) - 1$ **do**   ▷ Replicate cols
5:         $\tilde{A}[j]+ =$ Rot($\tilde{A}[j], -2^i$)

    // *Preprocess B*
6: **for** $j = 0$ to $d - 1$ **do**
7:      $\tilde{B}[j] \leftarrow$ cMult(B, ψ_j)   ▷ Splitting B rows
8:      $\tilde{B}[j] \leftarrow$ Rot($\tilde{B}[j], j \cdot d$)   ▷ Top align all rows
9:      **for** $i = 0$ to $\log_2(d) - 1$ **do**   ▷ Replicate the rows
10:         $\tilde{B}[j]+ =$ Rot($\tilde{B}[j], -2^i \cdot d$)

    // *Compute C*
11: **for** $j = 0$ to $d - 1$ **do**
12:      $C+ =$ cMult($\tilde{A}[j], \tilde{B}[j]$)

Since one ciphertext can only pack data from one row-wise or column-wise split, a post-multiplication transform is unnecessary, streamlining the process. Notably, no left alignment is required in the algorithms when $j = 0$. As a result, this technique demands $2 \cdot d$ pt-ct multiplications for the necessary transformations, d ct-ct multiplications to combine the results, and $2 \cdot d(1 + \log_2 d) - 2$ rotations to align the data correctly for the final multiplication. This efficient use of resources reduces the overall computational complexity. These details are clearly summarized in Table 3.

3.2 Technique for $2 \cdot d^2$ Packing

In the previous case, we focused on d^2 packing, where one ciphertext could only encode a single matrix. However, for smaller matrices, it is possible that more slots are available- specifically, $2 \cdot d^2$, which allows for packing two matrices within the same ciphertext. To accommodate this scenario, we modify the algorithm outlined in the previous section and describe it in Algorithm 3.

Instead of directly initiating row-wise or column-wise matrix decomposition, we propose packing two copies of the matrices within the same ciphertext. This

---

Algorithm 3 Optimized.Matrix.Mult_2 · d^2

Require: $A, B \leftarrow$ row_enc($\mathbf{A}_{d \times d}, \mathbf{B}_{d \times d}$)
Out: $C =$ row_enc($\mathbf{A}_{d \times d} \times \mathbf{B}_{d \times d}$)

    // *Preprocess A*
1:   $A+ = \text{Rot}(A, -d^2 + 1)$
2:   **for** $j = 0$ to $(d/2) - 1$ **do**
3:      $\tilde{A}[j] \leftarrow \text{cMult}(A, \pi_j)$ ▷ Splitting A cols
4:      $\tilde{A}[j] \leftarrow \text{Rot}(\tilde{A}[j], 2j)$ ▷ Right align all cols
5:      **for** $i = 0$ to $\log_2(d) - 1$ **do**   ▷ Replicate cols
6:         $\tilde{A}[j]+ = \text{Rot}(\tilde{A}[j], -2^i)$

    // *Preprocess B*
7:   $B+ = \text{Rot}(B, -d^2 + d)$
8:   **for** $j = 0$ to $(d/2) - 1$ **do**
9:      $\tilde{B}[j] \leftarrow \text{cMult}(B, \psi_j)$ ▷ Splitting B rows
10:     $\tilde{B}[j] \leftarrow \text{Rot}(\tilde{B}[j], 2 \cdot j \cdot d)$ ▷ Top align all rows
11:     **for** $i = 0$ to $\log_2(d) - 1$ **do**   ▷ Replicate the rows
12:        $\tilde{B}[j]+ = \text{Rot}(\tilde{B}[j], -2^i \cdot d)$

    // *Compute C*
13: **for** $j = 0$ to $(d/2) - 1$ **do**
14:     $C+ = \text{cMult}(\tilde{A}[j], \tilde{B}[j])$
15: $C+ = \text{Rot}(c, d^2)$

---

critical step (Steps 1 and 7) is where our technique diverges from all prior works. The key innovation here is that rather than simply duplicating matrix A, the second ciphertext is shifted by one value to the left, as illustrated in Fig. 2.

After performing multiplication with the plaintext mask, we achieve duplicated and left-aligned columns as required for the subsequent steps, eliminating the need for an additional rotation for left alignment. Thus, this adjustment allows us to obtain two columns with a single multiplication and rotation. Similarly, the B matrix is duplicated by keeping d values to the left, ensuring a seamless integration of the matrices. Consequently, this technique requires only d pt-ct multiplications, $\frac{d}{2}$ ct-ct multiplications, and $d \cdot (1 + \log_2 d) + 1$ rotations, as outlined in Table 3. A final rotation is necessary for accumulating the two partial results packed in the same ciphertext (Step 15).

3.3 Generalization to Arbitrarily High Packing Complexity ($> d^2$)

If we estimate the cost for $4 \cdot d^2$ packing availability, we can initially pack four matrices per ciphertext. Hence, the required pt-ct and ct-ct multiplications, as well as the rotations for duplication and alignment, will decrease by a factor of four. This complexity is also outlined in Table 3. For generalization we observe that we can quantify the complexities based on the required steps and packing availability. Let s be the number of matrices that can be packed in a ciphertext. We can formulate the complexity in five parts, as follows.

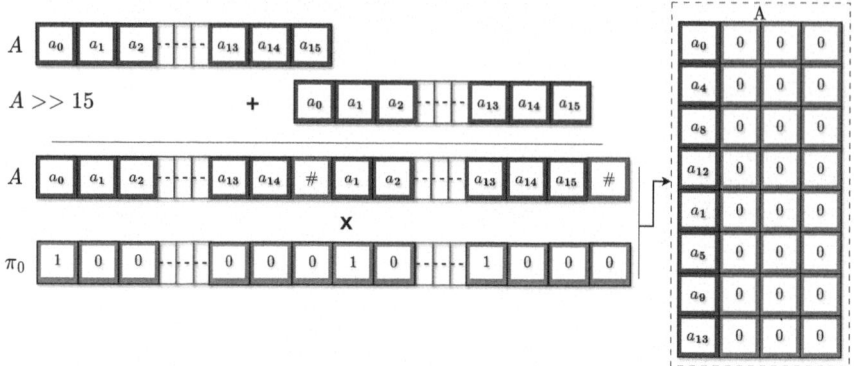

Fig. 2. Matrix duplication and alignment done in one step using rotation. First, Matrix A is rotated and added to the original A. Then, a mask is multiplied to remove the unwanted elements for the upcoming column-wise duplication step

Table 3. Complexity of the proposed secure d-dimensional matrix multiplications

Packing	# pt-ct Mult	# ct-ct Mult	# Rotations	Required Depth[†]
d^2	$2 \cdot d$	d	$2 \cdot d(1 + \log_2 d) - 2$	2
$2 \cdot d^2$	d	$\frac{d}{2}$	$d(1 + \log_2 d) + 1$	2
$4 \cdot d^2$	$\frac{d}{2}$	$\frac{d}{4}$	$\frac{d}{2}(1 + \log_2 d) + 4$	2
d^3	2	1	$5 \cdot \log_2 d$	2
$d^{3\,\ddagger}$	1	1	$3 \cdot \log_2 d$	2

† This includes pt-ct and ct-ct multiplications, which consume the same depth in Libraries like OpenFHE [2].
‡ If one of the matrices is unencrypted and can be defined in any form by the server.

1. Matrix Duplication. This part consists of matrix duplication depending on s. The duplication of this type has logarithmic complexity, as illustrated in Fig. 1. The cost of each matrix duplication is $\log_2 s$ rotations and additions. For example, when $s = 1$, no duplication is possible, resulting in a cost of 0. When $s = 2$, one duplication is possible per matrix, yielding a cost of 1. Therefore, the total cost for this step is $2 \log_2 s$ rotations and additions for both matrices.

2. RC Extraction. Once the matrices have been duplicated, their row/column-wise (RC) extraction requires plaintext multiplication with the appropriate masks, π_i and ψ_i. This extraction is performed $\frac{d}{s} \times$ per matrix, as duplication within the ciphertext offers s times more rows/columns per multiplication. Thus, the total pt-ct multiplication count amounts to $\frac{2 \cdot d}{s}$. While the first extraction does not require alignment, every subsequent extraction necessitates left alignment via rotations. Therefore, the total number of required rotations for alignment is $2(\frac{d}{s} - 1)$ for both the rows and columns of matrices A and B.

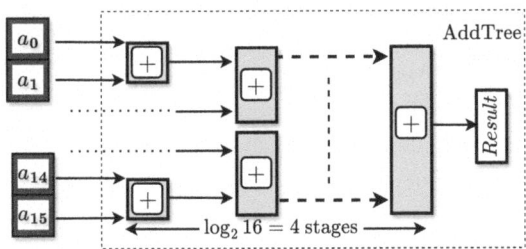

Fig. 3. The adder tree depicting final result accumulation within a ciphertext. Every stage with t elements performs an addition (in SIMD) and results in $\frac{t}{2}$ resultant elements, which are then added in the next stage. This process continues until all the elements are accumulated.

3. RC Duplication. After alignment, the rows and columns require duplication. Each duplication incurs a fixed cost of $\log_2 d$ rotations and additions. However, its frequency changes depending on the number of duplications needed, $\frac{2 \cdot d}{s}$. Thus, the total number of required rotations and additions for this step is $\frac{2 \cdot d}{s}(\log_2 d)$ for both row and column duplications.

4. Multiplication. We proceed to their multiplication once we have the row and column-wise duplicated matrices. This process is directly related to the number of ciphertexts that must be multiplied, resulting in $\frac{d}{s}$ ct-ct multiplications.

5. Accumulation. Finally, we must accumulate the results packed within a ciphertext. This accumulation process is not an issue for d^2 packing since no matrix copies can be stored in a single ciphertext. However, accumulation becomes necessary for any higher packing availability. The accumulation is analogous to duplication but in the reverse direction, as depicted in Fig. 3. It is important to note that elements cannot be accessed prior to rotation in an FHE ciphertext; hence, each $\log_2 s$ stage in the addition tree requires a rotation. Overall, the complexity of this stage is $\log_2 s$ rotations and additions.

Finally, we arrive at the following generalized operation requirements for matrix multiplication at depth two. The algorithm for this process is outlined in Algorithm 4. By substituting the appropriate value of s into the formulas below, we can derive the operation complexities listed in Table 3. As previously mentioned, Key Switching is only necessary following the ct-ct multiplications and rotations. Therefore, while pt-ct multiplication reduces the multiplicative depth, it does not significantly impact the overall time consumption.

- **# pt-ct Multiplications:** $\frac{2 \cdot d}{s}$
- **# ct-ct Multiplications:** $\frac{d}{s}$
- **# Rotations:** $\frac{2 \cdot d}{s}(\log_2 d) + 2(\frac{d}{s} - 1) + 3 \cdot \log_2 s$

Algorithm 4 `Generalized.Matrix.Mult` (for arbitrary s matrix packing)

Require: $A, B \leftarrow$ row_enc($\mathbf{A}_{d \times d}, \mathbf{B}_{d \times d}$)
Out: $C = $ row_enc($\mathbf{A}_{d \times d} \times \mathbf{B}_{d \times d}$)

   // Preprocess A
1: **for** $i = 0$ to $\log_2 s - 1$ **do**
2:     $A{+} = \text{Rot}(A, -d^2 \cdot 2^i + 2^i)$ ▷ Matrix Duplication

3: **for** $j = 0$ to $\frac{d}{s} - 1$ **do**
4:     $\tilde{A}[j] \leftarrow \text{cMult}(A, \pi_j)$ ▷ Extracting A column-wise
5:     $\tilde{A}[j] \leftarrow \text{Rot}(\tilde{A}[j], s \cdot j)$ ▷ Left align all ciphertexts
6:     **for** $i = 0$ to $\log_2(d) - 1$ **do**   ▷ Column Duplication
7:         $\tilde{A}[j]{+} = \text{Rot}(\tilde{A}[j], -2^i)$

   // Preprocess B
8: **for** $i = 0$ to $\log_2 s - 1$ **do**
9:     $B{+} = \text{Rot}(B, -d^2 \cdot 2^i + d \cdot 2^i)$ ▷ Matrix Duplication

10: **for** $j = 0$ to $\frac{d}{s} - 1$ **do**
11:     $\tilde{B}[j] \leftarrow \text{cMult}(B, \psi_j)$ ▷ Extracting B row-wise
12:     $\tilde{B}[j] \leftarrow \text{Rot}(\tilde{B}[j], s \cdot j \cdot d)$ ▷ Left align all ciphetexts
13:     **for** $i = 0$ to $\log_2(d) - 1$ **do**   ▷ Row Duplication
14:         $\tilde{B}[j]{+} = \text{Rot}(\tilde{B}[j], -2^i \cdot d)$

   // Compute C
15: **for** $j = 0$ to $\frac{d}{s} - 1$ **do**
16:     $C{+} = \text{cMult}(\tilde{A}[j], \tilde{B}[j])$ ▷ Matrix Multiplication

17: **for** $i = 0$ to $\log_2 s - 1$ **do**
18:     $C{+} = \text{Rot}(C, d^2 \cdot 2^i)$ ▷ Accumulation

The d^3 Packing Case

For d^3 packing, we note that $s = d$, resulting in two required pt-ct multiplications and one ct-ct multiplication, yielding a constant time complexity. Additionally, the number of rotations required is $5 \log_2 d$, which is logarithmic in terms of the matrix dimension d. This performance surpasses all prior works, including those operating at multiplicative depth three [21].

In specific scenarios where the server has control over the model, and only one matrix is encrypted, this operation count can be further reduced to $3 \log_2 d$ rotations (as shown in Table 3). If we assume that the client can format the data as needed during the encoding step before encryption, the rotation operations can be minimized to $\log_2 d$, and the multiplicative depth requirement becomes one. Therefore, the proposed technique is versatile and can be optimized for various scenarios.

A comparison of actual operation counts for encrypted matrix-matrix multiplication with prior works is illustrated in Fig. 4, specifically for d^3 packing availability. It underscores the efficiency and effectiveness of our approach in enhancing secure matrix multiplication tasks within untrusted environments.

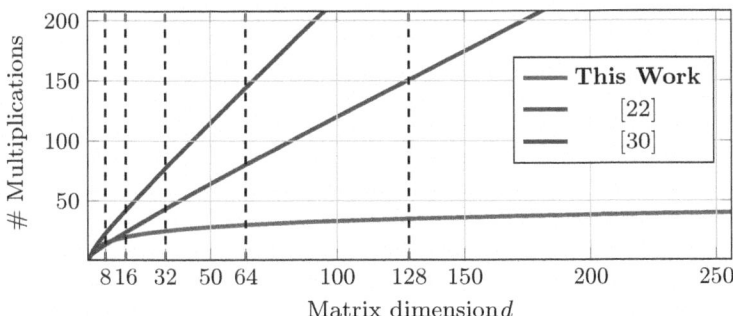

Fig. 4. Graph comparing the rotation count reduction for d^3 packing

4   Generalization to Rectangular Matrices

4.1   Padding Based Technique

The above technique can also be adapted for rectangular matrices of the form-$A_{l \times d} \cdot B_{d \times t}$. Previous works [29] employ zero-padding to the rows and columns of the matrices, transforming them into square matrices. This approach allows them to leverage techniques designed specifically for square matrices. We extend this idea to our method as well. By adding appropriate padding to the matrices A and B, we can ensure they fit the necessary dimensions for our algorithm.

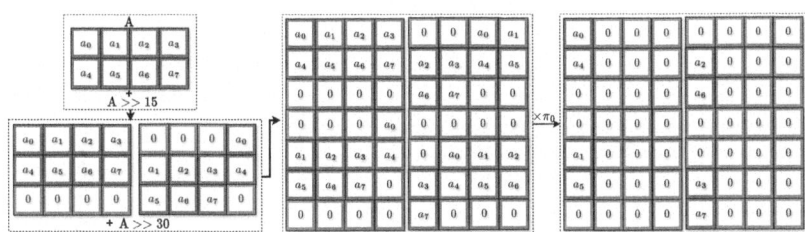

Fig. 5. The $A_{2 \times 4}$ matrix duplication transformation considering padding. The rectangular matrix undergoes two rotated accumulations to result in the middle ciphertext. This ciphertext behaves like a square matrix with zero padding. The grey elements are not needed and are removed via multiplication with the mask. After this step, column-wise duplication follows. (Color figure online)

This transformation can be incorporated into the application logic, ensuring that shifts consistently account for the factor d, as shown in Fig. 5. Notably, most complexities remain unchanged except for row-wise and column duplication steps. Specifically, the columns of A will now need to be duplicated t times, while the rows of B require l duplications. Thus, the complexity for the RC Duplication step is adjusted to $\frac{d}{s}(\log_2 l + \log_2 t)$. This allows us to perform the same efficient operations while accommodating the rectangular structure of the matrices. The zero-padding process does not significantly impact the overall complexity, as it effectively maintains the logarithmic characteristics of our approach.

The technique discussed can also be generalized for rectangular matrices by adopting the approach from [29]. Specifically, when $d < t$, $t - d$ zero-padding columns are added to matrix A, and conversely, when $d > t$, zero-padding columns are appended to matrix B. This method is bounded by the naive technique that ensures both matrices become square, with dimensions $k \times k$, where $k = \max(l, d, t)$. As a result, the final rotation complexity for this approach is $5 \cdot \log_2 k$, necessitating k^3 packing.

4.2   Common Divide-and-Conquer Technique

The previous technique necessitates high packing availability, a requirement that becomes increasingly challenging when there is a significant difference between d and l or t. To address this, we explore an alternate matrix-splitting and divide-and-conquer approach that can be applied in all cases, irrespective of whether $l \neq d$ or $t \neq d$. This method is illustrated in Fig. 6, showcasing two scenarios: one where $d < l, t$ and another where $d > l, t$.

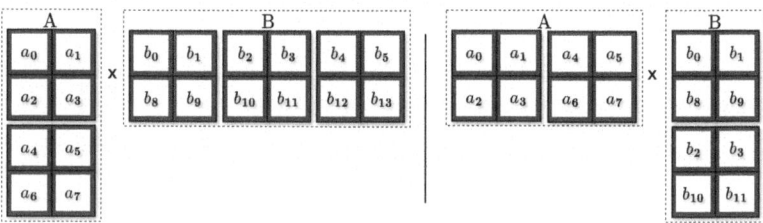

Fig. 6. Rectangular matrices division for the divide-and-conquer technique

The essence of this technique lies in dividing the larger matrix into smaller $d \times d$ matrices, which mitigates the need for excessive padding. For instance, in the first scenario (where $l = 4$, $d = 2$, and $t = 6$), instead of requiring 6^3 packings, we only need $6 \cdot 2^2$ slots per ciphertext. This reduces the overall packing requirement significantly. Additionally, the prior rotation complexity of $5 \cdot \log_2 6$ simplifies to $5 \cdot \log_2 2$, as no accumulation is necessary. In the second scenario (where $l, t = 2, d4$), instead of needing 4^3 packing, we again require only $4 \cdot 2^2$ slots per ciphertext. Here, the rotation requirement changes from $5 \cdot \log_2 4$

to $5 \cdot \log_2 2 + \log_2 2$, as one accumulation is required. Consequently, the overall packing requirement can be generalized as $k \cdot p^2$, where $k = \max(l, d, t)$ and $p = \min(l, d, t)$. This alternate approach reduces the need for extensive padding.

Next, the $p \times p$ matrix chunks are appended. These chunks need to be duplicated $\frac{t}{d}\times$ for A and $\frac{l}{p}\times$ for B. After the initial duplication, all the $p \times p$ matrices are packed in one ciphertext. Hence, $3 \cdot (\log_2 p)$ rotations are required for row-wise and column-wise duplication and result accumulation for a $p \times p$ matrix. $\log_2 \frac{d}{p}$ rotations are required for the final accumulation if d is greater than l, t (the second case in Fig. 6). There is no increase in ct-ct multiplications, and the multiplicative depth stays at two. The rotation complexity for this approach can be expressed as: $(\frac{l}{p} + \frac{t}{p}) \log_2 p + (\log_2 \frac{t}{p} + \log_2 \frac{l}{p}) + 3 \cdot (\log_2 p) + \log_2 \frac{d}{p}$. The breakdown of each component is as follows.

- *Initial Matrix Duplication.* The term $(\frac{l}{p} + \frac{t}{p}) \log_2 p$ accounts for the number of rotations required to duplicate the initial matrices. This arises from the need to replicate the smaller matrix chunks across the dimensions of l, t.
- *Duplication$_{p \times p}$.* The $p \times p$ matrix chunks need to be duplicated $\frac{t}{d}\times$ for matrix A and $\frac{l}{p}\times$ for matrix B. This contributes to the subsequent rotation complexity of $\log_2 \frac{t}{p} + \log_2 \frac{l}{p}$ operations.
- *Accumulation$_{p \times p}$.* Once the $p \times p$ matrices are packed into a single ciphertext, an additional $3 \cdot (\log_2 p)$ rotations are required for the row-wise and column-wise duplication and the accumulation of results for all the packed $p \times p$ matrix separately.
- *Final Accumulation.* The term $\log_2 \frac{d}{p}$ is included if d exceeds both l and t, indicating the rotations required for the final accumulation of results.

There is no increase in ct-ct multiplications throughout this process, and the multiplicative depth remains at two. This method effectively optimizes both the packing requirements and the rotation complexity.

5   Experimental Evaluation

We utilize the CKKS [11] FHE scheme for our experimental evaluation. Prior works also adopt this scheme due to its ability to perform computations over approximate arithmetic, making it suitable for various applications, including Neural Networks. For our benchmarks, we employ the open-source OpenFHE [2] library, ensuring compatibility with the setup provided by the FHERMA matrix multiplication challenge[2]. Our proposed matrix multiplication technique was tested in this challenge, where the available packing was $2 \cdot d^2$, yielding the best results[3]. Our artefacts for the general solution are available at[4].

We take ring-degree $N = 2^{16}$, which enables us to fully pack 32×32 matrices, duplicated 32 times. For our benchmarks, we evaluate matrix sizes ranging from

---

[2] https://fherma.io/challenges/652bf669485c878710fd020b/overview.
[3] Winner of the competition.
[4] https://anonymous.4open.science/r/MatMul-0568.

Table 4. Runtime evaluation of privacy-preserving matrix multiplication.

Matrix Dimension		Utilized Packing	Runtime (s)	Slot Usage
A	B			
2×2	2×2	$d^3 = 2^3$	1.54	0.02%
2×4	4×4	$d^3 = 2^6$	1.79	0.20%
4×4	4×2	$d^3 = 2^6$	1.80	0.20%
4×4	4×4	$d^3 = 2^6$	1.94	0.20%
8×8	8×4	$d^3 = 2^9$	2.21	1.56%
8×8	8×8	$d^3 = 2^9$	2.31	1.56%
16×16	16×4	$d^3 = 2^{12}$	3.12	12.5%
16×16	16×16	$d^3 = 2^{12}$	3.23	12.5%
32×32	32×8	$d^3 = 2^{15}$	7.54	100%
32×32	32×32	$d^3 = 2^{15}$	7.88	100%
64×64	64×64	$8 \cdot d^2 = 2^{15}$	17.42	100%
128×128	128×128	$2 \cdot d^2 = 2^{15}$	157.1	100%

2×2 to 128×128 to demonstrate the scalability and efficiency of the technique. We observe that the proposed method is highly parallelizable. Therefore, we leverage the available parallelization using the '*pragma omp parallel*' routine. The runtimes reported in Table 4 are based on benchmarks executed on a 12th Gen Intel® Core™i7-1260P processor with 16 threads and 32 GB RAM. We also provide benchmarks for the rectangular matrix case (Case 1) where padding is used. Additionally, the runtime for the divide-and-conquer approach can be extrapolated from the results reported for square matrices.

6   Discussion on the Case of Packed Multiplications

The proposed technique and results may raise the question: for an application requiring several simultaneous matrix multiplications (r), is it more efficient to duplicate a matrix within the same ciphertext or pack r distinct matrices for separate multiplications? In the case of d^2 packing, our technique necessitates $2 \cdot d(1 + \log_2 d - 2)$ rotations. If s packing slots are available, we could pack s distinct matrices into one ciphertext instead of duplicating the same matrix. This approach can be applied $\frac{r}{s}$ times if $r > s$. The ct-ct multiplication count remains d, and the number of rotations required for processing s packed matrix multiplications is $2 \cdot d(1 + \log_2 d) - 2$. However, packing s distinct matrices in one ciphertext requires $s - 1$ additional rotations and additions per matrix. Therefore, the total rotation requirement for $r > s$ becomes- $\frac{r}{s}(2 \cdot d(1 + \log_2 d) + 2 \cdot s - 4)$.

For $r \leq s$, the complexity simplifies to:

$$2 \cdot d(1 + \log_2 d) + 2 \cdot r - 4$$

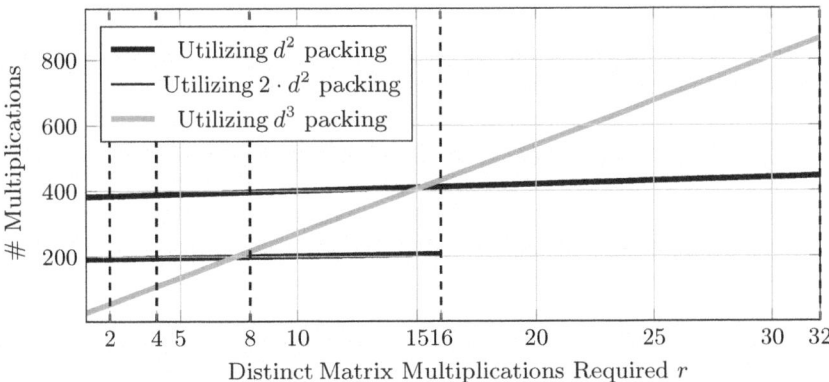

Fig. 7. Graph comparing the rotation count for $d^3(d = 32)$ packing available peer ciphertext with increasing distinct simultaneous matrix multiplication requirement r.

In the prior work [29], similar to our technique for d^3 packing, the approach involves packing (or duplicating) the same matrix s times within a single ciphertext. The total number of rotations required to process $r \leq s$ matrix multiplications using this technique are as follows.

$$2 \cdot r \cdot d(1 + \log_2 d) + 2 \cdot r(\frac{d}{s} - 1) + 3 \cdot r \cdot \log_2 s$$

For the case where $r = s = 1$, both our proposed technique and the technique from [29] offer the same computational complexity. However, as r increases, the complexity of the latter technique grows much more rapidly. This makes our proposed technique for d^2 packing more suitable for scenarios where multiple matrix multiplications need to be performed simultaneously, as it lowers computational costs. The choice of technique ultimately depends on the available matrix packing capacity (s) and the required number of matrix multiplications (r). Users can make informed decisions that align with their specific application requirements by providing a detailed complexity breakdown regarding these values.

In Fig. 7, we illustrate the number of required rotations for the above two techniques for matrices of dimension $d = 32$ and packing availability $d^3 = 2^{15}$. When required r is 15 or higher, the former technique, utilizing d^2 packed computation with distinct matrix packing, results in a lower rotation requirement. However, the d^3 packing technique per ciphertext is better for lower simultaneous matrix multiplication requirements. A similar analysis can be done for more combinations of s and r with different packing utilization per matrix multiplication to make it comprehensive.

7 Conclusion

In this work, we introduced a novel technique for secure matrix multiplication using homomorphic encryption that significantly outperforms all previous approaches. By optimizing the required key-switch operation complexity from $\mathcal{O}(d)$ to $\mathcal{O}(\log_2 d)$ for square matrices of dimension d, our approach enhances computational efficiency. We also extended this technique to handle rectangular matrices by proposing two methods: one based on padding and another employing a divide-and-conquer strategy. These methods generalize our approach beyond square matrices, making it adaptable to arbitrary packing between d^2 and d^3. This adaptability is particularly beneficial in secure neural network applications, where the matrix dimensions and the available packing per matrix change at each layer. By thoroughly analyzing the complexities and trade-offs of these techniques, we offer insights that can guide future research in privacy-preserving computation. This work lays a strong foundation for future exploration in optimizing secure matrix operations using homomorphic encryption.

Acknowledgement. This work was supported by the State Government of Styria, Austria - Department Zukunftsfonds Steiermark. We also extend our gratitude to the organizers of the FHERMA challenges for motivating this work.

References

1. Aikata, Mert, A.C., Kwon, S., Deryabin, M., Roy, S.S.: REED: chiplet-based scalable hardware accelerator for fully homomorphic encryption. In: IACR Cryptol. ePrint Arch, p. 1190 (2023). https://eprint.iacr.org/2023/1190
2. Al Badawi, A.: OpenFHE: open-source fully homomorphic encryption library. In: WAHC'22, Proceedings of the 10th Workshop on Encrypted Computing & Applied Homomorphic Cryptography, pp. 53–63. Association for Computing Machinery, New York, NY, USA (2022). https://doi.org/10.1145/3560827.3563379
3. angelolmg: Textile texture database (tilda) for defect detection (2023). https://www.kaggle.com/datasets/angelolmg/tilda-400-64x64-patches
4. Beirendonck, M.V., D'Anvers, J., Turan, F., Verbauwhede, I.: FPT: a fixed-point accelerator for torus fully homomorphic encryption. In: Meng, W., Jensen, C.D., Cremers, C., Kirda, E. (eds.) Proceedings of the 2023 ACM SIGSAC Conference on Computer and Communications Security, CCS 2023, Copenhagen, Denmark, November 26–30, 2023, pp. 741–755. ACM (2023). https://doi.org/10.1145/3576915.3623159
5. Bootland, C., Castryck, W., Vercauteren, F.: On the security of the multivariate ring learning with errors problem. In: IACR Cryptol. ePrint Arch, p. 966 (2018). https://eprint.iacr.org/2018/966
6. Bossuat, J., Mouchet, C., Troncoso-Pastoriza, J.R., Hubaux, J.: Efficient bootstrapping for approximate homomorphic encryption with non-sparse Keys. In: Canteaut, A., Standaert, F. (eds.) Advances in Cryptology - EUROCRYPT 2021 - 40th Annual International Conference on the Theory and Applications of Cryptographic Techniques, Zagreb, Croatia, October 17–21, 2021, Proceedings, Part I. Lecture Notes in Computer Science, vol. 12696, pp. 587–617. Springer (2021). https://doi.org/10.1007/978-3-030-77870-5_21
7. Brakerski, Z., Gentry, C., Vaikuntanathan, V.: Fully homomorphic encryption without bootstrapping. Electron. Colloquium Comput. Complex. 111 (2011). https://eccc.weizmann.ac.il/report/2011/111

8. Chen, H., Chillotti, I., Song, Y.: Improved bootstrapping for approximate homomorphic encryption. In: Ishai, Y., Rijmen, V. (eds.) Advances in Cryptology - EUROCRYPT 2019 - 38th Annual International Conference on the Theory and Applications of Cryptographic Techniques, Darmstadt, Germany, May 19-23, 2019, Proceedings, Part II. Lecture Notes in Computer Science, vol. 11477, pp. 34–54. Springer (2019). https://doi.org/10.1007/978-3-030-17656-3_2

9. Cheon, J.H., Han, K., Kim, A., Kim, M., Song, Y.: Bootstrapping for approximate homomorphic encryption. In: Nielsen, J.B., Rijmen, V. (eds.) Advances in Cryptology - EUROCRYPT 2018 - 37th Annual International Conference on the Theory and Applications of Cryptographic Techniques, Tel Aviv, Israel, April 29–May 3, 2018 Proceedings, Part I. Lecture Notes in Computer Science, vol. 10820, pp. 360–384. Springer (2018). https://doi.org/10.1007/978-3-319-78381-9_14

10. Cheon, J.H., Han, K., Kim, A., Kim, M., Song, Y.: A full RNS variant of approximate homomorphic encryption. In: Cid, C., Jr., M.J.J. (eds.) Selected Areas in Cryptography - SAC 2018 - 25th International Conference, Calgary, AB, Canada, August 15–17, 2018, Revised Selected Papers. Lecture Notes in Computer Science, vol. 11349, pp. 347–368. Springer (2018). https://doi.org/10.1007/978-3-030-10970-7_16

11. Cheon, J.H., Kim, A., Kim, M., Song, Y.S.: Homomorphic encryption for arithmetic of approximate numbers. In: Takagi, T., Peyrin, T. (eds.) Advances in Cryptology - ASIACRYPT 2017 - 23rd International Conference on the Theory and Applications of Cryptology and Information Security, Hong Kong, China, December 3–7, 2017, Proceedings, Part I. Lecture Notes in Computer Science, vol. 10624, pp. 409–437. Springer (2017). https://doi.org/10.1007/978-3-319-70694-8_15

12. Cheon, J.H., Kim, A., Yhee, D.: Multi-dimensional packing for HEAAN for approximate matrix arithmetics. In: IACR Cryptol. ePrint Arch, p. 1245 (2018). https://eprint.iacr.org/2018/1245

13. Chillotti, I., Gama, N., Georgieva, M., Izabachène, M.: TFHE: fast fully homomorphic encryption over the torus. J. Cryptol. **33**(1), 34–91 (2020)

14. Fan, J., Vercauteren, F.: Somewhat practical fully homomorphic encryption. In: IACR Cryptol. ePrint Arch, p. 144 (2012). http://eprint.iacr.org/2012/144

15. Garner, H.L.: The residue number system. IRE Trans. Electron. Comput. **8**(2), 140–147 (1959). https://doi.org/10.1109/TEC.1959.5219515

16. Geelen, R., et al.: BASALISC: programmable hardware accelerator for BGV fully homomorphic encryption. IACR Trans. Cryptogr. Hardw. Embed. Syst. **2023**(4), 32–57 (2023). https://doi.org/10.46586/TCHES.V2023.I4.32-57

17. Halevi, S., Shoup, V.: Algorithms in helib. In: Garay, J.A., Gennaro, R. (eds.) Advances in Cryptology - CRYPTO 2014 - 34th Annual Cryptology Conference, Santa Barbara, CA, USA, August 17–21, 2014, Proceedings, Part I. Lecture Notes in Computer Science, vol. 8616, pp. 554–571. Springer (2014). https://doi.org/10.1007/978-3-662-44371-2_31

18. Han, K., Hong, S., Cheon, J.H., Park, D.: Logistic regression on homomorphic encrypted data at scale. In: The Thirty-Third AAAI Conference on Artificial Intelligence, AAAI 2019, The Thirty-First Innovative Applications of Artificial Intelligence Conference, IAAI 2019, The Ninth AAAI Symposium on Educational Advances in Artificial Intelligence, EAAI 2019, Honolulu, Hawaii, USA, January 27–February 1, 2019, pp. 9466–9471. AAAI Press (2019). https://doi.org/10.1609/aaai.v33i01.33019466

19. Huang, H., Zong, H.: Secure matrix multiplication based on fully homomorphic encryption. J. Supercomput. **79**(5), 5064–5085 (2023). https://doi.org/10.1007/S11227-022-04850-4

20. Jang, J., et al.: Privacy-preserving deep sequential model with matrix homomorphic encryption. In: Suga, Y., Sakurai, K., Ding, X., Sako, K. (eds.) ASIA CCS '22: ACM Asia Conference on Computer and Communications Security, Nagasaki, Japan, 30 May 2022–3 June 2022, pp. 377–391. ACM (2022). https://doi.org/10.1145/3488932.3523253

21. Jiang, X., Kim, M., Lauter, K.E., Song, Y.: Secure outsourced matrix computation and application to neural networks. In: Lie, D., Mannan, M., Backes, M., Wang, X. (eds.) Proceedings of the 2018 ACM SIGSAC Conference on Computer and Communications Security, CCS 2018, Toronto, ON, Canada, October 15–19, 2018, pp. 1209–1222. ACM (2018). https://doi.org/10.1145/3243734.3243837

22. Ju, J.H., Park, J., Kim, J., Kim, D., Ahn, J.H.: Neujeans: private neural network inference with joint optimization of convolution and bootstrapping. CoRR arXiv:2312.04356 (2023)

23. Kim, A., Song, Y., Kim, M., Lee, K., Cheon, J.: Logistic regression model training based on the approximate homomorphic encryption. BMC Med. Genomics **11** (2018). https://doi.org/10.1186/s12920-018-0401-7

24. Lu, W., Kawasaki, S., Sakuma, J.: Using fully homomorphic encryption for statistical analysis of categorical, ordinal and numerical data. In: 24th Annual Network and Distributed System Security Symposium, NDSS 2017, San Diego, California, USA, February 26–March 1, 2017. The Internet Society (2017)

25. Mavi, A.: Sign language digits dataset (2017). https://doi.org/10.34740/KAGGLE/DSV/11071, https://www.kaggle.com/dsv/11071

26. Mert, A.C., et al.: Medha: microcoded hardware accelerator for computing on encrypted data. IACR Trans. Cryptogr. Hardw. Embed. Syst. **2023**(1), 463–500 (2023). https://doi.org/10.46586/tches.v2023.i1.463-500

27. Micciancio, D.: A first glimpse of cryptography's holy grail. Commun. ACM **53**(3), 96 (2010). https://doi.org/10.1145/1666420.1666445

28. Pedrouzo-Ulloa, A., Troncoso-Pastoriza, J.R., Pérez-González, F.: Multivariate lattices for encrypted image processing. In: 2015 IEEE International Conference on Acoustics, Speech and Signal Processing, ICASSP 2015, South Brisbane, Queensland, Australia, April 19–24, 2015, pp. 1707–1711. IEEE (2015). https://doi.org/10.1109/ICASSP.2015.7178262

29. Rizomiliotis, P., Triakosia, A.: On matrix multiplication with homomorphic encryption. In: Regazzoni, F., van Dijk, M. (eds.) Proceedings of the 2022 on Cloud Computing Security Workshop, CCSW 2022, Los Angeles, CA, USA, 7 November 2022, pp. 53–61. ACM (2022). https://doi.org/10.1145/3560810.3564267

30. Wang, S., Huang, H.: Secure outsourced computation of multiple matrix multiplication based on fully homomorphic encryption. KSII Trans. Internet Inf. Syst. **13**(11), 5616–5630 (2019). https://doi.org/10.3837/TIIS.2019.11.019

31. Zheng, X., Li, H., Wang, D.: A new framework for fast homomorphic matrix multiplication. In: IACR Cryptol. ePrint Arch, p. 1649 (2023). https://eprint.iacr.org/2023/1649

32. Zhu, L., Hua, Q., Chen, Y., Jin, H.: Secure outsourced matrix multiplication with fully homomorphic encryption. In: Tsudik, G., Conti, M., Liang, K., Smaragdakis, G. (eds.) Computer Security - ESORICS 2023 - 28th European Symposium on Research in Computer Security, The Hague, The Netherlands, September 25–29, 2023, Proceedings, Part I. Lecture Notes in Computer Science, vol. 14344, pp. 249–269. Springer (2023). https://doi.org/10.1007/978-3-031-50594-2_13

On Approx-SVP in Multiquadratic Ideal Lattices

Semen Aleksandrovich Novoselov[(✉)] [iD]

Immanuel Kant Baltic Federal University, A. Nevskogo street, 14,
Kaliningrad 236016, Russia
novsem@gmail.com

Abstract. Let $K = \mathbb{Q}(\sqrt{d_1}, \ldots, \sqrt{d_n})$ be a multiquadratic field of degree $m = 2^n$, and $D = d_1 \cdot \ldots \cdot d_n$. Assume also that $\log D = (\log m)^{O(1)}$. In this work, we prove that Ideal-SVP$_\gamma$ can be solved for ideals of multiquadratic fields in heuristic quasi-polynomial time in m with approximation factor $\gamma = e^{\widetilde{O}(m^{1/2})}$. Our work is an extension of the the algorithm (and the experiments) of Bauch, Bernstein, de Valence, Lange, and van Vredendaal for principal ideals in real multiquadratic fields to the case of non-principal ideals in real and imaginary multiquadratic fields.

Keywords: Multiquadratic fields · Approx-Ideal-SVP

1 Introduction

The shortest vector problem (SVP) is a classical hard problem in the complexity theory and lattice-based cryptography. Ideal lattices and other structured lattices can be used to build more efficient and practical cryptosystems in comparison with unstructured ones [9,12]. The security of such cryptographic schemes relies on the assumption that the corresponding structure does not lead to much better attacks than generic attacks on unstructured lattices. In our work we study the lattices which come from the ideals of multiquadratic fields.

A multiquadratic field is a field $K = \mathbb{Q}(\sqrt{d_1}, \ldots, \sqrt{d_n})$ where $d_1, \ldots, d_n$ are pairwise coprime integers. The field K is called real if all $d_1, \ldots, d_n$ are positive and imaginary otherwise. Any ideal I of the field K is a lattice under the canonical embedding. For real multiquadratic fields, it was shown in [2] that we can find short vectors for ideal lattices corresponding to principal ideals in quasi-polynomial time by using norm equations to reduce the problem to quadratic subfields. The natural question is to consider the problem of finding short elements in non-principal ideals and to extend all of these results from real to imaginary multiquadratic fields. This is the subject of our work. For this purpose we adopt known techniques for cyclotomic fields [4,5,10,11]. The main idea of these techniques is to reduce the problem to the case of principal ideals by solving for a target ideal I the discrete logarithm problem in the class group and reducing the result using specially crafted sublattices (having small generators) of the lattice of class group relations. The original algorithm with this

© The Author(s), under exclusive license to Springer Nature Switzerland AG 2025
S. Mukhopadhyay and P. Stănică (Eds.): INDOCRYPT 2024, LNCS 15495, pp. 75–93, 2025.
https://doi.org/10.1007/978-3-031-80308-6_4

technique was presented in [10,11] and it is called now CDW-algorithm. Other works [4,5,20] are modifications of this algorithm. These algorithms use special sublattices with small generators constructed either from a special ideal called Stickelberger ideal or via preprocessing. In our work we provide experimental evidences that the class group relations lifted from quadratic subfields of the field also provide short enough generators of the lattice.

Prior Work. In [6] the authors proposed a generalization of the original CDW-algorithm [11] for finding short vectors in ideals of cyclotomic fields to number fields with norm relations. These include multiquadratics as a special case. However, a complexity analysis was performed only for cyclotomic fields. The authors of [6] only removed quantum steps in the CDW-algorithm for cyclotomics without considering complexity of the algorithm for other fields. Neither does it include a description of short bases for relations (like it was done for cyclotomics in [4]) and nor does it contain an analysis for achieved approximation factor. Moreover, the original CDW-algorithm has the h_K^+ factor in the complexity. This is the class number of maximal real subfield of the field. While for cyclotomics the value of h_K^+ is conjectured to be small, it is large in practice for multiquadratic fields (see Table 1). To avoid this we have to consider a modified algorithm from [4].

Our Contribution. We propose an heuristic algorithm for finding short elements in non-principal ideals of real and imaginary multiquadratic fields. Our algorithm approximates shortest vectors by subexponential approximation factor $e^{\tilde{\mathcal{O}}(\sqrt{m})}$ and works in quasi-polynomial time in the degree of the field. This extends the results from [2] for principal ideals. As minor contribution, we also report new full class group computation (all S-units and matrices of relations) for degrees of the field equal to 256. We implemented our algorithm in Sage [23] and demonstrate finding the short elements in practice for degrees of the fields up to 64.

Organization of the Paper. In Sects. 2 and 3 we give notations and necessary background material. Section 4 contains description of our class group computations with modifications to the existing algorithm made for minimization of the size of S-units. Our work is based on two heuristic assumptions on the lengths of vectors in Log-unit and Log_S-unit lattices. The first one is presented in Sect. 5, where we describe an approximation factor of principal ideal generators returned by the algorithm from [2]. The second assumption concerns lengths of class group relations returned by the algorithm from [8]. In Sect. 6 we provide experimental data supporting the assumption that these lengths are small. In Sect. 7 we combine the two assumptions to obtain an heuristic algorithm that returns "mildly" short elements of a non-principal ideal, i.e. elements approximating shortest element of the ideal with approximation factor $e^{\tilde{\mathcal{O}}(\sqrt{m})}$. Section 8 contains results of the implementation of our algorithm and experiments with practical evaluation of the achieved approximation factor.

2   Notation

We use the following notation.

- $K = \mathbb{Q}(\sqrt{d_1}, \ldots, \sqrt{d_n})$ is a n-quadratic field where all d_i are pairwise coprime and square-free, $\mathcal{O}_K$ is its ring of integers.
- σ is the canonical embedding.
- K^+ is a maximal real subfield of K.
- $\mathrm{Cl}_K := \mathrm{I}_K / \mathrm{Princ}_K$ – class group of a field K, i.e. the factor group I_K of fractional ideals of K factored by principal ideals of $\mathcal{O}_K$.
- Cl_K^+ is the ideal class group of K^+.
- C_b is a cyclic group of order b.
- h_K, h_K^+ are the orders of Cl_K, and Cl_K^+ respectively.
- Δ_K is a discriminant of K.
- $G_K := \mathrm{Gal}(K/\mathbb{Q})$ is the absolute Galois group of K.
- $\lambda_1^{(2)}(I)$ is the length of a shortest vector (with respect to the Euclidean norm) in the lattice $\sigma(I)$.
- $\lambda_1^{(\infty)}(I)$ is the length of a shortest vector (with respect to infinity norm) in the lattice $\sigma(I)$.
- Log is Dirichlet's logarithm map.
- $N(I)$ is the absolute norm of an ideal I.
- $\mathcal{O}_{K,S}^\times$ is the S-unit group, where S is a set of prime ideals of the field K.
- $\mathcal{U}_K^\times$ is the set of all units from quadratic subfields of K.

3   Preliminaries

3.1   Bounds for Shortest Vectors in Ideal Lattices

Let K be a number field of degree m with signature (r_1, r_2), i.e. the field K has r_1 real and $2r_2$ complex embeddings to $\mathbb{C}$. If I is a fractional ideal of K then under the canonical embedding σ an ideal I can be considered as a full rank sublattice of $\mathbb{R}^m$. It has volume $\sqrt{\Delta_K} \cdot N(I)$ and the lengths of its shortest vectors are bounded (see Lemma 6.1 and Lemma 6.2 in [19]) as follows:

$$N(I)^{1/m} \leq \lambda_1^{(\infty)}(I) \leq \sqrt{|\Delta_K|}^{1/m} N(I)^{1/m} \left(\frac{2}{\pi}\right)^{r_2/m}, \tag{1}$$

$$\sqrt{m} \cdot N(I)^{1/m} \leq \lambda_1^{(2)}(I) \leq \sqrt{m} \cdot \sqrt{|\Delta_K|}^{1/m} N(I)^{1/m} \left(\frac{2}{\pi}\right)^{r_2/m}. \tag{2}$$

Here, $\lambda_1^{(\infty)}(I)$ and $\lambda_1^{(2)}(I)$ are the lengths of the shortest vector in $\sigma(I)$ with respect to the infinity-norm and to the Euclidean norm respectively. For ideal lattices we have the lower bound on the length of a shortest vector. Moreover, the upper and lower bounds are only differ by a value of size $\mathcal{O}(\sqrt{|\Delta_K|}^{1/m})$.

When the root discriminant $|\Delta_K|^{1/m}$ is small, these bounds significantly simplify the problem of determining the approximation factor for approximate closest/shortest vector problems. For instance, in the case of cyclotomic fields of prime power degree, we have $\sqrt{|\Delta_K|}^{1/m} = \mathcal{O}(m)$ and we know the value of $\lambda_1(I)$ up to a factor of polynomial size. For multiquadratic fields the situation is similar. Let $K = \mathbb{Q}(\sqrt{d_1}, \ldots, \sqrt{d_n})$ where all $d_1, \ldots, d_n$ are primes. Then the discriminant of the field K has [8, Lemma 2.2] a form

$$\Delta_K = (2^a D)^{2^{n-1}} = (2^a D)^{m/2},$$

for $a \in \{0, 2, 3\}$ and $D = d_1 \cdot \ldots \cdot d_n$. Therefore, $|\Delta_K|^{1/m} = \mathcal{O}(D^{1/2})$. We especially interested in the case when $\log D = (\log m)^{\mathcal{O}(1)}$, since we have here efficient algorithms for the short principal ideal problem (SPIP) [2] and the class group computation [8]. For this case, the D and the root-discriminant $|\Delta_K|^{1/m}$ are quasi-polynomial values in m. This means that we know the length of a shortest vector $\lambda_1(I)$ up to a quasi-polynomial in m factor $D^{1/4}$. And this will be enough for our proofs.

3.2   Finding Short Generators of Ideals Using Short Bases

The works of [4,10,11] exploit full-rank sublattices having short generators. In general, if we run the polynomial time Babai's nearest plane algorithm for a target vector t, we obtain an exponential approximation to the solution for the closest vector problem. However, if we have a short basis of the lattice with a known bound on their lengths, we can do better as shown in the following lemma proved in [10,11] (see also Regev's lectures [21, Claim 4]).

Lemma 1. *[11, Lemma 2.4] Let $\Lambda \subset \mathbb{R}^m$ be a lattice and let W be a set of k vectors in Λ such that W spans $\Lambda \otimes \mathbb{Q}$. Then there exists a classical randomized polynomial time algorithm that given a target vector $t \in \Lambda \otimes \mathbb{Q}$, outputs a vector $v \in \mathbb{Z}^k$ such that*

$$\|W \cdot v - t\|_\infty \leq \sqrt{2 \cdot \ln(8\,m)} \cdot \max_{w \in W} \|w\|$$

with probability at least $1/2$.

We can run the algorithm polynomial times to make probability of failure in Lemma 1 negligible (as it done in Algorithm 1 from [11]).

4   Class Group Computation

The computation of class group Cl_K can be done by collecting enough relations for a set $S = \{\mathfrak{p}_1, \ldots, \mathfrak{p}_\ell\}$ of prime ideals generating the class group. A relation is a pair $(g, \boldsymbol{a}) \in K \times \mathbb{Z}^\ell$ such that

$$g\mathcal{O}_K = \prod_{i=1}^{\ell} \mathfrak{p}_i^{a_i}.$$

The element g in the pair is called an S-unit. The matrix H whose rows are vectors $\boldsymbol{a}$ in a set of relations A is called a matrix of relations. Having a set of relations A we can compute a tentative regulator r^* of the field K and a tentative class number $h^* = \det H$ and compare the product $h^* r^*$ with the product $r_K h_K$ which can be computed in polynomial time. When $h^* r^* \approx r_K h_K$ we conclude that we have computed enough relations.

For multiquadratic fields the algorithm of Biasse-van Vredendaal [8] computes class group relations together with S-units by reducing the problem to the quadratic subfields of the field. Moreover, in this case the regulator r_K of the field can be computed in quasi-polynomial time using [2]. So, we do not have to compute a tentative regulator (we have the real one) and we can compute an approximation of h_K in polynomial time.

Remark 1. We note that for multiquadratics there is also a fast algorithm [7] for computing the class number and the class group structure. However, the current implementation [18] ("abelianbnf" package) does not compute S-units (and that's why it is fast). Moreover, the multiquadratic fields are Grunwald-Wang special fields and there can be infinite loop in computation (see details in [7]). Nevertheless, in most cases we are able to compute class group structure by this algorithm and use it for verification of output of Biasse-van-Vredendaal's algorithm. In other cases, we perform the verification by computing the product $h_K r_K$ using Hecke package [13].

We performed computations of the class group with generators of S-units for multiquadratic fields of degrees $64 - 256$. The results are presented in Table 1.

The fields in the table are defined as follows.

1. $L = \mathbb{Q}(\sqrt{d_1}, \ldots, \sqrt{d_n})$, where d_i for $i = 1, 2, \ldots$, are the first primes (taken with minuses for imaginary fields) such that $d_i \equiv 1 \bmod 4$.
2. L^+ is a maximal real subfield of the field L.

For computations in our work we used implementation "multiclass-im" [16] of Biasse-van Vredendaal algorithm with modifications to deal with obstructions in order to make representation of S-units smaller (see details in the subsection below). Computations were performed on one core of Intel Xeon Silver 4201R clocked at 2.40GHz on the machine with 629 GB RAM. For the degrees $64 - 128$ the timings are given for the computation with our modified implementation. For degree 256 the data is presented for computations with original "multiclass-im" implementation. The reference column contains the links to the respective works where full class group computation (together with S-units) was reported for the first time.

Remark 2. For the field L of degree $m = 128$ in the work [8, Table 1] the authors report the wrong size and structure of the class group. The correct class number differs from the reported one by the factor equal to 256 as showed in our table. We verified this using "abelianbnf" package which returns the same output as ours in the table. This also can be checked using the computation of the product $r_K h_K$. The product $r_K h_K$ computed using Hecke is equal to 632.80.

Table 1. Computation of the class group of the multiquadratic fields of degree m.

field	type	m	h^*/h_K	time (sec.)	Cl_K	ref.
L	real	64	1	$26.61 \cdot 10^2$	$C_2^9 \times C_4^3 \times C_8 \times C_{16}^4 \times C_{48} \times C_{240}$	[8]
	real	128	256	$10.62 \cdot 10^{4\ddagger}$	$C_2^{10} \times C_4^{16} \times C_8^{13} \times C_{16}^2 \times C_{48}^6 \times C_{96}^3 \times C_{480} \times C_{960}$	[8]
	real	128	1	$58.58 \cdot 10^3$	$C_2^2 \times C_4^{16} \times C_8^{13} \times C_{16}^2 \times C_{48}^6 \times C_{96}^3 \times C_{480} \times C_{960}$	†
	real	256	$\approx 1/128$	$61.34 \cdot 10^4$	$C_2^{38} \times C_4^{17} \times C_8^8 \times C_{16}^{35} \times C_{48}^6 \times C_{96}^5 \times C_{192}^5 \times C_{960} \times$ $\times C_{10560}^2 \times C_{781440}$	†
	imag.	64	1	$28.50 \cdot 10^2$	$C_2^2 \times C_4^9 \times C_8^3 \times C_{16} \times C_{48} \times C_{96}^2 \times C_{192}^2 \times$ $\times C_{6720}^2 \times C_{927360}$	[16]
	imag.	128	1	$22.78 \cdot 10^3$	$C_2^6 \times C_4^{11} \times C_8^{15} \times C_{16}^9 \times C_{32} \times C_{96}^4 \times C_{192}^2 \times$ $\times C_{960} \times C_{1920} \times C_{13440}^5 \times C_{443520} \times C_{20401920}$ $\times C_{554955114954240}$	[16]
	imag.	256	$\approx 1/32208$	$34.28 \cdot 10^{4\ddagger}$	$C_2^{15} \times C_4^{33} \times C_8^{19} \times C_{16}^{15} \times C_{32} \times C_{48}^{10} \times C_{96}^{19} \times C_{480} \times$ $\times C_{960}^7 \times C_{6720} \times C_{13440}^4 \times C_{40320} \times C_{80640}^2 \times C_{11531520}^3 \times$ $\times C_{219098880} \times C_{438197760} \times C_{514005972480} \times$ $\times C_{16648653448627200} \times C_{2788471376483444453514726400}$	[16]
	imag.	256	$\approx 1/2$	$11.16 \cdot 10^5$	$C_2^{15} \times C_4^{33} \times C_8^{16} \times C_{16}^{18} \times C_{48}^8 \times C_{96}^{20} \times C_{192} \times$ $\times C_{960}^6 \times C_{1920}^2 \times C_{13440}^2 \times C_{26880}^3 \times C_{80640}^3 \times C_{11531520}^3 \times$ $\times C_{219098880} \times C_{438197760} \times C_{514005972480} \times$ $\times C_{16648653448627200} \times C_{2788471376483444453514726400}$	†
L^+	real	64	1	$21.25 \cdot 10^2$	$C_4^9 \times C_8^2 \times C_{16}^3 \times C_{240}^2$	†
	real	128	1	$15.02 \cdot 10^3$	$C_2^{11} \times C_4^7 \times C_8^{12} \times C_{48}^4 \times C_{96}^6 \times C_{480}^2 \times C_{960}$	†

† Our work.
‡ Timings from the reference.

For our computations we have $\log h^* r^* \approx 632.84 \approx r_K h_K$. For the data from the work [8] we have $\log h^* r^* \approx 638.39$ and so h^* contains an extra factor of size ≈ 267.73.

Choice of the Factor Base. For the factor base S we chose unramified primes of norm p such that $p \nmid [\mathbb{Z}[\theta_L] : \mathcal{O}_L]$ for all subfields $L = \mathbb{Q}(\theta_L) \subseteq K$ from which the relations are lifted in the algorithm of Biasse-van Vredendaal. As we will see in the following, we have to choose p and $\#S$ as small as possible for keeping approximation factors small in the practical computations of short elements in ideals. So, we took p to be the first primes satisfied the conditions above and tried to select the minimal number of such primes.

Obstructions in the Biasse-Van Vredendaal Implementation. The original implementation of Biasse-van Vredendaal from [8] uses a slightly different algorithm that is described in their paper. It computes S-units in the compact representation of field elements, but with rational exponents. Namely, a computed S-unit g has the form $g = g_1^{e_1} \cdot \ldots \cdot g_k^{e_k}$, where $g_1, \ldots, g_k$ are small elements of the field K and $e_1, \ldots, e_k \in \mathbb{Z}[\frac{1}{2^{m-1}}]$. This is done to avoid expensive square root computations when we are recursively lifting relations from subfields. So, instead of computing square roots, the exponents in the compact representation of a field element are divided by 2. However, it introduces obstructions to the class group computation since it puts limitations on characters used in square root detection subroutine of the algorithm. And these constraints lead to false detection of non-squares as squares in some cases, which we call "obstructions".

To be more precise, the procedure of lifting relations from the subfield is the following. Let ν, μ are two automorphisms of the field K. Denote by K_ν, K_μ, $K_{\nu\mu}$ the corresponding fixed subfields by these automorphisms. Then for an S-unit g there is known norm equation

$$g^2 = \frac{N_{K/K_\nu}(g)N_{K/K_\mu}(g)}{\mu(N_{K/K_{\nu\mu}}(g))},$$

which allows us to join relations from subfields. To do this the algorithm computes generators of S-unit groups for subfields K_ν, K_μ, $K_{\nu\mu}$ and then applies the square-root detection algorithm to find generators of $(\mathcal{O}_{K,S}^\times)^2$, and then extracts square roots from these generators (in the case of Biasse-van Vredendaal implementation this is done formally by dividing exponents in compact representation by 2).

The square root detection algorithm works as follows. Let T be the union of generators of $\mathcal{O}_{K_\mu,S}^\times, \mathcal{O}_{K_\nu,S}^\times$, and $\mu(\mathcal{O}_{K_{\nu\mu},S}^\times)$. Note that the norm equation guaranties that $(\mathcal{O}_{K,S}^\times)^2$ is contained in a set generated by T. The algorithm computes the values of characters $\chi_q : g \mapsto \left(\frac{\phi_q(t_i)}{q}\right)$ for random prime numbers q (such that $\sqrt{d_1}, \ldots, \sqrt{d_n} \in \mathbb{F}_q$) and $i = 1, \ldots, \#T$. Here ϕ_q is a homomorphism $K \to \mathbb{F}_q^\times$. After that it computes a basis V of the left kernel of the matrix $A = \left(\log_{-1}\chi_q(t_i)\right)_{q,i}$ and the generators of $(\mathcal{O}_{K,S}^\times)^2$ as $f_i = \prod_j^{\#V} t_j^{v_{i,j}}$ for $i = 1, \ldots, \#V$.

This works under the assumption that if g is a non square in K and we have enough characters χ_q, then there is a prime number q such that $\log_{-1}\chi_q(g) = 1$. However, if we work with "formal square roots" this assumption is wrong, since instead of square roots we have to detect 2^r-powers (with additional knowledge that the elements in the set T are 2^{r-1}-powers). This requires to use r-power residue symbols and the maps $(\log_{\zeta_{2^r}}\chi_{q,r}(t_i))/2^{r-1}$ instead of $\log_{-1}(\chi_q(t_i))$, where $\chi_{q,r}(t_i)$ is a 2^r-power residue symbol computed at $\phi_q(t_i)$ and ζ_{2^r} is a primitive 2^r-th root of unity. Using of 2^r-symbols adds a constraint $q \equiv 1 \mod 2^r$ and the number of obstructions occur. For example, from the law of quadratic reciprocity we have for a prime a that $\left(\frac{a}{q}\right) = 1$ for all $a \equiv 1 \mod 4$ and $q \equiv 1 \mod 8$. Then all elements of K of the form $a^{2^{r-1}}$ with $a \equiv 1 \mod 4$ are obstructions and will always be detected as squares for $r \geq 3$, while they may not be squares in K.

Remark 3. Despite these obstructions, the original implementation of Biasse-van Vredendaal is able to compute class group for the fields $K = \mathbb{Q}(\sqrt{d_1}, \ldots, \sqrt{d_n})$ with $d_1, \ldots, d_n \equiv 1 \bmod 4$. For the case of the fields with $d_1, \ldots, d_n \equiv 3 \bmod 4$ we have a small extra multiplier in the outputted class number. However, we have to take more primes in the factor base and S-units have larger exponents in the compact representation. The obstructions are also do not alter the computation of short elements. We checked this on small degrees fields by ideal membership tests.

To avoid obstructions we computed two types of additional characters.

1. Quadratic characters without restrictions on primes q to ensure that 2^r-powers are at least squares. In practice, this simple computation rules out most of obstructions for the case when $d_i \equiv 1 \bmod 4$ for $i = 1, \ldots, n$. And leads to less memory consumption during computation and smaller S-units.
2. Schirokauer maps [22] as described in [3, §6.1] for detection of e-powers where e is a prime power.

Since we work with S-units, another set of characters can be defined using the valuations $v_{\mathfrak{p}}$ at prime ideals $\mathfrak{p} \in S$. These characters could be useful in case of occurrence of rational exponents at prime ideals in relations, but we did not experienced any such cases during computations (at least for the case of $d_i \equiv 1 \bmod 4$).

5   Short Principal Ideal Problem (SPIP) in Multiquadratics

Since we reduce the problem of finding short vectors in non-principal ideal lattices to the same problem in principal ideal lattices following the approaches from [4,11], we formulate here the basic assumptions for the lengths of vectors in principal ideal lattices required for obtaining the bounds for the lengths of vectors in non-principal ones. The main assumption is the following.

Assumption 1. Let $K = \mathbb{Q}(\sqrt{d_1}, \ldots, \sqrt{d_n})$ be a multiquadratic field of degree m such that $\log D = \log(d_1 \cdot \ldots \cdot d_n) = (\log m)^{\mathcal{O}(1)}$. Then there is an algorithm ShortGenerator that given a principal ideal $I = h\mathcal{O}_K$ returns an element $g \in I$ such that $h\mathcal{O}_K = g\mathcal{O}_K$ and $\|g\|_2 = e^{\widetilde{\mathcal{O}}(m^{1/2})} N(h)^{1/m}$. This algorithm takes quasi-polynomial time in m and $\log N(h)$. The element g approximates SVP in the lattice I with an approximation factor $e^{\widetilde{\mathcal{O}}(m^{1/2})}$.

This assumption is a multiquadratic analog of Theorem 3.7 from [11] for cyclotomic fields. Let us argue why the assumption is reasonable. We split the work in two parts. First, we describe the class of ideals for which we have an quasi-polynomial time algorithm for computing of the shortest generator. These are ideals that have generators of specific length. After that we prove that any ideal of the field falls into this class with probability $1 - \frac{1}{m}$ as m tends to ∞. So, we can find the shortest generator of a principal ideal with high probability of success and we heuristically assume that the shortest generator approximates SVP in the lattice $\sigma(I)$ with the claimed approximation factor (this is definitely true for ideals used in cryptography).

5.1   Finding Generators of Principal Ideals in Quasi-Polynomial Time

In [2] the authors proposed an (BBVLvV-)algorithm for finding a short generator of a given principal ideal in a real multiquadratic field. However, the authors did not analyze the achieved approximation factor in their work. The algorithm takes quasi-polynomial time in the degree of the field. It uses reduction in special sublattice of $\mathrm{Log}(\mathcal{O}_K^\times)$ with the orthogonal basis formed by units from quadratic subfields of the field K. This sublattice is called a multiquadratic unit lattice $\mathrm{Log}(\mathcal{U}_K^\times)$. Since the multiquadratic unit lattice has the orthogonal basis, we can solve closest vector problem (CVP) in this lattice in polynomial time using rounding [1]. This leads us to the following lemma (a restatement, refinement, and slight generalization of the result from [2, §8.4]).

Lemma 2. *Let I be a principal ideal of a multiquadratic field K of degree m, let $\varepsilon_1, \ldots, \varepsilon_{m-1}$ be the generators of multiquadratic units, and D be quasi-polynomial in m. If I has a generator g such that $\mathrm{Log}(g) = \sum_{j=1}^{m-1} c_j \mathrm{Log}(\varepsilon_j) + c \cdot \mathbf{1}$ where $|c_j| < \frac{1}{2m}$ and $c = \frac{\log|N(g)|}{m}$ then g is the unique generator (modulo torsion) of such kind in the ideal I. Moreover, this generator can be computed in quasi-polynomial time in m.*

Proof. Since the multiquadratic unit lattice is orthogonal, CVP can be solved here in polynomial time using rounding. Also, it follows from the orthogonality that we can compute c_j as $c_j = \frac{\langle \mathrm{Log}(g), \mathrm{Log}(\varepsilon_j)\rangle}{\|\mathrm{Log}(\varepsilon_j)\|^2}$. We will reduce the problem of finding the generator g to solving CVP in the multiquadratic unit lattice. A generator h of the ideal I can be computed in quasi-polynomial time using the algorithm from [2]. Since generators of ideals differ by a multiplication by a unit, the result of a generator computation will have the form $h = ug$ for some unit u. But there is no guarantee that u is a multiquadratic unit. However, it is known that u^m is a multiquadratic unit [2, Th. 5.2] for any unit u. Compute $m \mathrm{Log}(ug)$ and solve CVP (using rounding) for this vector in the multiquadratic unit lattice. We have $|c_j| < \frac{1}{2m}$, so the coefficients of the vector $m \mathrm{Log}(g)$ will be $< \frac{1}{2}$ in the multiquadratic unit lattice and thus the closest vector in the lattice to the vector $m \mathrm{Log}(g)$ is the zero vector. Then solving CVP for the target vector $m \mathrm{Log}(ug)$ produces $m \mathrm{Log}(u)$. This reveals the value of $\pm g^m$ and we can compute the element g with successive square root computations. The sign on every step is chosen in a such way that a square root exists (this is a character computation).

Let g' be another generator of I that satisfies the conditions of the lemma, i.e. $\mathrm{Log}(g') = \sum_{j=1}^{m-1} c_j' \mathrm{Log}(\varepsilon_j) + c' \cdot \mathbf{1}$ where $|c_j'| < \frac{1}{2m}$ and $c' = \frac{\log|N(g')|}{m}$. Then $g = u'g'$ for some unit $u' \in \mathcal{O}_K^\times$ and $h = uu'g'$. Since $|c_j'| < \frac{1}{2m}$, solving CVP (via rounding) in multiquadratic unit lattice for the target vector $\mathrm{Log}\, h^m =$

$\mathrm{Log}\,(uu'g)^m$ produces $\mathrm{Log}\,(uu')^m$. But we already know that $\mathrm{Log}\,u^m$ is a solution of CVP for $\mathrm{Log}\,h^m = \mathrm{Log}\,(u^m g^m)$. Therefore, $u'^m = 1$, so $u' \in \mu_K$, and $\|\mathrm{Log}(g')\| = \|\mathrm{Log}(g)\|$. Thus, the generator g of I is unique up to multiplication by an element from torsion.   □

5.2   Existence of Principal Ideal Generators of Bounded Length

In previous section we showed that we can find the shortest generator of a principal ideal in quasi-polynomial time in m if it fits the bounds in the lemma. Experiments from [2] show that for ideals in cryptosystems on multiquadratic fields, the conditions of the lemma are satisfied with high probability when $d_i > m^{1.03}$. In fact the conditions of the lemma are not so restrictive when m and D are big enough. The coefficients of $\mathrm{Log}\,\varepsilon_j$ have the size of regulators of the corresponding quadratic subfields, i.e. $\|\mathrm{Log}\,\varepsilon_j\|_\infty = D_j^{1/2+o(1)} = D^{1/2+o(1)}$. So, we have $0 \leq \|g\|_\infty < e^{D^{1/2+o(1)}} N(I)^{1/m}$ if g satisfies the conditions of the lemma or with respect to Euclidean norm:

$$0 \leq \|g\|_2 < \sqrt{m} \cdot e^{D^{1/2+o(1)}} N(I)^{1/m} \tag{3}$$

when $m \to \infty$. The upper bound in (3) is much higher than the upper bound for the shortest element of I (we have $\lambda_1(I) = \mathcal{O}(\sqrt{m}D^{1/4}N(I)^{1/m})$ by (2)) and we can expect that many ideals fit this value. To be more precise, the size of generators can be bounded using the covering radius of the lattice, which is defined as maximal distance between a vector from $\mathbb{R}^r$ and a vector from the lattice. For multiquadratic unit lattices the covering radius is bounded as follows.

Lemma 3. Let $K = \mathbb{Q}(\sqrt{d_1}, \dots, \sqrt{d_n})$ of degree m be a multiquadratic field and let $D = d_1 \cdot \dots \cdot d_n$. Then the covering radius $\mu_K^{(2)}$ (with respect to Euclidean norm) of the unit lattice $\sigma(\mathcal{O}_K)$ and the multiquadratic unit lattice $\sigma(\mathcal{U}_K)$ are bounded from above as:

$$\mu_K^{(2)}(\mathcal{O}_K^\times) \leq \mu_K^{(2)}(\mathcal{U}_K^\times) \leq \frac{m}{2} e^{D^{1/2+o(1)}},$$

when $D \to \infty$.

Proof. We have that $\mu_K^{(2)}(\mathcal{O}_K^\times) \leq \mu_K^{(2)}(\mathcal{U}_K^\times)$, since $\sigma(\mathcal{U}_K)$ is a sublattice of $\sigma(\mathcal{O}_K)$. Let $r = \dim \mathcal{U}_K^\times$. When K is real, we have $r = m$, while for imaginary K we have $r = \frac{m}{2}$. The covering radius can be bounded [14, Lemma 4.3] by using the r-th successive minimum:

$$\frac{1}{2}\lambda_r \leq \mu_K^{(2)}(\mathcal{U}_K^\times) \leq \frac{\sqrt{r}}{2}\lambda_r. \tag{4}$$

Since the generators $\varepsilon_1, \dots, \varepsilon_r$ of $\mathcal{U}_K^\times$ are multiquadratic units, we have

$$\lambda_r^\infty(\mathrm{Log}(\mathcal{U}_K^\times)) \leq \max_i \|\mathrm{Log}(\varepsilon_i)\|_\infty \leq \max_i R_i,$$

where R_i is a regulator corresponding to the quadratic subfield $\sqrt{D_i}$ of ε_i. It is known that regulators grow as $D_i^{1/2+o(1)}$ by Hua's bound [15, Theorem 13.4]. The maximal D_i is equal to D. Thus $\lambda_r^\infty(\text{Log}(\mathcal{U}_K^\times)) \leq D^{1/2+o(1)}$, and for the lattice $\sigma(\mathcal{U}_K^\times)$ this implies $\lambda_r^{(2)}(\mathcal{U}_K^\times) \leq \sqrt{r}e^{D^{1/2+o(1)}}$. Putting this in Eq.(4) gives us the claimed result, since $r \leq m$. $\square$

The bound in the lemma above is obtained by using known basis for the lattice $\sigma(\mathcal{U}_K^\times)$. Experimentally, the lattice $\text{Log}(\mathcal{U}_K^\times)$ has index of size $m^{\mathcal{O}(m)}$ in $\text{Log}(\mathcal{O}_K^\times)$. Moreover, it can be proved that this index is a power of 2 (see [2]). So, potentially this bound can be improved by finding a better basis. However it is an open problem to find such a basis. As a consequence of the lemma, we can bound now the generators of principal ideals.

Corollary 1. *Let I be a principal ideal of a multiquadratic field. Then there exists a generator g of I such that $\|g\|_2 \leq \frac{m}{2}e^{D^{1/2+o(1)}} N(I)^{1/m}$.*

As we can see all principal ideals of K have generators that differ from bounds in Lemma 2 by a factor of at most $\sqrt{m}$. Thus, if we take a principal ideal I and assume that lengths of its generators are distributed uniformly in the interval $(0, m \cdot e^{D^{1/2+o(1)}} N(I)^{1/m}]$, then with probability $\approx 1 - \frac{1}{\sqrt{m}}$ we obtain an ideal with generator of length $\leq \sqrt{m}e^{D^{1/2+o(1)}} N(I)^{1/m}$. Moreover, this generator is unique (so, it is the shortest) and it can be computed in quasi-polynomial time by Lemma 2.

Experiments from [2, §8.1] show that the secret generators in multiquadratic cryptosystems are much shorter. We have $|c_i| = \mathcal{O}\left(\frac{1}{\sqrt{m}\ln|\varepsilon_i|}\right) = \mathcal{O}\left(\frac{1}{\sqrt{m}D^{1/2+o(1)}}\right)$ with high probability. If we have such generators, then they are the shortest by Lemma 2 and we can find them in quasi-polynomial time. Since we know the length of the shortest vector upto a quasi-polynomial factor, it is easy to show that such generators approximate the shortest vector problem in I with approximation factor $\gamma = e^{\widetilde{\mathcal{O}}(\sqrt{m})}$. This explains a phenomena from the experiments in [2], when we have success probability for finding short elements in the ideal tending to 1 for $D \to \infty$.

For an arbitrary principal ideal, by resorting to probabilistic algorithms, we can also compute with high probability the generators of length as in Corollary 1.

Proposition 1. *Let I be a principal ideal of a multiquadratic field of degree m. Then there exists a classical randomized polynomial time algorithm that given a generator h of I returns a generator g such that $\|g\|_2 \leq me^{D^{1/2+o(1)}} N(I)^{1/m}$ with success probability $\to 1$ as $m \to \infty$.*

Proof. Let $u_1, \ldots, u_r$ be generators of multiquadratic unit lattice $\sigma(\mathcal{U}_K^\times)$. Then by Lemma 1 there is a classical randomized polynomial time algorithm that given a vector $t = \sigma(h)$ returns $\sigma(u) \in \mathcal{U}_K^\times$ such that $\|\sigma(u) - \sigma(h)\|_\infty \leq \sqrt{2 \cdot \ln(8r)} \cdot \max(\|\sigma(u_1)\|, \ldots, \|\sigma(u_r)\|) \leq \sqrt{2 \cdot \ln(8r)}\sqrt{r}e^{D^{1/2+o(1)}} \leq$

$\sqrt{2 \cdot \ln(8m)}\sqrt{m}e^{D^{1/2+o(1)}}$ with probability at least $1/2$. By running this algorithm m-times, we obtain desired result.   □

To summarize results of this section: we can find the shortest generator of a principal ideal of a multiquadratic field in quasi-polynomial time with probability that tends to 1 as $m \to \infty$. As a consequence, when the principal ideal has short generators (e.g. ideals in cryptographic applications) this gives us a short element of ideal in quasi-polynomial time.

6   Euclidean Norms of Class Group Relations

Here we present results of our experiments for lengths estimation of class group relations. For this purpose we used the class groups computed with the code from [16] modified as described in Sect. 4. Note that the relation matrices for all intermediate subfields and for the final output are LLL-reduced. For a set of relations W in the class group let $b_2 = \max_{w \in W}\|w\|_2$ and $b_\infty = \max_{w \in W}\|w\|_\infty$. These values determine (see Lemma 1) the efficiency of reduction of a vector using log-S-unit lattice, since they bound lengths of output vectors of polynomial time approx-CVP-solver – Babai's nearest plane algorithm. For the set W computed by Biasse-van Vredendaal algorithm the values b_2 and b_∞ are presented in Table 2. In addition to the field L defined in Sect. 4, we included the following fields.

1. $L_\nu = \mathbb{Q}(\sqrt{d_1}, \ldots, \sqrt{d_{n-2}}, \sqrt{d_n})$ is the fixed field of L by the automorphism $\nu \in G_K$ induced by the map $\sqrt{d_{n-1}} \mapsto -\sqrt{d_{n-1}}$.
2. $L_{\nu\mu} = \mathbb{Q}(\sqrt{d_1}, \ldots, \sqrt{d_n}, \sqrt{d_{n-1}}\sqrt{d_n})$ is the fixed field of L by composition of the automorphisms ν and μ, where μ is defined by the map $\sqrt{d_n} \mapsto -\sqrt{d_n}$.

The Biasse-van Vredendaal algorithm recursively reduces the problem of class group computation from the field L to subfields L_ν, L_μ and $L_{\nu\mu}$. So, we have the relations for all subfields of such type "for free" without running class group computation for them.

Remark 4. For our purpose of computing short elements of ideals it is enough to compute a sublattice of full S-unit lattice with small ratio h^*/h_K, i.e. we do not have to compute the full class group. For such a sublattice we can solve the discrete logarithm problem with high probability for a target ideal.

As we can see the relations have small lengths. Such lengths are sufficient for obtaining the "mildly" short vectors as it is done for cyclotomics in [11]. This motivates us to introduce the following heuristic.

Assumption 2. Let K be a multiquadratic field and S is a set of prime ideals generating the class group of K. Then the generators of the lattice $\mathrm{Log}_S \mathcal{O}_{K,S}^\times$ obtained by lifting relations from quadratic subfields of K have length $\mathcal{O}(\sqrt{m \log D})$.

Table 2. Euclidean lengths for class group relations computed using Biasse-van Vredendaal algorithm.

field	type	m	h^*/h_K	rank Λ_S	b_2	b_∞	$b_2 \leq \sqrt{m}$	$b_2 \leq 2\sqrt{m}$	$b_2 \leq \sqrt{m}\log_2 m$
L	imag.	32	1	128	4	1	✓	✓	✓
	imag.	64	1	512	7	2	✓	✓	✓
	imag.	128	1	1024	12.80	2	x	✓	✓
	imag.	256	$\approx 1/2$	2944	23.28	2	x	✓	✓
	real	32	1	112	3.16	1	✓	✓	✓
	real	64	1	448	5.29	1	✓	✓	✓
	real	128	1	1344	9.53	2	✓	✓	✓
	real	256	$\approx 1/128$	1664	15.03	2	✓	✓	✓
L_ν	imag.	32	1	368	4.24	1	✓	✓	✓
	imag.	64	1	736	8	2	✓	✓	✓
	imag.	128	1	1472	13.63	2	x	✓	✓
	real	32	1	424	3.31	1	✓	✓	✓
	real	64	1	864	6.08	1	✓	✓	✓
	real	128	1	1024	8.71	2	✓	✓	✓
$L_{\nu\mu}$	imag.	32	1	384	5.65	1	✓	✓	✓
	imag.	64	1	736	8.83	1	x	✓	✓
	imag.	128	$\approx 1/3$	1472	15.16	2	x	✓	✓
	real	32	2	424	4.12	1	✓	✓	✓
	real	64	2	848	7.28	1	✓	✓	✓
	real	128	32	1024	9.69	2	✓	✓	✓
L^+	real	64	1	560	5.56	1	✓	✓	✓
	real	128	1	1088	8.88	2	✓	✓	✓

We note that algorithm of Biasse-van Vredendaal lifts S-units from quadratic subfields of the field K using norm relations and saturation technique. The heuristic is built using analogy with multiquadratic unit group $\mathcal{U}_K^\times \subseteq \mathcal{O}_K^\times$ with replacing multiquadratic units by multiquadratic S-units. The quadratic subfield with largest discriminant has discriminant $|D|$. The heuristic claims in fact that the lengths of the vectors are bounded as $\mathcal{O}(\sqrt{\log h_K})$, since $h_K = \mathcal{O}(\sqrt{\Delta_K}) = \mathcal{O}(\sqrt{D^m})$. Note that we usually assume that $\log D = (\log m)^{O(1)}$ and then the bounds presented in Table 2 are more tight than in heuristic.

However, we do not know how to prove the claim in heuristic. If we put the relation matrix in HNF then it can be shown that $b_2 \leq h_K$ (see the proof of Proposition 3.10 in [4]), but h_K is expected to be huge (see class numbers in [8,16] or in Table 1). This can be improved using the fact the class group is highly non-cyclic. If $\mathrm{Cl}_K \sim C_{s_1} \times \ldots \times C_{s_k}$ (here $s_1, \ldots, s_k$ are diagonal entries from SNF of the matrix of relations) then it is often the case (heuristic) when

$b_2 \leq s_k$. Unfortunately, this does not give us bounds because s_k is typically higher then m.

The computation of the class number and the class group structure are relatively fast operation. It can be done with algorithm from [7]. However, the full class group computation (all relations, generators, S-units) is quite expensive (it takes weeks, see the timings in [16, Table 2] and [8, Table 1]) and requires a lot of memory. The algorithm computes S-units and this is what consumes memory, because they are represented in "compact" representation, i.e. as products $\prod_i g_i^{e_i}$ where g_i are small elements of the field K (in the algorithm these are S-units of quadratic subfields) and only exponents e_i and g_i are stored. These reduces amount of memory required to store an S-unit, but nevertheless e_i are huge. Moreover, the exponents e_i in the implementation of [8] are not integers. They have a form $e_i = a/2^r$ for some $r \geq 0$ and $a \in \mathbb{Z}$ (note that the existence of 2^r-roots is guarantied by algorithm). This makes the exponents even bigger.

So, it is not easy to check our heuristic for many fields. However, we already have examples of computations for the degrees of cryptographic size as presented in Table 2.

7   Finding Short Elements in Non-Principal Ideals

Now we have all necessary components to describe an algorithm for finding short elements in the ideals of multiquadratic fields. In Section §6 we showed that the algorithm [8] (with the usage of LLL-reduction as implemented in [16]) produces short generators of the lattice $\Lambda_S = \mathrm{Log}_S \mathcal{O}_{K,S}^\times$. To be more precise, Assumption 2 claims that the generators of the lattice have lengths $\mathcal{O}(\sqrt{m \log D})$ and there is an experimental evidence supporting this claim for big degrees of the fields as shown in Table 2. By Lemma 1 this gives us an heuristic polynomial time algorithm to solve approx-CVP in the lattice Λ_S with lengths of output vectors bounded by $\widetilde{\mathcal{O}}(\sqrt{m})$ (assuming that D is quasi-polynomial). So, we can build an algorithm which finds "mildly" short elements in integral ideals (as it is provided in [11] for cyclotomic fields).

Proposition 2. *Let* $K = \mathbb{Q}(\sqrt{d_1}, \ldots, \sqrt{d_n})$, $m = \deg K = 2^n$, *and* $D = d_1 \cdot \ldots \cdot d_n$ *is quasi-polynomial in* m. *Then Algorithm 1 is correct and under Assumptions 1, 2 it takes quasi-polynomial time in* m. *Moreover, it returns an element* g *of norm* $\|g\|_2 \leq e^{\widetilde{\mathcal{O}}(\sqrt{m})} N(I)^{\frac{1}{m}}$, *where* $m = \deg K = 2^n$. *This element approximate the shortest vector problem in* $\sigma(I)$ *with approximation factor* $e^{\widetilde{\mathcal{O}}(\sqrt{m})}$.

Proof. For correctness we need to show that the output value of g belongs to the ideal I. To ensure this a drift s is used. This is a standard technique for algorithms working with ideal lattices (see for example, proof of Theorem 3.3 in [5]). For completeness, we give the proof here. By definition of the decoding radius we have for any $\boldsymbol{\beta}$ that

$$\|\boldsymbol{\alpha} - (s, \ldots, s) - \boldsymbol{\beta}\|_\infty \leq s.$$

Data: An I integral ideal of $\mathcal{O}_K$, where $K = \mathbb{Q}(\sqrt{d_1}, \ldots, \sqrt{d_n})$.

- A set $S = \{\mathfrak{P}_i \mid i = 1, \ldots, d\}$ generating Cl_K.
- An Approx-CVP oracle for Log-S-unit lattice.
- A drift $s \in \mathbb{Z}$ such that it is bigger than decoding radius of the Approx-CVP algorithm.
- A short principal ideal problem oracle (ShortGenerator).

Result: A short element $g \in I$.

1 Compute the class group, the generators of the lattice $\Lambda_S = \mathrm{Log}_S \mathcal{O}_{K,S}^{\times}$, and the generators $\{t_i\}_i$ of $\mathcal{O}_{K,S}^{\times}$ corresponding to a basis matrix $W = B(\Lambda_S)$ (using Biasse-van Vrendendaal algorithm);

2 Find h and $\boldsymbol{\alpha} \in \mathbb{Z}^d$ s.t. $\langle h \rangle = I \cdot \prod_{i=1}^{d} \mathfrak{P}_i^{\alpha_i}$;

3 Compute $\boldsymbol{\beta} = \mathrm{CV}(\boldsymbol{\alpha} - (s, \ldots, s), \Lambda_S) = \boldsymbol{\delta} \cdot W$ using Approx-CVP algorithm;

4 $g = h / \left(\prod_i t_i^{\delta_i} \right)$;

5 **return** ShortGenerator(g);

Algorithm 1: mqASVP(I). Finding short elements in an ideal I.

This implies $|\alpha_i - s - \beta_i| \leq s$ and it follows that $0 \leq \alpha_i - \beta_i \leq 2s$. Now, we have $\langle g \rangle = I \cdot \prod_{i=1}^{d} \mathfrak{P}_i^{\alpha_i - \beta_i}$. Since the ideal I is integral and all $\alpha_i - \beta_i$ are positive, it holds that $g \in I$.

Let us find now an approximation factor for the output of the algorithm. By Assumption 1 there is a ShortGenerator algorithm that returns an element f such that

$$\|f\|_2 = e^{\tilde{\mathcal{O}}(\sqrt{m})} N(g)^{1/m} = e^{\tilde{\mathcal{O}}(\sqrt{m})} N(I \cdot \prod_{i=1}^{d} \mathfrak{P}_i^{\alpha_i - \beta_i})^{\frac{1}{m}} =$$

$$= e^{\tilde{\mathcal{O}}(\sqrt{m})} N(I)^{\frac{1}{m}} \cdot \prod_{i=1}^{d} N(\mathfrak{P}_i)^{(\alpha_i - \beta_i)/m} =$$

$$= e^{\tilde{\mathcal{O}}(\sqrt{m})} N(I)^{\frac{1}{m}} \cdot \ell^{\sum_i (\alpha_i - \beta_i)/m},$$

where ℓ is a bound for norms of prime ideals in S. By Bach's bound the class group is generated by ideals of norm:

$$N(\mathfrak{P}_i) \leq \ell = 12 \cdot \ln^2 \Delta_K = \mathcal{O}(m^2 \ln^2 D) = \mathcal{O}(m^2 (\log m)^{\mathcal{O}(1)}) = e^{\tilde{\mathcal{O}}(\log m)}.$$

Consider now the value of the sum $\sum_{i=1}^{d} (\alpha_i - \beta_i)/m$. For each a prime ideal $\mathfrak{P}_i$ there are up to m ideals in S of the same norm. These are G_K-conjugates of $\mathfrak{P}_i$. So, in total we have $d = m \log \ell = \mathcal{O}(m \log m \log \log m)$ prime ideals in S. By applying multiple times an algorithm in Lemma 1 we can obtain in polynomial time a vector $\boldsymbol{\beta}$ such that

$$\|\boldsymbol{\alpha} - \boldsymbol{s} - \boldsymbol{\beta}\|_{\infty} \leq \sqrt{2 \ln(8d)} \max_{\boldsymbol{w} \in W} \|\boldsymbol{w}\|_2.$$

By Assumption 2 we have $\max\limits_{w \in W} \|w\|_2 = \mathcal{O}(\sqrt{m \log D}) = \widetilde{\mathcal{O}}(\sqrt{m})$. So, s can be taken of size $\widetilde{\mathcal{O}}(\sqrt{m})$. Then $\frac{1}{m}\sum_i(\alpha_i - \beta_i) = \frac{1}{m}\|\alpha - \beta\|_1 \leq \frac{d}{m}\|\alpha - \beta\|_\infty = \widetilde{\mathcal{O}}(\sqrt{m})$.

Thus, we have $\|f\|_2 = e^{\widetilde{\mathcal{O}}(\sqrt{m})}N(I)^{\frac{1}{m}}$. The shortest vector in the lattice $\sigma(I)$ has the length quasipoly$(m) \cdot N(I)^{\frac{1}{m}}$ when D is quasi-polynomial (see Sect. 3.1). So, the approximation factor is $e^{\widetilde{\mathcal{O}}(\sqrt{m})}$.

It remains to prove the complexity claim. Biasse-van Vredendaal algorithm takes quasi-polynomial time by Proposition 5.1 from [8]. Step 2 is solving of discrete logarithm problem in the class group of the field. This task can also be solved in quasi-polynomial time using the algorithm from [17]. Step 3 is polynomial time by Lemma 1. Step 4 is polynomial time assuming that we use compact representation of the field elements. Step 5 is quasi-polynomial time if we use algorithm from [2]. □

8   Implementation and Experiments

We implemented[1] algorithm in SageMath 10.2 on top of the code from [2,8,16,17] for class group and discrete logarithm computation. The output elements are stored in compact representation, i.e. as the products of type $\prod_i g_i^{e_i}$ where g_i are small elements of the field K. This representation prevents coefficient explosion in the algorithm and allows us to compute approximation factors without explicit evaluation of the product.

8.1   Optimizing the Drift

Due to the presence of a drift s in Algorithm 1 the vector $\alpha - \beta$ is very dense in practice. Most of its elements are non zero. This leads to unnecessary big approximation factors in the output of the algorithm. This is a common problem for algorithms that use drifts [4,5,20] and the following methods to overcome the problem are presented in the literature.

1. Minimizing factor base: the extra factor occurs from the product of ideals in the factors base. So, we should select as small factor base S and norms of ideals in S as possible.
2. Dichotomic approach [5, §5.3]: a binary search for optimal value of a drift by selecting bigger s when output element of the algorithm does not belong to the ideal and smaller value when it does. However, as noted in [4] this approach does not scale to big dimensions.
3. Sampling drifts from ℓ_∞-balls of radius 1 (see [4, §5.4]). This method consist of two stages. First, we sample $\mathcal{O}(\deg K)$ points in $s + B_\infty(1)$ for a wide range of random s to find the value s_0 with the best output element g (i.e. with the minimal $\|g\|_2$) that belongs to target ideal. At the second stage we trying to refine result by taking $\mathcal{O}(\deg K)$ sample points in $[0.9\beta_0, 1.1\beta_0] \cdot \mathbf{1} + B_\infty(1)$, the neighborhood of s_0.

---

[1] The source code is available here: https://github.com/novoselov-sa/mqASVP.

In our computations we were able to obtain the best results by combining the first and the third methods with the following additional approach.

1. Let $\rho = 0$ be a drift vector which we will subtract from α instead of s.
2. Compute $\beta = CV(\alpha - \rho, \Lambda_S)$.
3. If all coefficients of $\alpha - \beta$ are positive then output ρ and stop.
4. Set $\rho_i = -(\alpha_i - \beta_i)$ for each $\alpha_i - \beta_i < 0$.
5. Output FAIL if the number of iterations exceed $\mathcal{O}(\deg K)$. Otherwise, go to step 2.

The method above never failed in our computations. We run our method followed by the third method and outputted the drift with the best value of $\|g\|_2$. In most cases our method returned the best result.

8.2   Evaluation of the Approximation Factor

In our experiments we selected 10 random ideals of prime norm $\leq 10^6$, computed elements of ideals using our implementation, and calculated the average value of approximation factor $\mathtt{af}_{\mathrm{gh}} = \|g\|_2/\lambda_1(I)$ with respect to the Gaussian Heuristic:

$$\lambda_1(I) \approx \sqrt{\frac{m}{2\pi e}} N(I)^{1/m} \sqrt{|\Delta_K|}^{1/m}.$$

Since the value of $\sqrt{|\Delta_K|}^{1/m}$ is quasi-polynomial (see §3.1) for our multi-quadratic fields, the Gaussian Heuristic differs from shortest vector only by a multiplier of quasi-polynomial size. We present in Table 3 the results of computations for the multiquadratic fields $K = \mathbb{Q}(\sqrt{d_1}, \ldots, \sqrt{d_n})$ where $|d_1|, \ldots, |d_n|$ are the first primes such that $d_i \equiv 1 \bmod 4$ for $i = 1, \ldots, n$. For small degrees ($m \leq 16$) we evaluated the elements in compact representation to obtain explicit elements of the field and checked that the elements belong to the target ideal.

Table 3. Approximation factors reached by Algorithm 1 (with respect to Gaussian Heuristic).

$[K : \mathbb{Q}]$	Field	$\ln(\mathtt{af}_{\mathrm{gh}})$	$\sqrt{2m \log m \log D}$	Cl_K
8	real	1.45	15.26	trivial
16	real	4.27	30.33	C_4^2
32	real	18.07	55.69	$C_2 \times C_4 \times C_8^4$
64	real	64.55	97.06	$C_2^9 \times C_4^3 \times C_8 \times C_{16}^4 \times C_{48} \times C_{240}$
8	imag.	1.00	13.45	C_3
16	imag.	3.09	27.27	$C_8 \times C_{48}$
32	imag.	11.30	50.55	$C_2 \times C_4^3 \times C_{24} \times C_{48}^2 \times C_{3360}$
64	imag.	49.59	89.22	$C_2^2 \times C_4^9 \times C_8^3 \times C_{16} \times C_{48} \times C_{96}^2 \times C_{192}^2 \times C_{6720}^2 \times C_{927360}$

To achieve better approximation factors, we also performed optimizations for the drift parameter from the previous section. Computations were done on one core of Intel Xeon Silver 4201R clocked at 2.40GHz on the machine with 629 GB RAM and they took less than a week.

Under Assumptions 1, 2 we expect the value of $\ln(\mathsf{af_{gh}})$ to be bounded by $\mathcal{O}(\sqrt{2\ln(8d)m\log D})$, where $d = \mathcal{O}(m\log m\log\log m)$ is the size of the factor base. So, under the assumptions we have $\ln(\mathsf{af_{gh}}) = \mathcal{O}(\sqrt{(2+o(1))m\log m\log D})$ and we added the value of $\sqrt{2m\log m\log D}$ to the table as the expected bound for the logarithm of the approximation factor.

As we can see for both imaginary and real fields the resulting approximation factors perfectly fit into the expected bound.

Acknowledgement. The research was funded by the Russian Science Foundation (project No. 22-41-04411, https://rscf.ru/en/project/22-41-04411/).

Disclosure of Interests. The author has no competing interests to declare that are relevant to the content of this article.

References

1. Babai, L.: On Lovász' lattice reduction and the nearest lattice point problem. Combinatorica **6**, 1–13 (1986)
2. Bauch, J., Bernstein, D.J., de Valence, H., Lange, T., van Vredendaal, C.: Short generators without quantum computers: the case of multiquadratics. In: Coron, J.-S., Nielsen, J.B. (eds.) EUROCRYPT 2017, pp. 27–59. Springer, Cham (2017). https://doi.org/10.1007/978-3-319-56620-7_2
3. Bernard, O., Fouque, P.A., Lesavourey, A.: Computing e-th roots in number fields. In: 2024 Proceedings of the Symposium on Algorithm Engineering and Experiments (ALENEX), pp. 207–219. SIAM (2024)
4. Bernard, O., Lesavourey, A., Nguyen, T.H., Roux-Langlois, A.: Log-S-unit lattices using explicit Stickelberger generators to solve approx ideal-SVP. Cryptology ePrint Archive, Report 2021/1384 (2021). https://ia.cr/2021/1384
5. Bernard, O., Roux-Langlois, A.: Twisted-PHS: using the product formula to solve approx-SVP in ideal lattices. In: International Conference on the Theory and Application of Cryptology and Information Security, pp. 349–380. Springer (2020)
6. Biasse, J.F., Erukulangara, M.R., Fieker, C., Hofmann, T., Youmans, W.: Mildly short vectors in ideals of cyclotomic fields without quantum computers. Math. Cryptol. **2**(1), 84–107 (2022)
7. Biasse, J.F., Fieker, C., Hofmann, T., Page, A.: Norm relations and computational problems in number fields. J. Lond. Math. Soc. **105**(4), 2373–2414 (2022)
8. Biasse, J.F., Van Vredendaal, C.: Fast multiquadratic S-unit computation and application to the calculation of class groups. Open Book Ser. **2**(1), 103–118 (2019)
9. Bos, J., et al.: CRYSTALS-Kyber: a CCA-secure module-lattice-based KEM. In: 2018 IEEE European Symposium on Security and Privacy (EuroS&P), pp. 353–367. IEEE (2018)

10. Cramer, R., Ducas, L., Wesolowski, B.: Short Stickelberger class relations and application to Ideal-SVP. In: Annual International Conference on the Theory and Applications of Cryptographic Techniques, pp. 324–348. Springer (2017)
11. Cramer, R., Ducas, L., Wesolowski, B.: Mildly short vectors in cyclotomic ideal lattices in quantum polynomial time. J. ACM **68**(2) (2021)
12. Ducas, L., et al.: CRYSTALS-Dilithium: a lattice-based digital signature scheme. IACR Trans. Cryptographic Hardw. Embed. Syst. **2018**(1), 238–268 (2018)
13. Fieker, C., Hart, W., Hofmann, T., Johansson, F.: Nemo/Hecke: computer algebra and number theory packages for the Julia programming language. In: Proceedings of the 2017 ACM on International Symposium on Symbolic and Algebraic Computation, pp. 157–164 (2017)
14. Guruswami, V., Micciancio, D., Regev, O.: The complexity of the covering radius problem. Comput. Complexity **14**, 90–121 (2005)
15. Hua, L.K.: Introduction to Number Theory. Springer Science & Business Media (1982)
16. Novoselov, S.A.: On ideal class group computation of imaginary multiquadratic fields. Prikl. Diskr. Mat **58**, 22–30 (2022)
17. Novoselov, S. A.: On the discrete logarithm problem in the ideal class group of multiquadratic fields. In: International Conference on Cryptology and Information Security in Latin America, pp. 192–211. Springer (2023)
18. Page, A.: abelianbnf (2020). https://hal.inria.fr/hal-02961482
19. Peikert, C., Rosen, A.: Lattices that admit logarithmic worst-case to average-case connection factors. In: Proceedings of the Thirty-ninth Annual ACM Symposium on Theory of Computing, pp. 478–487 (2007)
20. Pellet-Mary, A., Hanrot, G., Stehlé, D.: Approx-SVP in ideal lattices with preprocessing. In: Advances in Cryptology–EUROCRYPT 2019: 38th Annual International Conference on the Theory and Applications of Cryptographic Techniques, Darmstadt, Germany, May 19–23, 2019, Proceedings, Part II 38, pp. 685–716. Springer (2019)
21. Regev, O.: Lattices in computer science: CVP algorithm (2004). https://cims.nyu.edu/regev/teaching/lattices_fall_2004/ln/cvp.pdf
22. Schirokauer, O.: Discrete logarithms and local units. Philos. Trans. Roy. Soc. Lond. Ser. A Phys. Eng. Sci. **345**(1676), 409–423 (1993)
23. The Sage Developers: SageMath, the Sage Mathematics Software System (Version 10.2) (2023). https://www.sagemath.org

Symmetric-Key Cryptography

ChakraVyuha - A Fast Self Synchronizing Stream Cipher

Rajeeva Laxman Karandikar$^{(\boxtimes)}$

Chennai Mathematical Institute, Chennai 603103, India
rlk@cmi.ac.in

Abstract. We propose a Self Synchronizing Stream Cipher ChakraVyuha: this is like a block cipher in CFB mode, where each block is encrypted using a new key, in turn generated by a fast stream cipher Chakra. The later uses a suitable non-linear transformation on the bitstream generated by a Linear Feedback Shift Register (LFSR) acting on 64-bit integers with degree 32 and a high order (over 2^{238}). The focus has been on efficient computation. The efficiency of the LFSR acting on 64-bit integers is achieved by putting together 64 LFSR's (usual, acting on single bits) of orders between 17 and 32 (4 LFSR's for each order).

The stream cipher ChakraVyuha is able to encrypt (or decrypt) over 3 giga bits of plaintext per second on Mac with M1 chip.

Keywords: Symmetric Key Cryptography · Stream Cipher · Linear Feedback Shift Register · Self Synchronizing Stream Cipher

1 Introduction

Linear Feedback Shift Registers (LFSR) have been in use as a pseudo random number generator in various applications. In cryptography it is used as a building block, for a stream cipher. A simple example of a stream cipher that uses 3 LFSRs and combines the output of the LFSRs via a non-linear function is the Geffe generator (see [9]). It is common to have several LFSRs and combine the output via a non-linear function. For computational efficiency, it is common to choose LFSR's with small number of taps. However, when using a 64 bit-processor, we could simultaneously use 64 LFSR's acting independently of each other and then suitably transform the output of the LFSR's- instead of using 3 LFSR's as in Geffe generator, we use 64 LFSR's and instead of output being a single bit, here the output is also 64-bits.

Thus we propose a Synchronous Stream Cipher algorithm, named Chakra. This combines 64 LFSR's, with orders varying between 17 and 32 and a suitable non-linear combining function.

It is common to use primitive polynomial with large t but a small number of non-zero terms in the connection polynomials (same as number of taps in the LFSR) for fast computation while generating output of a LFSR. However, we use "balanced" polynomials, where the number of non-zero terms in the connection

S. Mukhopadhyay and P. Stănică (Eds.): INDOCRYPT 2024, LNCS 15495, pp. 97–116, 2025.
https://doi.org/10.1007/978-3-031-80308-6_5

polynomials is roughly half of its order and where the maximum number of consecutive 1's is at most 3.

The computational efficiency is achieved by ensuring that the 64 LFSR's are put together in such a way that as a whole it yields a LFSR acting on 64 bit integers and in 32 steps, we get one new 64-bit random number. This yields a 64-bit PRNG. To generate the stream cipher, we proceed as follows: At each step, we generate a 64-bit random number and along with carefully chosen 8 out of the last 255 random numbers generated, we generate an array (of size 9) of 64 bit integers, and transform this array (of size 9) of 64 bit integers using a non-linear non-invertible function, to generate an array (of size 8) of 64-bit random numbers. The array of size 8 of 64-bit integers generated above is the output of Chakra. The algorithm is optimised for a 64-bit processor and is very fast, more than 7 giga bits per second on Macbook with M1 chip. The same idea can be adopted for 32-bit processors.

Self Synchronizing Stream Ciphers have been in use for long (see [1,3–5,8,10]) and here we introduce a variant of this Chakra algorithm by combining it with a suitable Mixing algorithm (a combination of "confusion" and "diffusion" steps in a block cipher). This yields a good Self Synchronizing Stream Cipher, we have called it ChakraVyuha. This is like a Block Cipher used in CFB mode, and it is also fast, can encrypt (or decrypt) over 3 giga bits data per second.

We also extend the randomness criterion for Self Synchronizing Stream Ciphers proposed in [1] and the proposed algorithm satisfies the extended criterion.

2   The Design of a New Stream Cipher

Recall that if $p(x)$ is a primitive polynomial of degree t over $GF(2)$, given by

$$p(x) = 1 + c_1 x + \ldots + c_{t-1} x^{t-1} + c_t x^t$$

where $c_k = 0$ or 1 for $1 \le k < t$ and $c_t = 1$. Then given $s_0, s_1, \ldots, s_{t-1}$, with each s_j being 0 or 1, let us define $\{s_j : j \ge t\}$ by

$$s_j = (c_1 s_{j-1} + c_2 s_{j-2} + \ldots + c_t s_{j-t}) \bmod 2,$$

or equivalently, writing $\oplus$ for addition mod 2 (same as bitwise XOR),

$$s_j = c_1 s_{j-1} \oplus c_2 s_{j-2} \oplus \ldots \oplus c_t s_{j-t}.$$

When $t \ge 16$ say, then the period of the output sequence $\{s_j : j \ge t\}$ is large. Using the assumption that $p(x)$ is a primitive polynomial, it can be shown that for large t, the output has period $(2^t - 1)$ and can be considered to be a random bit stream (of length less than 2^t) and used as a pseudo random number generator (PRNG).

This algorithm that takes t bits as input and generates output of arbitrary length, is easy to implement on a computing device and thus is popularly used as a PRNG and is called Linear Feedback Shift Register, LFSR. Here $s_0, s_1, \ldots, s_{t-1}$ is called the *initial state*, t is the length of the LFSR, $p(x)$ is called the *connection polynomial* and $\{s_j : j \ge t\}$ is the *output sequence* of the LFSR.

2.1   LFSR on 64 Bit Integers

For $0 \leq i < m$, let a polynomial p_i be given by

$$p_i(x) = 1 \oplus c_{i,1}x \oplus c_{i,2}x^2 \oplus \ldots \oplus c_{i,t_i}x^{t_i}, \tag{1}$$

where $c_{i,j} \in \{0,1\}$, for $0 < j < t_i$ and $c_{i,t_i} = 1$. Suppose p_i is a primitive polynomial over GF(2) with degree t_i for each i, $0 \leq i < m$. We assume that the polynomials are distinct- $i.e.$ for $i \neq k$, p_i and p_k and are distinct. Let $\{s_{i,j} : 0 \leq j < t_i \ 0 \leq i < m\}$ be fixed. Then $\{s_{i,j} : j \geq t_i \ 0 \leq i < m\}$ defined by

$$s_{i,j} = c_{i,1}s_{i,j-1} \oplus c_{i,2}s_{i,j-2} \oplus \ldots \oplus c_{i,t_i}s_{i,j-t_i} \tag{2}$$

are outcomes of the m LFSR's. Let us write $\mathcal{U} = \{0,1\}^m$ - the set of all m-bit integers. For $j \geq 0$, let

$$\mathbf{s}[\mathbf{j}] = (s_{0,j}, s_{1,j}, \ldots, s_{m-1,j}) \in \mathcal{U}. \tag{3}$$

Let $T = \max\{t_i : 0 \leq i < m\}$. Given $\{\mathbf{s}[\mathbf{j}] : 0 \leq j < T\}$, we can use (2) to recursively generate $\{\mathbf{s}[\mathbf{j}] : j \geq T\}$. It is another matter that $\{\mathbf{s}[\mathbf{j}] : j \geq T\} \subseteq \mathcal{U}$ do not depend upon $\{s_{i,j-k} : 0 \leq i < m, \ t_i \leq k < T\}$. Note that the size of $\{\mathbf{s}[\mathbf{j}] : 0 \leq j < T\}$ is $m \times T$.

Let $\theta =$ g.c.d.$(s_0, \ldots, s_{m-1})$, where $s_i = 2^{t_i} - 1$. It can be seen that the period of the sequence $\{\mathbf{s}[\mathbf{j}] : j \geq T\}$ is θ.

Let us define $c_{i,j} = 0$ if $t_i < j \leq T$. Then (2) can be written as,

$$s_{i,j} = c_{i,1}s_{i,j-1} \oplus c_{i,2}s_{i,j-2} \oplus \ldots \oplus c_{i,T}s_{i,j-T} \ , \ 0 \leq i < m. \tag{4}$$

For $0 \leq j < T$, let

$$\mathbf{c}[\mathbf{j}] = (c_{1,j}, c_{2,j}, \ldots, c_{m,j}) \in \mathcal{U}. \tag{5}$$

Using (3) and (5), it can be checked that (4) is same as

$$\mathbf{s}[\mathbf{j}] = (\mathbf{c}[\mathbf{1}]\&\mathbf{s}[\mathbf{j-1}]) \oplus (\mathbf{c}[\mathbf{2}]\&\mathbf{s}[\mathbf{j-2}]) \oplus \ldots \oplus (\mathbf{c}[\mathbf{T}]\&\mathbf{s}[\mathbf{j-T}]) \tag{6}$$

where $\&$ is bitwise multiplication and $\oplus$ is bitwise addition mod 2, or XOR. Indeed, it can be seen that the ith component of $\{\mathbf{s}[\mathbf{j}] : j \geq T\}$ is simply the sequence generated by (2).

In the proposed algorithm designed for use with a hardware having $m = 64$-bit processor, we have chosen 64 LFSR's. We have chosen $T = 32$ and then for each integer i in the interval $[17, 32]$, 4 (distinct) LFSRs with length i have been chosen. While in most such constructions, primitive polynomials with a small number of non-zero terms are chosen for computational efficiency, we are looking for LFSRs such that little over half the coefficients are 1, rest being 0, and the maximum length of consecutive 1's is at most 4. The equation (6) yields a 64-bit integer $\mathbf{s}[\mathbf{j}]$ in $T = 32$-steps!

A list of all primitive polynomials of orders upto 32 is available on the web, see [13]. A code for generating such a list is also available at [14]. We have randomly chosen 4 LFSR's from the list of length i for each $i \in [17, 32]$ satisfying the conditions mentioned above. The reason for choosing the LFSR's randomly from the list as opposed to choosing them with some specified criterion is to ensure that the LFSR's do not have any algebraic relation among them. If one chooses LFSR's so that they are not related to each other based on some known criterion, it leaves the possibility that some algebraic relationship may be found among the chosen ones. The other reason is to ensure that for actual useage, an entity can generate the LFSR's and do the relevant testing to ensure that the chosen LFSR's are different but as good as the one presented in this paper.

The coefficients c[j] : $0 \leq j < m = 64$ based on the chosen LFSRs are then constructed using (6). The list of $\{c[j] : 0 \leq j < 64\}$ used in our algorithm is given in the source code ChakraVyuha_lib.cpp in the Appendix. The source code includes various other choices of permutations used and all the functions described in the later sections.

We will follow the following convention: a[j], b[j] for some sequence of integers j will represent an array of bytes for all lower case characters a,b,... while A[j], B[j] for some sequence of integers j will represent an array of 64-bit integers for all upper case characters A,B,.... This convention will be used for the first character of the name: Thus Cx[j] will represent an array of 64-bit integers while p256[j] will represent an array of bytes.

For the chosen primitive polynomials, $T = 32$ and it can be verified that the period θ of the proposed 64-bit generator is over 2^{238}. Having chosen $\{c[j] : 0 \leq j < 64\}$, given $\{s[j] : 0 \leq j < 32\}$, we can generate recursively

$$\{s[j] : 32 \leq j < 32 + M\}$$

for any M and these can be treated as output of a random number generator.

2.2 Key Expansion

As a first step for construction of the stream cipher, one needs a good algorithm for expanding a "key" to the required key, $\{s[j] : 0 \leq j < T = 32\}$ called here "expanded key" of size $64 \times 32 = 2048$ bits. In any case, it is a good idea to suitably transform "key" to "expanded key" which along with expanding, mixes the bits of the key. For even if the size of the given key is 2048, we know that 480 bits out of 2048 bits of the expanded key do not contribute to the output!

Let us choose a permutation p256 acting on $\{0,1,2,...,255\}$ and a permutation p32 acting on $\{0,1,2,...,31\}$. These choices are given in the source code ChakraVyuha_lib.cpp in the Appendix.. The choice of p256 has been carefully made to ensure that the Nonlinearity, Differential Uniformity and The Diffusion Order are same as that in the AES S-box : 112, 4 and 2 respectively. The choice of p32 has been made so as to yield good mixing properties. And of course, the extensive randomness testing of resulting stream cipher through Maurer's test.

We define a function Mix32 from $\mathcal{U}^{32}$ onto itself, which transforms an array of 64-bit integers of size 32 as follows: For $x \in \mathcal{U}^{32}$,

Mix32(X)=Y $\in \mathcal{U}^{32}$

where
```
Y[p32[0]]=((X[0])<<17)|((X[0])>>47)
```
and for i=0,1,2,...,30,
```
Y[p32[i+1]]=(((Y[p32[i]])<<17)|((Y[p32[i]])>>47))+X[i+1] .
```
Here, we are using c++ notations. Thus `((X[0])<<17)|((X[0])>>47)` is left circular shift by 17, operating on the bits of `X[0]` and so on.

Let the key be represented as an array of bytes: `KY[j]`, j=0,1,...,(N-1) where N is an integer between 8 and 256 (thus the key size, in bits, is between 64 and 2048). Let `iv[j]`, j=0,1,...,255 be an arbitrarily chosen and fixed array of bytes, called "initial vector", used in key expansion.

Let `S[i]` for i=0,1,2,...,31 (an array of 64-bit integers) be initialised to 0. First step is to fill these 32 elements using the key `KY`:
```
S=ExpandKey(KY)
```
is defined by
```
a[0]=p256[KY[0]∧iv[0]]
for i in 1,2,...,255
{
     ij=i%N
     a[i]=p256[a[i-1]∧KY[ij]∧iv[i]]
}
b[0]=p256[a[255]]∧a[0]
for i in 1,2,...,255
     b[i]=p256[b[i-1]]∧a[i]
memcpy(U,b,256)
V=Mix32(U)
U=Mix32(V)
V=Mix32(U)
U=Mix32(V)
V=Mix32(U)
S=Mix32(V)
```
Here i%N is the remainder when i is divided by N.

The important point to be noted is that in all functions used in key expansion, only bitwise XOR, bitwise AND, rotating the bits and composition with a fixed permutation have been used. Multiplication has been comp. The same applies to the other functions that follow. This is what gives the algorithm the speed.

2.3   The Underlying PRNG

Having generated the expanded "key" : $\{S[j] : 0 \leq j < 32\}$, for any $M < \infty$, we can generate the PRNG $\{S[i] : 32 \leq i < 32 + M\}$ by recursion as follows: Recall that c[j], j=1,2,...,32 are the coefficients of 64-LFSR's. For any i, $32 \leq i < 32 + M$, let `S[i]` be defined by:
```
A=0
for j=0,1,... ,31
{
     ij=i-j-1
```

```
        A=A∧(S[ij]&C[j])
    }
    S[i]=A
```

Here, & is the bitwise AND (same as product of the bits) and ∧ is bitwise XOR (same as bitwise addition mod 2).

Note that for any $i \geq 32$, S[i] is defined in terms of S[k], $i - 32 \leq k < i$ (and the array of Coefficients).

By construction, {S[i] : $32 \leq i < 32 + M$} is generated via a LFSR action on 64-bit integers with bitwise XOR and bitwise AND. It is suitable as a Pseudo Random Number Generator. Let us note that, if we can observe for some t, {S[i] : $t \leq i < t + 32$} then we can get {S[i] : $t + 32 \leq i < 32 + M$} *without knowing the "key"* ! This makes it unsuitable to be used as a stream cipher. One approach to deal with this to use a non-linear transformation, preferably non-invertible. This is what is achieved in the next section.

2.4   Stream Cipher Chakra

The output of the stream cipher Chakra would be generated sequentially, using the PRNG {S[i] : $32 \leq i < 32 + M$} described above as follows: at each time t, the 64-bit array Z[t] is the output of the function Get_Next(t,KY) described below (here KY is the key).

Let us choose a permutation p9 acting on {0,1,2,...,8}. The choice made is given in the Appendix.

Let Mix9 : $\mathcal{U}^9 \mapsto \mathcal{U}^9$ be defined by : (for x∈ $\mathcal{U}^9$)

```
        Mix9(X)=Y ∈ 𝒰⁹
```

where

```
        Y[p9[0]]=((X[0])<<43)|((X[0])>>21)
```

and for i=0,1,2,...,7,

```
        Y[p9[i+1]]=(((Y[p9[i]])<<43)|((Y[p9[i]])>>21))+X[i+1] .
```

Let Next9 from $\mathcal{U}^9$ to $\mathcal{U}^9 \times \mathcal{U}^9$ be defined by for A ∈ $\mathcal{U}^9$,

```
        Next9(A)=(W,V) ∈ 𝒰⁹ × 𝒰⁹
```

where

```
        B=Mix9(A)
        A=Mix9(B)
        B=Mix9(A)
        A=Mix9(B)
        B=Mix9(A)
        V=Mix9(B)
        W=Mix9(V)
```

Let Aux ∈ $\mathcal{U}^9$ be an arbitrarily chosen array of 64-bit integers of length 9 (used in the first step and updated sequentially). For each $t \geq 0$, we first get X[t] by choosing 9 out of the last 256 elements in the PRNG {S[i] : $t - 255 \leq i \leq t$} along with Aux, and then using Next9(X[t]), we generate the output Y[t] as well as update Aux:

Let id[9]={153,58,255,89,0,217,122,189,23}.

For $t \geq 0$, let

X[t], Y[t]$\in \mathcal{U}^9$

be sequentially defined by, for j=0,1,2,...,8

X[t][j]=S[t+256-id[j]]$\wedge$Aux[j]

and

(Y[t],Aux)=Next9(X[t]).

Note that in the last step, the auxiliary array Aux is updated.

For each $0 \leq t < M$, Z[t]=Get_Next(t,KY) is defined as follows: Using KY, first one initialises $\{$S[i] $: 0 \leq i < 32)\}$ and then generates $\{$S[i] $: 32 \leq i < (32+M)\}$ as described above. Then using Next9, one generates Y[t] and updates Aux at each step t. Note at step t, one needs $\{$S[k] $: t < k \leq t+256\}$ and Aux, and we generate Y[t] and update Aux.

Finally, the output

Z[t]=Get_Next(t,KY)$\in \mathcal{U}^8$

is defined by

Z[t][j]=Y[t][j] for j=0,1,2,...,7.

Note that Z[t] depends only on $\{$S[i] $: t < i \leq t + 256\}$ and Aux.

Now it can be seen that this function is non-invertible, since at the last step we are dropping the last element in the array $\{$Y[t][j] $: 0 \leq j \leq 8\}$.

The output of the Stream cipher Chakra is given by $\{$Z[t] $: 0 \leq t < M\}$.

3   Self Synchronizing Stream Cipher

The Self Synchronizing Stream Cipher, called ChakraVyuha will be defined via the stream cipher Chakra described above. The ChakraVyuha cipher is like a block cipher, taking 64 bytes (512 bits) as block size. A plaintext is represented as array P[t] of elements in $\mathcal{U}^8$, say of length M. If the given plaintext length is not a multiple of 64 bytes, the plaintext is padded as usual and thus the plaintext is expressed as an array of elements of $\mathcal{U}^8$.

Using the "key", one first initialises the stream cipher Chakra. For each t=1,2,...,M, one obtains S[t] and then get Z[t] using Get_Next and treating Z[t] as key, transform plaintext P[t] to generate ciphertext C[t]. Indeed, for each t, the algorithm will take last ciphertext Prev_CipherText = C[t-1] along with plaintext P[t] and key Z[t] to generate ciphertext C[t]. Of course we will need arbitrarily chosen but fixed Prev_CipherText to begin with and after each step t, we will update Prev_CipherText by setting it equal to C[t]. Thus this algorithm resembles block cipher operating in CFB mode.

The non-linear non-invertible transformation of the plaintext is achieved as follows. We introduce a function similar to the Mix9 and Mix32 described earlier, with one twist- it also has an additional parameter resebmling a "key". The function Chakra8 plays the role played by "confusion" and "diffusion" in block ciphers and same is true of the inverse Chakra8Inv. The algorithm also needs a permutation p8 on $\{0,1,2,...,7\}$ (The choice made of p8 is given in the Appendix) and an array Prev_CipherText[j] j=0,1,...,7 of length 8 of 64-bit integers. This is chosen arbitrarily to begin with and then is updated in each step.

The functions Chakra8 and Chakra8Inv map $\mathcal{U}^8 \times \mathcal{U}^8$ to $\mathcal{U}^8$:

```
Chakra8(X,H)=Y
```

where

```
Y[p8[0]]=X[0]∧H[0];
Y[p8[1]]=(((Y[p8[0]])<<17)|((Y[p8[0]])>>47))+X[1]+H[1];
Y[p8[2]]=(((Y[p8[1]])<<25)|((Y[p8[1]])>>39)) ∧X[2]∧H[2];
Y[p8[3]]=(((Y[p8[2]])<<7)|((Y[p8[2]])>>57))+X[3]+H[3];
Y[p8[4]]=(((Y[p8[3]])<<47)|((Y[p8[3]])>>17)) ∧X[4]∧H[4];
Y[p8[5]]=(((Y[p8[4]])<<29)|((Y[p8[4]])>>35))+X[5]+H[5];
Y[p8[6]]=(((Y[p8[5]])<<13)|((Y[p8[5]])>>51)) ∧X[6]∧H[6];
Y[p8[7]]=(((Y[p8[6]])<<33)|((Y[p8[6]])>>31))+X[7]+H[7];
```

and

```
Chakra8Inv(Y,H)=X
```

where

```
X[0]=Y[p8[0]]∧H[0];
X[1]=Y[p8[1]]-(((Y[p8[0]])<<17)|((Y[p8[0]])>>47))-H[1];
X[2]=Y[p8[2]]∧ (((Y[p8[1]])<<25)|((Y[p8[1]])>>39)) ∧H[2];
X[3]=Y[p8[3]]-(((Y[p8[2]])<<7)|((Y[p8[2]])>>57))-H[3];
X[4]=Y[p8[4]]∧ (((Y[p8[3]])<<47)|((Y[p8[3]])>>17)) ∧H[4];
X[5]=Y[p8[5]]-(((Y[p8[4]])<<29)|((Y[p8[4]])>>35))-H[5];
X[6]=Y[p8[6]]∧ (((Y[p8[5]])<<13)|((Y[p8[5]])>>51)) ∧H[6];
X[7]=Y[p8[7]]-(((Y[p8[6]])<<33)|((Y[p8[6]])>>31))-H[7];
```

The first step in the encryption (as well as decryption) is to initialise Chakra using the "key" and obtain $\{S[j] : 0 \leq j < 256\}$. At step $t \geq 1$, one obtains Z[t] and using this as key along with Prev_CipherText and the plaintext P[t], one generates ciphertext C[t] using Encrypt_Next, updates Prev_CipherText and if need be, move onto t+1.

The same steps are followed for decryption - using Decrypt_Next. The two functions are described below: P[t] is the plaintext and C[t] is the ciphertext. Each call to these functions also updates Prev_CipherText.

```
Encrypt_Next(P[t])=C[t]
```

where

```
Z[t]=Get_Next(t,KY)
V=Chakra8(P[t],Z[t])
for i=1,2,3,4,5
{
    U=Chakra8(V,Z[t])
    V=Chakra8(U,Z[t])
}
C[t]=Chakra8(V,Prev_CipherText)
Prev_CipherText=C[t].
```

Likewise we have

```
Decrypt_Next(C[t])=P[t]
```
where
```
Get_Next(Z[t])
V=Chakra8Inv(C[t],Prev_CipherText)
Prev_CipherText=C[t].
for i=1,2,...,5
{
    U=Chakra8Inv(V,Z[t])
    V=Chakra8Inv(U,Z[t])
}
P[t]=Chakra8Inv(V,Z[t]).
```

Thus, starting with a "key" of length N bytes, and plaintext P[t], $1 \leq t \leq M$, an integer $M \geq 1$, we can generate Z[t] for each t, and using Encrypt_Next, we can get the cipher text C[t].

4   Efficient Implementation of ChakraVyuha

The stream cipher algorithm ChakraVyuha is based on the coefficients of 64 LFSR's as explained above and expressed as C[j], $j = 1, 2, .., T = 32$.

The "key" is assumed to be of N bytes, $8 \leq N \leq 256$. The algorithm also depends upon the following choices (arbitrarily chosen):

an array iv[j] j=0,1,...,255 of bytes,
an array Aux[j] of length 9 of 64-bit integers,
an array Prev_CipherText[j] of length 8 of 64-bit integers,

The algorithm needs following permutations

a permutation p9 on $\{0, 1, 2, \ldots, 8\}$,
a permutation p32 on $\{0, 1, 2, \ldots, 31\}$,
a permutation p256 on $\{0, 1, 2, \ldots, 255\}$.

The permutations need to be carefully chosen. The choices made are given in the Appendix. Also we will need functions

Mix9 - a one one onto transformation on $\mathcal{U}^9$.
Mix32 - a one-one onto transformation on $\mathcal{U}^{32}$.

along with

Chakra8 : $\mathcal{U}^8 \times \mathcal{U}^8 \mapsto \mathcal{U}^8$ and Chakra8Inv: $\mathcal{U}^8 \times \mathcal{U}^8 \mapsto \mathcal{U}^8$

such that for all T $\in \mathcal{U}^8$, one has for all P $\in \mathcal{U}^8$

Chakra8(P,T)=C $\Rightarrow$ Chakra8Inv(C,T)=P.

The functions Mix9, Mix32, Chakra8, Chakra8Inv have been defined in earlier sections.

For efficient implementation, we introduce State of the cypher- it is an array of length 256 of 64-bit integers along with an integer indx ("index"). The choice of 256 is made since at each t we needed a subset of the the the previous 255 elements in the Last S[j].

To begin with we first generate $\{S[j] : 0 \leq j < 32\}$ based on the "key" and "initial vector" and then go no to define $\{S[j] : 32 \leq j < 256\}$ and then copy $\{S[j] : 0 \leq j < 256\}$ to $\{State[j] : 0 \leq j < 256\}$. This is the initialisation phase.

Once we have next "block" P[t] of plaintext at "time" t, and we have updated Aux, Prev_CipherText, we first "update" the "state" as follows:

```
indx= t (mod 256).
State[indx]=0
for j=0,1,... ,31
{
        ij=indx+(256-j-1) (mod 256)
        State[indx]=State[indx]∧(State[ij]&C[j])
}
```

and then go on to define X[t] by

```
for j=0,1,...,8
{
        ij=indx+(256-id[j]) (mod 256)
        X[t][j]=State[ij]∧Aux[j]
}
```

and then generate Y[t] and update Aux as :

```
(Y[t],Aux)=Next9(X[t]).
```

and then get Z[t] by dropping the last element in the array Y[t].

Finally, using Z[t] along with Prev_CipherText, we obtain ciphertext C[t] and update Prev_CipherText. This implementation yields encryption at the rate of about 3 giga bits per second.

5   Randomness Tests for Chakra and ChakraVyuha

One requirement of a stream cipher is that if an adversary gets hold of one chunk of the bit-stream generated by the stream cipher, she/he should not be able to recover the subsequent parts of the bit-stream. This algorithm ensures this by first generating a stream X[t], transforming it to Y[t], via a non-linear function and then ignoring last 8 bits of Y[t] to get Z[t].

Another requirement for a stream cipher that the output stream should be indistinguishable from the output of repeated coin tosses, *i.e.* satisfy all statistical tests for a PRNG, say as described in [12].

To facilitate discussion, let us fix notation for the rest of this section. K will denote a key, with N bytes, where N is an integer. $\{Z[t] : 1 \leq t \leq M\}$ will denote the stream cipher Chakra, expressed as an array of 512-bit integers of size M. For each t, $Z[t]$ is itself an array of 64-bit integers, of size 8, *i.e.* $Z[t] \in \mathcal{U}^8$. We will express the stream cipher generated by key K as

$$Z = \mathcal{S}(K).$$

Of course this depends upon the length M, but we will not express the same in notation.

Coming to ChakraVyuha,

$$\{P[t] : 1 \leq t \leq M\}$$

will denote a plaintext, also expressed as an array of 512-bit integers of size M, where for each t, $P[t] \in \mathcal{U}^8$.

The cipher text generated by the algorithm ChakraVyuha, corresponding to the plain text P and key K will be denoted by $\mathcal{E}(P, K) = C$ which itself will be an array of size M of elements in $\mathcal{U}^8$. Denoting the decryption algorithm by $\mathcal{D}$, we have $\mathcal{D}(C, K) = P$.

For an integer $\alpha < 512$, let $0 \le j < 8$, $0 < i < 64$ be such that $\alpha = 64j + i$, and let $\Phi_\alpha = (\phi_\alpha(k) : 0 \le j < 8) \in \mathcal{U}^8$, where $\phi_\alpha(k) = 0$ for $k \ne j$ and $\phi_\alpha(j) = 2^i$, so that if we interpret Φ_α as a 512-bit word, then the αth bit is 1 and all other bits are zero. Let $\mathcal{T}_\alpha : \mathcal{U}^8 \mapsto \mathcal{U}^8$ be defined by $\mathcal{T}_\alpha(A) = A \wedge \Phi_\alpha$ (recall $\wedge$ stands for bitwise XOR), so that A and $\mathcal{T}_\alpha(A)$ differ exactly at α^{th} bit. Let $\widetilde{P}[t] = \mathcal{T}_\alpha(P[t])$, $0 \le t < M$.

Likewise, let K' denote a key such that K and K' agree at all bits other than β^{th}, $\beta < 8N$. When we need to stress β, we will express it as $K' = \mathcal{T}_\beta(K)$.

Given key K, plain text P (of length M) and constants α, β, let us define

$$G_1 = \mathcal{S}(K)$$
$$G_2 = \mathcal{S}(K')$$
$$G_3 = \mathcal{S}(K) \wedge \mathcal{S}(K')$$
$$H_1 = \mathcal{E}(P, K)$$
$$H_2 = \mathcal{E}(P, K) \wedge \mathcal{E}(\widetilde{P}, K)$$
$$H_3 = \mathcal{E}(P, K) \wedge \mathcal{E}(P, K')$$
$$H_4 = \mathcal{D}(\mathcal{E}(P, K), K').$$
$$H_5 = \mathcal{D}(P, K)$$
$$H_6 = \mathcal{D}(P, K) \wedge \mathcal{D}(\widetilde{P}, K)$$
$$H_7 = \mathcal{D}(P, K) \wedge \mathcal{D}(P, K')$$
$$H_8 = \mathcal{E}(\mathcal{D}(P, K), K').$$

As mentioned above, it is a standard requirement for a stream cipher that the output stream should be indistinguishable from the output of repeated coin tosses. Thus, G_1, G_2 should satisfy all statistical tests as described in [12]. The same applies also to the Self Synchronizing Stream Cipher and hence H_1 should be indistinguishable from the output of repeated coin tosses.

In [1], the authors presented a general framework for the application of the ideas of differential cryptanalysis to stream ciphers. These roughly translate to the following: the bit streams G_3, H_2, H_3 should have roughly half the bits as 1. Or if we take any, say 8-bit stream, for a large M, this 8-bit stream should occur roughly $\frac{1}{256}$ proportion of times.

We make a stronger requirement that the bit streams G_3, H_2, H_3 should themselves be indistinguishable from the output of repeated coin tosses. We will add H_4 to this list, for if H_4 is not indistinguishable from the output of repeated coin tosses, if decryption of a cipher text by a key K^* does not show any pattern, it means K^* and other keys closer to K^* are ruled out, reducing the number of trials for a brute force attack.

The decryption algorithm should also be equally strong and this translates to the requirement that H_5, H_6, H_7, H_8 should also be indistinguishable from the output of repeated coin tosses.

Thus we have the requirement that G_1, G_2, G_3 as well as H_1, H_2, H_3, H_4, H_5, H_6, H_7, H_8 all should be indistinguishable from the output of repeated coin tosses.

How large should M be? Recall, M is the number of 512 bit "blocks" in the output bitstream for G_1, G_2, G_3 and for H_1, H_2, H_3, H_4, H_5, H_6, H_7, H_8 it is also the number of 512 bit "blocks" in the plaintext P as well as cipher text C. The choice of M depends upon the randomness test being used. Some of the tests listed under diehard tests may require over 1 gegabits. And unless implemented efficiently, may take long time to run for each choice of K, P, α and β.

And how does one choose K, P, α and β? We should test with as many choices of K, P, α, β as feasible. Let us note that if we choose say 10 keys K and 10 plain texts P and for each of these 100 choices, choose α and β randomly, then we will have 1100 tests to perform. So the tests need efficient implementation.

6   Maurer's Universal Test of Randomness

Maurer [8] proposed Test of Randomness which is very different from most of the tests which depend essentially upon frequency of L bit patterns, for a suitable L. Maurer's test is based on (products of) gaps between L bit patterns. This test had been added to the battery of NIST tests. Also see [2,9,12]. If one is looking for gaps between $L = 8$ bit patterns, then the test needs little over 2 million bits (2068480 to be precise) while for $L = 16$ bit patterns, the test needs little over 1 billion bits (1059061760). The test statistic follows Normal distribution. The mean and variance of the test statistic is available in [9].

The Maurer's test detects departure from successive toss of fair coin very effectively as the test statistic is based on gaps between successive occurrence of L-bit patterns rather than frequency of these occurrences.

We can efficiently run Maurer's randomness tests for $L = 16$. We randomly choose 10 keys K_i (each of 512 bits) and 10 plain texts P_j (each of 1059061760 bits) and for each of these 100 choices, choose α_{ij} and β_{ij} randomly. Let X_{ijt} denote the Maurer's test value for t^{th} test for plaintext P_j, key K_i, chosen values of α_{ij} and β_{ij}. Let $Z_t = \max_{1 \leq j \leq 100} |X_{ijt}|$. We propose Z_t as the test statistic for the tth test.

Table 1 below gives distribution of $Z = \max_{1 \leq j \leq 100} |X_j|$ when $\{X_j : 1 \leq j \leq 100\}$ are i.i.d. random variables with Standard Normal distribution.

Table 1. Distribution of the test statistic based on 100 runs of Maurer's test

-	a=2	a=2.5	a=3	a=3.5	a=3.88
$\mathbf{P}(\max_{1 \leq j \leq 100} |X_j| \leq a)$	0.009	0.287	0.763	0.955	0.990

We have run Maurer's test (with $L = 8$) thousands of time and we have seen that when the test fails, the test statistic typically exceeds 5 and can cross 100. Noting that $P(Z = \max_{1 \leq j \leq 100} |X_j| \geq 3.88) = 0.01$ (when $\{X_j : 1 \leq j \leq 100\}$ are i.i.d. N(0,1)), we choose $\{Z \geq 3.88\}$ as the critical region.

We run the Maurer test and compute Z for each of the 11 tests, as described above. Each row in the following table gives the 11 values. The test is repeated 20 times and observed values given in a row in Table 2. It should be noted that each of the 220 values in the table represents statistics computed out of 100 gigabits.

Table 2. Test statistic values for ChakraVyuha

G1	G2	G3	H1	H2	H3	H4	H5	H6	H7	H8
2.90	2.18	2.10	1.94	2.66	2.49	2.07	2.23	2.34	3.24	2.20
2.61	3.46	2.83	2.61	2.58	2.14	2.56	2.44	2.67	3.01	3.53
3.47	1.80	2.58	2.67	2.22	2.44	2.30	2.47	2.42	2.34	2.17
2.23	3.00	2.23	3.00	2.64	2.60	3.14	2.52	2.23	2.50	2.45
2.46	3.31	2.65	3.14	2.15	2.53	2.45	2.64	2.22	2.50	2.26
2.02	2.76	2.73	1.72	2.89	**3.98**	2.01	2.75	2.24	2.75	2.31
3.25	2.55	2.87	1.96	2.88	2.08	2.49	2.64	2.80	2.09	2.43
2.56	1.91	2.28	2.62	**4.07**	2.06	2.24	2.49	1.88	2.44	1.94
2.21	2.14	2.61	3.15	2.32	2.49	2.51	2.89	2.65	2.78	2.55
2.37	2.38	2.80	2.69	2.90	2.47	3.02	2.67	2.45	2.32	2.24
2.14	1.93	3.84	2.45	2.66	2.66	3.00	2.60	2.43	2.07	2.68
2.18	2.27	2.13	2.83	2.22	2.80	2.90	2.40	2.36	2.58	2.71
2.29	2.32	3.01	2.83	2.54	2.90	2.68	2.73	2.50	2.52	2.01
2.64	2.35	3.24	3.08	2.15	2.72	2.66	2.89	2.32	2.37	2.89
3.24	2.09	2.44	3.40	1.98	2.54	2.66	2.78	3.04	3.17	2.37
2.26	2.60	2.59	1.98	2.44	2.06	2.74	2.54	2.88	2.42	2.60
2.64	2.23	2.18	2.14	1.96	2.96	2.28	2.21	3.16	2.89	2.95
2.40	2.81	2.61	2.80	2.56	2.67	3.17	2.52	2.23	2.45	2.61
2.97	1.90	2.14	2.21	2.19	2.43	2.54	2.71	2.22	3.25	3.47
1.94	2.88	2.64	2.77	2.35	2.47	2.52	2.24	3.28	3.00	3.20

Let Y denote the maximum observed value of the 11 test statistics over 20 runs. Then under Null hypothesis that the algorithm is good, Y is the maximum of absolute value of 22000 observations from standard Normal distribution. As can be seen from table, the observed value of Y is 4.07. If we treat Y as the overall test statistic, the p-value of this observation is 0.644.

It is worth mentioning that the choices of various constants as well as the design of various functions have been done using Maurer's test for $L = 8$ which can be run very fast.

The code for running the randomness tests along with the full output is available on my personal page:
https://www.cmi.ac.in/~rlk/Indocrypt2024/

Appendix

ChakraVyuha_lib.cpp

```cpp
#include<iostream>
#include<fstream>
#include<cstdlib>
#include<cstring>
#include<cmath>
#include<new>
using namespace std;
extern const unsigned int NK=256;
extern const unsigned int NB=64;
const unsigned int NB8=8;
u_int64_t Prev_CipherText[8]={0};
u_int64_t State[256]={0};
u_int8_t ix={0};
const u_int8_t id[9]={153,58,234,83,0,209,119,189,27};
const u_int64_t Coef32[32]={0xffffffffffffffff, 0x5b7bfef801f1f531,
0x99256a7af9ec748, 0x14842119fa4b0823, 0xc77c36663deebbbf,
0xe2431c6013b4557c, 0x9fea9b37c6bf3cc0, 0x6db5ef813d510805,
0x933894cd7ac2c1bf, 0x7683093ac21eba1d, 0xf9fbde3cde0d7a62,
0x8c5789c3b8e18da4, 0xfb4871c86728528a, 0x73a6a94ea5687f8e,
0x8d9d789918d12e5d, 0xbcebd693e0037493, 0x6c6761f7349710d7,
0x72aeb94ecfeebd6e, 0xdc5ce55c715c43dc, 0x34380ab89738feb8,
0xe0707e70ec70ad70, 0x5ee096e058e0d2e0, 0x99c065c0bdc025c0,
0xab8063800380c380, 0x2700d700d700f700, 0xae00ae00ee006e00,
0x5c00dc005c009c00, 0x380038003800b800, 0x7000700070007000,
0xe000e000e000e000, 0xc000c000c000c000, 0x8000800080008000};
const u_int8_t  p256[256]={165,138,236,74,28,120,227,253,146,
48,213,75,36,156,144,29,240,214,184,127,7,246,108,250,133,168,
151,124,85,179,251,15,215,140,11,180,232,238,230,123,158,97,64,
125,45,95,37,9,61,162,92,231,241,69,89,33,80,234,154,157,0,26,
21,55,6,78,222,117,98,82,212,13,58,41,219,134,109,102,216,155,
242,14,81,192,225,20,220,209,113,31,221,99,205,223,181,84,38,
172,186,40,182,88,200,70,122,10,18,130,187,131,76,104,174,178,
142,150,51,35,4,103,203,128,137,152,83,121,90,67,47,153,175,
196,247,195,87,27,105,177,208,188,198,116,197,96,136,34,149,224,
12,24,254,167,5,68,32,30,19,166,42,132,79,228,159,207,185,145,52,
2,44,73,16,17,106,46,173,56,114,72,252,170,115,57,91,235,39,60,54,
233,171,194,176,1,62,126,147,229,244,112,110,43,53,22,226,101,141,
93,191,71,245,201,160,218,169,100,248,8,183,63,189,202,255,139,
163,210,3,193,237,118,217,86,239,25,204,111,135,206,94,249,129,
50,143,190,119,161,107,199,211,148,59,243,65,164,23,77,49,66};
```

```
void Mix9(const u_int64_t X[], u_int64_t Y[])
{
      Y[p9[0]]=((X[0])<<43)|((X[0])>>21);
      for(u_int8_t i=0;i<8;i++)
      {
            Y[p9[i+1]]=(((Y[p9[i]])<<43)|((Y[p9[i]])>>21))+X[i+1];
      }
}
void Mix32(const u_int64_t X[], u_int64_t Y[])
{
      Y[p32[0]]=((X[0])<<17)|((X[0])>>47);
      for(u_int8_t i=0;i<31;i++)
      {
            Y[p32[i+1]]=(((Y[p32[i]])<<17)|((Y[p32[i]])>>47))+X[i+1];
      }
}
u_int64_t Ky[8]={0};
const u_int8_t iv[256]={225,6,157,197,37,127,128,175,64,230,
50,45,162,188,30,36,20,108,78,1,66,114,122,168,49,172,221,
217,42,228,155,138,192,216,98,106,148,7,201,40,147,67,22,
150,92,240,166,14,95,12,223,23,164,91,68,133,59,81,21,44,
117,54,239,140,32,10,174,247,69,193,233,196,94,243,28,248,
71,101,209,9,220,198,145,33,89,208,176,169,151,96,29,134,
161,207,86,211,70,190,75,0,46,253,118,56,214,85,137,206,
241,31,125,41,142,252,238,84,102,109,48,242,194,195,245,
3,87,60,149,160,210,77,222,80,139,107,143,76,61,250,111,
163,170,200,90,79,132,191,119,126,110,219,181,255,93,121,
144,57,185,173,237,124,184,15,17,178,53,129,202,130,136,
58,236,212,231,234,156,229,141,39,35,186,51,63,52,43,62,
227,159,182,99,47,73,120,189,112,213,218,199,8,235,25,
154,187,183,88,105,146,254,244,205,123,5,11,179,226,167,
152,224,215,116,24,55,74,177,13,19,249,180,104,165,72,
153,27,135,16,4,2,113,246,131,34,232,18,38,83,82,100,203,
65,204,103,171,97,251,158,115,26};
u_int64_t Aux[9]={0x5ed0b5b19f1eb641, 0x6a03816a4d599fe6,
0x65790ad0e3806762, 0x564735d25b3d9c0a, 0xe9911ed5cac06417,
0xe88c8a8da66fe1a, 0x26f6e54ab78f3a1c, 0x3eb5f75c418bd022,
0xb26e1fa3eefeae7c};
void make_init_State(const u_int8_t *key)
{
      u_int64_t i,jj;
      u_int8_t ij;
      u_int64_t UU[32]={0};
      u_int64_t VV[32]={0};
      u_int8_t A[256]={0};
```

```
u_int8_t B[256]={0};
A[0]=p256[key[0]^iv[0]];
State[0]=0;
for(i=1;i<256;i++)
{
    State[i]=0;
    ij=i%NK;
    A[i]=p256[A[i-1]^key[ij]^iv[i]];
}
B[0]=p256[A[255]]^A[0];
for(i=1;i<256;i++)
{
    B[i]=p256[B[i-1]]^A[i];
}
memcpy(UU,B,256);
Mix32(UU,VV);
Mix32(VV,UU);
Mix32(UU,VV);
Mix32(VV,UU);
Mix32(UU,VV);
Mix32(VV,State);
indx=32;
for(jj=0;jj<224;jj++)
{
    State[indx]=0;
    for(u_int16_t j=0;j<32;j++)
    {
        ij=indx-j-1;
        State[indx]=State[indx]^(State[ij]&Coef32[j]);
    }
    indx++;
}
memcpy(Last,p256+3,72);
memcpy(Prev_CipherText,p256+9,64);
}
void Get_Next()
{
    u_int8_t ij;
    u_int64_t A[9];
    u_int64_t B[9];
    State[indx]=0;
    for(u_int8_t j=0;j<32;j++)
    {
        ij=indx-j-1;
        State[indx]=State[indx]^(State[ij]&Coef32[j]);
```

```
        }
        for(u_int8_t j=0;j<9;j++)
        {
              ij=indx-id[j];
              A[j]=(State[ij]^Last[j]);
        }
        indx++;
        Mix9(A,B);
        Mix9(B,A);
        Mix9(A,B);
        Mix9(B,Last);
        Mix9(Last,B);
        memcpy(Ky,B,64);
}
void Stream_Crypt(const char Out[],const u_int8_t key[],const u_int32_t &N)
{
        make_init_State(key);
        ofstream out(Out, ios_base::out |ios_base::binary|ios_base::trunc);
        for(u_int64_t j=0;j<N;j=j+NB)
        {
              Get_Next();
              out.write( reinterpret_cast<char *>(Ky),NB);
        }
        out.close();
}
const u_int8_t p8[8]={7,4,1,6,0,3,5,2};
void Chakra8(const u_int64_t X[], u_int64_t Y[],u_int64_t H[])
{
        Y[p8[0]]=X[0]^H[0];
        Y[p8[1]]=(((Y[p8[0]])<<17)|((Y[p8[0]])>>47))+X[1]+H[1];
        Y[p8[2]]=(((Y[p8[1]])<<25)|((Y[p8[1]])>>39))^X[2]^H[2];
        Y[p8[3]]=(((Y[p8[2]])<<7)|((Y[p8[2]])>>57))+X[3]+H[3];
        Y[p8[4]]=(((Y[p8[3]])<<47)|((Y[p8[3]])>>17))^X[4]^H[4];
        Y[p8[5]]=(((Y[p8[4]])<<29)|((Y[p8[4]])>>35))+X[5]+H[5];
        Y[p8[6]]=(((Y[p8[5]])<<13)|((Y[p8[5]])>>51))^X[6]^H[6];
        Y[p8[7]]=(((Y[p8[6]])<<33)|((Y[p8[6]])>>31))+X[7]+H[7];
}
void Chakra8Inv(const u_int64_t Y[], u_int64_t X[],u_int64_t H[])
{
        X[0]=Y[p8[0]]^H[0];
        X[1]=Y[p8[1]]-(((Y[p8[0]])<<17)|((Y[p8[0]])>>47))-H[1];
        X[2]=Y[p8[2]]^(((Y[p8[1]])<<25)|((Y[p8[1]])>>39))^H[2];
        X[3]=Y[p8[3]]-(((Y[p8[2]])<<7)|((Y[p8[2]])>>57))-H[3];
        X[4]=Y[p8[4]]^(((Y[p8[3]])<<47)|((Y[p8[3]])>>17))^H[4];
        X[5]=Y[p8[5]]-(((Y[p8[4]])<<29)|((Y[p8[4]])>>35))-H[5];
```

```
      X[6]=Y[p8[6]]^(((Y[p8[5]])<<13)|((Y[p8[5]])>>51))^H[6];
      X[7]=Y[p8[7]]-(((Y[p8[6]])<<33)|((Y[p8[6]])>>31))-H[7];
}
void Encr_Next(u_int64_t X[])
{
      u_int64_t U[8],V[8];
      Get_Next();
      Chakra8(X,V,Ky);
      Chakra8(V,U,Ky);
      Chakra8(U,V,Ky);
      Chakra8(V,U,Ky);
      Chakra8(U,V,Ky);
      Chakra8(V,U,Ky);
      Chakra8(U,V,Ky);
      Chakra8(V,U,Ky);
      Chakra8(U,V,Ky);
      Chakra8(V,U,Ky);
      Chakra8(U,V,Ky);
      Chakra8(V,X,Prev_CipherText);
      memcpy(Prev_CipherText,X,64);
}
void Decr_Next(u_int64_t X[])
{
      u_int64_t U[8],V[8];
      Get_Next();
      Chakra8Inv(X,V,Prev_CipherText);
      memcpy(Prev_CipherText,X,64);
      Chakra8Inv(V,U,Ky);
      Chakra8Inv(U,V,Ky);
      Chakra8Inv(V,U,Ky);
      Chakra8Inv(U,V,Ky);
      Chakra8Inv(V,U,Ky);
      Chakra8Inv(U,V,Ky);
      Chakra8Inv(V,U,Ky);
      Chakra8Inv(U,V,Ky);
      Chakra8Inv(V,U,Ky);
      Chakra8Inv(U,V,Ky);
      Chakra8Inv(V,X,Ky);
}
void Encrypt(const char In[], const char Out[],const unsigned char key[])
{
      u_int64_t Z[NB8];
      make_init_State(key);
      ifstream in(In);
      if(!in)
```

```
        {
            cerr<<"Could Not Open the file to encrypt - "<<In<<endl;
            exit(1);
        }
        ofstream out(Out, ios_base::out |ios_base::binary|ios_base::trunc);
        u_int64_t CC;
        while(!in.eof())
        {
            in.read( reinterpret_cast<char *>(Z), NB);
            CC=in.gcount();
            if (CC==NB)
            {
                Encr_Next(Z);
                out.write( reinterpret_cast<char *>(Z),NB);
            }
        }
        in.close();
        out.close();
}
void Decrypt(const char In[], const char Out[],const unsigned char key[])
{
        u_int64_t Z[NB8];
        make_init_State(key);
        u_int64_t CC;
        ifstream in(In, ios_base::in |ios_base::binary);
        if(!in)
        {
            cerr<<"Could Not Open the file to decrypt - "<<In<<endl;
            exit(1);
        }
        ofstream out(Out);
        while(!in.eof())
        {
            in.read(reinterpret_cast<char *>(Z),NB);
            CC=in.gcount();
            if (CC==NB)
            {
                Decr_Next(Z);
                out.write( reinterpret_cast<char *>(Z),NB);
            }
        }
        in.close();
        out.close();
}
```

References

1. Biham, E., Dunkelman, O.: Differential Cryptanalysis in Stream Ciphers Cryptology ePrint Archive, Paper 2007/218 https://eprint.iacr.org/2007/218
2. Coron, J.S., Naccache, D.: An accurate evaluation of Maurer's universal test International Workshop on Selected Areas in Cryptography. Springer, Berlin (1998)
3. Daemen, J., Govaerts, R..Vandewalle, J.: On the design of high speed self-synchronizing stream ciphers. Singapore ICCS-ISITA 1992 Conference Proceedings, pp 279-283 (1992)
4. Fontaine, C.: Self-Synchronizing Stream Cipher. In: van Tilborg, H.C.A., Jajodia, S. (eds.) Encyclopedia of Cryptography and Security. Springer, Boston, MA (2011). https://doi.org/10.1007/978-1-4419-5906-5_371
5. Francq, J., Besson, L., Huynh, P., Guillot, P., Millerioux, G., et al.: Non-triangular self-synchronizing stream ciphers. IEEE Trans. Comput. **71**(1), 134–145 (2022)
6. Lamba, C.S.: Design and analysis of stream cipher for network security. In: 2010 Second International Conference on Communication Software and Networks, Singapore, pp. 562-567 (2010)
7. Maurer, U.M.: New approaches to the design of self-synchronizing stream ciphers. In: Advances in Cryptology - Eurocrypt 1991. LNCS, vol. 547., pp 458-471. Springer (1991). https://doi.org/10.1007/3-540-46416-6_39
8. Maurer, U.M.: A universal statistical test for random bit generators. J. Cryptol. **5**, 89–105 (1992)
9. Menezes, A.J., van Oorschot, P., Vanstone, S.A.: Handbook of Applied Cryptography. CRC Press (1997)
10. Millerioux, G., Guillot, P.: Self-synchronizing stream ciphers and dynamical systems: state of the art and open issues. Int. J. Bifurcat. Chaos Appl. Sci. Eng. **20**(9), 2979–2991 (2010)
11. Sarkar, P.: Hiji-bij-bij: a new stream cipher with a self-synchronizing mode of operation. In: Johansson, T., Maitra, S. (eds.) INDOCRYPT 2003. LNCS, vol. 2904, pp. 36–51. Springer, Heidelberg (2003). https://doi.org/10.1007/978-3-540-24582-7_3
12. Bassham, L.E.,, et al.: A Statistical Test Suite for Random and Pseudorandom Number Generators for Cryptographic Applications. Revised: April 2010, NIST Special Publication 800-22, Revision 1a, National Institute of Standards and Technology, United States https://www.nist.gov/publications/statistical-test-suite-random-and-pseudorandom-number-generators-cryptographic?pub_id=906762https://www.nist.gov/publications/statistical-test-suite-random-and-pseudorandom-number-generators-cryptographic?pub-id=906762
13. http://www.ece.cmu.edu/koopman/lfsr/index.html
14. https://github.com/hayguen/mlpolygen

UFLM: A Unified Framework for Feistel Structure and Lai-Massey Structure

Zhengyi Dai[1], Chun Guo[2,3,4(✉)], and Chao Li[1(✉)]

[1] College of Computers, National University of Defense Technology,
Changsha 410073, Hunan, China
lichao_nudt@sina.com

[2] School of Cyber Science and Technology, Shandong University,
Qingdao, Shandong, China
chun.guo@sdu.edu.cn

[3] Key Laboratory of Cryptologic Technology and Information Security of Ministry of
Education, Shandong University, Qingdao 266237, Shandong, China

[4] Shandong Research Institute of Industrial Technology,
Jinan 250102, Shandong, China

Abstract. Feistel structure and Lai-Massey structure are 2-branch structures with the property that encryption is similar to decryption. Many cryptographers study these two structures individually regarding design, provable security, and cryptanalysis. This paper adopts a unified perspective to research these two structures. We propose a framework called **UFLM**. Feistel structure and Lai-Massey structure are instances of **UFLM**. The differences between these two structures stem from the varying orders of branch permutation and orthomorphic permutation as viewed from the perspective of **UFLM**. Specifically, the order of the branch permutation is 2, while the order of an orthomorphic permutation is at least 3. We further investigate the number of rounds of impossible differentials, zero correlation linear hulls, and integral distinguishers of **UFLM** instances with bijective f-functions based on different orders of the linear transformation adopted. Finally, we prove the CCA security of 4-round **UFLM** construction using secret random f-functions in two cases. Interestingly, the CCA security of 4-round Lai-Massey construction is superior to that of 4-round Feistel construction when utilizing the same f-function in each round. The results can be easily transferred to random permutation-based **UFLM** constructions. With these, the Lai-Massey structure does benefit from orthomorphic permutation in both cryptanalysis and provable security settings. We further provide a **UFLM** instance that is better than Feistel structure regarding several distinguishers, which may be beneficial for future block cipher designs.

Keywords: Feistel structure · Lai-Massey structure · UFLM · Permutation · Distinguishers · CCA security

© The Author(s), under exclusive license to Springer Nature Switzerland AG 2025
S. Mukhopadhyay and P. Stănică (Eds.): INDOCRYPT 2024, LNCS 15495, pp. 117–142, 2025.
https://doi.org/10.1007/978-3-031-80308-6_6

1   Introduction

The design of modern block ciphers is based on the principles of confusion and diffusion proposed by Shannon [25]. Most block ciphers such as DES [6], AES [7], IDEA [14], and CLEFIA [26] are designed by calling a keyed round transformation multiple times to gain enough security. The round transformation generally consists of an overall iterative structure and local f-functions.

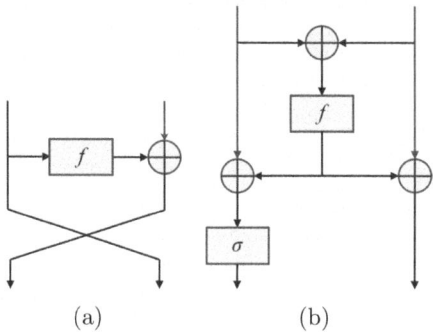

(a)   (b)

Fig. 1. Feistel structure and Lai-Massey structure.

There are several types of iterative structures used to design block ciphers, such as Feistel structure [9], SPN structure [7], Lai-Massey structure [12,29], and generalized Feistel structures [23,34]. The advantage of these structures, except for SPN structure, is that the decryption process does not require the calculation of the inverse of f-functions. Therefore, f-functions may not necessarily need to be invertible, which reduces costs in both software and hardware implementations and enhances the versatility, flexibility, and economics for the design of cipher structures. Furthermore, compared with the generalized Feistel structure, Feistel structure and Lai-Massey structure divide the state into two equal-sized branches. Feistel structure and Lai-Massey structure are shown in Fig. 1.

There are remarkable differences between the Feistel and Lai-Massey structures in design. On one hand, the input and output of the f-function are related to only one branch for the Feistel structure, while they are associated with both branches for the Lai-Massey structure. On the other hand, the Feistel structure employs a branch permutation, while the Lai-Massey structure utilizes an orthomorphic permutation acting on one branch. There are also similarities between the Feistel and Lai-Massey structures. Yun et al. propose the concept of quasi-Feistel construction and prove that these two structures are instances of quasi-Feistel construction [33].

There are similar results regarding provable security in both chosen-plaintext attack (CPA) and chosen-ciphertext attack (CCA) for the Feistel and Lai-Massey structures. Luby and Rackoff prove that 3-round and 4-round Feistel constructions are pseudorandom permutation (PRP) and strong PRP (SPRP), respectively [18]. Later, Vaudenay proves that 3-round and 4-round group-oriented

Lai-Massey constructions are PRP and SPRP in the Luby-Rackoff model, respectively [29]. Additionally, the quasi-Feistel construction sheds light on the consistency between the Feistel and Lai-Massey constructions regarding CPA and CCA security results [33]. However, the Lai-Massey construction enjoy stronger security against certain cryptanalytic methods [20,21]. Besides, Guo et al. prove the sequential indifferentiability for 6-round Lai-Massey construction using independent random f-functions [10].

There is an observation regarding impossible differential cryptanalysis [1] for the Feistel and Lai-Massey ciphers. Assuming that f-functions are bijective, 5-round impossible differentials exist for the Feistel structure [13]. Moreover, there will always be longer impossible differentials for block ciphers that adopt the Feistel structure when considering the details of f-functions. For instance, 8-round impossible differentials exist for the block cipher Camellia [30]. However, for the FOX block cipher family [12] using the Lai-Massey structure, 4-round impossible differentials exist [16,32]. Therefore, the Lai-Massey structure may enjoy better security against impossible differential cryptanalysis.

In the 2023 issue of TIT, Liu et al. proposed a unified structure [17] for characterizing Feistel-like structures that utilize a single f-function. Subsequently, in Crypto 2024 [28], Sun et al. presented conditions for the equivalence of Feistel-like structures and provided provable security evaluations of the unified structure against impossible differential [1] and zero correlation linear cryptanalysis [2]. Notably, the Lai-Massey structure, which employs an orthomorphic permutation, is not encompassed within this unified structure.

Inspired by the quasi-Feistel construction and unified structure, this paper adopts another unified perspective to research the Feistel and Lai-Massey structures from two aspects: distinguishers and provable security.

Our Results: Firstly, we propose a framework **UFLM** that builds on the properties of the Lai-Massey structure and orthomorphic permutation. Furthermore, we attribute differences in the number of rounds of distinguishers and provable security between the Feistel and Lai-Massey structures to the varying orders of permutations employed. Specifically, the order of the branch permutation is 2, while the order of an orthomorphic permutation is at least 3.

Secondly, we research the number of rounds of impossible differentials, zero correlation linear hulls, and integral distinguishers for **UFLM** instances with a mild constraint. **UFLM** instances employ linear transformations with different orders. Considering that f-functions are bijective, there exist 5-round (4-round) impossible differentials, 5-round (4-round) zero correlation linear hulls, and 5-round (4-round) integral distinguishers for **UFLM** instances where the order of the linear transformation is 2 (3), respectively. Furthermore, there exist 3-round impossible differentials, 3-round zero correlation linear hulls, and 3-round integral distinguishers for **UFLM** instances regardless of the order of the linear transformation. By this, the Lai-Massey structure may enjoy better security regarding the number of rounds of these distinguishers due to the use of orthomorphic permutation.

Finally, we investigate the CCA security of 4-round **UFLM** construction in two cases: (i) the same (secret) random f-function is adopted in each round; (ii) independent (secret) random f-functions are utilized in each round. Compared to case (ii), case (i) requires the linear transformation to satisfy an extra mild requirement, which is stronger than the swapping operation in Feistel construction. We prove that 4-round **UFLM** construction is CCA secure up to the birthday bound in these two cases. The number of rounds is tight by Nandi's lower bound on SPRP constructions. Note that Feistel construction instantiates an involution when using the same function in each round. By this, the Lai-Massey construction does benefit from stronger linear transformation in provable settings. These results can be easily extended to the cases where each f-function is a PRP. In addition, we provide a **UFLM** instance that is better than Feistel structure for several distinguishers, which may be utilized in future block cipher designs.

Organization: The remainder of this paper is organized as follows. Section 2 introduces some notations and concepts. Section 3 presents the equivalent structure of the Lai-Massey structure and the properties of orthomorphic permutation. Section 4 proposes the framework **UFLM** and researches the number of rounds of different distinguishers. Section 5 provides the CCA-security proofs for 4-round **UFLM** construction and presents a **UFLM** instance. Section 6 concludes the paper.

2   Preliminaries

Let E_r be an r-round branch-oriented block cipher with block size $m = b \cdot n$ bits, where b is the number of branches and n is the size of each branch. E_r employs a nonlinear cryptographic keyed function f_i in each round and the input and output size of f_i equals the branch size. Generally, each f-function is selected as an S-box. Due to the effect of the round key, functions $f_1, f_2, \cdots, f_r$ may be different. The Feistel and Lai-Massey ciphers correspond to the case of $b = 2$.

2.1   Impossible Differential and Zero Correlation Linear Hull

Impossible Differential. Given a vectorial Boolean function $F : \mathbb{F}_2^n \mapsto \mathbb{F}_2^n$. Let $\alpha \in \mathbb{F}_2^n$ and $\beta \in \mathbb{F}_2^n$ be the input difference and output difference of F, respectively. Then, the differential probability of $\alpha \to \beta$ is defined as:

$$\Pr(\alpha \to \beta) := \frac{\#\{x \in \mathbb{F}_2^n \mid F(x) \oplus F(x \oplus \alpha) = \beta\}}{2^n}.$$

If $\Pr(\alpha \to \beta) = 0$, then $\alpha \to \beta$ is an impossible differential of F. Otherwise, $\alpha \to \beta$ is a possible differential of F. Furthermore, let $S \subseteq \mathbb{F}_2^n$ and $T \subseteq \mathbb{F}_2^n$. If for all $\alpha \in S, \beta \in T, \Pr(\alpha \to \beta) = 0$, then $S \to T$ is a (truncated) impossible differential of F. In the case that F is a permutation on $\mathbb{F}_2^n$, $\alpha \to 0$ and $0 \to \beta$ are trivial impossible differentials of F where $\alpha \neq 0$ and $\beta \neq 0$.

Zero Correlation Linear Hull. Let $\lambda \in \mathbb{F}_2^n$ and $\gamma \in \mathbb{F}_2^n$ be the input mask and output mask of F, respectively. Then, the linear correlation of $\lambda \to \gamma$ is defined as:

$$\text{Cor}(\lambda \to \gamma) := \frac{1}{2^n} \sum_{x \in \mathbb{F}_2^n} (-1)^{\lambda \cdot x \oplus \gamma \cdot F(x)}.$$

If $\text{Cor}(\lambda \to \gamma) = 0$, then $\lambda \to \gamma$ is a zero correlation linear hull of F. Otherwise, $\lambda \to \gamma$ is a non-zero correlation linear hull of F. In addition, let $\Lambda \subseteq \mathbb{F}_2^n$ and $\Gamma \subseteq \mathbb{F}_2^n$. If for all $\lambda \in \Lambda, \gamma \in \Gamma, \text{Cor}(\lambda \to \gamma) = 0$, then $\Lambda \to \Gamma$ is a (truncated) zero correlation linear hull of F. In the case that F is a permutation on $\mathbb{F}_2^n$, $\lambda \to 0$ and $0 \to \gamma$ are trivial zero correlation linear hulls of F where $\lambda \neq 0$ and $\gamma \neq 0$.

2.2   Cipher Structure

An r-round cipher structure $\mathcal{E}^{(r)}$ is a set of all r-round block ciphers $E_r : \mathbb{F}_2^m \mapsto \mathbb{F}_2^m$. $\mathcal{E}^{(r)}$ is the same as E_r except that the S-boxes (non-linear components) can take all possible transformations on the corresponding domains [27]. We focus on the cipher structures with bijective S-boxes, specifically bijective f-functions, and view each bijective f-function as a bijective S-box.

Let $a, b \in \mathbb{F}_2^m$. If for any $E_r' \in \mathcal{E}^{(r)}, a \to b$ is an impossible differential (zero correlation linear hull) of E_r', then $a \to b$ is an impossible differential (zero correlation linear hull) of $\mathcal{E}^{(r)}$. In addition, if there exists $E_r' \in \mathcal{E}^{(r)}$, $a \to b$ is a possible differential (non-zero correlation linear hull) of E_r', then $a \to b$ is a possible differential (non-zero correlation linear hull) of $\mathcal{E}^{(r)}$. Note that $a \to 0$ and $0 \to b$ are structural impossible differentials (zero correlation linear hulls) where $a \neq 0$ and $b \neq 0$, which are key points for finding impossible differentials (zero correlation linear hulls) of a cipher structure based on the miss-in-the-middle technique [13,19,31].

2.3   CCA Security and H-Coefficient Method

Let $\mathcal{F}(\mathbb{F}_2^n, \mathbb{F}_2^n)$ be a set of all functions $f : \mathbb{F}_2^n \mapsto \mathbb{F}_2^n$ and $\mathcal{P}(\mathbb{F}_2^m)$ be a set of all permutations $\Pi : \mathbb{F}_2^m \mapsto \mathbb{F}_2^m$. Let $\mathcal{E}^{(r)}[f_1, \cdots, f_r] : \mathbb{F}_2^m \mapsto \mathbb{F}_2^m$ be an r-round idealized block cipher construction defined upon *secret* functions $f_1, \cdots, f_r \xleftarrow{\$} \mathcal{F}(\mathbb{F}_2^n, \mathbb{F}_2^n)$. The construction $\mathcal{E}^{(r)}[f_1, \cdots, f_r]$ is CCA-secure if it is resistant to computationally unbounded attackers who make a bounded number of queries to the construction. Formally, we consider the ability of an adversary D to distinguish two worlds: the "real world", in which it is given oracle access to $\mathcal{E}^{(r)}[f_1, \cdots, f_r]$ and its inverse $(\mathcal{E}^{(r)}[f_1, \cdots, f_r])^{-1}$, and the "ideal world", in which it has access to a random permutation $\Pi \xleftarrow{\$} \mathcal{P}(\mathbb{F}_2^m)$ and its inverse Π^{-1}. With these, the

SPRP advantage against the r-round construction $\mathcal{E}^{(r)}[f_1, \cdots, f_r]$ is defined as:

$$\mathsf{Adv}^{\mathrm{CCA}}_{\mathcal{E}^{(r)}}(D) := \left| \Pr\left(f_1, \cdots, f_r \overset{\$}{\leftarrow} \mathcal{F}(\mathbb{F}_2^n, \mathbb{F}_2^n) : D^{\mathcal{E}^{(r)}[f_1, \cdots, f_r], (\mathcal{E}^{(r)}[f_1, \cdots, f_r])^{-1}} = 1\right) \right. \tag{1}$$

$$\left. - \Pr\left(\Pi \overset{\$}{\leftarrow} \mathcal{P}(\mathbb{F}_2^m) : D^{\Pi, \Pi^{-1}} = 1\right) \right|$$

for an adversary D. Furthermore, the *SPRP security* of $\mathcal{E}^{(r)}[f_1, \cdots, f_r]$ is defined as:

$$\mathsf{Adv}^{\mathrm{CCA}}_{\mathcal{E}^{(r)}}(q) := \max_{D} \left\{ \mathsf{Adv}^{\mathrm{CCA}}_{\mathcal{E}^{(r)}}(D) \right\},$$

where the maximum is taken over all adversaries that make at most q queries to the two oracles in total.

The H-coefficient method. We will use Patarin's H-coefficient technique [24] to prove the SPRP security of a construction in Sect. 5. Here, we provide a brief overview of its main ingredients, which are borrowed from [4].

Fix an adversary D that makes at most q queries to its oracles. As in the security definition presented above, D aims to distinguish between a "real world" and an "ideal world". Assume that D is deterministic. The execution of D defines a *transcript* that includes a sequence of queries and answers received from its oracles; D's output is a deterministic function of its transcript. Let μ, ν denote the probability distributions on transcripts induced by the real world and ideal world, respectively. The distinguishing advantage of D is upper bounded by the statistical distance

$$\mathsf{Dist}(\mu, \nu) := \frac{1}{2} \sum_{\mathcal{Q}} \left| \mu(\mathcal{Q}) - \nu(\mathcal{Q}) \right|, \tag{2}$$

where the sum is taken over all possible transcripts $\mathcal{Q}$.

Denote $\mathcal{T}$ as a set of all transcripts such that $\nu(\mathcal{Q}) > 0$ for all $\mathcal{Q} \in \mathcal{T}$. We aim to partition $\mathcal{T}$ into two sets $\mathcal{T}_1$ and $\mathcal{T}_2$, representing "good" and "bad" transcripts, respectively, while also identifying a constant $\epsilon_1 \in [0, 1)$, such that

$$\mathcal{Q} \in \mathcal{T}_1 \implies \mu(\mathcal{Q})/\nu(\mathcal{Q}) \geq 1 - \epsilon_1. \tag{3}$$

It is then possible to show (see [4] for details)

$$\mathsf{Dist}(\mu, \nu) \leq \epsilon_1 + \Pr(\mathcal{Q} \in \mathcal{T}_2), \tag{4}$$

which is an upper bound of the distinguishing advantage.

3 Properties of Lai-Massey Structure

3.1 Lai-Massey Structure and Its Equivalent Structure

The Lai-Massey structure was proposed by Lai and Massey when designing the block cipher International Data Encryption Algorithm (IDEA) [14,15]. Subsequently, this structure was adopted to design FOX block cipher family [12].

The process of the i-th round of the Lai-Massey structure is illustrated in Fig. 2(a). The input is initially divided into two equal-sized branches, L_{i-1} and R_{i-1}. Then, an f-function is applied on the XOR value of two branches, and the output of f-function is XORed to both branches. Finally, a linear orthomorphic permutation σ is applied to one branch to produce the output, L_i and R_i. The steps are formalized as follows:

$$\begin{cases} z_{i-1} = L_{i-1} \oplus R_{i-1}, \\ L_i = \sigma(L_{i-1} \oplus f(z_{i-1})), \\ R_i = R_{i-1} \oplus f(z_{i-1}). \end{cases}$$

Moreover, another representation of the Lai-Massey structure can be derived, as shown in Fig. 2(b).

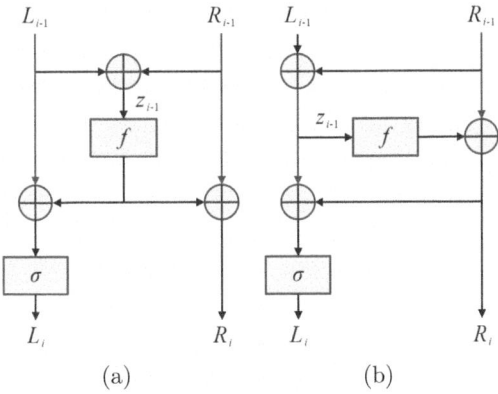

(a)   (b)

Fig. 2. Lai-Massey structure and its another representation.

Considering the multiple-round iteration process of the Lai-Massey structure, a significant observation is that it is possible to merge linear transformations between two consecutive f-functions. This refers to the part enclosed by the red rectangle, as shown in Fig. 3(a). The matrix representation of the merged linear transformation is

$$\sigma' = \begin{pmatrix} I & I \\ O & I \end{pmatrix} \begin{pmatrix} \sigma & O \\ O & I \end{pmatrix} \begin{pmatrix} I & I \\ O & I \end{pmatrix} = \begin{pmatrix} \sigma & \sigma \oplus I \\ O & I \end{pmatrix}.$$

where I and O are the identity matrix and zero matrix, respectively. Here, linear transformations σ, σ' and their corresponding matrix representations adopt the same symbols. Moreover, the r-round iteration process of the Lai-Massey structure is shown in Fig. 3(b).

Based on the above observation, an equivalent structure $\mathcal{ELM}$ of the Lai-Massey structure $\mathcal{LM}$ is derived, as shown in Fig. 4. Denote the r-round Lai-Massey structure as $\mathcal{LM}^{(r)}$, and denote the r-round equivalent structure as

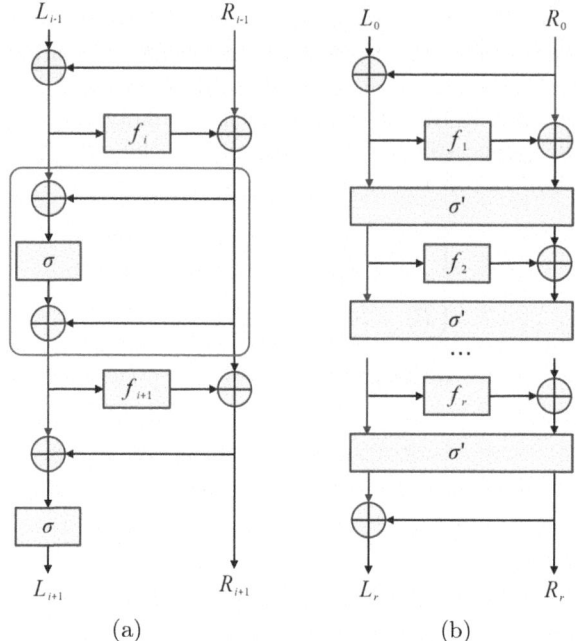

$$(a) \qquad\qquad (b)$$

Fig. 3. The 2-round iteration and r-round iteration of the Lai-Massey structure.

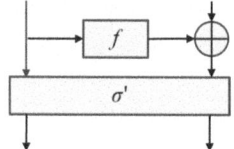

Fig. 4. The equivalent structure of Lai-Massey structure.

$\mathcal{ELM}^{(r)}$. Notably, $\mathcal{LM}^{(r)}$ is the same as $\mathcal{ELM}^{(r)}$ with linear transformations on the input of the first round and the output of the r-th round, respectively, i.e.,

$$\mathcal{LM}^{(r)} = \begin{pmatrix} I\ I \\ O\ I \end{pmatrix} \circ \mathcal{ELM}^{(r)} \circ \begin{pmatrix} I\ I \\ O\ I \end{pmatrix}.$$

Consequently, $\mathcal{LM}^{(r)}$ and $\mathcal{ELM}^{(r)}$ are affine equivalent. Please refer to [28] for the definition of affine equivalence. Therefore, the cryptanalysis of the Lai-Massey structure is equivalent to that of its equivalent structure.

Furthermore, the differences between the Lai-Massey and Feistel structures in design and security are attributed to different properties of orthomorphic permutation and branch permutation. We further investigate the properties of orthomorphic permutation in the following.

3.2   Orthomorphic Permutation σ and Linear Transformation σ'

Let $(G, +)$ be a finite abelian group. Define a mapping $\sigma : G \mapsto G$. If σ and $x \mapsto \sigma(x) - x$ are both permutations, then σ is called an orthomorphic permutation. This paper sets G as $\mathbb{F}_2^n$, the group operation as bit-wise XOR operation $\oplus$, and the mapping σ as a linear orthomorphic permutation. The linear orthomorphic permutation σ can be represented by an invertible matrix over $\mathbb{F}_2$. The order of σ, denoted as $\text{ord}(\sigma)$, is the smallest positive integer t such that σ^t is the identity mapping. The properties for σ are presented in the following.

Property 1. For a linear orthomorphic permutation σ, we have $\text{ord}(\sigma) \geq 3$.

Proof. Since the identity mapping is not an orthomorphic permutation, we have $\text{ord}(\sigma) \neq 1$. Assume that $\text{ord}(\sigma) = 2$, namely, $\sigma^2(x) \oplus x \equiv 0$, thus $\sigma^2(x) \oplus \sigma(x) \equiv \sigma(x) \oplus x$ holds for any $x \in \mathbb{F}_2^n$. Due to $\sigma(x) \oplus x$ being a permutation, then $\sigma(x) \oplus x \equiv 0$, which leads to a contradiction. Consequently, $\text{ord}(\sigma) \geq 3$.

Example 1. There are six 2×2 invertible matrices over $\mathbb{F}_2$, i.e.,

$$M_1 = \begin{pmatrix} 1 & 0 \\ 0 & 1 \end{pmatrix}, M_2 = \begin{pmatrix} 1 & 1 \\ 0 & 1 \end{pmatrix}, M_3 = \begin{pmatrix} 0 & 1 \\ 1 & 0 \end{pmatrix},$$

$$M_4 = \begin{pmatrix} 1 & 0 \\ 1 & 1 \end{pmatrix}, M_5 = \begin{pmatrix} 0 & 1 \\ 1 & 1 \end{pmatrix}, M_6 = \begin{pmatrix} 1 & 1 \\ 1 & 0 \end{pmatrix}.$$

Note that matrices M_5 and M_6 are orthomorphic permutations, and the order of these two matrices is 3. Other matrices are not orthomorphic permutations. The order of M_2, M_3, M_4 is 2, and that of M_1 is 1.

Property 2. The linear mapping $x \mapsto \sigma^2(x) \oplus x$ is a permutation.

Proof. Assume that the linear mapping $x \mapsto \sigma^2(x) \oplus x$ is not a permutation. Then there exists a non-zero vector $y \in \mathbb{F}_2^n$, such that $\sigma^2(y) \oplus y = 0$, thus $\sigma(y) \oplus y = 0$. We have $y = 0$, which is a contradiction.

Corollary 1. *Suppose the order of a linear orthomorphic permutation σ is k, then linear mappings $x \mapsto \sigma^i(x) \oplus x$ are permutations for $i = 1, 2, \cdots, k - 1$.*

Following [3,8], we propose the following definition to establish a link between orthomorphic permutation σ and linear transformation σ'.

Definition 1. *Suppose M, N are $n \times n$ invertible matrices over $\mathbb{F}_2$, if there exists an $n \times n$ invertible matrix P over $\mathbb{F}_2$, such that $P^{-1}MP = N$, then matrix M is said to be conjugated equivalent to N, denoted as $M \sim N$.*

The conjugated equivalence of $n \times n$ invertible matrices over $\mathbb{F}_2$ constitutes an equivalence relationship, which satisfies reflexivity, symmetry, and transitivity. The following property presents a necessary condition for conjugated equivalence according to the order.

Property 3. Suppose M, N are $n \times n$ invertible matrices over $\mathbb{F}_2$, if M is conjugated equivalent to N, then $\text{ord}(M) = \text{ord}(N)$.

According to Property 3, if the orders of matrices M and N are different, then these two matrices must not be conjugated equivalent. However, if different invertible matrices have the same order, these matrices may be conjugated equivalent and belong to the same equivalence class.

Example 2. For a linear orthomorphic permutation σ, the invertible matrix $\sigma' = \begin{pmatrix} \sigma & \sigma \oplus I \\ O & I \end{pmatrix}$ is conjugated equivalent to the invertible matrix $\begin{pmatrix} \sigma & O \\ O & I \end{pmatrix}$ and $\text{ord}(\sigma') = \text{ord}(\sigma)$.

Example 3. As shown in Example 1, there are three equivalence classes, i.e.,

$$\{M_1\}, \{M_2, M_3, M_4\}, \{M_5, M_6\}.$$

4 Design and Cryptanalysis of Framework UFLM

In this section, we propose a new framework called **UFLM**. The framework **UFLM** is a collection of cipher structures that includes the Feistel and Lai-Massey structures, as illustrated in Fig. 5. Then, we research the number of rounds of impossible differentials, zero correlation linear hulls, and integral distinguishers of **UFLM** instances.

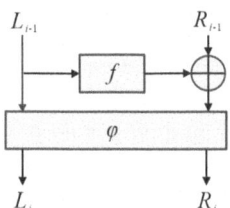

Fig. 5. The framework **UFLM**.

Let the input and output of the i-th round be (L_{i-1}, R_{i-1}) and $(L_i, R_i) \in \mathbb{F}_2^{2n}$, respectively, where n is the size of each branch. Let f be a mapping over $\mathbb{F}_2^n$, and φ be an invertible linear transformation over $\mathbb{F}_2^{2n}$. Then, the process of a single round of the framework **UFLM** is defined as a mapping over $\mathbb{F}_2^{2n}$, namely,

$$\begin{pmatrix} L_i \\ R_i \end{pmatrix} := \varphi \begin{pmatrix} L_{i-1} \\ R_{i-1} \oplus f(L_{i-1}) \end{pmatrix}.$$

Generally, the f-function is selected as a nonlinear bijective mapping in the design of block ciphers. However, the details of the f-function do not need to be taken into account when designing cipher structures. If the linear transformation

φ is chosen, then an **UFLM** instance is obtained. An **UFLM** instance $\mathcal{U}_\varphi$ is defined as:

$$\mathcal{U}_\varphi := \{E_{f,\varphi} \mid f \in \mathcal{F}(\mathbb{F}_2^n, \mathbb{F}_2^n)\},$$

where $E_{f,\varphi}$ is a (single-round) block cipher employing the instance $\mathcal{U}_\varphi$. If φ is chosen as the swapping branch permutation, then the **UFLM** instance is Feistel structure. If $\varphi = \sigma'$, then the **UFLM** instance is equivalent Lai-Massey structure. Furthermore, the (single-round) **UFLM** construction $\mathcal{UFLM}$ is defined as: $\mathcal{UFLM} := \{\mathcal{U}_\varphi \mid \varphi \in \Pi\}$.

In addition, an r-round **UFLM** instance is defined as $\mathcal{U}_\varphi^{(r)}$, namely, the r-fold composition of $\mathcal{U}_\varphi$ with itself. Moreover, f-functions adopted in $\mathcal{U}_\varphi^{(r)}$ in each round are considered as (secret) random functions. Furthermore, the r-round **UFLM** construction $\mathcal{UFLM}^{(r)}$ is defined as:

$$\mathcal{UFLM}^{(r)} := \{\mathcal{U}_\varphi^{(r)} \mid \varphi \in \Pi\}.$$

This paper treats each f-function as a bijective mapping. Note that any **UFLM** instance is invertible. Therefore, the construction $\mathcal{UFLM}$ is also invertible. For any number of rounds r, denote $A, B, \mathcal{A}^{(r)}$ and $\mathcal{B}^{(r)}$ as the following matrices:

$$A = \begin{pmatrix} I & O \end{pmatrix}, B = \begin{pmatrix} O & I \end{pmatrix}, \mathcal{A}^{(r)} = \begin{pmatrix} A \\ A\varphi \\ \vdots \\ A\varphi^{r-1} \end{pmatrix}, \mathcal{B}^{(r)} = \begin{pmatrix} B \\ B\varphi^{\mathrm{T}} \\ \vdots \\ B(\varphi^{\mathrm{T}})^{r-1} \end{pmatrix},$$

where φ^{T} is the transpose of φ. The symbol for the matrix representation of φ is also represented by φ.

If $\mathrm{ord}(\varphi) = 1$, then the construction $\mathcal{UFLM}$ is not secure. In the following, we investigate the number of rounds for three types of distinguishers of **UFLM** instances adopting linear transformations with $\mathrm{ord}(\varphi) \geq 2$. Moreover, we further constrain that both $\mathcal{A}^{(2)}$ and $\mathcal{B}^{(2)}$ are full-rank to filter some **UFLM** instances, namely, there will be longer distinguishers for these instances.

Property 4. If $\mathcal{A}^{(2)}$ is full-rank, then there exists at least one differentially active f-function covering two consecutive rounds for **UFLM** instances.

Property 5. If $\mathcal{B}^{(2)}$ is full-rank, then there exists at least one linearly active f-function covering two consecutive rounds for **UFLM** instances.

4.1 Impossible Differential Cryptanalysis

The following presents the number of rounds of impossible differentials for **UFLM** instances in three cases.

Theorem 1. *Assume that $\mathcal{A}^{(2)}$ and $\mathcal{B}^{(2)}$ are full-rank. There exists a 5-round impossible differential for **UFLM** instances where $\mathrm{ord}(\varphi) = 2$.*

Proof. According to the fact that $\mathcal{A}^{(2)}$ is full-rank, there exist non-zero solutions for equation $\mathcal{A}^{(1)}x = 0$, and there only exist zero solution for equation $\mathcal{A}^{(2)}x = 0$.

For any bijective f-functions, consider the possible differential propagation from the encryption direction. Let the non-zero input difference be α such that $\mathcal{A}^{(1)}\alpha = 0$, namely, $A\alpha = 0$ and $A\varphi\alpha \neq 0$, then the input and output difference of the f-function in the first round are both 0. Thus, the output difference of the first round is $\varphi\alpha$.

Since $A\varphi\alpha \neq 0$, i.e., the input difference of the f-function in the second round is non-zero, then the corresponding output difference is non-zero, denoted by β_1. Thus, the output difference of the second round is $\alpha \oplus \varphi B^{\mathrm{T}}\beta_1$.

The input difference of the f-function in the third round is $A\varphi B^{\mathrm{T}}\beta_1$. The corresponding output difference is denoted as β_2. Thus, the output difference of the third round is $\varphi\alpha \oplus B^{\mathrm{T}}\beta_1 \oplus \varphi B^{\mathrm{T}}\beta_2$.

Similarly, consider the possible differential propagation from the decryption direction. Let the output difference of the fifth round be $\varphi\alpha$. Then the output differences of the fourth round and third round are α and $\varphi\alpha \oplus B^{\mathrm{T}}\beta_3$, respectively, where β_3 is the output difference of the f-function in the fourth round.

Therefore, we have $\varphi\alpha \oplus B^{\mathrm{T}}\beta_1 \oplus \varphi B^{\mathrm{T}}\beta_2 = \varphi\alpha \oplus B^{\mathrm{T}}\beta_3$, namely,

$$\begin{pmatrix} B^{\mathrm{T}} & \varphi B^{\mathrm{T}} \end{pmatrix} \begin{pmatrix} \beta_1 \oplus \beta_3 \\ \beta_2 \end{pmatrix} = 0.$$

According to $\mathcal{B}^{(2)}$ is full-rank, there exists only zero solution in the above equation, namely, $\beta_1 \oplus \beta_3 = \beta_2 = 0$. Thus, we have

$$\begin{pmatrix} A \\ A\varphi \end{pmatrix} B^{\mathrm{T}}\beta_1 = 0.$$

Since $\mathcal{A}^{(2)}$ is full-rank, we have $B^{\mathrm{T}}\beta_1 = 0$. Moreover, $\varphi B^{\mathrm{T}}\beta_1 = 0$, therefore $\beta_1 = 0$, which leads to a contradiction.

Consequently, $\alpha \to \varphi\alpha$ is a 5-round impossible differential for **UFLM** instances where $\mathrm{ord}(\varphi) = 2$.

Similarly, we have the following corollaries.

Corollary 2. *Assume that $\mathcal{A}^{(2)}$ and $\mathcal{B}^{(2)}$ are full-rank. There exists a 4-round impossible differential $\alpha \to \varphi\alpha$ for* **UFLM** *instances where α is a non-zero solution for equation $\mathcal{A}^{(1)}x = 0$ and $\mathrm{ord}(\varphi) = 3$.*

Corollary 3. *Assume that $\mathcal{A}^{(2)}$ and $\mathcal{B}^{(2)}$ are full-rank. There exists a 3-round impossible differential $\alpha \to \varphi^3\alpha$ for* **UFLM** *instances where α is a non-zero solution for equation $\mathcal{A}^{(1)}x = 0$ and $\mathrm{ord}(\varphi) > 3$.*

4.2   Zero Correlation Linear Cryptanalysis

Based on $\mathrm{ord}(\varphi^{\mathrm{T}}) = \mathrm{ord}(\varphi)$, the following theorem and corollaries for zero correlation linear hulls hold, which are projected by impossible differentials.

Theorem 2. *Assume that $\mathcal{A}^{(2)}$ and $\mathcal{B}^{(2)}$ are full-rank. There exists a 5-round zero correlation linear hull for* **UFLM** *instances where* $\operatorname{ord}(\varphi) = 2$.

Proof. According to the fact that $\mathcal{B}^{(2)}$ is full-rank, there exist non-zero solutions for equation $\mathcal{B}^{(1)}x = 0$, and there only exist zero solution for equation $\mathcal{B}^{(2)}x = 0$.

For any bijective f-functions, consider possible mask propagation from encryption direction. Let the non-zero input mask be γ such that $\mathcal{B}^{(1)}\gamma = 0$, namely, $B\gamma = 0$ and $B\varphi^{\mathrm{T}}\gamma \neq 0$, then the input and output mask of the f-function in the first round are both 0. Thus, the output mask of the first round is $\varphi^{\mathrm{T}}\gamma$.

Since $B\varphi^{\mathrm{T}}\gamma \neq 0$, i.e., the output mask of the f-function in the second round is non-zero, then the corresponding input mask is non-zero, denoted by λ_1. Thus, the output mask of the second round is $\gamma \oplus \varphi^{\mathrm{T}}A^{\mathrm{T}}\lambda_1$.

Similarly, consider possible mask propagation from the decryption direction. Let the output mask of the fifth round be $\varphi^{\mathrm{T}}\gamma$. Then the output masks of the fourth round and third round are γ and $\varphi^{\mathrm{T}}\gamma \oplus A^{\mathrm{T}}\lambda_2$, respectively, where λ_2 is the input mask of the f-function in the fourth round.

The output mask of the f-function in the third round is $B\varphi^{\mathrm{T}}A^{\mathrm{T}}\lambda_2$. The corresponding input mask is denoted as λ_3. Thus, the output mask of the second round is $\gamma \oplus \varphi^{\mathrm{T}}A^{\mathrm{T}}\lambda_2 \oplus A^{\mathrm{T}}\lambda_3$.

Therefore, we have $\gamma \oplus \varphi^{\mathrm{T}}A^{\mathrm{T}}\lambda_1 = \gamma \oplus \varphi^{\mathrm{T}}A^{\mathrm{T}}\lambda_2 \oplus A^{\mathrm{T}}\lambda_3$, namely,

$$\begin{pmatrix} A^{\mathrm{T}} & \varphi^{\mathrm{T}}A^{\mathrm{T}} \end{pmatrix} \begin{pmatrix} \lambda_3 \\ \lambda_1 \oplus \lambda_2 \end{pmatrix} = 0.$$

According to $\mathcal{A}^{(2)}$ is full-rank, there exists only zero solution in the above equation, namely, $\lambda_1 \oplus \lambda_2 = \lambda_3 = 0$. Thus, we have

$$\begin{pmatrix} B \\ B\varphi^{\mathrm{T}} \end{pmatrix} A^{\mathrm{T}}\lambda_1 = 0.$$

Since $\mathcal{B}^{(2)}$ is full-rank, we have $A^{\mathrm{T}}\lambda_1 = 0$. Moreover, $\varphi^{\mathrm{T}}A^{\mathrm{T}}\lambda_1 = 0$, therefore $\lambda_1 = 0$, which leads to a contradiction.

Consequently, $\gamma \to \varphi^{\mathrm{T}}\gamma$ is a 5-round zero correlation linear hull for **UFLM** instances where $\operatorname{ord}(\varphi) = 2$.

Corollary 4. *Assume that $\mathcal{A}^{(2)}$ and $\mathcal{B}^{(2)}$ are full-rank. There exists a 4-round zero correlation linear hull $\gamma \to (\varphi^{\mathrm{T}})^2\gamma$ for* **UFLM** *instances where γ is a non-zero solution for equation $\mathcal{B}^{(1)}x = 0$ and $\operatorname{ord}(\varphi) = 3$.*

Corollary 5. *Assume that $\mathcal{A}^{(2)}$ and $\mathcal{B}^{(2)}$ are full-rank. There exists a 3-round zero correlation linear hull $\gamma \to (\varphi^{\mathrm{T}})^{k-3}\gamma$ for* **UFLM** *instances where γ is a non-zero solution for equation $\mathcal{B}^{(1)}x = 0$ and $\operatorname{ord}(\varphi) = k > 3$.*

4.3   Integral Cryptanalysis

Sun et al. have shown the link between a zero correlation linear hull and an integral distinguisher: *a nontrivial zero correlation linear hull of a block cipher always implies the existence of an integral distinguisher* [27]. Thus, the following theorem holds based on the Theorem 2, Corollary 4 and Corollary 5.

Theorem 3. *Assume that $\mathcal{A}^{(2)}$ and $\mathcal{B}^{(2)}$ are full-rank. If $\mathrm{ord}(\varphi) = 2$, then there exists a 5-round integral distinguisher for* **UFLM** *instances. If $\mathrm{ord}(\varphi) = 3$, then there exists a 4-round integral distinguisher for* **UFLM** *instances. If $\mathrm{ord}(\varphi) > 3$, then there exists a 3-round integral distinguisher for* **UFLM** *instances.*

Table 1 summarizes the number of rounds of different distinguishers for **UFLM** instances in three cases under the condition that $\mathcal{A}^{(2)}$ and $\mathcal{B}^{(2)}$ are full-rank. The Feistel structure is typically an instance with $\mathrm{ord}(\varphi) = 2$. The FOX64 structure, known as the Lai-Massey structure used in the FOX64 block cipher [12], is an instance with $\mathrm{ord}(\varphi) = 3$. Moreover, the decryption structure is the same as the encryption structure for **UFLM** instances with $\mathrm{ord}(\varphi) = 2$, but it requires more rounds to achieve sufficient security. On the contrary, the number of iteration rounds may be fewer for **UFLM** instances with $\mathrm{ord}(\varphi) \geq 3$. However, it is necessary to implement φ^{-1} for decryption.

Table 1. Distinguishers for **UFLM** instances in three cases.

$\mathrm{ord}(\varphi)$	Distinguishers	Rounds	Structures
2	Impossible differential	5	Feistel structure
	Zero correlation linear hull	5	
	Integral distinguisher	5	
3	Impossible differential	4	FOX64 structure
	Zero correlation linear hull	4	
	Integral distinguisher	4	
> 3	Impossible differential	3	—
	Zero correlation linear hull	3	
	Integral distinguisher	3	

5   CCA Security of $\mathcal{UFLM}$

Following [18], we prove that 4-round construction $\mathcal{UFLM}^{(4)}$ is CCA-secure up to the birthday bound in this section. The number of rounds is tight by Nandi's lower bound on SPRP constructions [22]. Consider the following two cases.

(i) The 4-round construction $\mathcal{UFLM}^{(4)}[f]$ adopts the same f-functions in each round, i.e., $f_1 = f_2 = f_3 = f_4 = f$.
(ii) The 4-round construction $\mathcal{UFLM}^{(4)}[f_1, f_2, f_3, f_4]$ utilizes independent f-functions in each round.

The first three rounds map (L_{i-1}, R_{i-1}) to $(L_i, R_i) = \varphi\big(L_{i-1}, f_i(L_{i-1}) \oplus R_{i-1}\big)$ for $i = 1, 2, 3$. The last round omits φ and simply maps (L_3, R_3) to $(L_4, R_4) = \big(L_3, f_4(L_3) \oplus R_3\big)$. Next, we first introduce some notations and then present security proofs.

5.1   Additional Notations

We will frequently write $\varphi : \mathbb{F}_2^{2n} \mapsto \mathbb{F}_2^{2n}$ in the block form of 4 submatrices over $\mathbb{F}_2^{n \times n}$. For this, we follow the convention using U, B, L, R to represent *upper*, *bottom*, *left*, and *right*, respectively, i.e.,

$$\varphi = \begin{pmatrix} \varphi_{\mathrm{UL}} & \varphi_{\mathrm{UR}} \\ \varphi_{\mathrm{BL}} & \varphi_{\mathrm{BR}} \end{pmatrix}.$$

In other words, the i-th round process maps (L_{i-1}, R_{i-1}) to

$$\begin{cases} L_i = \varphi_{\mathrm{UL}} \cdot L_{i-1} \oplus \varphi_{\mathrm{UR}} \cdot \big(f_i(L_{i-1}) \oplus R_{i-1}\big), \\ R_i = \varphi_{\mathrm{BL}} \cdot L_{i-1} \oplus \varphi_{\mathrm{BR}} \cdot \big(f_i(L_{i-1}) \oplus R_{i-1}\big). \end{cases}$$

The "right halves" $f_i(L_{i-1}) \oplus R_{i-1}$ and R_i can be derived from the "left halves" L_i and L_{i-1} according to the above equations, i.e.,

$$\begin{cases} f_i(L_{i-1}) \oplus R_{i-1} = \varphi_{\mathrm{UR}}^{-1} \cdot \big(L_i \oplus \varphi_{\mathrm{UL}} \cdot L_{i-1}\big), \\ R_i = \varphi_{\mathrm{BL}} \cdot L_{i-1} \oplus \varphi_{\mathrm{BR}} \cdot \big(\varphi_{\mathrm{UR}}^{-1} \cdot \big(L_i \oplus \varphi_{\mathrm{UL}} \cdot L_{i-1}\big)\big), \end{cases}$$

which will be used in the security proofs later. We use brackets, i.e., $(\varphi^{-1})_{\mathrm{POS}}$, $\mathrm{POS} \in \{\mathrm{UL}, \mathrm{UR}, \mathrm{BL}, \mathrm{BR}\}$, to distinguish submatrices of φ^{-1} (the inverse of φ) from $\varphi_{\mathrm{POS}}^{-1}$ (the inverse of φ_{POS}).

Recall that matrices $\mathcal{A}^{(2)}$ and $\mathcal{B}^{(2)}$ are full-rank. It holds that two submatrices φ_{UR} and $(\varphi^{-1})_{\mathrm{UR}}$ are full-rank. Furthermore, we characterize extra requirements on φ and φ^{-1} to obtain an excellent CCA security bound for $\mathcal{UFLM}^{(4)}[f]$.

Definition 2 (Good Linear Transformation). *A linear transformation*

$$\varphi = \begin{pmatrix} \varphi_{\mathrm{UL}} & \varphi_{\mathrm{UR}} \\ \varphi_{\mathrm{BL}} & \varphi_{\mathrm{BR}} \end{pmatrix}$$

over $\mathbb{F}_2^{2n \times 2n}$ *is said to be good if the three matrices* φ_{UR}, $(\varphi^{-1})_{\mathrm{UR}}$ *and* $\varphi_{\mathrm{UR}} \oplus (\varphi^{-1})_{\mathrm{UR}}$ *are full-rank.*

An example of constructing good linear transformations is as follows.

Example 4. Let σ be a linear orthomorphic permutation over $\mathbb{F}_2^{n \times n}$ and $\varphi = \begin{pmatrix} \sigma & \sigma \oplus I \\ O & I \end{pmatrix}$. Then $\varphi^{-1} = \begin{pmatrix} \sigma^{-1} & \sigma^{-1} \oplus I \\ O & I \end{pmatrix}$. According to Corollary 1, $\sigma \oplus I$, $\sigma^{-1} \oplus I$ and $\sigma \oplus \sigma^{-1}$ are full-rank. Thus, φ is a good linear transformation.

Next, we will provide detailed security proof for CCA of the construction $\mathcal{UFLM}^{(4)}[f]$. By employing a similar method, we will obtain the CCA security proof for the construction $\mathcal{UFLM}^{(4)}[f_1, f_2, f_3, f_4]$.

5.2   CCA Security for $\mathcal{UFLM}^{(4)}[f]$

The CCA-security results of 4-round construction $\mathcal{UFLM}^{(4)}[f]$ are formally stated as follows.

Theorem 4. *Assume $q \leq 2^n/2$. Then, for the 4-round idealized construction $\mathcal{UFLM}^{(4)}[f]$ defined upon a secret random function f and a good linear transformation φ (see Definition 2), it holds*

$$\mathsf{Adv}^{\mathrm{CCA}}_{\mathcal{UFLM}^{(4)}}(q) \leq \frac{6q^2}{2^n} + \frac{q^2}{2^{2n}}. \tag{5}$$

All this subsection is devoted to proving Theorem 4. We first establish an interaction between an adversary and oracles. Then, we employ the H-coefficient method without using bad transcripts. Finally, we bound the ratio $\mu(\mathcal{Q})/\nu(\mathcal{Q})$ with attainable transcripts to complete the proof.

Proof Setup. Fix a deterministic adversary D. We assume D makes exactly q (non-redundant) forward/inverse queries to its left oracle: either $\mathcal{UFLM}^{(4)}[f]$ or Π.

The interaction between D and oracles is recorded as a list of records $\mathcal{Q} \subseteq \mathbb{F}_2^{2n} \times \mathbb{F}_2^{2n}$. Among them, $\mathcal{Q} = \{((L_0^{(1)}, R_0^{(1)}), (L_4^{(1)}, R_4^{(1)})), \cdots, ((L_0^{(q)}, R_0^{(q)}), (L_4^{(q)}, R_4^{(q)}))\}$ lists the queries-responses of D, where the i-th record $((L_0^{(i)}, R_0^{(i)}), (L_4^{(i)}, R_4^{(i)}))$ indicates the query is either a forward query $(L_0^{(i)}, R_0^{(i)})$ that was answered by $(L_4^{(i)}, R_4^{(i)})$ or an inverse query $(L_4^{(i)}, R_4^{(i)})$ that was answered by $(L_0^{(i)}, R_0^{(i)})$. Significantly, the interaction between D and oracles can be unambiguously reconstructed from $\mathcal{Q}$.

Bounding the Ratio $\mu(\mathcal{Q})/\nu(\mathcal{Q})$. Let $\mathcal{Q}'$ be a transcript. Given a function $f^* \in \mathcal{F}(\mathbb{F}_2^n, \mathbb{F}_2^n)$, we say that the 4-round construction $\mathcal{UFLM}^{(4)}[f^*]$ *extends* $\mathcal{Q}'$ and denote $\mathcal{UFLM}^{(4)}[f^*] \vdash \mathcal{Q}'$, if $\mathcal{UFLM}^{(4)}[f^*](L_0, R_0) = (L_4, R_4)$ for all $((L_0, R_0), (L_4, R_4)) \in \mathcal{Q}'$. Similarly, a permutation $\Pi^* \in \mathcal{P}(\mathbb{F}_2^{2n})$ *extends* $\mathcal{Q}'$ (denoted $\Pi^* \vdash \mathcal{Q}'$), if $\Pi^*(L_0, R_0) = (L_4, R_4)$ for all $((L_0, R_0), (L_4, R_4)) \in \mathcal{Q}'$.

For the rest of the proof, we fix an attainable transcript $\mathcal{Q} \in \mathcal{T}$. It is easy to see that

$$\mu(\mathcal{Q}) = \Pr(f \leftarrow \mathcal{F}(\mathbb{F}_2^n, \mathbb{F}_2^n) : \mathcal{UFLM}^{(4)}[f] \vdash \mathcal{Q}),$$
$$\nu(\mathcal{Q}) = \Pr(\Pi \leftarrow \mathcal{P}(\mathbb{F}_2^{2n}) : \Pi \vdash \mathcal{Q}).$$

Thus,

$$\frac{\mu(\mathcal{Q})}{\nu(\mathcal{Q})} = \frac{\Pr(f \leftarrow \mathcal{F}(\mathbb{F}_2^n, \mathbb{F}_2^n) : \mathcal{UFLM}^{(4)}[f] \vdash \mathcal{Q})}{\Pr(\Pi \leftarrow \mathcal{P}(\mathbb{F}_2^{2n}) : \Pi \vdash \mathcal{Q})}.$$

By these and $|\mathcal{Q}| = q$, it is immediate that

$$\Pr\big(\Pi \leftarrow \mathcal{P}(\mathbb{F}_2^{2n}) : \Pi \vdash \mathcal{Q}\big) = \prod_{i=0}^{q-1} \frac{1}{2^{2n} - i}.$$

It remains to lower bound the real-world probability $\mu(\mathcal{Q})$. For this, we proceed in two steps. First, based on $\mathcal{Q}$ and some entries in f that define the relevant first and fourth round f-function values, we derive the second and the third round intermediate values: these constitute a special transcript $\mathcal{Q}_{mid}$ in the middle two rounds. We characterize conditions on f that will ensure specific good properties in the derived $\mathcal{Q}_{mid}$. Therefore, in the second step, we analyze such "good" $\mathcal{Q}_{mid}$ to yield the final bounds. Each of the two steps will take a paragraph as follows.

Predicate on f. We follow a clean "predicate" approach from [5] that defines a "bad" predicate $\mathsf{Bad}(f)$ on $f : \mathbb{F}_2^n \to \mathbb{F}_2^n$. If $\mathsf{Bad}(f)$ does not hold (the probability of which has a lower bound), then the event $\mathcal{UFLM}^{(4)}[f] \vdash \mathcal{Q}$ is equivalent with $2q$ distinct equations on the f-function. For convenience, define

$$\mathsf{ExtF} := \big\{ X \in \mathbb{F}_2^n \,|\, ((X, R_0), (L_4, R_4)) \in \mathcal{Q} \text{ for some } R_0, L_4, R_4 \text{ or}$$
$$((L_0, R_0), (X, R_4)) \in \mathcal{Q} \text{ for some } L_0, R_0, R_4 \big\}.$$

Clearly, $|\mathsf{ExtF}| \leq 2q$.

Then, given a random function f, we let $\mathsf{Bad}(f)$ be a predicate that holds if any of the following conditions is met:

- (B-1) There exists a record $((L_0, R_0), (L_4, R_4)) \in \mathcal{Q}$ such that $\varphi_{\mathrm{UL}} \cdot L_0 \oplus \varphi_{\mathrm{UR}} \cdot R_0 \oplus \varphi_{\mathrm{UR}} \cdot f(L_0) \in \mathsf{ExtF}$ or $(\varphi^{-1})_{\mathrm{UL}} \cdot L_4 \oplus (\varphi^{-1})_{\mathrm{UR}} \cdot R_4 \oplus (\varphi^{-1})_{\mathrm{UR}} \cdot f(L_4) \in \mathsf{ExtF}$;
- (B-2) There exist distinct records $((L_0, R_0), (L_4, R_4)), ((L'_0, R'_0), (L'_4, R'_4)) \in \mathcal{Q}$ such that $L_0 \neq L'_0$, but $\varphi_{\mathrm{UL}} \cdot L_0 \oplus \varphi_{\mathrm{UR}} \cdot R_0 \oplus \varphi_{\mathrm{UR}} \cdot f(L_0) = \varphi_{\mathrm{UL}} \cdot L'_0 \oplus \varphi_{\mathrm{UR}} \cdot R'_0 \oplus \varphi_{\mathrm{UR}} \cdot f(L'_0)$;
- (B-3) There exist distinct records $((L_0, R_0), (L_4, R_4)), ((L'_0, R'_0), (L'_4, R'_4)) \in \mathcal{Q}$ such that $L_4 \neq L'_4$, but $(\varphi^{-1})_{\mathrm{UL}} \cdot L_4 \oplus (\varphi^{-1})_{\mathrm{UR}} \cdot R_4 \oplus (\varphi^{-1})_{\mathrm{UR}} \cdot f(L_4) = (\varphi^{-1})_{\mathrm{UL}} \cdot L'_4 \oplus (\varphi^{-1})_{\mathrm{UR}} \cdot R'_4 \oplus (\varphi^{-1})_{\mathrm{UR}} \cdot f(L'_4)$;
- (B-4) There exist two records $((L_0, R_0), (L_4, R_4)), ((L'_0, R'_0), (L'_4, R'_4)) \in \mathcal{Q}$ (not necessarily distinct) such that $\varphi_{\mathrm{UL}} \cdot L_0 \oplus \varphi_{\mathrm{UR}} \cdot R_0 \oplus \varphi_{\mathrm{UR}} \cdot f(L_0) = (\varphi^{-1})_{\mathrm{UL}} \cdot L'_4 \oplus (\varphi^{-1})_{\mathrm{UR}} \cdot R'_4 \oplus (\varphi^{-1})_{\mathrm{UR}} \cdot f(L'_4)$.

The conditions capture collisions of the inputs of the second and the third round f-functions from (distinct) queries. It then holds

$$\Pr\big(f \leftarrow \mathcal{F}(\mathbb{F}_2^n, \mathbb{F}_2^n) : \mathcal{UFLM}^{(4)}[f] \vdash \mathcal{Q}\big)$$
$$\geq \Pr_f\big(\mathcal{UFLM}^{(4)}[f] \vdash \mathcal{Q} \mid \neg\mathsf{Bad}(f)\big) \times \big(1 - \Pr_f\big(\mathsf{Bad}(f)\big)\big). \qquad (6)$$

Hence, all that remains to determine the lower bound of the two terms in the product of (6).

Lemma 1. *When $q \leq 2^n/2$, we have*

$$\Pr_f\big(\mathsf{Bad}(f)\big) \leq \frac{6q^2}{2^n}. \tag{7}$$

Proof. Consider the conditions in turn. First, the two function values $f(L_0)$ and $f(L_4)$ are uniformly distributed in $\mathbb{F}_2^n$ for any record $\big((L_0, R_0), (L_4, R_4)\big) \in \mathcal{Q}$. In addition, they are independent of the values in the set ExtF, which are given by $\mathcal{Q}$. Therefore, the probability of having $\varphi_{\mathrm{UL}} \cdot L_0 \oplus \varphi_{\mathrm{UR}} \cdot R_0 \oplus \varphi_{\mathrm{UR}} \cdot f(L_0) \in \mathsf{ExtF}$ or $(\varphi^{-1})_{\mathrm{UL}} \cdot L_4 \oplus (\varphi^{-1})_{\mathrm{UR}} \cdot R_4 \oplus (\varphi^{-1})_{\mathrm{UR}} \cdot f(L_4) \in \mathsf{ExtF}$ is at most $2|\mathsf{ExtF}|/2^n \leq 4q/2^n$. Summing over the q choices of $\big((L_0, R_0), (L_4, R_4)\big) \in \mathcal{Q}$ yields

$$\Pr\big((\text{B-1})\big) \leq \frac{4q^2}{2^n}. \tag{8}$$

Second, given any pair $\big((L_0, R_0), (L_4, R_4)\big), \big((L_0', R_0'), (L_4', R_4')\big) \in \mathcal{Q}$ such that $L_0 \neq L_0'$, the two function values $f(L_0)$ and $f(L_0')$ are independent and uniformly distributed in $\mathbb{F}_2^n$. Thus, the probability of having $\varphi_{\mathrm{UL}} \cdot L_0 \oplus \varphi_{\mathrm{UR}} \cdot R_0 \oplus \varphi_{\mathrm{UR}} \cdot f(L_0) = \varphi_{\mathrm{UL}} \cdot L_0' \oplus \varphi_{\mathrm{UR}} \cdot R_0' \oplus \varphi_{\mathrm{UR}} \cdot f(L_0')$ is $1/2^n$. Since the number of choices of such pairs is $\binom{q}{2} \leq q^2/2$, it holds

$$\Pr\big((\text{B-2})\big) \leq \frac{q^2}{2 \cdot 2^n}. \tag{9}$$

The argument for (B-3) is similar by symmetry and yields the same bound.

$$\Pr\big((\text{B-3})\big) \leq \frac{q^2}{2 \cdot 2^n}. \tag{10}$$

Finally, for any two records $\big((L_0, R_0), (L_4, R_4)\big), \big((L_0', R_0'), (L_4', R_4')\big) \in \mathcal{Q}$, we distinguish two cases:

- Case 1: $L_0 \neq L_4'$. Then, $f(L_0)$ and $f(L_4')$ are independent and the probability of having $\varphi_{\mathrm{UL}} \cdot L_0 \oplus \varphi_{\mathrm{UR}} \cdot R_0 \oplus \varphi_{\mathrm{UR}} \cdot f(L_0) = (\varphi^{-1})_{\mathrm{UL}} \cdot L_4' \oplus (\varphi^{-1})_{\mathrm{UR}} \cdot R_4' \oplus (\varphi^{-1})_{\mathrm{UR}} \cdot f(L_4')$ is also $1/2^n$.
- Case 2: $L_0 = L_4'$. Then the equality $\varphi_{\mathrm{UL}} \cdot L_0 \oplus \varphi_{\mathrm{UR}} \cdot R_0 \oplus \varphi_{\mathrm{UR}} \cdot f(L_0) = (\varphi^{-1})_{\mathrm{UL}} \cdot L_4 \oplus (\varphi^{-1})_{\mathrm{UR}} \cdot R_4 \oplus (\varphi^{-1})_{\mathrm{UR}} \cdot f(L_4)$ is equivalent with $\big(\varphi_{\mathrm{UR}} \oplus (\varphi^{-1})_{\mathrm{UR}}\big) \cdot f(L_0) = \varphi_{\mathrm{UL}} \cdot L_0 \oplus \varphi_{\mathrm{UR}} \cdot R_0 \oplus (\varphi^{-1})_{\mathrm{UL}} \cdot L_4 \oplus (\varphi^{-1})_{\mathrm{UR}} \cdot R_4$, which holds with probability $1/2^n$ since $\big(\varphi_{\mathrm{UR}} \oplus (\varphi^{-1})_{\mathrm{UR}}\big)$ is full-rank.

The number of such two (not necessarily distinct) records is bounded by q^2. Therefore,

$$\Pr\big((\text{B-4})\big) \leq \frac{q^2}{2^n}. \tag{11}$$

Summing over equalities (8), (9), (10) and (11) yields the claim.

The probability of $\mathcal{UFLM}^{(4)}[f] \vdash \mathcal{Q}$ *for good* f. The step is determining the lower bound of the term $\Pr_f(\mathcal{UFLM}^{(4)}[f] \vdash \mathcal{Q} \mid \neg\mathsf{Bad}(f))$ in (6). To this end, for any f such that $\mathsf{Bad}(f)$ does not hold, we define an "extended transcript"

$$\mathcal{Q}^{out}(f) := \{((L_0, f(L_0)), (L_4, f(L_4))) \mid ((L_0, R_0), (L_4, R_4)) \in \mathcal{Q}\}.$$

We further define $\mathcal{T}^{out}$ as the set of all such extended transcripts, i.e.,

$$\mathcal{T}^{out} := \{\mathcal{Q}^{out}(f) \mid f \in \mathcal{F}(\mathbb{F}_2^n, \mathbb{F}_2^n)\},$$

and a set of "good" extended transcripts based on f that do not fulfill the predicate Bad, i.e.,

$$\mathcal{T}_{good}^{out} := \{\mathcal{Q}^{out}(f) \mid f \in \mathcal{F}(\mathbb{F}_2^n, \mathbb{F}_2^n), \neg\mathsf{Bad}(f)\}.$$

For any extended transcript $\mathcal{Q}^{out} \in \mathcal{T}^{out}$, we define another extended transcript $\mathcal{Q}^{mid}(\mathcal{Q}^{out})$ that captures the intermediate values of the two rounds in the middle of the construction. Formally, let f^* be a function such that $\mathcal{Q}^{out}(f^*) = \mathcal{Q}^{out}$, then

$$\mathcal{Q}^{mid}(\mathcal{Q}^{out}) = \{(L_1^{(i)}, Y_1^{(i)}), (L_2^{(i)}, Y_2^{(i)}) \mid i = 1, 2, \cdots, q\}, \tag{12}$$

where the values $L_1^{(i)}, Y_1^{(i)}, L_2^{(i)}$ and $Y_2^{(i)}$ are defined via

$$L_1^{(i)} = \varphi_{\mathrm{UL}} \cdot L_0^{(i)} \oplus \varphi_{\mathrm{UR}} \cdot R_0^{(i)} \oplus \varphi_{\mathrm{UR}} \cdot f^*(L_0^{(i)}),$$
$$R_1^{(i)} = \varphi_{\mathrm{BL}} \cdot L_0^{(i)} \oplus \varphi_{\mathrm{BR}} \cdot R_0^{(i)} \oplus \varphi_{\mathrm{BR}} \cdot f^*(L_0^{(i)}),$$
$$Y_1^{(i)} = \varphi_{\mathrm{UR}}^{-1} \cdot (L_2^{(i)} \oplus \varphi_{\mathrm{UL}} \cdot L_1^{(i)}) \oplus R_1^{(i)},$$
$$L_2^{(i)} = (\varphi^{-1})_{\mathrm{UL}} \cdot L_4^{(i)} \oplus (\varphi^{-1})_{\mathrm{UR}} \cdot R_4^{(i)} \oplus (\varphi^{-1})_{\mathrm{UR}} \cdot f^*(L_4^{(i)}),$$
$$Y_2^{(i)} = \varphi_{\mathrm{BR}} \cdot \varphi_{\mathrm{UR}}^{-1} \cdot (L_2^{(i)} \oplus \varphi_{\mathrm{UL}} \cdot L_1^{(i)}) \oplus \varphi_{\mathrm{BL}} \cdot L_1^{(i)}$$
$$\oplus (\varphi^{-1})_{\mathrm{BL}} \cdot L_4^{(i)} \oplus (\varphi^{-1})_{\mathrm{BR}} \cdot (f^*(L_4^{(i)}) \oplus R_4^{(i)}). \tag{13}$$

For all f^* with $\mathcal{Q}^{out}(f^*) = \mathcal{Q}^{out}$, the transcripts $\mathcal{Q}^{mid}(\mathcal{Q}^{out})$ defined above are the same since the condition $\mathcal{Q}^{out}(f^*) = \mathcal{Q}^{out}$ ensures that f^* is consistent with the input-output relations defined in $\mathcal{Q}^{out}$, which will fully characterize $\mathcal{Q}^{mid}(\mathcal{Q}^{out})$.

Further: for any $\mathcal{Q}^{out} \in \mathcal{T}^{out}$, we say f *extends* $\mathcal{Q}^{out}$ and denote $f \vdash \mathcal{Q}^{out}$, if $f(L) = Y$ for all $(L, Y) \in \mathcal{Q}^{out}$; we say f *extends* $\mathcal{Q}^{mid}(\mathcal{Q}^{out})$ and denote $f \vdash \mathcal{Q}^{mid}(\mathcal{Q}^{out})$, if $f(L) = Y$ for all $(L, Y) \in \mathcal{Q}^{mid}(\mathcal{Q}^{out})$. It is easy to see that $f \vdash \mathcal{Q}^{out}$ if and only if $\mathcal{Q}^{out}(f) = \mathcal{Q}^{out}$.

With the above notations and the definition of $\mathcal{UFLM}^{(4)}[f]$, we have

$$
\begin{aligned}
&\Pr_f\left(\mathcal{UFLM}^{(4)}[f] \vdash \mathcal{Q} \mid \neg\mathsf{Bad}(f)\right) \\
&= \sum_{\mathcal{Q}^{out} \in \mathcal{T}^{out}} \Pr_f\left(f \vdash \mathcal{Q}^{out} \mid \neg\mathsf{Bad}(f)\right) \\
&\qquad\qquad \cdot \Pr_f\left(f \vdash \mathcal{Q}^{mid}(\mathcal{Q}^{out}) \mid f \vdash \mathcal{Q}^{out} \wedge \neg\mathsf{Bad}(f)\right) \\
&\geq \underbrace{\sum_{\mathcal{Q}^{out} \in \mathcal{T}^{out}_{good}} \Pr_f\left(f \vdash \mathcal{Q}^{out} \mid \neg\mathsf{Bad}(f)\right)}_{=1} \\
&\qquad\qquad \cdot \Pr_f\left(f \vdash \mathcal{Q}^{mid}(\mathcal{Q}^{out}) \mid f \vdash \mathcal{Q}^{out} \wedge \neg\mathsf{Bad}(f)\right)
\end{aligned}
$$

For any $\mathcal{Q}^{out} \in \mathcal{T}^{out}_{good}$, the condition $\neg$(B-1) ensures that

$$
\{L|\exists Y : (L,Y) \in \mathcal{Q}^{mid}(\mathcal{Q}^{out})\} \bigcap \underbrace{\{L'|\exists Y' : (L',Y') \in \mathcal{Q}^{out}\}}_{=\mathsf{ExtF}} = \emptyset.
$$

Thus

$$
\Pr_f\left(f \vdash \mathcal{Q}^{mid}(\mathcal{Q}^{out}) \mid f \vdash \mathcal{Q}^{out} \wedge \neg\mathsf{Bad}(f)\right) = \frac{1}{(2^n)^{|\mathcal{Q}^{mid}(\mathcal{Q}^{out})|}}.
$$

In the next paragraph, we show $|\mathcal{Q}^{mid}(\mathcal{Q}^{out})| = 2q$ to complete the proof.

$\mathcal{Q}^{mid}(\mathcal{Q}^{out})$ *has $2q$ values.* By the definitions, for any $\mathcal{Q}^{out} \in \mathcal{T}^{out}_{good}$, there exists f such that $\mathsf{Bad}(f)$ does not hold, and $\mathcal{Q}^{out}(f) = \mathcal{Q}^{out}$. It remains to be proven that the values $L_1^{(1)}, \cdots, L_1^{(q)}, L_2^{(1)}, \cdots, L_2^{(q)}$ in $\mathcal{Q}^{mid}(\mathcal{Q}^{out})$ (see equations (12) and (13)) are $2q$ distinct values. Conditioned on $\mathsf{Bad}(f)$, it holds:

(i) The q inputs $L_1^{(1)}, ..., L_1^{(q)}$ of the second round function are distinct. Consider any two distinct indices $i, j \in \{1, \cdots, q\}$. If $L_0^{(i)} \neq L_0^{(j)}$, then $L_1^{(i)} = L_1^{(j)}$ would imply (B-2) and contradict the assumption that $\mathsf{Bad}(f)$ is not fulfilled. If $L_0^{(i)} = L_0^{(j)}$, then $L_1^{(i)} = L_1^{(j)}$ would imply $\varphi_{\mathrm{UR}} \cdot R_0^{(i)} = \varphi_{\mathrm{UR}} \cdot R_0^{(j)}$ and contradicts the invertibility of φ_{UR}.

(ii) Similarly to the claim (i), the q inputs $L_2^{(1)}, \cdots, L_2^{(q)}$ of the third round function are distinct by $\neg$(B-3) and by the invertibility of $(\varphi^{-1})_{\mathrm{UR}}$.

(iii) Finally, every pair of $L_2^{(i)}$ and $L_3^{(j)}$ are distinct, otherwise, (B-4) is fulfilled.

Furthermore, conditioned on $f \vdash \mathcal{Q}^{out}$, the $2q$ function values

$$
f(L_1^{(1)}), \cdots, f(L_1^{(q)}), f(L_2^{(1)}), \cdots, f(L_2^{(q)})
$$

remain undetermined, otherwise, (B-1) is fulfilled. Therefore, for each $i \in \{1, 2, \cdots, q\}$ we have $\Pr\left(f(L_1^{(i)}) = Y_1^{(i)} \wedge f(L_2^{(i)}) = Y_2^{(i)}\right) = \frac{1}{(2^n)^2}$. For any good transcript $\mathcal{Q}$,

$$\frac{\mu(\mathcal{Q})}{\nu(\mathcal{Q})} = \Pr\Big(f \leftarrow \mathcal{F}(\mathbb{F}_2^n, \mathbb{F}_2^n) : \mathcal{UFLM}^{(4)}[f] \vdash \mathcal{Q}\Big) \Big/ \left(\prod_{i=0}^{q-1} \frac{1}{2^{2n} - i}\right)$$

$$\geq \frac{\Pr_f\Big(\mathcal{UFLM}^{(4)}[f] \vdash \mathcal{Q} \mid \neg\mathsf{Bad}(f)\Big) \cdot \Big(1 - \Pr_f\big(\mathsf{Bad}(f)\big)\Big)}{\prod_{i=0}^{q-1} \frac{1}{2^{2n}-i}}$$

$$\geq \left(1 - \frac{6q^2}{2^n}\right) \left(\frac{1}{(2^n)^{2q}}\right) \Big/ \left(\prod_{i=0}^{q-1} \frac{1}{2^{2n} - i}\right)$$

$$\geq \left(1 - \frac{6q^2}{2^n}\right) \left(1 - \frac{q^2}{2^{2n}}\right) \geq 1 - \frac{6q^2}{2^n} - \frac{q^2}{2^{2n}}.$$

Gathering this and (4) yields the claim of (5), which completes the proof.

5.3 CCA Security for $\mathcal{UFLM}^{(4)}[f_1, f_2, f_3, f_4]$

Using a similar method, the CCA-security result of the 4-round construction $\mathcal{UFLM}^{(4)}[f_1, f_2, f_3, f_4]$ is formally stated as follows.

Theorem 5. *Assume $q \leq 2^n/2$. Then, for the 4-round idealized construction $\mathcal{UFLM}^{(4)}[f_1, f_2, f_3, f_4]$ defined upon four independent secret random functions f_1, f_2, f_3, f_4 and an invertible linear transformation φ, it holds*

$$\mathsf{Adv}_{\mathcal{UFLM}^{(4)}}^{\mathrm{CCA}}(q) \leq \frac{q^2}{2^n} + \frac{q^2}{2^{2n}}. \tag{14}$$

We have the following corollaries based on the Theorem 4 and Theorem 5.

Corollary 6. *The CCA security of the 4-round Lai-Massey construction is superior to that of the 4-round Feistel construction when utilizing the same f-function in each round.*

Corollary 7. *If the linear transformation φ of a 4-round* **UFLM** *instance adopts $O - I$ block matrix, then its CCA security is identical to the 4-round Feistel construction.*

5.4 Using Random Permutations: Security of $\mathcal{UFLM}^{(4)}[p]$ and $\mathcal{UFLM}^{(4)}[p_1, p_2, p_3, p_4]$

While Theorems 4 and 5 focused on **UFLM** constructions defined upon random (non-invertible) functions, the results can be easily transferred to random permutation-based **UFLM** constructions using the PRP/PRF Switching Lemma. Concretely, the PRP/PRF Switching Lemma states that the advantage of a q-query distinguisher against a random permutation and a random function is at most $\frac{q(q-1)}{2 \cdot 2^n} \leq \frac{q^2}{2 \cdot 2^n}$ [11]. By this, we consider the following two scenarios.

- Consider the 4-round idealized construction $\mathcal{UFLM}^{(4)}[p]$ defined upon **a secret random permutation** p and a good linear transformation φ in the sense of Definition 2). When an adversary makes q queries, the construction makes at most $4q$ queries to p. Therefore, replacing p with a secret random function f yields a gap of at most $\frac{(4q)^2}{2 \cdot 2^n} \leq \frac{8q^2}{2^n}$. Therefore, when $q \leq 2^n/2$, the 4-round random permutation-based construction $\mathcal{UFLM}^{(4)}[p]$ has

$$\mathsf{Adv}^{\mathrm{CCA}}_{\mathcal{UFLM}^{(4)}}(q) \leq \frac{14q^2}{2^n} + \frac{q^2}{2^{2n}}. \tag{15}$$

- Consider the 4-round idealized construction $\mathcal{UFLM}^{(4)}[p_1, p_2, p_3, p_4]$ defined upon **four independent secret random permutations** p_1, p_2, p_3, p_4 and an invertible linear transformation φ. When an adversary makes q queries, the construction makes at most q queries to each of p_1, p_2, p_3, p_4. Therefore, replacing p_1, p_2, p_3, p_4 with four independent secret random functions yields a gap of at most $4 \times \frac{q^2}{2 \cdot 2^n} \leq \frac{2q^2}{2^n}$. Therefore, when $q \leq 2^n/2$, the 4-round random permutation-based construction $\mathcal{UFLM}^{(4)}[p_1, p_2, p_3, p_4]$ has

$$\mathsf{Adv}^{\mathrm{CCA}}_{\mathcal{UFLM}^{(4)}}(q) \leq \frac{3q^2}{2^n} + \frac{q^2}{2^{2n}}. \tag{16}$$

5.5   Proposal for a UFLM Instance

We provide a **UFLM** instance **kite** that employs an orthomorphism permutation acting on two branches. This instance is superior to the Feistel structure regarding the number of rounds of existing impossible differentials and zero correlation linear hulls. Besides, the CCA security of **kite** is identical to that of the Feistel structure. Additionally, the implementation cost of **kite** may be higher than that of the Feistel structure but lower than that of the Lai-Massey structure, assuming that these three structures adopt the same f-function.

Denote the input and output of the i-th round of **kite** as (L_{i-1}, R_{i-1}) and $(L_i, R_i) \in \mathbb{F}_2^{2n}$, respectively, where n is the size of each branch. Let f be a bijective mapping over $\mathbb{F}_2^n$. Then a single-round process of **kite** is defined as a mapping over $\mathbb{F}_2^{2n}$, as shown in Fig. 6, i.e.,

$$\begin{cases} L_i = R_{i-1} \oplus f(L_{i-1}), \\ R_i = L_{i-1} \oplus R_{i-1} \oplus f(L_{i-1}). \end{cases}$$

According to the results of the framework **UFLM** regarding cryptanalysis and CCA security presented in the previous subsections, the following propositions hold for the instance **kite**.

Proposition 1. *There exists a 4-round impossible differential* $(0, \alpha) \to (\alpha, \alpha)$ *for **kite** where* $\alpha \neq 0$.

Proposition 2. *There exists a 4-round zero correlation linear hull* $(\gamma, 0) \to (\gamma, \gamma)$ *for **kite** where* $\gamma \neq 0$, *which leads to a 4-round integral distinguisher.*

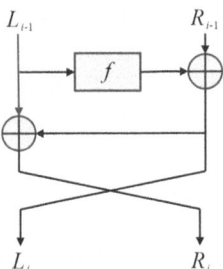

Fig. 6. The UFLM instance **kite**

Proposition 3. *The 4-round construction* **kite**$^{(4)}$ *is CCA-secure when utilizing different f-functions in each round.*

Based on the $\mathcal{U}$-method [13], we cannot find 5-round impossible differentials and 5-round zero correlation linear hulls for **kite**. Consequently, both the longest impossible differential and the longest zero correlation linear hull for **kite** are limited to 4 rounds, which is an improvement over the Feistel structure.

Proposition 4. *There do not exist 5-round impossible differentials and 5-round zero correlation linear hulls for* **kite** *based on the* $\mathcal{U}$-*method.*

6   Conclusion and Future Work

This paper proposes a framework **UFLM** for reassessing the security of Feistel and Lai-Massey structures. This framework demonstrates that the linear transformation employed in a cipher structure is directly related to its security, which provides guidance for the design and cryptanalysis of 2-branch cipher structures. Furthermore, we analyze the security of **UFLM** instances from two perspectives: cryptanalysis and provable security settings. The results indicate that the Lai-Massey structure does benefit from the orthomorphic permutation in both aspects. Additionally, we present a **UFLM** instance better than the Feistel structure regarding the number of rounds of several distinguishers, which may be beneficial for future block cipher designs.

Nevertheless, several issues deserve further consideration. This paper focuses on the case where **UFLM** instances employ bijective f-functions when evaluating the number of rounds of distinguishers. The case of non-invertible f-functions remains a topic for subsequent investigation. Furthermore, consider a scenario where an f-function is composed of multiple smaller components, such as S-boxes. In this case, it is feasible to convert a **UFLM** instance into an alternative structure with several smaller-scale f-functions, for example, a Type-II generalized Feistel structure employing a linear transformation. This paper concentrates on structures with two branches and one f-function. The security evaluation of structures with multiple branches, multiple f-functions, and a linear transformation will be explored in future research.

Acknowledgments. The authors sincerely thank the anonymous reviewers for their valuable comments and suggestions. Chun Guo was supported by the National Natural Science Foundation of China (Grant 62372274). Chao Li was supported by the National Natural Science Foundation of China (Grant 62172427).

References

1. Biham, E., Biryukov, A., Shamir, A.: Cryptanalysis of skipjack reduced to 31 rounds using impossible differentials. In: Stern, J. (ed.) EUROCRYPT 1999. LNCS, vol. 1592, pp. 12–23. Springer (1999). https://doi.org/10.1007/3-540-48910-X_2
2. Bogdanov, A., Rijmen, V.: Linear hulls with correlation zero and linear cryptanalysis of block ciphers. Des. Codes Cryptogr. **70**(3), 369–383 (2014). https://doi.org/10.1007/s10623-012-9697-z
3. Cauchois, V., Gomez, C., Thomas, G.: General diffusion analysis: how to find optimal permutations for generalized type-ii Feistel schemes. IACR Trans. Symmetric Cryptol. **2019**(1), 264–301 (2019). https://doi.org/10.13154/tosc.v2019.i1.264-301
4. Chen, S., Steinberger, J.P.: Tight security bounds for key-alternating ciphers. In: Nguyen, P.Q., Oswald, E. (eds.) EUROCRYPT 2014. LNCS, vol. 8441, pp. 327–350. Springer (2014). https://doi.org/10.1007/978-3-642-55220-5_19
5. Cogliati, B., et al.: Provable security of (tweakable) block ciphers based on substitution-permutation networks. In: Shacham, H., Boldyreva, A. (eds.) CRYPTO 2018. LNCS, vol. 10991, pp. 722–753. Springer (2018). https://doi.org/10.1007/978-3-319-96884-1_24
6. Coppersmith, D.: The data encryption standard (DES) and its strength against attacks. IBM J. Res. Dev. **38**(3), 243–250 (1994). https://doi.org/10.1147/rd.383.0243
7. Daemen, J., Rijmen, V.: The Design of Rijndael. Information Security and Cryptography, 1st edn. Springer, Berlin, Heidelberg (2002). https://doi.org/10.1007/978-3-662-04722-4
8. Derbez, P., Fouque, P., Lambin, B., Mollimard, V.: Efficient search for optimal diffusion layers of generalized Feistel networks. IACR Trans. Symmetric Cryptol. **2019**(2), 218–240 (2019). https://doi.org/10.13154/tosc.v2019.i2.218-240
9. Feistel, H., Notz, W.A., Smith, J.L.: Some cryptographic techniques for machine-to-machine data communications. Proc. IEEE **63**(11), 1545–1554 (1975). https://doi.org/10.1109/PROC.1975.10005
10. Guo, C., Luo, Y., Xiao, C.: On the sequential indifferentiability of the Lai-Massey construction. Des. Codes Cryptogr. **92**(6), 1637–1661 (2024). https://doi.org/10.1007/s10623-024-01361-6
11. Hoang, V.T., Tessaro, S.: Key-alternating ciphers and key-length extension: exact bounds and multi-user security. In: Robshaw, M., Katz, J. (eds.) CRYPTO 2016. LNCS, vol. 9814, pp. 3–32. Springer (2016). https://doi.org/10.1007/978-3-662-53018-4_1
12. Junod, P., Vaudenay, S.: FOX : a new family of block ciphers. In: Handschuh, H., Hasan, M.A. (eds.) SAC 2004. LNCS, vol. 3357, pp. 114–129. Springer (2004). https://doi.org/10.1007/978-3-540-30564-4_8
13. Kim, J., Hong, S., Sung, J., Lee, C., Lee, S.: Impossible differential cryptanalysis for block cipher structures. In: Johansson, T., Maitra, S. (eds.) INDOCRYPT 2003. LNCS, vol. 2904, pp. 82–96. Springer (2003). https://doi.org/10.1007/978-3-540-24582-7_6

14. Lai, X., Massey, J.L.: A proposal for a new block encryption standard. In: Damgård, I. (ed.) EUROCRYPT '90. LNCS, vol. 473, pp. 389–404. Springer (1990). https://doi.org/10.1007/3-540-46877-3_35

15. Lai, X., Massey, J.L., Murphy, S.: Markov ciphers and differential cryptanalysis. In: Davies, D.W. (ed.) EUROCRYPT '91. LNCS, vol. 547, pp. 17–38. Springer (1991). https://doi.org/10.1007/3-540-46416-6_2

16. Li, X., Jin, C., Fu, F.: Improved results of impossible differential cryptanalysis on reduced FOX. Comput. J. **59**(4), 541–548 (2016). https://doi.org/10.1093/comjnl/bxv073

17. Liu, J., et al.: New wine old bottles: Feistel structure revised. IEEE Trans. Inf. Theory **69**(3), 2000–2008 (2023). https://doi.org/10.1109/TIT.2022.3223139

18. Luby, M., Rackoff, C.: How to construct pseudorandom permutations from pseudorandom functions. SIAM J. Comput. **17**(2), 373–386 (1988). https://doi.org/10.1137/0217022

19. Luo, Y., Lai, X., Wu, Z., Gong, G.: A unified method for finding impossible differentials of block cipher structures. Inf. Sci. **263**, 211–220 (2014). https://doi.org/10.1016/j.ins.2013.08.051

20. Luo, Y., Lai, X., Zhou, Y.: Generic attacks on the Lai-Massey scheme. Des. Codes Cryptogr. **83**(2), 407–423 (2017). https://doi.org/10.1007/s10623-016-0235-2

21. Mao, S., Guo, T., Wang, P., Hu, L.: Quantum attacks on Lai-Massey structure. In: Cheon, J.H., Johansson, T. (eds.) PQCrypto 2022. LNCS, vol. 13512, pp. 205–229. Springer (2022). https://doi.org/10.1007/978-3-031-17234-2_11

22. Nandi, M.: On the optimality of non-linear computations of length-preserving encryption schemes. In: Iwata, T., Cheon, J.H. (eds.) ASIACRYPT 2015. LNCS, vol. 9453, pp. 113–133. Springer (2015). https://doi.org/10.1007/978-3-662-48800-3_5

23. Nyberg, K.: Generalized Feistel networks. In: Kim, K., Matsumoto, T. (eds.) ASIACRYPT '96. LNCS, vol. 1163, pp. 91–104. Springer (1996). https://doi.org/10.1007/BFb0034838

24. Patarin, J.: The coefficients H technique. In: Avanzi, R.M., Keliher, L., Sica, F. (eds.) SAC 2008. LNCS, vol. 5381, pp. 328–345. Springer (2008). https://doi.org/10.1007/978-3-642-04159-4_21

25. Shannon, C.E.: Communication theory of secrecy systems. Bell Syst. Tech. J. **28**(4), 656–715 (1949). https://doi.org/10.1002/j.1538-7305.1949.tb00928.x

26. Shirai, T., Shibutani, K., Akishita, T., Moriai, S., Iwata, T.: The 128-Bit Blockcipher CLEFIA (extended abstract). In: Biryukov, A. (ed.) FSE 2007. LNCS, vol. 4593, pp. 181–195. Springer (2007). https://doi.org/10.1007/978-3-540-74619-5_12

27. Sun, B., et al.: Links among impossible differential, integral and zero correlation linear cryptanalysis. In: Gennaro, R., Robshaw, M. (eds.) CRYPTO 2015. LNCS, vol. 9215, pp. 95–115. Springer (2015). https://doi.org/10.1007/978-3-662-47989-6_5

28. Sun, B., et al.: Feistel-like structures revisited: classification and cryptanalysis. In: Reyzin, L., Stebila, D. (eds.) CRYPTO 2024. LNCS, vol. 14923, pp. 275–304. Springer (2024). https://doi.org/10.1007/978-3-031-68385-5_9

29. Vaudenay, S.: On the Lai-Massey scheme. In: Lam, K., Okamoto, E., Xing, C. (eds.) ASIACRYPT '99. LNCS, vol. 1716, pp. 8–19. Springer (1999). https://doi.org/10.1007/978-3-540-48000-6_2

30. Wei, Y., Li, P., Sun, B., Li, C.: Impossible differential cryptanalysis on Feistel Ciphers with SP and SPS round functions. In: Zhou, J., Yung, M. (eds.) ACNS 2010. LNCS, vol. 6123, pp. 105–122 (2010). https://doi.org/10.1007/978-3-642-13708-2_7

31. Wu, S., Wang, M.: Automatic search of truncated impossible differentials for word-oriented block Ciphers. In: Galbraith, S.D., Nandi, M. (eds.) INDOCRYPT 2012. LNCS, vol. 7668, pp. 283–302. Springer (2012). https://doi.org/10.1007/978-3-642-34931-7_17

32. Wu, Z., Luo, Y., Lai, X., Zhu, B.: Improved cryptanalysis of the FOX block Cipher. In: Chen, L., Yung, M. (eds.) INTRUST 2009. LNCS, vol. 6163, pp. 236–249. Springer (2009). https://doi.org/10.1007/978-3-642-14597-1_15

33. Yun, A., Park, J.H., Lee, J.: On Lai-Massey and quasi-Feistel ciphers. Des. Codes Cryptogr. **58**(1), 45–72 (2011). https://doi.org/10.1007/s10623-010-9386-8

34. Zheng, Y., Matsumoto, T., Imai, H.: On the Construction of Block Ciphers provably secure and not relying on any unproved hypotheses. In: Brassard, G. (ed.) CRYPTO '89. LNCS, vol. 435, pp. 461–480. Springer (1989). https://doi.org/10.1007/0-387-34805-0_42

ASURA: An Efficient Large-State Tweakable Block Cipher for ARM Environment

Atsushi Tanaka[1], Rentaro Shiba[2,3], Kosei Sakamoto[1,2], Mostafizar Rahman[1], Takuro Shiraya[1], and Takanori Isobe[1(✉)]

[1] University of Hyogo, Kobe, Japan
atsushi.mayh1105@gmail.com, takanori.isobe@ai.u-hyogo.ac.jp
[2] Mitsubishi Electric Corporation, Kamakura, Japan
sakamoto.kosei@dc.mitsubishielectric.co.jp
[3] Nagoya University, Nagoya, Japan
shiba.rentaro.k7@s.mail.nagoya-u.ac.jp

Abstract. In this paper, we propose a large-state tweakable block cipher dubbed ASURA, which is optimized for software implementation using AES instructions on AArch64 processors. ASURA takes a 256-bit plaintext, 256-bit secret key, and 128-bit tweak as inputs. Unlike existing AES-based ciphers such as Ghidle and Pholkos, the ASURA round function integrates AES rounds and their inverse operations. For the linear layer of the round function, we employ an efficient byte-wise shuffle using only a single vuzpq_u8 instruction, enabling faster diffusion than the 32-bit shuffle in Pholkos-256 and the matrix multiplication over $GF(2^{64})$ in Ghidle-256. Leveraging NEON instructions in AArch64, we optimize the balance of AES and shuffle layers for both security and performance. Finally, we show that ASURA outperforms Pholkos-256 and Ghidle-256 by 64% and 74%, respectively, in single-block encryption and by over 40% in parallel processing on AArch64, and remains competitive even in x86 environments.

Keywords: tweakable block cipher · AES-NI · wide-block encryption · NEON · SIMD

1 Introduction

1.1 Background

With the expansion of the Internet of Things (IoT), not only general-purpose and personal computers, but also mobile devices such as smartphones and tablets, as well as household appliances, medical devices, sensing equipment, and other edge devices connected to the internet, are becoming increasingly prevalent. ARM architectures are often used for CPUs of these edge devices. Particularly in mobile devices that require advanced processing, processors of the AArch64 architecture are often embedded in their system-on-chips (SoCs). The AArch64

© The Author(s), under exclusive license to Springer Nature Switzerland AG 2025
S. Mukhopadhyay and P. Stănică (Eds.): INDOCRYPT 2024, LNCS 15495, pp. 143–164, 2025.
https://doi.org/10.1007/978-3-031-80308-6_7

architecture, which is a 64-bit variant of ARM architectures, has attracted attention because of its high customizability and its high-end performance. In recent years, even Apple Inc.'s personal computers employ original architectures based on the AArch64 instead of x86_64 architectures.

In symmetric cryptography, the AES block cipher is widely employed. On the AArch64, dedicated circuits for executing AES operations allow for high-speed processing. AES is a block cipher with a block length of 128 bits and supports key lengths of 128, 192, or 256 bits. While there are no significant security concerns with key lengths of 128 bits and above so far, the 128-bit block length has been identified as a practical limitation when used with certain modes of operation. Specifically, encryption modes such as CBC, GCM, and OCB, which are subject to Birthday bound security, cannot securely encrypt more than 2^{64} blocks of data with a single key. Amazon Web Services has reported encrypting 2^{64} blocks in just two weeks [14], and with the anticipated growth in users and the corresponding increase in data processing, a 128-bit block size is becoming insufficient. As a result, there is a demand for block ciphers with larger block sizes (e.g., 256 bits or more) to provide stronger Birthday bound security and support long-term use.

Two large-state block ciphers, Pholkos [8] and Ghidle [19], have been proposed. Both employ the AES New Instructions Set (AES-NI), a set of instructions capable of executing the AES round function and integrated into modern CPUs for acceleration through hardware design. Pholkos and Ghidle both feature a substitution-permutation (SP) structure composed of an AES layer and a linear layer. In the AES layer, the input state, either 256-bit or 512-bit, is divided into 128-bit subblocks, and the AES round function is applied to each subblock in parallel. The linear layer involves shuffle operations or matrix operations using SIMD instructions.

1.2   Limitations

Linear layers of existing large-state block ciphers are optimized for Intel's x86 (x86_64) architectures and not for ARM. Additionally, the types of AES instructions differ between x86 and AArch64, making it unclear whether the existing AES layers and linear layers are optimal for AArch64. Another issue with existing large-state block ciphers is that 32-bit wise operations are employed in their linear layers, which may lead to distinguishers exploiting insufficient diffusion, such as impossible differentials, over extended rounds. Consequently, the optimal structure for large-state block ciphers that balance security and implementation efficiency on AArch64 remains unclear.

1.3   Our Contribution

In this paper, we propose a large-state tweakable block cipher, dubbed ASURA, optimized for software implementation on AArch64 processors. ASURA is a tweakable block cipher taking a 256-bit block, a 256-bit secret key, and a 128-bit tweak, as the inputs. The round function of ASURA comprises AES round functions and a linear layer. Although the round function is similar to existing software-oriented algorithms such as Pholkos, Ghidle, Haraka v2 [17], and AESQ [7], it differs significantly in that its AES layer incorporates inverse operations of the AES round function.

Besides, we employ a byte-wise shuffle operation in the linear layer of the ASURA round function, which can be efficiently implemented using a single vuzpq_u8 operation. Compared to the 32-bit or 64-bit shuffles used in Pholkos-256 and Ghidle-256, ASURA's linear layer offers faster diffusion and performs more efficiently due to the fact that it can be realized with only one instruction. Furthermore, unlike AES-NI in x86, the NEON instruction set in AArch64 includes a specific instruction for executing only the MixColumns operation. This allows for consideration of AES layer operations that are not possible in x86 architectures.

To determine the optimal combination of these AES and linear layers, we derive the number of instructions needed to achieve security against differential attacks, impossible differential attacks, and integral attacks. As a result, we demonstrate that the AES layer, including three decryption AES round functions and one encryption AES round function, is optimal in terms of implementation. We demonstrate that the single-block encryption speed of ASURA is approximately 64% and 74% faster than Pholkos-256 and Ghidle-256, respectively, on AArch64. In parallel processing, the ASURA encryption speed is over 40% faster than that of both Pholkos-256 and Ghidle-256. Even in x86, which is not the primary target environment, the performance of ASURA is competitive with or outperforms Pholkos-256 and Ghidle-256.

2 Preliminaries

2.1 AES Instructions

SIMD (Single Instruction / Multiple Data) is a processing model that processes multiple data with a single instruction. Modern processors often support SIMD instruction sets that enable implementing complex operations with few instructions. A well-known operation supported by SIMD instructions is the computation for the most widely used block cipher, AES.

In the x86 (x86_64) architecture, various SIMD instructions are supported, but the instructions related to AES are specifically defined in a group of instructions called AES-NI (AES New Instructions set). AES-NI includes instructions for AES round functions and some operations of key scheduling functions. Specifically, AES-NI includes AESENC executing the round function, AESENCLAST executing the final encryption round. AES-NI also includes inverse operations such as AESDEC, AESDECLAST, and AESIMC. Let SubBytes, ShiftRows, MixColumns and AddRoundKey be SB, SR, MC, AK, respectively. In addition, let InvSubBytes, InvShiftRows, and InvMixColumns, be SB^{-1}, SR^{-1}, and MC^{-1}, respectively. Each of these instructions executes the following operation:

$$\mathsf{AESENC} = \mathrm{AK} \circ \mathrm{MC} \circ \mathrm{SB} \circ \mathrm{SR}, \quad \mathsf{AESENCLAST} = \mathrm{AK} \circ \mathrm{SB} \circ \mathrm{SR},$$
$$\mathsf{AESDEC} = \mathrm{AK} \circ \mathrm{MC}^{-1} \circ \mathrm{SB}^{-1} \circ \mathrm{SR}^{-1}, \quad \mathsf{AESDECLAST} = \mathrm{AK} \circ \mathrm{SB}^{-1} \circ \mathrm{SR}^{-1},$$
$$\mathsf{AESIMC} = \mathrm{MC}^{-1}.$$

The AArch64 architecture supports a SIMD instruction set called NEON and includes instructions realizing AES operations like AES-NI. Instructions related to AES are included in a specific group of NEON instructions called the Cryptography Extension, along with instructions for implementing cryptographic algorithms such as SHA1 and SHA256. Although the AES instructions in AES-NI in x86 and NEON Cryptography Extension in AArch64 are equipped for the common purpose of AES software implementation, they differ in the types and configurations of instructions. Unlike AES-NI, NEON cryptography extension does not include instructions related to the key scheduling function and only has four instructions related to the round function and its inverse.

The AES instructions defined in AArch64's NEON Cryptography Extension are AESE, AESMC, AESD, and AESIMC. The realized operation of AESIMC in NEON is the same as the AESIMC in AES-NI. Other instructions can be expressed as follows:

$$\mathsf{AESE} = \mathrm{SB} \circ \mathrm{SR} \circ \mathrm{AK} \tag{1}$$
$$\mathsf{AESMC} = \mathrm{MC} \tag{2}$$
$$\mathsf{AESD} = \mathrm{SB}^{-1} \circ \mathrm{SR}^{-1} \circ \mathrm{AK} \tag{3}$$
$$\mathsf{AESIMC} = \mathrm{MC}^{-1}. \tag{4}$$

AESE and AESMC are used for the encryption round function, and AESD and AESIMC are used for the decryption round function. AESE and AESD are similar to AESENCLAST and AESDECLAST, respectively. However, there is no instruction similar to AESMC in x86. Thus, in AArch64, it is possible to execute a MC independently of other operations by AESMC, while it is impossible in x86.

2.2   Existing Large-State Block Ciphers

Pholkos. Pholkos [8] is a family of large-state tweakable block ciphers that employ 256-bit secret key. Pholkos-256, Pholkos-512 and Pholkos-1024 support 256/512/1024-bit block lengths, respectively. It consists of parallel executions of 2-round AES and a 32-bit word-wise shuffle operation. The construction is dedicated to the software implementation using AES-NI on x86. Besides, the shuffle operation can be implemented by vpblendd instructions, which is a class of shuffle instructions on x86. The designers claim that vpblendd is more efficient than other shuffle instructions such as punpckldq and punpckhdq for the linear layer of Haraka v2 permutation [17] if the number of instances for parallel execution is large.

Ghidle. Ghidle [19] is a family of large-state block ciphers, which is designed so that the performance for the encryption and decryption are equivalent and both are efficient. Ghidle-256 and Ghidle-512 support a 256 and a 512-bit block, respectively. Both variants use a 256-bit secret key. The AES layer consists of the operations which can be realized by AESENC or AESENCLAST, *i.e.*, it includes the AES round function for final rounds. In the linear layer, an almost MDS matrix employed in Midori is used. The linear layer for Ghidle-512, a 512-bit block variant, can be implemented by only 6 XOR instructions. The number of ports that accept micro-operations generated by XOR instruction is larger than other shuffle instructions. Thus, in many cases, the Ghidle-512 is faster than Pholkos-512 on x86 processors. Ghidle-256 is not efficient because it uses punpckhdq and punpckldq in addition to XOR instructions in the linear layers. However, since the round function for both variants is friendly with the software implementation with NEON instructions, both are efficient on AArch64 processors, even though they are not optimized for such environments.

2.3   Limitations of Existing Schemes

The existing schemes are efficient when implemented using SIMD instructions, including AES instructions. However, there remain some challenges or areas for improvement.

Pholkos. Pholkos uses 32-bit shuffle instructions, resulting in slow diffusion property and requiring many rounds to guarantee sufficient security. Besides, the round function is not optimized for implementation with NEON instructions. When implementing on AArch64, the second key addition requires additional XOR instructions to match the x86 implementation shown in Sect. 1. As a result, on mobile devices with AArch64 processors, Pholkos-256 performs more slowly than Ghidle-256, as demonstrated in [19]. Additionally, the decryption process is significantly slower than the encryption.

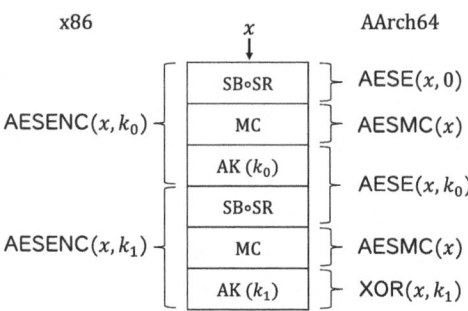

Fig. 1. Differences between Pholkos-256's AES instructions implemented on x86 and AArch64

Ghidle. The AES layer of the Ghidle round function can be implemented using 2 or 4 parallel instances of sequential executions of two AESE and one AESMC instructions, resulting in a total of three AES instructions. In contrast, Pholkos, which includes two MixColumns operations, requires the sequential execution of two AESE and two AESMC instructions, for a total of four instructions. Therefore, Ghidle is more efficient than Pholkos on AArch64 processors.

However, the Ghidle construction is not fully optimized for AArch64 processors, leaving room for improvements. Specifically, it is worth exploring the adoption of a linear layer that incorporates operations optimized for AArch64-specific instructions or a design that further reduces the number of instructions in the AES layer on target processors.

Moreover, Ghidle-256 is not efficient on x86 processors due to the number of instructions required for its linear layer and is not a tweakable block cipher. These limitations motivate the development of new software-oriented tweakable block ciphers.

3   Specification

This section specifies a large-state tweakable block cipher ASURA. ASURA is 256-bit block cipher employing a 256-bit key K and a 128-bit tweak T.

3.1   The Round Function

The ASURA's round function $R : \mathbb{F}_2^{256} \times \mathbb{F}_2^{256} \times \mathbb{F}_2^{256} \rightarrow \mathbb{F}_2^{256}$ operates 256-bit input $x = x_0 || x_1 \in \mathbb{F}_2^{256}$ and two round keys $k_0, k_1 \in \mathbb{F}_2^{256}$ as follows:

$$R(x, k_0, k_1) = perm_{256}(F_\gamma(x_0, k_0), F_\delta(x_1, k_1)).$$

The input is divided into two 128-bit substates $x_0, x_1 \in \mathbb{F}_2^{128}$. After that, these two substates are processed by two functions : $F_\gamma : \mathbb{F}_2^{128} \times \mathbb{F}_2^{256} \rightarrow \mathbb{F}_2^{128}$ and $F_\delta : \mathbb{F}_2^{128} \times \mathbb{F}_2^{256} \rightarrow \mathbb{F}_2^{128}$ in the AES layer. Finally, the output of the AES layer is processed by $perm_{256}$, which is a byte shuffle operation in the linear layer. In the last round, $perm_{256}$ is not applied. We refer to the round function for the last round as $R_{last} : \mathbb{F}_2^{256} \times \mathbb{F}_2^{256} \times \mathbb{F}_2^{256} \rightarrow \mathbb{F}_2^{256}$, and it is defined as

$$R_{last}(x, k_0, k_1) = F_\gamma(x_0, k_0) || F_\delta(x_1, k_1).$$

In the following, we explain the detail for these two layers constituting the ASURA round function.

The AES Layer. The AES layer is realized by the parallel execution of the following two operations.

$$\text{MC} \circ \text{SB} \circ \text{SR} \circ \text{AK} \circ \text{MC}^{-1} \circ \text{SB}^{-1} \circ \text{SR}^{-1} \circ \text{AK} \tag{5}$$

$$\text{MC}^{-1} \circ \text{SB}^{-1} \circ \text{SR}^{-1} \circ \text{AK} \circ \text{MC}^{-1} \circ \text{SB}^{-1} \circ \text{SR}^{-1} \circ \text{AK} \tag{6}$$

In the final round, MC and MC^{-1}, which are linear transformations, are not needed. Using operations realized by NEON instructions shown in Eq. (1)–Eq.

(4) as functions such that $\mathsf{AESE} : \mathbb{F}_2^{128} \times \mathbb{F}_2^{128} \to \mathbb{F}_2^{128}$, $\mathsf{AESMC} : \mathbb{F}_2^{128} \to \mathbb{F}_2^{128}$, $\mathsf{AESD} : \mathbb{F}_2^{128} \times \mathbb{F}_2^{128} \to \mathbb{F}_2^{128}$ and $\mathsf{AESIMC} : \mathbb{F}_2^{128} \to \mathbb{F}_2^{128}$, we can rewrite Eq. (5) and Eq. (6) as functions F_γ and F_δ, respectively. F_γ and F_δ are as follows:

$$F_\gamma(x,k) = \mathsf{AESMC}(\mathsf{AESE}(\mathsf{AESIMC}(\mathsf{AESD}(x,k_0)),k_1)) \tag{7}$$

$$F_\delta(x,k) = \mathsf{AESIMC}(\mathsf{AESD}(\mathsf{AESIMC}(\mathsf{AESD}(x,k_0)),k_1)). \tag{8}$$

And their inverse are as follows:

$$F_\gamma^{-1}(x,k) = \mathsf{XOR}(\mathsf{AESE}(\mathsf{AESMC}(\mathsf{XOR}(\mathsf{AESD}(\mathsf{AESIMC}(x),0),k_1)),k_0) \tag{9}$$

$$F_\delta^{-1}(x,k) = \mathsf{XOR}(\mathsf{AESE}(\mathsf{AESMC}(\mathsf{XOR}(\mathsf{AESE}(\mathsf{AESMC}(x),0),k_1)),k_0). \tag{10}$$

where $\mathsf{XOR} : \mathbb{F}_2 \times \mathbb{F}_2^{128} \to \mathbb{F}_2^{128}$ is a simple XOR operation which takes two 128-bit inputs. Thus, we can implement the AES layer for the encryption using only AES instructions in NEON. The round function is shown in Fig. 2.

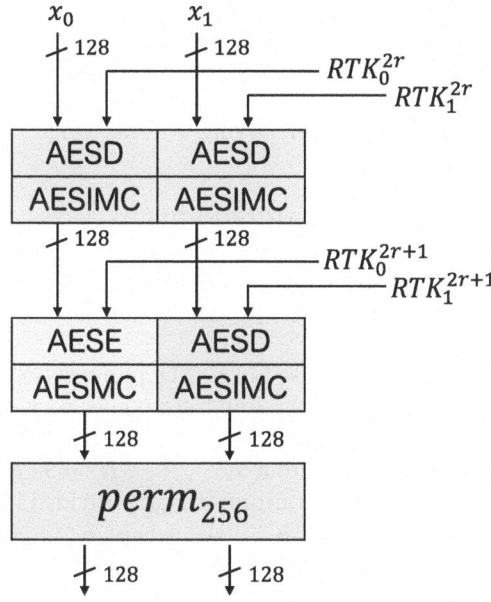

Fig. 2. The round function of ASURA

The Linear Layer. The linear layer is a byte shuffle operation that can be implemented by one vuzpq_u8 instruction on AArch64 processors. In this layer each byte in the input is permuted as a manner shown in Fig. 3. The inverse operation of $perm_{256}$ can be realized by vizpq_u8.

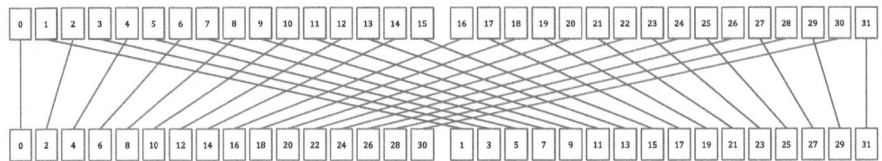

Fig. 3. $perm_{256}$ (vuzpq_u8)

3.2   The Tweakey Schedule Function

The tweakey schedule of ASURA is shown in Fig. 4. The tweakey schedule of ASURA is a modified version of the Pholkos tweakey schedule in which the τ function is replaced by another function.

To enhance security against related-tweak differential attacks, we employ the AES round function for the tweak update function θ alternative to τ of Pholkos. The details for each function used in the ASURA tweakey schedule are following:

RC^r : The tweakey schedule uses round constants $(RC^0, RC^1, \ldots, RC^{2r})$. We adopted the round constants proposed in Haraka v2 [17]. This is also used as a constant in Pholkos-256's tweakey schedule.

θ : θ is a AES-based function for mixing a tweak value. The tweak update function θ is as follows:

$$\theta(T) = \text{MC}(\text{SB}(\text{SR}(T))).$$

π_{256} : The 32-bit word permutation. The permutation manner is shown in Table 1a. The i-th word of the input state goes to $\pi_{256}(i)$-th word of the output state.

π_τ : The byte-wise permutation used in κ. The permutation manner is shown in Table 1b. The i-th byte of the input state goes to $\pi_\tau(i)$-th word of the output state.

κ : The function κ obtains K^I from K_{I-1}, i.e., $K^i = \kappa(K^{i-1})$. π_{256} is first applied to K^{I-1}, and then the output of π_{256} is divided into 128-bit states, i.e., $\pi_{256}(K^{I-1}) = (K^{I-1'}[0]||K^{I-1'}[1])$ where $K^{I-1'}[j] \in GF(2)$. Each byte of the round key is doubled in $\mathbb{F}_{2^8}$. Lastly, π_τ is applied to each $K^{I-1'}[j]$.

γ : The function γ obtains RTK^i from T^i, K^i, and RC^i, i.e. $RTK^i = \gamma(T^i, K^i, RC^i)$. 0 and 1 are indexes of 128-bit substate, γ has been computed as follows:

$$RTK_0^i = T \oplus K_0^i \oplus RC^i, \quad RTK_1^i = T \oplus K_1^i, \quad RTK^i = RTK_0^i||RTK_1^i.$$

These functions are iterated until generating RTK. The pseudo code for the tweakey schedule is shown in Algorithm 1.

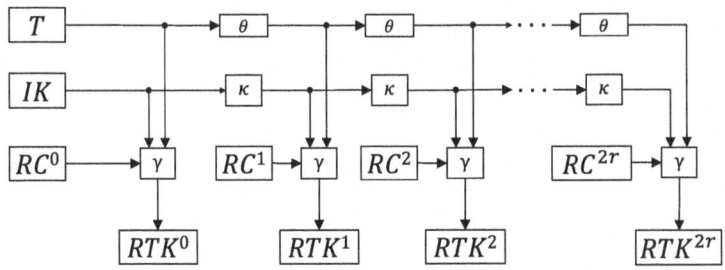

Fig. 4. Tweakey schedule of ASURA

Table 1. 32-bit-wise permutation π_{256} and Byte-wise permutation π_τ.

(a) π_{256}

i	0 1 2 3 4 5 6 7
$\pi_{256}(i)$	0 5 2 7 4 1 6 3

(b) π_τ

i	0 1 2 3 4 5 6 7 8 9 10 11 12 13 14 15
$\pi_\tau(i)$	11 12 1 2 15 0 5 6 3 4 9 10 7 8 13 14

Algorithm 1. ASURA Tweakey Schedule

procedure TWEAKEYSCHEDULE(IK, T^0)
  $K^0 \leftarrow IK$
  $RTK^0 \leftarrow \gamma(RC^0, K^0, T^0)$
  for $i = 1$ to $2R$ **do**
    $T^i \leftarrow \theta(T^{i-1})$
    $K^i \leftarrow \kappa(K^{i-1})$
    $RTK^i \leftarrow \gamma(RC^i, K^i, T^i)$
  end for
  return $(RTK^0, RTK^1, \ldots, RTK^{2R})$
end procedure

---

procedure $\theta(T^{i-1})$
  $T^i \leftarrow \mathrm{MC}(\mathrm{SB}(\mathrm{SR}(T^{i-1})))$
  return T^i
end procedure

---

procedure $\kappa(K^{i-1})$
  $(K_0^i, K_1^i, \ldots, K_7^i) \leftarrow K^i$
  for $j = 0$ to 7 **do**
    $K_j^i \leftarrow K_{\pi_{256}(j)}^{i-1}$
  end for
  for $j = 0$ to 1 **do**
    $(K^i[j][0], K^i[j][1], \ldots, K^i[j][15]) \leftarrow K^i[j]$
    for $b = 0$ to 15 **do**
      $K^i[j][b] \leftarrow 2 \cdot K^i[j][b]$
    end for
  end for
  $K^i \leftarrow \pi_\tau(K^i)$
  return K^i
end procedure

---

procedure $\gamma(RC^i, K^i, T^i)$
  $K_0^i \| K_1^i\} \leftarrow K^i$
  $RTK_0^i \leftarrow T^i \oplus K_0^i \oplus RC^i$
  $RTK_1^i \leftarrow T^i \oplus K_1^i$
  return RTK^i
end procedure

3.3 Encryption and Decryption

The pseudo code for the encryption is shown in Algorithm 2. The round function iterates between $0 \leq r \leq R - 1$, while the tweakey schedule function iterates between $0 \leq r \leq 2R$ for generating keys for whitening in addition to round keys. Let the round key of the r-th round be RTK^{2r} and RTK^{2r+1}. Note that the value of RTK^r is $RTK^r = RTK_0^r \| RTK_1^r$. For the whitening, RTK^{2R} is XORed to the state after the final round x. We recommend setting R to 6 based on the security analysis which is described in detail in the next section.

Algorithm 2. ASURA Encryption

---

procedure ASURAENCRYPT(PlainText, SecretKey)
    $x_0 \| x_1 \leftarrow$ PlainText
    $\{RTK^0, \cdots, RTK^{2R}\} \leftarrow$ TweakeySchedule(SecretKey)
    for $r = 0, \cdots, R - 2$ **do**
        $x \leftarrow R(x, RTK_0^{2r} \| RTK_0^{2r+1}, RTK_1^{2r} \| RTK_1^{2r+1})$
    end for
    $x \leftarrow R_{\mathsf{last}}(x, RTK_0^{2(R-1)} \| RTK_0^{2(R-1)+1}, RTK_1^{2(R-1)} \| RTK_1^{2(R-1)+1})$
    $x_0 \leftarrow x_0 \oplus RTK_0^{2R}$
    $x_1 \leftarrow x_1 \oplus RTK_1^{2R}$
    CipherText $\leftarrow x_0 \| x_1$
    return CipherText
end procedure

---

4   Design Rationale

4.1   Target Construction

The target construction in this paper is a tweakable block cipher with an SPN structure, having a block length and key length of 256 bits and supporting 128-bit tweak, similar to Pholkos-256 and Ghidle-256 as shown in Fig. 5. The target construction consists of an AES layer and a linear layer. For the linear layer, we employ a simple shuffle operation, which can be efficiently implemented using NEON instructions.

For the AES layer, we adopt an approach that selects the optimal combination of AES instructions. Since the NEON implementations of Ghidle-256 and Pholkos-256 require four consecutive AES instructions for their AES layer, we search for an optimal AES layer consisting of two parallel instances of four or fewer successive AES instructions to search for a more efficient construction than existing schemes. Specifically, we construct block ciphers with all target AES layers and evaluate their security against differential, impossible differential, and integral attacks. We then select an optimal AES layer that makes the entire encryption algorithm sufficiently secure against these attacks with the lowest implementation.

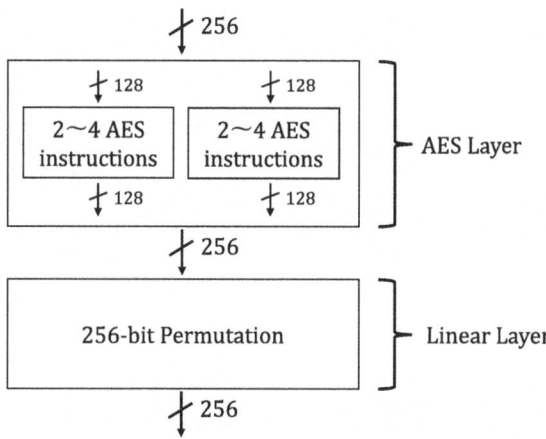

Fig. 5. The round function of large block cipher

4.2 Linear Layer

In AArch64, NEON includes instructions like vizpq_u8 and vuzpq_u8, which operate on a structure composed of two NEON registers. These instructions shuffle the elements within each lane and then write the results into another structure consisting of two NEON registers. Specifically, the vuzpq_u8 instruction enables byte-wise shuffling, and the vizpq_u8 instruction performs the inverse shuffle operation. Consequently, they can be applied in the round functions for both encryption and decryption, respectively.

Since these instructions allow for byte-wise shuffling in a single instruction, they offer faster diffusion than existing 32-bit shuffles and the binary matrix of Ghidle-256. Therefore, we select the byte-wise shuffle operation, which can be realized by only one vuzpq_u8 for two lanes as $perm256$ shown in Fig. 3.

4.3 AES Layer

Since the goal of this paper is to design a block cipher that is efficient on AArch64 processors, we consider the AES layer, which consists of the AES instructions in NEON: AESE, AESMC, AESD, and AESIMC, defined in Eq. (1)–Eq. (4). This paper examines constructions that have two parallel executions of sequential AES instructions grouped in sets of 2, 3, and 4. The target AES layers are shown in Fig. 6.

In the following, we refer to the classes of AES layers using two consecutive instructions as AL-2, three as AL-3, and four as AL-4. There are four types of AES instructions, and thus AL-2, AL-3, and AL-4 require 4, 6, and 8 instructions, respectively. The number of all possible AES layers in AL-2, AL-3 and AL-4 is 56 ($= 4^4$), 4096 ($= 4^6$), and 65536 ($= 4^8$), respectively. However, since the shuffle instructions in the linear layer are symmetric left and right, equivalent instruction sets can be eliminated by excluding those with swapped left and

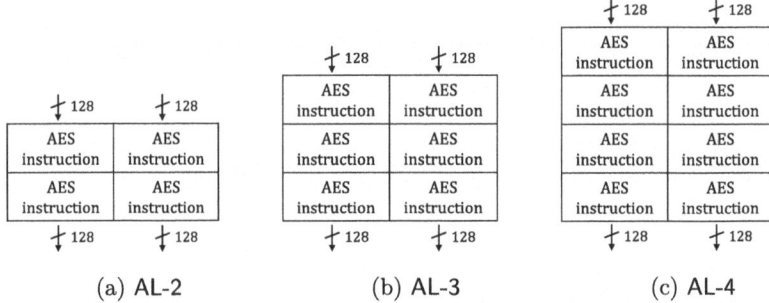

Fig. 6. Target AES layers

right 128-bit halves. This reduces the number of candidate AES layers to 136, 2080, and 33024 AES layers in AL-2, AL-3, and AL-4, respectively. Thus, after eliminating equivalent classes, we have a total of 35,240 candidate AES layers.

Our goal is to identify the optimal AES layer from the 35,240 candidates depicted in Fig. 6. To achieve this, we reduce the number of candidate AES layers by security evaluations against several attacks. Specifically, we choose an optimal AES layer that minimizes the number of rounds such that the block cipher with the round function indicated in Fig. 5 using the target AES layer is sufficiently secure against differential attacks, impossible differential attacks, and integral attacks. Since we evaluate all schemes byte-wise, MC^{-1} and MC are considered equivalent for this analysis.

Step 1 : Differential Attacks. First, we reduce the number of candidate AES layers based on the minimum number of rounds required to achieve 256-bit security against differential attacks. We use SAT-based automatic search to search the minimum number of active S-boxes for all candidate AES layers. Since the maximum differential probability of AES S-box is 2^{-6}, it requires at least 43 active S-boxes to make the differential characteristic probability below 2^{-256} ($2^{-6 \times 43} < 2^{-256}$). When the number of active S-boxes is at least 43, the 256-bit security is achieved. Since our search is byte-wise, MC and MC^{-1} are treated as equivalent instructions due to the MDS property. Therefore, we consider only MC instruction. This reduces the number of instruction candidates to three instructions: AESE, AESD, and AESMC. Consequently, the number of AES layer candidates is 45, 378, and 3,321 AES layers in AL-2, AL-3, and AL-4, respectively.

As a result of our active S-boxes search, 18 AES layers remain. Specifically, we found 6, 3, and 9 candidate AES layers with a minimal number of rounds in each class of AL-2, AL-3, and AL-4, respectively. The constructions with remaining candidates AES layers in AL-2, AL-3, and AL-4 require 6, 4, and 3 rounds, respectively. All remaining candidates require 12 AES instructions when they are used as AES layers for block ciphers with their minimal number of rounds. Table 2 shows the remaining AES layers in each class.

Table 2. The optimal number of instruction patterns for the AES layer explored through differential attack

(a) AL-2 (6)

AL-2			
AESE	AESE	AESE	AESD
AESMC	AESMC	AESMC	AESMC
AESD	AESD	AESMC	AESMC
AESMC	AESMC	AESE	AESE
AESMC	AESMC	AESMC	AESMC
AESD	AESE	AESD	AESD

(b) AL-3 (3)

AL-3	
AESE	AESE
AESE	AESE
AESMC	AESMC
AESD	AESD
AESD	AESD
AESMC	AESMC
AESE	AESD
AESE	AESD
AESMC	AESMC

(c) AL-4 (9)

AL-4					
1		2		3	
AESE	AESE	AESE	AESE	AESE	AESE
AESMC	AESMC	AESMC	AESMC	AESMC	AESMC
AESE	AESE	AESE	AESD	AESD	AESD
AESMC	AESMC	AESMC	AESMC	AESMC	AESMC
4		5		6	
AESD	AESD	AESD	AESD	AESD	AESD
AESMC	AESMC	AESMC	AESMC	AESMC	AESMC
AESE	AESE	AESE	AESD	AESD	AESD
AESMC	AESMC	AESMC	AESMC	AESMC	AESMC
7		8		9	
AESE	AESD	AESE	AESD	AESE	AESD
AESMC	AESMC	AESMC	AESMC	AESMC	AESMC
AESE	AESE	AESE	AESD	AESD	AESD
AESMC	AESMC	AESMC	AESMC	AESMC	AESMC

Step2 : Impossible Differential Attacks. For all 18 AES layers remaining in Step 1, we search for byte-wise impossible differentials where both input and output differences have only one active byte. As a result, when all AES layers in AL-2, AL-3, and AL-4, which remain in Step 1, are used in a round function with $perm256$ as the linear layer, 6, 6, and 3 rounds are required to erase all impossible differentials, respectively. Also, unlike Step 1, the number of AES instructions to achieve sufficient security against impossible differential attacks depends on the class of the AES layer. The number of rounds and AES instructions required for each class are shown in Table 3. Consequently, only the AES layers in AL-3 require 18 AES instructions, while those in AL-2 and AL-4 require 12 AES instructions. However, to achieve sufficient security against impossible differentials, the block ciphers with AES layers in AL-2 and AL-4 require 6 and 3 instructions for $perm256$, *i.e.*, vuzpq_u8, respectively. Therefore, the total number of instructions for 6-round block ciphers with AES layers in AL-2 is 18, while that for 3-round block ciphers with AES layers in AL-4 is 15. Thus, in this step, we select 9 AES layers in AL-4 shown in Table 2c, which requires the lowest number of AES instructions of all target constructions.

Table 3. The minimum number of rounds and total number of AES instructions required to ensure 256-bit security against differential and impossible differential attacks

	AL-2	AL-3	AL-4
round	6	6	3
total of AES instructions	12	18	12

Step3 : Integral Attack. In this step, we evaluate the security against integral attacks [11,16] for the remaining 9 AES layers selected in Step 2 to determine the optimal AES layer. Here, we use the notations which are often used for security against integral attacks such as ALL, CONSTANT, and the division property $\mathcal{D}$. For detail on these, see [16,22].

In this evaluation, we assume that the available amount of data for the attacker is less than 2^{127}. We search for the longest integral distinguishers for all block ciphers with the nine AES layers using the MILP model of the division property. Specifically, to obtain the longest integral distinguishers that could be constructed with a data volume of 2^{127}, ALL is assigned to 127 bits of the input of either the left or right AES instruction, and CONSTANT was assigned to the remaining 129 bits. That is D_{128}^{127} was given to the input of either AES instruction, and the remaining bits were set to CONSTANT. For this evaluation, since the evaluation is performed in D_k^8 units, the input space to be searched is 25 (= 2×16).

From Table 4, it is evident that constructions 3, 4, 5, and 6 do not have integral distinguisher in three or more rounds, indicating the highest level of security against integral attacks. Based on the results in Table 4 and the outcomes from Steps 1 and 2, it is clear that these four candidates are optimal in terms of both security and implementation performance.

Here, constructions 3, 4, 5, and 6 are equivalent in terms of the number of instructions and security perspective. In this evaluation, since MC and MC^{-1} are considered equivalent, and given that x86 lacks instructions to process MC alone, MC is treated as MC^{-1}. Therefore, in this paper, we decide to switch MC in construction 5 to MC^{-1} as the final construction.

Table 4. Integral attack

	Round			
Construction	2	3	4	5
1	-	✓	×	-
2	-	✓	×	-
3	✓	×	-	-
4	✓	×	-	-
5	✓	×	-	-
6	✓	×	-	-
7	-	✓	×	-
8	-	-	✓	×
9	-	-	✓	×

✓: BALANCE
×: UNKNOWN
−: not evaluated

4.4 Tweakey Schedule Function

In this section, we explain the design rationale behind the tweakey schedule function of ASURA. First, we directly adopt the tweakey schedule from Pholkos-256 and evaluate its security against related-tweak differential attacks. We realize that it requires 4 rounds to ensure 256-bit security, which is one more round than in the single-key setting.

To prevent this, we utilize AES round function in the tweak update function to enhance security. Consequently, as shown in Table 5, we are able to achieve 256-bit security with the same three rounds as in the single-key evaluation. Note that the overhead is negligible due to AES-NI support.

Table 5. Lower bounds of the number of active S-boxes for differential attack

	Round Number					
Attack Setting	1	2	3	4	5	6
Single	5	25	**45**	61	80	96
Related-tweak (Pholkos-256's tweakey schedule)	2	14	33	**49**	63	77
Related-tweak (ASURA's tweakey schedule)	5	25	**45**	61	80	96

5 Security Evaluation

In this section, we evaluate the security against differential/linear, truncate differential, impossible-differential, integral, boomerang, rectangle, yoyo and mixture differential attack for ASURA. The summary of our evaluation is shown in Table 6.

Table 6. Summary of security evaluation.

	linear	Differential	Impossible	Integral	Boomerang	Yoyo
ASURA	3	3	3	3	$3^{\#}$	2
Pholkos-256	4	4	4	4	$3^{\#}$	-
Ghidle-256	3	3	5	3	$2^{\$}$	2

$^{\#}$: Related-tweak setting
$^{\$}$: Secret-key setting

5.1 Differential and Linear Attacks

Differential and linear attacks are among the most fundamental cryptanalytic techniques applied to block ciphers. In this paper, we employ the SAT (satisfiability problem)-based method proposed in [21] to evaluate the number of active

S-boxes. As a result, the lower bound of active S-boxes in ASURA reaches 45 after 3 rounds. Given that the differential and linear probabilities of the AES S-box are 2^{-6} and the available data complexity in ASURA is limited to 2^{256}, ASURA is secure against differential and linear attacks after 3 rounds in the single-key setting. In the related-tweak setting, the lower bound of active S-boxes reaches 46 across 3 rounds as shown in Table 5.

5.2   Impossible Differential Attack

Impossible differential attack [5] exploits the differences holding with probability 0. For ASURA, our SAT-aided search verify that there is no byte-wise impossible differential in the case where only one byte is active in each of the input and output after 3 round. We find 2 round impossible differentials on ASURA. In addition, we evaluate in the related-tweak setting, and find impossible differentials for up to 1 round of ASURA, where the tweak must have a difference. Examples of these in both the single-key and related-tweak settings are shown in Table 7.

Table 7. Impossible differential distinguisher on ASURA

Setting	Round	Impossible differential	
Single Key	2	input:	(1000000000000000, 0000000000000000)
		Out:	(0000000000000000, 1000000000000000)
Related Tweak	1	input:	(0000000000000000, 0000000000000000)
		Tweak:	(1000000000000000)
		Out:	(1000000000000000, 0000000000000000)

5.3   Integral Attack

Integral attacks utilize a set of plaintexts where one part of the input has a constant value at a specific bit, while the other part assumes all possible values. We apply the MILP-based method proposed by Xiang [25] to analyze this scenario. Specifically, we investigate cases where only one input byte remains constant while the remaining bytes are active, in order to estimate the upper bounds for the number of rounds an integral distinguisher can cover in ASURA. As a result, we identify 2-round integral distinguishers for ASURA. Furthermore, given that the available data complexity is limited to 2^{128}, ASURA is secure against integral attack after 3 rounds.

5.4   Boomerang and Rectangle Attacks

In boomerang attack, a cipher E is analyzed as a composition of two sub-ciphers, E_0 and E_1. The method combines two independent differential trails through these sub-ciphers to form a boomerang quartet for E [23]. Specifically, if there exists a differential trail $\alpha \to \beta$ with probability p through E_0, and another trail $\gamma \to \delta$ with probability q through E_1, the probability of finding a boomerang quartet is $p^2 q^2$. Later, it was demonstrated that the upper and lower trails are not independent [18]. To account for these dependencies, a middle layer E_m is introduced between the existing two layers. Corresponding to these dependencies several analytical tools are also introduced, such as *Boomerang Connectivity Table* (BCT) [9], *Boomerang Difference Table* (BDT) [24], etc. In the context of related tweakey settings, ASURA was analyzed for potential boomerang attacks by considering a single s-box layer as E_m (which is considered as 0.5 round). Table 8 shows the number of active s-boxes for various combinations of E_0 and E_1. The attack for 2 rounds (consider either $r_{E_0} = 1$, $r_{E_1} = 1.5$ or $r_{E_0} = 1.5$, $r_{E_1} = 1$) activates 12 s-boxes. The best attack on 3-round requires 30 active s-boxes when $r_{E_0} = 2$, $r_{E_1} = 1.5$ or vice versa. For attacking 4-round, minimum 72 s-boxes are required to be activated. Hence, we believe, ASURA is secure against boomerang attack after 3 rounds.

Table 8. Lower bounds on the number of active S-boxes for boomerang attacks. Here, r_{E_0} and r_{E_1} correspond to the total number of rounds in E_0 and E_1, respectively (E_m is 0.5 round). Hence, the total round is $(r_{E_0} + r_{E_1} - 0.5)$.

r_{E_0} \ r_{E_1}	1	1.5	2	2.5	3	3.5	4
1	4	12	22	52	76	92	110
1.5	12	20	30	60	82	100	114
2	22	30	44	74	97	111	-
2.5	52	60	72	100	122	-	-
3	76	84	97	126?	-	-	-
3.5	92	100	111?	-	-	-	-
4	110	118?	-	-	-	-	-

The Rectangle attack [6,15] modifies the boomerang attack by shifting the context from adaptive chosen plaintext/ciphertext to chosen plaintext, though this adjustment comes with a reduction in the probability of discovering quartets. The concept is akin to the boomerang attack, where two shorter differential trails are combined to form a longer trail over more rounds. For an n-bit block cipher E, the rectangle attack identifies a quartet with a probability of $2^{-n} p^2 q^2$, making the likelihood of finding a quartet significantly lower compared to the boomerang attack. As the block size is 256 bits, ASURA is secure against rectangle attacks after one round.

5.5   Yoyo Attack

The Yoyo attack, as introduced in [4], consider a pair of plaintexts that meet a specific criterion. After encryption, words between the resulting ciphertexts are swapped, and the new ciphertexts are decrypted to generate a new pair of plaintexts. The expectation is that this new pair will retain the original property. Rønjom *et al.*expanded on the Yoyo attack for SPN ciphers and proposed a deterministic distinguisher for a generalized 2-round SPN [20]. The attack scenario of yoyo relies on the underlying non-linear structures (supersbox and megasbox). In brief, the attack exploits a generic structure consisting of two non-linear layers with a linear layer in between.

Due to the use of two complete AES rounds before the application of perm256, the 32-bit supersboxes cannot span over two rounds, which restricts the deterministic yoyo attack to two rounds. Because perm256 is applied immediately following a mixing layer (MC and MC^{-1}), even 64-bit megasboxes are confined within each round. Although 128-bit megasboxes span more than one round (as shown in Fig. 7), they cannot be exploited to attack more than two rounds, as the subsequent linear and non-linear layers cannot span beyond the first MC^{-1} of the third round. Therefore, we believe ASURA is immune to yoyo attacks.

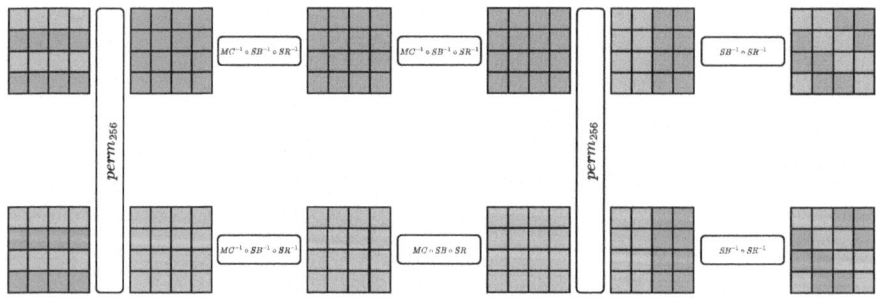

Fig. 7. 128-bit megasbox of ASURA. The same colored bytes are part of a single megasbox. (Color figure online)

5.6   Mixture Differential Attack

Mixture differential attack [12], considers the relation between the ciphertext differences generated by a pair of plaintexts and their mixture (created by swapping certain words between the original plaintexts). Initially a deterministic version of this attack was proposed to mount an attack on 4-round AES. Later, probabilistic variants of this attack have been introduced [1–3,13]. Similar to the yoyo attack, this technique also relies on the underlying non-linear structures of the cipher. As discussed in the case of yoyo, use of two mixing layers in each round and the following perm256 makes it difficult to attack more than two rounds.

6 Performance Evaluation

In this section, we evaluate the software performance of ASURA on AArch64 and x86 processors. For all evaluations on AArch64 and x86, we measure the gigabits per second (Gbps) for both single-threaded and 2 to 4 parallel block encryption and decryption. This is done by executing the target algorithm 1.0×10^8 times and measuring the average Gbps.

For the evaluation on AArch64, we use a MacBook Air (2024 model) with the Apple M3 chip. The M3 chip, developed by Apple, features custom cores and implements the AArch64 architecture, which supports NEON SIMD instructions. For the evaluations on x86, we use the Alder Lake platform, equipped with an Intel(R) Core(TM) i9-12900K CPU running at 3.20 GHz on performance cores (P-cores) and 2.40 GHz on efficiency cores (E-cores). Turbo Boost technology was disabled for all evaluations.

6.1 Performance on AArch64

We ran benchmarks on ASURA, Pholkos-256, and Ghidle-256, and compared their performance. The results are presented in Table 9.

Comparison. The single block encryption speed of ASURA is about 64% and 74% faster than Pholkos-256 and Ghidle-256, respectively. Similarly, the decryption speed of ASURA is 47% and 27% faster than Pholkos-256 and Ghidle-256, respectively. This result can be attributed to the smaller number of full rounds. Pholkos-256 and Ghidle-256 have 8/7 full rounds, respectively, while ASURA has only 6 rounds due to its strong and efficient round function in terms of security. Additionally, ASURA's decryption is slower than its encryption because two XOR instructions are required at the AES layer during decryption, as shown in Eq. (10). In parallel processing, we confirmed that it is possible to process two blocks concurrently, which is due to the latency and throughput of the AES instructions on the AArch64 architecture [10]. Overall, ASURA's encryption speed is more than 40% faster than that of both Pholkos-256 and Ghidle-256, while its decryption speed is more than 10% faster than theirs.

On the Number of Instructions. ASURA requires a lower number of instructions than Pholkos-256 in terms of the AES layer and than Ghidle-256 in terms of the linear layer. Since Pholkos-256 is optimized for the x86 environment, its implementation in the AArch64 environment requires a single XOR instruction (AddRoundKey) within each round, as shown in Sect. 1. In contrast, although ASURA can implement the AES layer with only four AES instructions, they cannot be parallelized, resulting in a delay of one XOR instruction in each round.

Ghidle-256 has the same number of instructions in the AES layer as ASURA but requires more instructions in the linear layer. Specifically, the linear layer of Ghidle-256 requires two vextq_u8 instructions and three XOR instructions, whereas the linear layer of ASURA can shuffle the entire 256 bits with a single instruction.

6.2   Performance on X86

We ran benchmarks on ASURA, Pholkos-256, Ghidle-256 and compared their performance. In the x86 environment, a single AESMC instruction is required during the AES layer of the decryption, but the instruction does not exist in x86. Therefore, we emulate a single AESMC by combining AESENC and AESDECLAST instructions. We can emulate AESMC by applying AESDECLAST and AESENC sequentially to the input with all zeros as the round key because of the relation of AESMC(x) = AESENC(AESDECLAST($x, 0), 0$). Regarding the linear layer, four each of punpckh_epi8 and punpckl_epi8 instructions are required for the encryption, and one each for the decryption.

Comparison. The results are presented in Table 10. The single block encryption speed of ASURA is about 15% slower than Pholkos-256, but 40% faster than Ghidle-256. In parallel processing, the relative speed difference remains the same, with Pholkos-256, ASURA, and Ghidle-256 being faster in that order. Also, the single decryption speed of ASURA is 53% faster than Pholkos-256 and same speed asGhidle-256. In parallel processing, the relative speed difference remains the same, with ASURA, Ghidle-256, and Pholkos-256 being faster in that order.

On the Number of Instructions. In encryption, ASURA requires four punpckh_epi8 and four punpckl_epi8 instructions in the linear layer. On the other hand, in decryption, ASURA requires one punpckh_epi8 and one punpckl_epi8 instructions in the linear layer. This difference in the number of instructions directly affects the respective speed differences between encryption and decryption.

Table 9. AArch64 (all values are given as Gbps)

	Sequential		Parallel					
			2		3		4	
Algorithm	Enc	Dec	Enc	Dec	Enc	Dec	Enc	Dec
ASURA	21.10	13.45	42.45	25.59	56.91	34.16	59.74	41.83
Pholkos-256	12.89	9.132	25.19	17.51	34.85	23.91	41.28	26.74
Ghidle-256	12.16	10.58	23.66	20.12	32.47	28.41	41.66	37.21

Table 10. x86 (all values are given as Gbps)

| | Sequential | | Parallel | | | | | |
| | | | 2 | | 3 | | 4 | |
Algorithm	Enc	Dec	Enc	Dec	Enc	Dec	Enc	Dec
ASURA	9.241	6.690	15.91	12.34	20.20	16.73	21.25	17.01
Pholkos-256	10.83	4.364	20.27	8.340	25.98	9.661	32.73	9.643
Ghidle-256	6.624	6.605	11.82	12.07	15.64	15.63	15.90	15.90

7 Conclusion

In this paper, we proposed ASURA, a 256-bit tweakable block ciphers optimized for the AArch64 environment. ASURA incorporates the vuzpq_u8 instruction for efficient byte-wise shuffling and enables high-speed processing through the parallel execution of four consecutive AES instructions. ASURA demonstrates faster encryption speeds, sequentially and in parallel, compared to Pholkos-256 and Ghidle-256 on AArch64.

Acknowledgements. This result is obtained from the commissioned research (JPJ012368C05801) by the National Institute of Information and Communications Technology (NICT), Japan. This work was also supported by JSPS KAKENHI Grant Number JP24H00696.

References

1. Bardeh, N.G.: A key-independent distinguisher for 6-round AES in an adaptive setting. Cryptology ePrint Archive, Paper 2019/945 (2019)
2. Bardeh, N.G., Rønjom, S.: The exchange attack: how to distinguish six rounds of AES with $2^{88.2}$ chosen plaintexts. In: Galbraith, S.D., Moriai, S. (eds.) ASIACRYPT 2019. LNCS, vol. 11923, pp. 347–370. Springer, Cham (2019). https://doi.org/10.1007/978-3-030-34618-8_12
3. Bardeh, N.G., Rønjom, S.: Practical attacks on reduced-round AES. In: AFRICACRYPT. LNCS, vol. 11627, pp. 297–310. Springer (2019)
4. Biham, E., Biryukov, A., Dunkelman, O., Richardson, E., Shamir, A.: Initial observations on skipjack: cryptanalysis of skipjack-3xor. In: Selected Areas in Cryptography. LNCS, vol. 1556, pp. 362–376. Springer (1998)
5. Biham, E., Biryukov, A., Shamir, A.: Cryptanalysis of skipjack reduced to 31 rounds using impossible differentials. J. Cryptol. **18**(4), 291–311 (2005)
6. Biham, E., Dunkelman, O., Keller, N.: The rectangle attack - rectangling the serpent. In: EUROCRYPT. LNCS, vol. 2045, pp. 340–357. Springer (2001)
7. Biryukov, A., Khovratovich, D.: PAEQ: parallelizable permutation-based authenticated encryption. In: ISC. LNCS, vol. 8783, pp. 72–89. Springer (2014)
8. Bossert, J., List, E., Lucks, S., Schmitz, S.: Pholkos - efficient large-state tweakable block ciphers from the AES round function. In: CT-RSA. LNCS, vol. 13161, pp. 511–536. Springer (2022)

9. Cid, C., Huang, T., Peyrin, T., Sasaki, Y., Song, L.: Boomerang connectivity table: a new cryptanalysis tool. In: Nielsen, J., Rijmen, V. (eds.) Advances in Cryptology - EUROCRYPT 2018 - 37th Annual International Conference on the Theory and Applications of Cryptographic Techniques, Tel Aviv, Israel, April 29 - May 3, 2018 Proceedings, Part II. LNCS, vol. 10821, pp. 683–714. Springer (2018)

10. ARM Corporation: Arm intrinsics guide, official webpage. https://developer.arm.com/architectures/instruction-sets/intrinsics/

11. Daemen, J., Knudsen, L., Rijmen, V.: The block cipher square. In: Biham, E. (ed) FSE. LNCS, vol. 1267, pp. 149–165. Springer (1997)

12. Grassi, L.: Mixture differential cryptanalysis: a new approach to distinguishers and attacks on round-reduced AES. IACR Trans. Symmetric Cryptol. **2018**(2), 133–160 (2018)

13. Grassi, L.: Probabilistic mixture differential cryptanalysis on round-reduced AES. In: Selected Areas in Cryptography. LNCS, vol. 11959, pp. 53–84. Springer (2019)

14. Kampanakis, P., Campagna, M., Crocket, E., Petcher, A., Gueron, S.: Practical challenges with AES-GCM and the need for a new cipher. In: The Third NIST Workshop on Block Cipher Modes of Operation (2023)

15. Kelsey, J., Kohno, T., Schneier, B.: Amplified boomerang attacks against reduced-round MARS and serpent. In: FSE. LNCS, vol. 1978, pp. 75–93. Springer (2000)

16. Knudsen, L., Wagner, D.: Integral cryptanalysis. In: Daemen, J., Rijmen, V. (eds.) FSE. LNCS, vol. 2365, pp. 112–127. Springer (2002)

17. Kölbl, S., Lauridsen, M.M., Mendel, F., Rechberger, C.: Haraka v2 - efficient short-input hashing for post-quantum applications. IACR Trans. Symmetric Cryptol. **2016**(2), 1–29 (2016)

18. Murphy, S.: The return of the cryptographic boomerang. IEEE Trans. Inf. Theory **57**(4), 2517–2521 (2011)

19. Nakahashi, M. et al.: Ghidle: efficient large-state block ciphers for post-quantum security. In: ACISP. LNCS, vol. 13915, pp. 403–430. Springer (2023)

20. Rønjom, S., Bardeh, N.G., Helleseth, T.: Yoyo tricks with AES. In: ASIACRYPT (1). LNCS, vol. 10624, pp. 217–243. Springer (2017)

21. Sun, L., Wang, W., Wang, M.: Accelerating the search of differential and linear characteristics with the SAT method. IACR Trans. Symmetric Cryptol. **2021**(1), 269–315 (2021)

22. Todo, Y.: Structural evaluation by generalized integral property. In: EUROCRYPT (1). LNCS, vol. 9056, pp. 287–314. Springer (2015)

23. Wagner, D.A.: The boomerang attack. In: FSE. LNCS, vol. 1636, pp. 156–170. Springer (1999)

24. Wang, H., Peyrin, T.: Boomerang switch in multiple rounds: application to AES variants and deoxys. IACR Trans. Symmetric Cryptol. **2019**(1), 142–169 (2019)

25. Xiang, Z., Zhang, W., Bao, Z., Lin, D.: Applying MILP method to searching integral distinguishers based on division property for 6 lightweight block ciphers. In: ASIACRYPT (1). LNCS, vol. 10031, pp. 648–678 (2016)

Proving the Security of the Extended Summation-Truncation Hybrid

Avijit Dutta[1,2(✉)] and Eik List[3]

[1] Institute for Advancing Intelligence, TCG-CREST, Kolkata, India
`avirocks.dutta13@gmail.com`
[2] Academy of Scientific and Innovative Research (AcSIR), Ghaziabad, India
[3] Weimar, Germany
`e.List@posteo.de`

Abstract. Since designing a dedicated secure symmetric PRF is difficult, various works studied optimally secure PRFs from the sum of independent permutations (SoP). At CRYPTO'20, Gunsing and Mennink proposed the Summation-Truncation Hybrid (STH). While based on SoP, STH releases additional $a \leq n$ bits of the permutation calls and sums $n-a$ bits of them. Thus, it produces $n+a$ bits at $O(n-a/2)$-bit PRF security. Both SoP or STH can be used directly in encryption schemes or MACs in place of permutation calls for higher security. However, simply replacing every call as in GCM-SIVr would demand more calls.

For encryption schemes, Iwata's XORP scheme is long known to provide a better trade-off between efficiency and security. It extends SoP to variable-length-outputs by using $r + 1$ calls to a block cipher where the output of one call is added to each of the other r outputs. A similar extension can be conducted for STH that we call XTH, the XORP-Truncation Hybrid. Such an extension was already suggested in the final discussion by Gunsing and Mennink, but left as an open problem.

This work fills the gap by formalizing and proving the security of XTH. For a rate of $r/(r + 1)$ as in XORP, we show $O(n - a/2 - 1.5\log(r))$-bit security for XTH.

Keywords: Secret-key cryptography · provable security · encryption · sum of permutations

1 Introduction

Since dedicated symmetric-key pseudorandom functions (PRFs) are hard to construct, the cryptographic community has been devoting sophisticated efforts towards designing PRFs from block ciphers and permutations. Research on the design of more secure PRFs from permutations came from truncation and summation. Hall et al. [11] truncated the output of an n-bit permutation from n bits

---

E. List—Independent researcher.

S. Mukhopadhyay and P. Stănică (Eds.): INDOCRYPT 2024, LNCS 15495, pp. 165–187, 2025.
https://doi.org/10.1007/978-3-031-80308-6_8

to a bits, which yielded security for up to $O(2^{n-a/2})$ queries [1,9], i.e., $(n-a/2)$-bit security. On the other hand, Bellare et al. [2] studied the security of the sum of permutations SoP. Initially, they studied two domain-separated instances of the same permutation Π. That is, they fed an $(n-1)$-bit value x, appended different domain bits and summed the outputs from $\Pi(x\|0) \oplus \Pi(x\|1)$. Alternatively, one could also consider the sum of two independent n-bit permutations, i.e. $\Pi_1(x) \oplus \Pi_2(x)$ (also called SoP2). After a long series of works, the PRF security of SoP and SoP2 is well-understood to be about $O(n)$ [3,4,6,8,16].

Summation-Truncation Hybrid. In [10], Gunsing and Mennink proposed a trade-off between output length and PRF security by reconsidering truncation. They introduced the *Summation-Truncation Hybrid* (STH), which filled the range between those extremes. STH outputs a bits of each permutation call and the sum of the remaining $(n-a)$-bit outputs from both permutations:

$$\mathsf{STH}[a](x) \stackrel{\Delta}{=} \Pi_1(x)[n-1..n-a]\| (0^a \|\Pi_1(x)[n-a-1..0]) \oplus \Pi_2(x) .$$

They showed that STH provides PRF security for up to $O(2^{n-a/2})$ queries.

SoP has proven highly useful for a number of designs, e.g. as a finalization of MACs or authentication parts of authenticated encryption schemes, e.g. in PMAC$^+$ [18], 3kf9 [19], Lightmac$^+$ [17], DBHtS [5], or Deoxys [15]. It is still an interesting question of finding good applications for STH. Efficient extensions to variable output lengths (VOL) could be one avenue towards more applications. Simply plugging in SoP or STH as a replacement for a block cipher in encryption or authenticated encryption can already suffice to increase a scheme's security, e.g. in GCM-SIVr [14]. However, such in-place instantiations would double the number of primitive calls compared to a usual block-cipher-based construction. For SoP, a more efficient extension is long known. In [12], Iwata had extended SoP to a VOL-PRF XORP, which takes an m-bit input and produces a sequence of r n-bit outputs as

$$\mathsf{XORP}[r](x) \stackrel{\Delta}{=} \|_{i=1}^{r} \Pi(x\|\langle 0 \rangle_s) \oplus \Pi(x\|\langle i \rangle_s) ,$$

where $s = \lceil \log_2(r+1) \rceil$ and $\langle i \rangle_s$ denotes the s-bit binary representation of the integer i and $m+s = n$. XORP achieved $O(n - \log_2(r^2))$-bit PRF security at a rate of $r/(r+1)$ [13].

Extending STH. In the concluding thoughts of their work, Gunsing and Mennink [10] suggested an extension of STH to more outputs, but left it as an open problem. We formalize such an extension as XTH, the XORP-Truncation Hybrid, which takes n-bit inputs x and produces $(a+rn)$-bit outputs as

$$\mathsf{XTH}[a,r](x) \stackrel{\Delta}{=} \Pi_0(x)[n-1..n-a]\|\Big\|_{i=1}^{r} (0^a \|\Pi_0(x)[n-a-1..0]) \oplus \Pi_i(x) .$$

Thus, it uses the first permutation's $(n-a)$ bits to mask the other outputs, increasing the output size by a bits compared to that of XORP. Thus, XTH seems interesting, but has not a security proof yet.

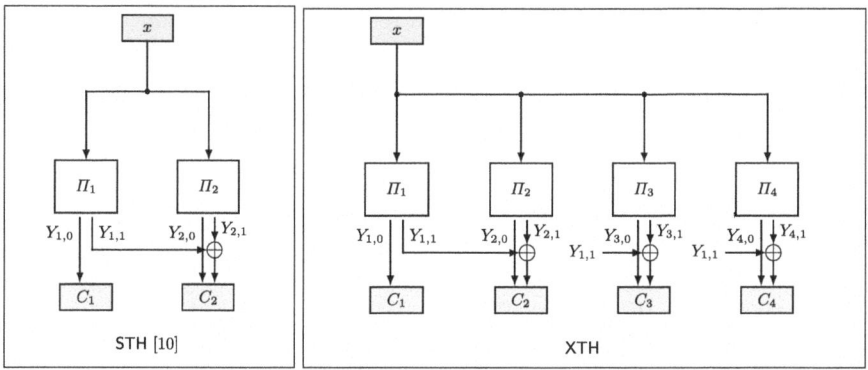

Fig. 1. Gunsing and Mennink's Summation-truncation hybrid and the XORP extension considered in this work.

Outline. In this work, we study the provable security of XTH and show that it achieves $O(n - a/2 - 1.5\log(r))$-bit PRF security. The remainder of this work is structured as follows. After preliminaries, Sect. 3 briefly recalls the definitions of STH and XTH before Sects. 4 and 5 analyze the security of the latter. Section 6 concludes (Fig. 1).

2   Preliminaries

For positive integers x, y, we write $[x] = \{1, \ldots, x\}$, $[0..x] = \{0, 1, \ldots, x\}$, and $[x..y] = \{x, x+1, \ldots, y\}$. We write $\{0, 1\}^n$ for n-bit strings, and $X\|Y$ for the concatenation of two bitstrings X and Y. By $\langle i \rangle_s$, we denote the s-bit binary representation of a non-negative integer i. For a bitstring X, $|X|$ denotes the length of the bitstring X in terms of the number of bits. For integers x, n and bitstring $X \in \{0, 1\}^n$, we use $X_1, \ldots, X_m \xleftarrow{x} X$ for the splitting of X into segments of $\leq x$ bits s. t. $|X_1| = \cdots = |X_{m-1}| = x$ and $|X_m| \leq x$. $(X_1, X_2) \xleftarrow{x, n-x} X$ indicates that $|X_1| = x$, $|X_2| = n - x$ and $X_1\|X_2 = X$. We write $X_1, X_2, \ldots \xleftarrow{\$} \mathcal{X}$ for the uniform and pairwise independent sampling with replacement of $X_1, X_2, \ldots$ from $\mathcal{X}$. Thus, $X_i \xleftarrow{\$} \mathcal{X}$, independent of the values X_j for $i \neq j$. For non-empty sets or spaces $\mathcal{T}$ and $\mathcal{X}$, $\mathsf{Perm}(\mathcal{X})$ is the set of permutations over $\mathcal{X}$ and $\widetilde{\mathsf{Perm}}(\mathcal{T}, \mathcal{X})$ the set of tweakable permutations over $\mathcal{X}$ with tweak space $\mathcal{T}$, that is, the functions $\widetilde{\Pi}(T, \cdot)$ that, for each tweak $T \in \mathcal{T}$, $\widetilde{\Pi}(T, \cdot)$, is a bijection over $\mathcal{X}$.

Distinguishers. A distinguisher $\mathbf{D}$ is an algorithm that interacts with one of several worlds that it shall distinguish between. Prior, the challenger samples a random bit $b \xleftarrow{\$} \{0, 1\}$ and presents $\mathbf{D}$ with one of two sets of oracles depending on the value of b. We use $b = 1$ for the real world. Moreover, the challenger uses internal secrets. $\mathbf{D}$ interacts with the individual oracles and collects the responses. At the end, $\mathbf{D}$ outputs a guess b' and wins iff $b = b'$. We write

$$\Delta_{\mathbf{D}}(\mathcal{R}_K; \mathcal{I}) \triangleq \left| \Pr_K \left[\mathbf{D}^{\mathcal{R}_K} = 1 \right] - \Pr \left[\mathbf{D}^{\mathcal{I}} = 1 \right] \right|$$

for the advantage of $\mathbf{D}$ in distinguishing a real keyed construction $\mathcal{R}_K$ from an ideal construction $\mathcal{I}$, where the probability is over the key K, the randomness of $\mathcal{I}$, the coins of $\mathbf{D}$ and that of the challenger, if any.

PRF Security. Given two non-empty sets or spaces $\mathcal{X}, \mathcal{Y}$, let $F : \mathcal{K} \times \mathcal{X} \to \mathcal{Y}$, $\rho \leftarrow_{\$} \mathsf{Func}(\mathcal{X}, \mathcal{Y})$, and let $K \leftarrow_{\$} \mathcal{K}$ be a secret key. The PRF advantage of a distinguisher $\mathbf{D}$ on F_K is defined as $\mathbf{Adv}_{F_K}^{\mathsf{PRF}}(\mathbf{D}) \overset{\Delta}{=} \Delta_{\mathbf{D}}(F_K; \rho)$.

The χ^2 Method. We will employ the χ-square method by Dai et al. [4]. For this purpose, we briefly recall its main theorem. For each $i \in [q]$ and each vector $\mathbf{W}^{i-1} = (W_2^{i-1}, \ldots, W_r^{i-1})$ with $\mathbf{W}_j^{i-1} = (W_j^1, W_j^2, \ldots, W_j^{i-1})$, define

$$\chi^2(\mathbf{W}^{i-1}) \overset{\text{def}}{=} \sum_{W \in (\mathbb{F}_2^b)^{r-1}} \frac{\left(\mathrm{Pr}_{\mathcal{O}_{\mathrm{real}}}[\mathbf{W}^i = W | \mathbf{W}^{i-1}] - \mathrm{Pr}_{\mathcal{O}_{\mathrm{ideal}}}[\mathbf{W}^i = W | \mathbf{W}^{i-1}]\right)^2}{\mathrm{Pr}_{\mathcal{O}_{\mathrm{ideal}}}[\mathbf{W}^i = W | \mathbf{W}^{i-1}]}.$$

Theorem 1 (χ^2 Method [4]). Consider two systems $\mathcal{O}_{\mathrm{real}}$ and $\mathcal{O}_{\mathrm{ideal}}$. Suppose that for any vector $\mathbf{W}$, it holds that $\mathrm{Pr}_{\mathcal{O}_{\mathrm{ideal}}}[\mathbf{W}^i] > 0$ whenever $\mathrm{Pr}_{\mathcal{O}_{\mathrm{real}}}[\mathbf{W}^i] > 0$. Then

$$\left| \mathop{\mathrm{Pr}}_{\mathcal{O}_{\mathrm{real}}}[\mathbf{W}^i] - \mathop{\mathrm{Pr}}_{\mathcal{O}_{\mathrm{ideal}}}[\mathbf{W}^i] \right| \leq \sqrt{\frac{1}{2} \sum_{i=1}^{q} \mathbb{E}_{\mathcal{O}_{\mathrm{real}}}[\chi^2(\mathbf{W}^{i-1})]}.$$

3   XTH

STH. We note that two versions of the Summation-Truncation-Hybrid [10] exist: based on a single n-bit secret full-round permutation Π, and based on a pair of independent permutations Π_1 and Π_2. Here, we focus on the second version and refer to it as STH.

It feeds the n-bit input x into two independent full round n-bit secret permutations Π_1 and Π_2, and splits each of their outputs Y_i, where $Y_i = \Pi_i(x)$, into an a-bit part $Y_{i,0}$ and an $(n-a)$-bit part $Y_{i,1}$, for $i \in \{1,2\}$, i.e., $(Y_{i,0}, Y_{i,1}) \xleftarrow{a, n-a} Y_i$, for $i \in \{1,2\}$. The a-bit parts $Y_{1,0}$ and $Y_{2,0}$ are output in plain; the $(n-a)$-bit parts are summed and output as $Y_{1,1} \oplus Y_{2,1}$. Gunsing and Mennink have shown that STH achieves roughly $(n - a/2)$-bit security.

XTH. We define the XORP-based Summation-Truncation Hybrid (XTH) as follows. For $a, r \in \mathbb{N}$ with $a \leq n$, $\mathsf{XTH}[a,r]$ feeds an n-bit input x into r independent n-bit secret permutations $\Pi_1, \ldots, \Pi_r$. Thereupon, it partitions each permutation output $Y_i = \Pi_i(x)$ into an a-bit part $Y_{i,0}$ and an $(n - a)$-bit part $Y_{i,1}$, for $i \in \{1, \ldots, r\}$. The a-bit parts $Y_{1,0}, Y_{2,0}, \ldots, Y_{r,0}$ are then returned as outputs. The $(n - a)$-bit parts $Y_{2,1}$, $Y_{3,1}$ etc., in contrast, are XORed to the $(n - a)$-bit output of the first permutation call, and the sum is returned for each block: $Y_{1,1} \oplus Y_{2,1}, Y_{1,1} \oplus Y_{3,1}, \ldots, Y_{1,1} \oplus Y_{r,1}$. Algorithm 1 lists formal definitions for both STH and XTH.

Algorithm 1. Definition of $\mathsf{XTH}[a,r]_{\Pi_0,\ldots,\Pi_r}$.

11: **function** $\mathsf{STH}[a]_{\Pi_1,\Pi_2}(x)$	21: **function** $\mathsf{XTH}[a,r]_{\Pi_1,\Pi_2,\ldots,\Pi_r}(x)$
12: $(Y_{1,0},Y_{1,1}) \xleftarrow{a,n-a} \Pi_1(x)$	22: **for** $i \leftarrow 1 \ldots r$ **do**
13: $(Y_{2,0},Y_{2,1}) \xleftarrow{a,n-a} \Pi_2(x)$	23: $(Y_{i,0},Y_{i,1}) \xleftarrow{a,n-a} \Pi_i(x)$
14: **return** $Y_{1,0} \| Y_{2,0} \oplus Y_{2,1}$	24: **return** $Y_{1,0} \| \left(\|_{i=2}^{r} Y_{i,0} \| Y_{1,1} \oplus Y_{i,1} \right)$

4 Security Analysis of XTH

In this section, we state and prove the following main security result of XTH.

Theorem 2. Let r, n, a, b and q be positive integers with $r \geq 2$, $a + b = n$, and $q < 2^{b-2}$ and $q \leq 2^n/(2r)$. Let $\Pi_1,\ldots,\Pi_r \xleftarrow{\$} \mathsf{Perm}(\{0,1\}^n)$ be independent random permutations. Let $\mathbf{D}$ be a PRF distinguisher on the construction $\mathsf{XTH}[a,r]_{\Pi_1,\Pi_2,\ldots,\Pi_r}$. Then

$$\mathbf{Adv}^{\mathsf{PRF}}_{\mathsf{XTH}[a,r]}(\mathbf{D}) \leq \left(\frac{4}{3}\right)^r \left(\frac{rq}{2^{n-a/3}}\right)^{3/2} + 2^{a-1} \cdot \left(\frac{16rq}{2^n}\right)^{2^{b-2}} + \mathbf{Adv}^{\mathsf{PRF}}_{\mathsf{trunc}_a}(rq).$$

By substituting $r = 2$ into Theorem 2, we recover the PRF advantage of STH using a pair of independent permutations as follows:

Corollary 1. Let n, a, b and q be positive integers such that $a + b = n$, and $q < 2^{b-2}$ and $q \leq 2^n/4$. Let $\Pi_1,\Pi_2 \xleftarrow{\$} \mathsf{Perm}(\{0,1\}^n)$ be independent random permutations. Let $\mathbf{D}$ be a PRF distinguisher on $\mathsf{STH}[a]_{\Pi_1,\Pi_2}$. Then

$$\mathbf{Adv}^{\mathsf{PRF}}_{\mathsf{STH}[a]}(\mathbf{D}) \leq 8 \cdot \left(\frac{q}{2^{n-a/3}}\right)^{3/2} + 2^{a-1} \cdot \left(\frac{32q}{2^n}\right)^{2^{b-2}} + \mathbf{Adv}^{\mathsf{PRF}}_{\mathsf{trunc}_a}(2q).$$

Proof (Proof of Theorem 2). The general proof strategy will follow that by [10]. Let $\Pi_1, \ldots, \Pi_r \xleftarrow{\$} \mathsf{Perm}(\mathbb{F}_2^n)$ such that all permutations Π_j are pairwise independent. We consider two oracles, $\mathcal{O}_{\mathrm{ideal}}$ and $\mathcal{O}_{\mathrm{real}}$. Let $\mathbf{D}$ be a distinguisher that is given access to one of them, chosen uniformly at random. $\mathbf{D}$ shall distinguish between both worlds, given the transcript τ of queries of $\mathbf{D}$ to the oracle, the corresponding responses, and intermediate variables. We define by $\mathbf{I}_n$ the identity permutation over $\mathbb{F}_2^n$. For integers $n = a + b$ and $X \in \mathbb{F}_2^n$ with $X = V \| Y$ and $V \in \mathbb{F}_2^a$, $Y \in \mathbb{F}_2^b$, we define $\mathsf{msb}_a(X) = V$ to always return the leftmost a bits of X and $\mathsf{lsb}_b(X) = Y$ to return the b least significant b bits of X, and $(V,Y) \xleftarrow{a,n-a} X$ splits X into an a-bit part V and an $(n-a)$-bit part Y.

On message input M^i, the real world $\mathcal{O}_{\mathrm{real}}$ uses $\mathsf{XTH}[a,r]_{\Pi_1,\ldots,\Pi_r}(M^i)$ and produces and outputs $V_1^i, V_2^i, W_2^i, \ldots, V_r^i, W_r^i$, where for each $j \in [r]$, $(V_j^i, Y_j^i) \xleftarrow{a,b} \Pi_j(M^i)$ and $W_j^i = Y_1^i \oplus Y_j^i$ for all $j \in [2..r]$. The values are collected in vectors $\mathbf{V} = (\mathbf{V}^1,\ldots,\mathbf{V}^q)$, $\mathbf{Y} = (\mathbf{Y}^1,\ldots,\mathbf{Y}^q)$, and $\mathbf{W} = (\mathbf{W}^1,\ldots,\mathbf{W}^q)$ with $\mathbf{V}^i = (V_1^i,\ldots,V_r^i)$, $\mathbf{Y}^i = (Y_1^i,\ldots,Y_r^i)$, and $\mathbf{W}^i = (W_2^i,\ldots,W_r^i)$ for all $i \in [q]$. Let $\tau = (\mathbf{V},\mathbf{W})$ be the transcript. Over all queries, we define the short-hand notation $\mathbf{V}_j = (V_j^1,\ldots,V_j^q)$ for some $j \in [r]$.

Algorithm 2. Real-world oracles from the analysis of $\mathsf{XTH}[a,r]_{\Pi_1,\ldots,\Pi_r}$.

11: **function** $\mathcal{O}_1(\mathbf{M})$	31: **function** $\mathcal{O}_2(\mathbf{M})$	51: **function** $\mathrm{PSoP}[r](\mathbf{M},\mathbf{V})$
12: $\quad \Pi_1,\Pi_2,\ldots\Pi_r \twoheadleftarrow \mathrm{Perm}(\mathbb{F}_2^n)$	32: $\quad \mathbf{V} \leftarrow \mathrm{PTrunc}[r](\mathbf{M})$	52: $\quad$ **for** $j \leftarrow 1$ **to** r **do**
13: $\quad M^1,\ldots,M^q \leftarrow \mathbf{M}$	33: $\quad \mathbf{W} \leftarrow \mathrm{PSoP}[r](\mathbf{M},\mathbf{V})$	53: $\quad\quad$ **if** $\mathrm{Perm}_{\mathrm{comp}}(\mathbf{V}_j) = \emptyset$
14: $\quad$ **for** $i \leftarrow 1$ **to** q **do**	34: $\quad$ **return** $\tau = (\mathbf{V},\mathbf{W})$	$\quad\quad$ **then**
15: $\quad\quad$ **for** $j \leftarrow 1$ **to** r **do**		54: $\quad\quad\quad \Pi_j \twoheadleftarrow \mathbf{I}_n$
16: $\quad\quad\quad (V_j^i,Y_j^i) \xleftarrow{a,b} \Pi_j(M^i)$	41: **function** $\mathrm{PTRUNC}[r](\mathbf{M})$	55: $\quad\quad$ **else**
17: $\quad\quad\quad$ **if** $j \geq 2$ **then**	42: $\quad$ **for** $i \leftarrow 1$ **to** q **do**	56: $\quad\quad\quad \Pi_j \twoheadleftarrow \mathrm{Perm}_{\mathrm{comp}}(\mathbf{V}_j)$
18: $\quad\quad\quad\quad W_j^i \leftarrow Y_1^i \oplus Y_j^i$	43: $\quad\quad$ **for** $j \leftarrow 1$ **to** r **do**	57: $\quad$ **for** $i \leftarrow 1$ **to** q **do**
19: $\quad\quad \mathbf{V}^i \leftarrow (V_1^i,V_2^i,\ldots,V_r^i)$	44: $\quad\quad\quad V_j^i \twoheadleftarrow \mathbb{F}_2^a$	58: $\quad\quad$ **for** $j \leftarrow 1$ **to** r **do**
20: $\quad\quad \mathbf{W}^i \leftarrow (W_2^i,\ldots,W_r^i)$	45: $\quad\quad \mathbf{V}^i \leftarrow (V_1^i,\ldots,V_r^i)$	59: $\quad\quad\quad Y_j^i \leftarrow \mathrm{lsb}_b(\Pi_j(\langle i \rangle))$
21: $\quad \mathbf{V} \leftarrow (\mathbf{V}^1,\ldots,\mathbf{V}^q)$	46: $\quad$ **return** $\mathbf{V} = (\mathbf{V}^1,\ldots,\mathbf{V}^q)$	60: $\quad\quad\quad$ **if** $j \geq 2$ **then**
22: $\quad \mathbf{W} \leftarrow (\mathbf{W}^1,\ldots,\mathbf{W}^q)$		61: $\quad\quad\quad\quad W_j^i \leftarrow Y_1^i \oplus Y_j^i$
23: $\quad \tau \leftarrow (\mathbf{V},\mathbf{W})$		62: $\quad\quad \mathbf{W}^i \leftarrow (W_2^i,\ldots,W_r^i)$
24: $\quad$ **return** τ		63: $\quad$ **return** $\mathbf{W} = (\mathbf{W}^1,\ldots,\mathbf{W}^q)$

The ideal world $\mathcal{O}_{\mathrm{ideal}}$ samples all outputs $V_j^i \twoheadleftarrow \mathbb{F}_2^a$, for all $i \in [q]$ and $j \in [r]$ and samples $W_2^i,\ldots,W_r^i \twoheadleftarrow \mathbb{F}_2^b$, for all $i \in [q]$. We denote the real-world oracle as $\mathcal{O}_1$ since we will modify it stepwise in the following. It holds that

$$\mathbf{Adv}_{\mathsf{XTH}[a,r]}^{\mathrm{PRF}}(\mathcal{A}) \leq |\Pr[\mathcal{O}_{\mathrm{ideal}}] - \Pr[\mathcal{O}_{\mathrm{real}}]| \,.$$

Next, we separate the a-bit values, $(V_1^i,\ldots,V_r^i)$, given out in clear from the results of the sums, $(W_2^i,\ldots,W_r^i)$. This yields the modified real world $\mathcal{O}_2$. Internally, $\mathcal{O}_2$ uses a function $\mathrm{PTrunc}[r]$ that samples the values $\mathbf{V} = (V_1,\ldots,V_r)$ as a-bit values sampled independently uniformly at random from $\mathbb{F}_2^a$ each. This is given in Algorithm 2. Moreover, we define $\mathrm{PSoP}[r]$, which takes $(V_1,\ldots,V_r)$ and samples $r-1$ permutations compatible to it (if they exist) and computes the vector of sum values, $\mathbf{W} = (W_2^i,\ldots,W_r^i)$, from it. For all $j \in [r]$ and given vectors of a-bit strings $\mathbf{V}_j = (V_j^1,\ldots,V_j^q) \in (\mathbb{F}_2^a)^q$, we define $\mathrm{Perm}_{\mathrm{comp}}(\mathbf{V}_j) \subseteq \mathrm{Perm}(\mathbb{F}_2^{n-a})$ as the set of all n-bit permutations that would produce $\mathbf{V}_j$ in their most significant a-bit outputs for the inputs in $\mathbf{V}_j$. The difference between both worlds is upper bounded by

$$|\Pr[\mathcal{O}_2] - \Pr[\mathcal{O}_{\mathrm{real}}]| \leq \mathbf{Adv}_{\mathrm{trunc}_a}^{\mathrm{PRF}}(rq) \,.$$

From the triangle inequality, the difference in the setting is at most

$$|\Pr[\mathcal{O}_{\mathrm{ideal}}] - \Pr[\mathcal{O}_{\mathrm{real}}]| \leq |\Pr[\mathcal{O}_{\mathrm{ideal}}] - \Pr[\mathcal{O}_2]| + \mathbf{Adv}_{\mathrm{trunc}_a}^{\mathrm{PRF}}(rq) \,.$$

We want to upper bound the distance between the multi-sum of pairwise independent permutations and the function that produces random bits. For the values $V_1, V_2, \ldots, V_r$, we define counters

$$C_{\mathbf{V},j}(i) \stackrel{\mathrm{def}}{=} \left| \left\{ V_j^{i'} : V_j^{i'} = V_j^i \right\} \right| \,, \text{ for all } j \in [r] \,.$$

Those counters will later have to remain below 2^{b-2}. For the case that one of them exceeds this amount, we define a set **bad** of vectors $\mathbf{V}$ such that there exists

$k \in [r]$ with $C_{\mathbf{V},k}(i) \geq 2^{b-2}$, which we denote as bad. Given a transcript τ that contains $\mathbf{V}$, we see that

$$\mathbb{E}_\tau[\Pr[\mathcal{O}_{\text{ideal}} = \tau] - \Pr[\mathcal{O}_2 = \tau]] \leq \mathbb{E}_\tau[\Pr[\mathcal{O}_{\text{ideal}} = \tau] - \Pr[\mathcal{O}_2 = \tau|\overline{\text{bad}}]] + \Pr[\text{bad}].$$

Multi-collision. We can upper bound $\Pr[\text{bad}]$ first, which requires a (2^{b-2})-collision of values $V_j^{i_1} = \cdots = V_j^{i_{2^{b-2}}}$ inside any one of r vectors $\mathbf{V}_j$ in $\mathbf{V}$. Since the values V_j^i are chosen independently and uniformly at random each, the probability for a t-collision is upper bounded by

$$\frac{(rq)^t}{2^{a(t-1)} \cdot t!}.$$

By using Stirling's approximation and substituting $t = 2^{b-2}$

$$\Pr[\text{bad}] \leq \frac{1}{\sqrt{2\pi}} \cdot \frac{(rq)^t}{2^{a(t-1)}} \cdot \left(\frac{1}{2^{3/2} \cdot t}\right)^t \leq \frac{2^a}{\sqrt{2\pi}} \cdot \left(\frac{rq}{2^{a-3/2} \cdot t}\right)^t$$

$$\leq \frac{2^a}{\sqrt{2\pi}} \cdot \left(\frac{rq}{2^{a-2} \cdot 2^{b-2}}\right)^{2^{b-2}} \leq 2^{a-1} \cdot \left(\frac{16rq}{2^n}\right)^{2^{b-2}}.$$

We have to upper bound the expectation of the difference of the probabilities of the realized good transcripts in world $\mathcal{O}_2$ and $\mathcal{O}_{\text{ideal}}$. Since for good transcripts, the vectors $\mathbf{V}$ are sampled equally in both worlds, we can focus on the vectors $\mathbf{W}$. We obtain the following.

Theorem 3. Let a, b, q, r be positive integers and $\tau = (\mathbf{V}, \mathbf{W})$ be a good transcript such that $C_{\mathbf{V},j}(i) < 2^{b-2}$ holds for all $i \in [q]$ and $j \in [r]$ and $q \leq 2^n/(3r)$. Then, for $r \geq 2$

$$\mathbb{E}_\tau\left[|\Pr[\mathcal{O}_2 = \tau] - \Pr[\mathcal{O}_{\text{ideal}} = \tau]|\right] \leq \left(\frac{4}{3}\right)^r \cdot \left(\frac{rq}{2^{n-a/3}}\right)^{3/2}.$$

5   Proof of Theorem 3

We can easily see that $\Pr_{\mathcal{O}_{\text{ideal}}}[W^i = W|\mathbf{W}^{i-1}] = 2^{-(r-1)b}$. Though, it remains to determine the probability in the real world. We denote the outputs $(Y_1^i, Y_2^i, \ldots, Y_r^i)$ also as $(y_1^i, y_2^i, \ldots, y_r^i)$ and the fixed sum values at the i-th step $(W_2^i, \ldots, W_r^i)$ also as $(w_2^i, \ldots, w_r^i)$. We consider r independent permutations π_1, $\ldots \pi_r$. We have to determine the probability

$$\Pr_{\mathcal{O}_{\text{real}}}[\mathbf{W}^i = (w_2^i, \ldots, w_r^i)|\mathbf{Y}^{i-1}],$$

where $\mathbf{Y}^{i-1} = (Y_1^1, \ldots, Y_r^1, \ldots, Y_1^{i-1}, \ldots, Y_r^{i-1})$. Fix a tuple $\mathbf{W}^i = (w_2^i, \ldots, w_r^i) \in (\mathbb{F}_2^b)^{r-1}$. We define $q \times r$ sets $\mathcal{S}_j^i = \{y_j^1, \ldots, y_j^{i-1}\}$ for all $i \in [q]$ and $j \in [r]$. Furthermore, we propose sets of translated values $\mathcal{S}_{y_j \to w_j}^i = \mathcal{S}_j^i \oplus w_j \triangleq$

$\{Y_j \in \mathcal{S}_j^i : Y_j \oplus w_j\}$ to denote the elementwise translation of $\mathcal{S}_j^i$ for the fixed scalar $w_j \in \mathbb{F}_2^b$ for all $j \in \{2, \ldots, r\}$. For consistency, we introduce $w_1^i = 0^b$ for all $i \in [q]$ so we can define $\mathcal{S}_{y_1 \to w_1}^i = \mathcal{S}_1^i$. We define cardinalities $s_j^{i,w_j} = |\mathcal{S}_{y_j \to w_j}^i| = |\mathcal{S}_j^i|$ for all $j \in [r]$, and will use the short form $s_j^i = s_j^{i,w_j}$ hereafter. We have to find the number of possible solutions $Y^i = (Y_1^i, \ldots, Y_r^i)$ for the next fixed tuple $W^i = (w_2^i, \ldots, w_r^i)$. For $Y_1^i \oplus Y_2^i = w_2^i$, $Y_1^i \oplus Y_3^i = w_3^i$, $\ldots$, it must hold that

$$Y_1^i \in \mathbb{F}_2^b \setminus \left(\mathcal{S}_1^i \cup \bigcup_{j=2}^r (\mathcal{S}_{y_j \to w_j}^i) \right).$$

Let n^i denote the number of choices for Y_1^i. From the inclusion-exclusion principle

$$
\begin{aligned}
n^i = {}& 2^b - \left(|\mathcal{S}_{y_1 \to w_1}^i| + |\mathcal{S}_{y_2 \to w_2}^i| + \cdots + |\mathcal{S}_{y_r \to w_r}^i| \right) \\
& + \left(|\mathcal{S}_{y_1 \to w_1}^i \cap \mathcal{S}_{y_2 \to w_2}^i| + |\mathcal{S}_{y_1 \to w_1}^i \cap \mathcal{S}_{y_3 \to w_3}^i| + \cdots + |\mathcal{S}_{y_{r-1} \to w_{r-1}}^i \cap \mathcal{S}_{y_r \to w_r}^i| \right) \\
& - \left(|\mathcal{S}_{y_1 \to w_1}^i \cap \mathcal{S}_{y_2 \to w_2}^i \cap \mathcal{S}_{y_3 \to w_3}^i| \right) + \cdots \right) + \cdots \\
= {}& 2^b - \left(\sum_{j=1}^r |\mathcal{S}_{y_j \to w_j}^i| \right) + \left(\sum_{j_1 < j_2} |\mathcal{S}_{y_{j_1} \to w_{j_1}}^i \cap \mathcal{S}_{y_{j_2} \to w_{j_2}}^i| \right) \\
& - \left(\sum_{1 \le j_1 < j_2 < j_3 \le r} |\mathcal{S}_{y_{j_1} \to w_{j_1}}^i \cap \mathcal{S}_{y_{j_2} \to w_{j_2}}^i \cap \mathcal{S}_{y_{j_3} \to w_{j_3}}^i| \right) + \cdots \\
= {}& 2^b - \left(\sum_{j=1}^r s_j^i \right) + \left(\sum_{1 \le j_1 < j_2 \le r} s_{j_1,j_2}^{i,w_{j_1},w_{j_2}} \right) - \left(\sum_{1 \le j_1 < j_2 < j_3 \le r} s_{j_1,j_2,j_3}^{i,w_{j_1},w_{j_2},w_{j_3}} \right) \\
& + \cdots + (-1)^r \left(\sum_{1 \le j_1 < \cdots < j_r \le r} s_{j_1,\ldots,j_r}^{i,w_{j_1},w_{j_2},\ldots,w_{j_r}} \right),
\end{aligned}
\tag{1}
$$

where we define $s_{1,2}^{i,w_1,w_2}$, $s_{1,2,3}^{i,w_1,w_2,w_3}$, $\ldots$ for the cardinalities of the corresponding intersection sets in a natural manner. We call the terms $s_{1,2}^{i,w_1,w_2}$ 2-tuple-related, $s_{1,2,3}^{i,w_1,w_2,w_3}$ 3-tuple-related, and so on. For each, we have to upper bound its expectation and variance.

Expectation and Variance of 2-Tuple-Related Terms. We can use the knowledge about $s_{1,2}^{i,w_1,w_2} = s_{1,2}^{i,0,w_2} = D_{i,w}$ from [4,10]. Thus, the expectation and variance of all cardinalities of two-component intersections can be taken from Equations (34), (35) in [10] as

$$
\mathbb{E}\left[s_{j_1,j_2}^{i,w_{j_1},w_{j_2}} \right] = \frac{s_{j_1}^i s_{j_2}^i}{2^b} \quad \text{and} \quad \mathbf{Var}\left[s_{j_1,j_2}^{i,w_{j_1},w_{j_2}} \right] \le \frac{2\, s_{j_1}^i s_{j_2}^i}{2^b}.
\tag{2}
$$

For independent permutations $\pi_1, \ldots, \pi_r$, and independent Binomial variables, we can derive them more precisely.

Lemma 1. For distinct $j_1, j_2 \in [r]$, it holds that

$$\mathbb{E}\left[s_{j_1,j_2}^{i,w_{j_1},w_{j_2}}\right] = \frac{s_{j_1}^i s_{j_2}^i}{2^b} \quad \text{and} \quad \mathbf{Var}\left[s_{j_1,j_2}^{i,w_{j_1},w_{j_2}}\right] = \frac{s_{j_1}^i s_{j_2}^i}{2^b} - \frac{(s_{j_1}^i s_{j_2}^i)^2}{2^{3b}}.$$

Proof. Let us focus on $s_{1,2}^{i,w_1,w_2}$; the remaining 2-tuple-related terms $s_{j_1,j_2}^{i,w_{j_1},w_{j_2}}$ behave similarly, for all $j_1 \neq j_2$, $j_1, j_2 \in [r]$. Given fixed $w_2 \in \mathbb{F}_2^b$, for each $y_1 \in \mathbb{F}_2^b$, we define Bernoulli variables I_{y_1} as

$$I_{y_1} \triangleq \begin{cases} 1 & y_1 \in \mathcal{S}_1^i \wedge y_1 \oplus w_2 \in \mathcal{S}_2^i \\ 0 & \text{otherwise.} \end{cases}$$

Then, we derive

$$\mathbb{E}\left[s_{1,2}^{i,w_1,w_2}\right] = \sum_{y_1 \in \mathbb{F}_2^b} \Pr[I_{y_1}].$$

To obtain

$$\mathbf{Var}\left[\mathsf{x}\right] = \mathbb{E}\left[\mathsf{x}^2\right] - \left(\mathbb{E}\left[\mathsf{x}\right]\right)^2,$$

we have to determine $\mathbb{E}\left[\mathsf{x}^2\right]$. For a sum of n independent Bernoulli variables I_{y_1}, with $\Pr[I_{y_1} = 1] = p$ for all y_1,

$$\mathsf{x} = \sum_{y_1} \Pr[I_{y_1} = 1],$$

it holds that

$$\mathbb{E}\left[\mathsf{x}^2\right] = \mathbb{E}\left[\left(\sum_{j=1}^{n} I_j\right)^2\right] = n(n-1)p^2 + np.$$

In our case, $n = 2^b$ and $p = s_1^i s_2^i \cdot 2^{-2b}$, for all $y_1 \in \mathbb{F}_2^b$. Given that $\left(\mathbb{E}\left[\mathsf{x}\right]\right)^2 = (2^b p)^2$, we obtain

$$\mathbf{Var}\left[s_{1,2}^{i,0,w_2}\right] \leq \frac{s_1^i s_2^i}{2^b} - \frac{(s_1^i s_2^i)^2}{2^{3b}} \quad \text{and in general}$$

$$\mathbf{Var}\left[s_{j_1,j_2}^{i,w_{j_1},w_{j_2}}\right] \leq \frac{s_{j_1}^i s_{j_2}^i}{2^b} - \frac{(s_{j_1}^i s_{j_2}^i)^2}{2^{3b}}.$$

Expectation and Variance of 3-Tuple-Related Terms. Next, we consider the expectation and variance of $s_{j_1,j_2,j_3}^{i,w_{j_1},w_{j_2},w_{j_3}}$.

Lemma 2. For distinct $j_1, j_2, j_3 \in [r]$, it holds that

$$\mathbb{E}\left[s_{j_1,j_2,j_3}^{i,w_{j_1},w_{j_2},w_{j_3}}\right] = \frac{s_{j_1}^i s_{j_2}^i s_{j_3}^i}{2^{2b}}, \quad \mathbf{Var}\left[s_{j_1,j_2,j_3}^{i,w_{j_1},w_{j_2},w_{j_3}}\right] = \frac{s_{j_1}^i s_{j_2}^i s_{j_3}^i}{2^{2b}} - \frac{(s_{j_1}^i s_{j_2}^i s_{j_3}^i)^2}{2^{5b}}.$$

Proof. Again, the remaining 3-tuple-related terms $s_{j_1,j_2,j_3}^{i,w_{j_1},w_{j_2},w_{j_3}}$ behave similarly, for all distinct $j_1, j_2, j_3 \in [r]$. Given fixed $w_1 = 0^b$ and $w_2, w_3 \in \mathbb{F}_2^b$, for each $y_1 \in \mathbb{F}_2^b$, we define Bernoulli variables I_{y_1} as

$$I_{y_1} \triangleq \begin{cases} 1 & y_1 \in \mathcal{S}_1^i \wedge y_1 \oplus w_2 \in \mathcal{S}_2^i \wedge y_1 \oplus w_3 \in \mathcal{S}_3^i \\ 0 & \text{otherwise.} \end{cases}$$

Then, it holds that

$$\mathbb{E}\left[s_{1,2,3}^{i,w_1,w_2,w_3}\right] = \mathbb{E}\left[\sum_{y_1 \in \mathbb{F}_2^b} I_{y_1}\right] = \sum_{y_1 \in \mathbb{F}_2^b} \mathbb{E}\left[I_{y_1}\right].$$

Since the expectations for a fixed value $y_1 \in \mathbb{F}_2^b$ and its translations to be in the list of all three permutations are mutually independent, the probability is 2^{-3b}. Over all elements of the sets $|\mathcal{S}_{y_1 \to w_1}^i| = |\mathcal{S}_1^i|$, $|\mathcal{S}_{y_2 \to w_2}^i|$, and $|\mathcal{S}_{y_3 \to w_3}^i|$, it holds that

$$\mathbb{E}\left[I_{y_1}\right] = \frac{s_1^i s_2^i s_3^i}{2^{3b}} \quad \text{and therefore} \quad \mathbb{E}\left[s_{y_1,y_2,y_3}^{i,w_1,w_2,w_3}\right] = \frac{s_1^i s_2^i s_3^i}{2^{2b}}. \tag{3}$$

It remains to determine its variance

$$\mathbf{Var}\left[s_{y_1,y_2,y_3}^{i,w_1,w_2,w_3}\right] = \mathbb{E}\left[\left(s_{y_1,y_2,y_3}^{i,w_1,w_2,w_3}\right)^2\right] - \left(\mathbb{E}\left[s_{y_1,y_2,y_3}^{i,w_1,w_2,w_3}\right]\right)^2$$

$$= \mathbf{Var}\left[\sum_{y_1 \in \mathbb{F}_2^b} I_{y_1}\right] = \sum_{y_1 \in \mathbb{F}_2^b} \mathbf{Var}\left[I_{y_1}\right] + \sum_{y_1 \neq y_1'} \mathbf{Cov}\left[I_{y_1}, I_{y_1'}\right],$$

with the covariance

$$\mathbf{Cov}\left[I_{y_1}, I_{y_1'}\right] = \mathbb{E}\left[I_{y_1} \cdot I_{y_1'}\right] - \mathbb{E}\left[I_{y_1}\right]\mathbb{E}\left[I_{y_1'}\right]$$

$$= \mathbb{E}\left[I_{y_1}\right] \cdot \Pr[I_{y_1'} = 1 | I_{y_1} = 1] - \mathbb{E}\left[I_{y_1}\right]\mathbb{E}\left[I_{y_1'}\right].$$

For the variance of the Bernoulli variables, it holds that

$$\mathbf{Var}\left[I_{y_1}\right] = \mathbb{E}\left[(I_{y_1})^2\right] - (\mathbb{E}\left[I_{y_1}\right])^2 = \mathbb{E}\left[I_{y_1}\right] - (\mathbb{E}\left[I_{y_1}\right])^2 = \frac{s_u^i s_v^i s_w^i}{2^{3b}} - \left(\frac{s_u^i s_v^i s_w^i}{2^{3b}}\right)^2.$$

For their covariance, we need to determine the conditional probability. We consider the case that $y_1' \notin \{y_1 \oplus w_2, y_1 \oplus w_3\}$. Since $y_1' \neq y_1$, it holds that all values differ mutually

$$\Pr[I_{y_1'} = 1 | I_{y_1} = 1] = \Pr[(y_1' \in \mathcal{S}_1^i) \wedge (y_1' \oplus w_2 \in \mathcal{S}_2^i) \wedge (y_1' \oplus w_3 \in \mathcal{S}_3^i) |$$

$$(y_1 \in \mathcal{S}_1^i) \wedge (y_1 \oplus w_2 \in \mathcal{S}_2^i) \wedge (y_1 \oplus w_3 \in \mathcal{S}_3^i)]$$

$$\leq \frac{(s_1^i - 1)(s_2^i - 1)(s_3^i - 1)}{(2^b - 1)^3}.$$

We conduct it for $y_1' = y_1 \oplus w_2$ exemplarily. From the requirement of the covariance that $y_1' \neq y_1$, we must exclude $w_2 = 0$.

$$\Pr[I_{y_1 \oplus w_2} = 1 | I_{y_1} = 1]$$
$$\leq \Pr[(y_1 \oplus w_2 \in \mathcal{S}_1^i) \wedge (y_1 \in \mathcal{S}_2^i) \wedge (y_1 \oplus w_2 \oplus w_3 \in \mathcal{S}_3^i) | I_{y_1} = 1]$$
$$\leq \frac{(s_u^i - 1)(s_v^i - 1)(s_w^i - 1)}{(2^b - 1)^3}.$$

From $s_1^i, s_2^i, s_3^i < 2^b$, it follows that

$$\Pr[I_{y_1 \oplus w_2} = 1 | I_{y_1} = 1] \leq \mathbb{E}[I_{y_1 \oplus w_2}],$$

and therefore $\mathbf{Cov}[I_{y_1}, I_{y_1 \oplus w_2}] \leq 0$ in this case. A similar argument holds for $y_1' = y_1 \oplus w_3$, $w_2 \neq w_3$. It remains to consider $y_1' = y_1 \oplus w_2$ with $w_2 = w_3$.

$$\Pr[I_{y_1 \oplus w_2} = 1 | I_{y_1} = 1, w_2 = w_3]$$
$$\leq \Pr[(y_1 \oplus w_2 \in \mathcal{S}_1^i) \wedge (y_1 \in \mathcal{S}_2^i) \wedge (y_1 \in \mathcal{S}_3^i) | I_{y_1} = 1]$$
$$\leq \frac{(s_1^i - 1)(s_2^i - 1)(s_3^i - 1)}{(2^b - 1)^3}.$$

Again, $s_1^i, s_2^i, s_3^i < 2^b$ implies

$$\Pr[I_{y_1 \oplus w_2} = 1 | I_{y_1} = 1] \leq \mathbb{E}[I_{y_1 + w_2}],$$

and therefore, $\mathbf{Cov}[I_{y_1}, I_{y_1 \oplus w_2}] \leq 0$. Thus, it holds that $\mathbf{Cov}[I_{y_1}, I_{y_1'}] \leq 0$ over all cases of y_1', and it follows that

$$\mathbf{Var}\left[s_{1,2,3}^{i,w_1,w_2,w_3}\right] \leq \sum_{y_1 \in \mathbb{F}_2^b} \mathbf{Var}[I_{y_1}]$$

$$= 2^b \cdot \left(\frac{s_1^i s_2^i s_3^i}{2^{3b}} - \left(\frac{s_1^i s_2^i s_3^i}{2^{3b}}\right)^2\right) = \frac{s_1^i s_2^i s_3^i}{2^{2b}} - \frac{(s_1^i s_2^i s_3^i)^2}{2^{5b}}.$$

Expectation and Variance of Terms for General Tuples

Lemma 3. Let $t \leq r$ and $\mathcal{I} = \{j_1, \ldots, j_t\} \subseteq \{1, \ldots, r\}$. Then, it holds for the expectation and variance that

$$\mathbb{E}\left[s_{j_1, j_2, \ldots, j_t}^{i, w_{j_1}, w_{j_2}, \ldots, w_{j_t}}\right] = \frac{\prod_{j \in \mathcal{I}} s_j^i}{2^{(t-1)b}}, \quad \mathbf{Var}\left[s_{j_1, j_2, \ldots, j_t}^{i, w_{j_1}, w_{j_2}, \ldots, w_{j_t}}\right] = \frac{\prod_{j \in \mathcal{I}} s_j^i}{2^{(t-1)b}} - \frac{\left(\prod_{j \in \mathcal{I}} s_j^i\right)^2}{2^{(2t-1)b}}.$$

Proof. Given fixed $w_{j_2}, \ldots w_{j_t} \in \mathbb{F}_2^b$, for each $y_1 \in \mathbb{F}_2^b$, we define Bernoulli variables I_{y_1} as

$$I_{y_1} \triangleq \begin{cases} 1 & y_1 \in \mathcal{S}_1^i \wedge y_1 \oplus w_{j_2} \in \mathcal{S}_{j_2}^i \wedge \ldots \wedge y_1 \oplus w_{j_t} \in \mathcal{S}_{j_t}^i \\ 0 & \text{otherwise.} \end{cases}$$

Then, it holds that

$$\mathbb{E}\left[s_{j_1,j_2,\cdots,j_t}^{i,w_{j_1},w_{j_2},\ldots,w_{j_t}}\right] = \mathbb{E}\left[\sum_{y_1\in\mathbb{F}_2^b} I_{y_1}\right] = \sum_{y_1\in\mathbb{F}_2^b} \mathbb{E}\left[I_{y_1}\right].$$

Since the expectations for a fixed value $y_1 \in \mathbb{F}_2^b$ and its translations to be in the list of all three permutations are mutually independent, the probability is 2^{-tb}. Over all elements of the sets, it holds that

$$\mathbb{E}\left[I_{y_1}\right] = \frac{\prod_{j\in\mathcal{I}} s_j^i}{2^{tb}} \quad \text{and therefore} \quad \mathbb{E}\left[s_{j_1,j_2,\cdots,j_t}^{i,w_{j_1},w_{j_2},\ldots,w_{j_t}}\right] = \frac{\prod_{j\in\mathcal{I}} s_j^i}{2^{(t-1)b}}.$$

For $\mathsf{x} = \sum_{y_1} \Pr[I_{y_1} = 1]$, as a sum of n independent Bernoulli variables I_{y_1}, with $\Pr[I_{y_1} = 1] = p$ for all y_1, it holds that

$$\mathbb{E}\left[\mathsf{x}^2\right] = \mathbb{E}\left[\left(\sum_{j=1}^n I_j\right)^2\right] = n(n-1)p^2 + np.$$

In our case, $n = 2^b$ and $p = \prod_{j\in\mathcal{I}} s_j^i \cdot 2^{-tb}$, for all $y_1 \in \mathbb{F}_2^b$. Given that $(\mathbb{E}[\mathsf{x}])^2 = (2^b p)^2$, we obtain

$$\mathbf{Var}\left[s_{j_1,j_2,\cdots,j_t}^{i,w_{j_1},w_{j_2},\ldots,w_{j_t}}\right] \leq \frac{\prod_{j\in\mathcal{I}} s_j^i}{2^{(t-1)b}} - \frac{\left(\prod_{j\in\mathcal{I}} s_j^i\right)^2}{2^{(2t-1)b}}.$$

Determining the Ratio. In the real and ideal worlds, it holds that

$$\Pr_{\mathcal{O}_{\text{real}}}\left[W^i = (w_2^i,\ldots,w_r^i)|\mathbf{Y}^{i-1}\right] = \mathbb{E}\left[\frac{n^i}{d^i}\right] \quad \text{and}$$

$$\Pr_{\mathcal{O}_{\text{ideal}}}\left[W^i = (w_2^i,\ldots,w_r^i)|\mathbf{W}^{i-1}\right] = \frac{1}{2^{(r-1)b}},$$

respectively, with n^i given in Eq. (1). The number of all choices of Y^i, that represents the denominator d^i, is given by

$$d^i = (2^b - s_1^i) \cdot (2^b - s_2^i) \cdots (2^b - s_r^i) = \prod_{j=1}^r (2^b - s_j^i)$$

$$= 2^{rb} - 2^{(r-1)b}\left(\sum_{j=1}^r s_j^i\right) + 2^{(r-2)b}\left(\sum_{1\leq j_1<j_2\leq r} s_{j_1}^i s_{j_2}^i\right) -$$

$$2^{(r-3)b}\left(\sum_{1\leq j_1<j_2<j_3\leq r} s_{j_1}^i s_{j_2}^i s_{j_3}^i\right) + \cdots + (-1)^r\left(\sum_{1\leq j_1<\cdots<j_r\leq r} s_{j_1}^i \cdots s_{j_r}^i\right),$$

$$\tag{4}$$

which yields

$$
\mathbb{E}\left[\left(\Pr_{\mathcal{O}_{\text{real}}}[W^i = (w_2^i, \dots, w_r^i)|\mathbf{Y}^{i-1}] - \Pr_{\mathcal{O}_{\text{ideal}}}[W^i = (w_2^i, \dots, w_r^i)|\mathbf{W}^{i-1}]\right)^2\right]
$$

$$
= \mathbb{E}\left[\left(\frac{n^i}{d^i} - \frac{1}{2^{(r-1)b}}\right)^2\right] = \mathbb{E}\left[\left(\frac{2^{(r-1)b} \cdot n^i - d^i}{2^{(r-1)b} \cdot d^i}\right)^2\right]
$$

$$
\leq \left(\frac{4}{3}\right)^{2r} \cdot \frac{1}{2^{(4r-2)b}} \cdot \mathbb{E}\left[\left(2^{(r-1)b} \cdot n^i - d^i\right)^2\right], \tag{5}
$$

where we used the assumption of $s_j^i < 2^{b-2}$, for all $j \in [r]$, to upper bound $d^i \geq \left(\frac{3}{4} \cdot 2^b\right)^r$. In the following, we focus on the rightmost term of Eq. (5), i.e., the expectation of the squared difference. We observe that the two leftmost terms of $2^{(r-1)b} \cdot n^i$, that we call $\underline{n}^i$ for short,

$$
\underline{n}^i = 2^{(r-1)b} \cdot \left(2^b - \sum_{j=1}^r s_j^i\right) = 2^{rb} - 2^{(r-1)b}\left(\sum_{j=1}^r s_j^i\right),
$$

are identical to the two leftmost terms in d^i as in Eq. (4) and cancel in the difference. We define

$$
\overline{n}^i = n^i - \left(2^b - \sum_{j=1}^r s_j^i\right) = \left(\sum_{1 \leq j_1 < j_2 \leq r} s_{j_1, j_2}^{i, w_{j_1}, w_{j_2}}\right)
$$

$$
- \left(\sum_{1 \leq j_1 < j_2 < j_3 \leq r} s_{j_1, j_2, j_3}^{i, w_{j_1}, w_{j_2}, w_{j_3}}\right) + \cdots + (-1)^r \left(s_{1,\dots,r}^{i, w_1, w_2, \dots, w_r}\right). \tag{6}
$$

We define $\overline{d}^i = d^i - \underline{n}^i$. We substitute the extended formulation of d^i from Eq. (4) into Eq. (6) and factor out $(2^{(r-1)b})^2$:

$$
\mathbb{E}\left[\left(2^{(r-1)b} \cdot n^i - d^i\right)^2\right] = \mathbb{E}\left[2^{2(r-1)b} \cdot \left(n^i - \frac{d^i}{2^{(r-1)b}}\right)^2\right]
$$

$$
= 2^{2(r-1)b} \cdot \mathbb{E}\left[\left(\overline{n}^i - \frac{\overline{d}^i}{2^{(r-1)b}}\right)^2\right]. \tag{7}
$$

We can write the rightmost term as

$$
\frac{\overline{d}^i}{2^{(r-1)b}} = \left(\sum_{1 \leq j_1 < j_2 \leq r} \frac{s_{j_1}^i s_{j_2}^i}{2^b}\right) - \left(\sum_{1 \leq j_1 < j_2 < j_3 \leq r} \frac{s_{j_1}^i s_{j_2}^i s_{j_3}^i}{2^{2b}}\right)
$$

$$
+ \cdots + (-1)^r \cdot \frac{s_1^i \cdots s_r^i}{2^{(r-1)b}}. \tag{8}
$$

From Eq. (1) for n^i, we can observe that for the sum of terms x in $\overline{n}^i$, Eq. (8) consists of exactly the sum of terms $\mathbb{E}[x]$.

$$(7) = 2^{(2r-2)b} \cdot \mathbb{E}\left[\left(\overline{n}^i - \mathbb{E}\left[\overline{n}^i\right]\right)^2\right] = 2^{(2r-2)b} \cdot \mathbf{Var}[\overline{n}^i].$$

Inserting it into Eq. (5) yields

$$\left(\frac{4}{3}\right)^{2r} \cdot \frac{1}{2^{(4r-2)b}} \cdot \mathbb{E}\left[\left(2^{(r-1)b} \cdot n^i - d^i\right)^2\right] \leq \left(\frac{4}{3}\right)^{2r} \cdot \frac{1}{2^{2rb}} \cdot \mathbf{Var}[\overline{n}^i].$$

For the sum of random variables x_i, it holds that

$$\mathbf{Var}[\overline{n}^i] = \sum_i \sum_j \mathbf{Cov}[x_i, x_j] = c^i,$$

where c^i is the sum of the pairwise covariances of all combinations of two addends in $\mathbf{Var}[\overline{n}^i]$, which includes the (always positive) variance terms:

$$
c^i = \left[\left(\sum_{1\leq j_1<j_2\leq r} \sum_{1\leq j_1'<j_2'\leq r} \mathbf{Cov}[s_{j_1,j_2}^{i,w_{j_1},w_{j_2}}, s_{j_1',j_2'}^{i,w_{j_1'},w_{j_2'}}]\right)\right.
$$
$$
- \left(\sum_{1\leq j_1<j_2\leq r} \sum_{1\leq j_1'<j_2'<j_3'\leq r} \mathbf{Cov}[s_{j_1,j_2}^{i,w_{j_1},w_{j_2}}, s_{j_1',j_2',j_3'}^{i,w_{j_1'},w_{j_2'},w_{j_3'}}]\right) + \cdots
$$
$$
\left. + (-1)^r \left(\sum_{j_1,j_2} \sum_{j_1',\ldots,j_r'} \mathbf{Cov}[s_{j_1,j_2}^{i,w_{j_1},w_{j_2}}, s_{j_1',\ldots,j_r'}^{i,w_{j_1'},w_{j_2'},\ldots,w_{j_r'}}]\right)\right]
$$
$$
- \left[\left(\sum_{1\leq j_1<j_2<j_3\leq r} \sum_{1\leq j_1'<j_2'<j_3'\leq r} \mathbf{Cov}[s_{j_1,j_2,j_3}^{i,w_{j_1},w_{j_2},w_{j_3}}, s_{j_1',j_2',j_3'}^{i,w_{j_1'},w_{j_2'},w_{j_3'}}]\right) - \cdots\right.
$$
$$
\left. + (-1)^r \left(\sum_{j_1,j_2,j_3} \sum_{j_1',\ldots,j_r'} \mathbf{Cov}[s_{j_1,j_2,j_3}^{i,w_{j_1},w_{j_2},w_{j_3}}, s_{j_1',\ldots,j_r'}^{i,w_{j_1'},w_{j_2'},\ldots,w_{j_r'}}]\right)\right] + \cdots.
$$

Covariance. Recall that the covariance of a term with itself equals its variance and is always positive: $\mathbf{Cov}[x_i, x_i] = \mathbf{Var}[x_i]$. Thus, we need to compute the covariance for all pairs of different variables. From the definition of the covariance,

$$\mathbf{Cov}[x_i, x_j] = \mathbb{E}[x_i \cdot x_j] - \mathbb{E}[x_i] \cdot \mathbb{E}[x_j], \tag{9}$$

we can compute the products of expectations, but have to find the expectations of the products $\mathbb{E}[x_i \cdot x_j]$, with dependent variables x_i and x_j.

Lemma 4 considers the expectation of products. We use $\mathcal{I}, \mathcal{J} \subseteq \{j_1,\ldots,j_r\}$ as distinct index sets and overload the notations so that for each set $\mathcal{I} = \{j_1',\ldots,j_t'\} \subseteq \{j_1,\ldots,j_r\}$, we define $s_{\mathcal{I}}^i = s_{j_1',\ldots,j_t'}^i$. Moreover, we define $p_{\mathcal{I}} = \prod_{j\in\mathcal{I}} p_j$. Note that

$$\mathbb{E}[s_{\mathcal{I}}^i] \cdot \mathbb{E}[s_{\mathcal{J}}^i] = np_{\mathcal{I}} \cdot np_{\mathcal{J}}, \quad \mathbb{E}[s_{\mathcal{I}}^i \cdot s_{\mathcal{J}}^i] = \mathbb{E}[s_{\mathcal{I}}^i] \cdot \mathbb{E}[s_{\mathcal{J}}^i] + \mathbf{Cov}[s_{\mathcal{I}}^i, s_{\mathcal{J}}^i].$$

If $\mathcal{I} \cap \mathcal{J} = \emptyset$, it follows that $p_{\mathcal{I} \cup \mathcal{J}} = p_{\mathcal{I}} \cdot p_{\mathcal{J}}$; thus, $\mathbf{Cov}[s_{\mathcal{I}}^i, s_{\mathcal{J}}^i] = 0$ and

$$\mathbb{E}[s_{\mathcal{I}}^i \cdot s_{\mathcal{J}}^i] = \mathbb{E}[s_{\mathcal{I}}^i] \cdot \mathbb{E}[s_{\mathcal{J}}^i].$$

Though, for the cases when $\mathcal{I} \cap \mathcal{J} \neq \emptyset$, we have to find $\mathbf{Cov}[s_{\mathcal{I}}^i, s_{\mathcal{J}}^i]$ in Lemma 4.

Lemma 4. It holds that $\mathbf{Cov}\left[s_{\mathcal{I}}^i, s_{\mathcal{J}}^i\right] = n p_{\mathcal{I} \cup \mathcal{J}} - n p_{\mathcal{I}} \cdot p_{\mathcal{J}}$.

Proof. Let us focus examplarily on $\mathcal{I} = \{1, 2, 3\}$ and $\mathcal{J} = \{1, 2, 4\}$. Thus, we consider $s_{1,2,3}^{i, w_1, w_2, w_3}$ and $s_{1,2,4}^{i, w_1, w_2, w_4}$. The remaining tuples behave similarly, for all $\mathcal{I} \neq \mathcal{J}$. Given fixed $w_1 = 0^b$ and $w_2, w_3, w_4 \in \mathbb{F}_2^b$, for each $y_1 \in \mathbb{F}_2^b$, we define Bernoulli variables I_{y_1} as

$$I_{y_1} \triangleq \begin{cases} 1 & y_1 \in \mathcal{S}_1^i \wedge y_1 \oplus w_2 \in \mathcal{S}_2^i \wedge y_1 \oplus w_3 \in \mathcal{S}_3^i \\ 0 & \text{otherwise} \end{cases}$$

and

$$J_{y_1} \triangleq \begin{cases} 1 & y_1 \in \mathcal{S}_1^i \wedge y_1 \oplus w_2 \in \mathcal{S}_2^i \wedge y_1 \oplus w_4 \in \mathcal{S}_4^i \\ 0 & \text{otherwise}. \end{cases}$$

Then, we derive

$$\mathbb{E}\left[s_{\mathcal{I}}^{i, w_1, w_2, w_3}\right] = \sum_{y_1 \in \mathbb{F}_2^b} \Pr[I_{y_1}]$$

$$\mathbb{E}\left[s_{\mathcal{J}}^{i, w_1, w_2, w_4}\right] = \sum_{y_1 \in \mathbb{F}_2^b} \Pr[J_{y_1}]$$

We want to bound

$$\mathbf{Cov}\left[s_{\mathcal{I}}^i, s_{\mathcal{J}}^i\right] = \mathbb{E}\left[s_{\mathcal{I}}^i \cdot s_{\mathcal{J}}^i\right] - \mathbb{E}\left[s_{\mathcal{I}}^i\right] \cdot \mathbb{E}\left[s_{\mathcal{J}}^i\right]. \tag{10}$$

We know the latter expectations from the proof of Lemma 3 to be bounded by

$$\mathbb{E}\left[s_{\mathcal{I}}^i\right] = 2^b \cdot \prod_{j \in \mathcal{I}} \frac{s_j^i}{2^b} \quad \text{and} \quad \mathbb{E}\left[s_{\mathcal{J}}^i\right] = 2^b \cdot \prod_{j \in \mathcal{J}} \frac{s_j^i}{2^b}.$$

Thus, it remains to bound the expectation of the product. It holds that

$$\mathbb{E}\left[s_{\mathcal{I}}^i \cdot s_{\mathcal{J}}^i\right] = \mathbb{E}\left[\left(\sum_{y_1 \in \mathbb{F}_2^b} I_{y_1}\right) \cdot \left(\sum_{y_1 \in \mathbb{F}_2^b} J_{y_1}\right)\right]$$

$$= \rho_{I_{y_1}, J_{y_1}} \cdot \sqrt{2^b \cdot 2^b \cdot \mathbf{Var}\left[I_{y_1}\right] \cdot \mathbf{Var}\left[J_{y_1}\right]}$$

$$+ \left(\sum_{y_1 \in \mathbb{F}_2^b} \mathbb{E}\left[I_{y_1}\right]\right) \cdot \left(\sum_{y_1 \in \mathbb{F}_2^b} \mathbb{E}\left[J_{y_1}\right]\right). \tag{11}$$

where

$$\rho_{I_{y_1}, J_{y_1}} = \frac{\mathbb{E}\left[I_{y_1} \cdot J_{y_1}\right] - \mathbb{E}\left[I_{y_1}\right] \cdot \mathbb{E}\left[J_{y_1}\right]}{\sqrt{\mathbf{Var}\left[I_{y_1}\right] \cdot \mathbf{Var}\left[J_{y_1}\right]}}. \tag{12}$$

We know

$$\mathbb{E}\left[I_{y_1}\right] = \prod_{j \in \mathcal{I}} \frac{s_j^i}{2^b} \quad \text{and} \quad \mathbb{E}\left[s_{\mathcal{J}}^i\right] = \prod_{j \in \mathcal{J}} \frac{s_j^i}{2^b}.$$

and need

$$\mathbb{E}\left[I_{y_1} \cdot J_{y_1}\right] = \mathbb{E}\left[I_{y_1}\right] \cdot \Pr\left[J_{y_1} | I_{y_1}\right].$$

We know that

$$\Pr\left[J_{y_1} | I_{y_1}\right] = \Pr\left[y_1 \in \mathcal{S}_1^i \wedge y_1 \oplus w_2 \in \mathcal{S}_2^i \wedge y_1 \oplus w_4 \in \mathcal{S}_4^i \,\middle|\, \right.$$
$$\left. y_1 \in \mathcal{S}_1^i \wedge y_1 \oplus w_2 \in \mathcal{S}_2^i \wedge y_1 \oplus w_3 \in \mathcal{S}_3^i\right]$$
$$= \Pr\left[y_1 \oplus w_4 \in \mathcal{S}_4^i \,\middle|\, y_1 \in \mathcal{S}_1^i \wedge y_1 \oplus w_2 \in \mathcal{S}_2^i \wedge y_1 \oplus w_3 \in \mathcal{S}_3^i\right]$$
$$= \frac{s_4^i}{2^b}$$

since the individual terms w_i are independent from each other and $\prod_{j \in \mathcal{J} \setminus \mathcal{I}} \frac{s_j^i}{2^b}$ in general. Thus, it holds that

$$\mathbb{E}\left[I_{y_1} \cdot J_{y_1}\right] = \prod_{j \in \mathcal{I}} \frac{s_j^i}{2^b} \cdot \prod_{j \in \mathcal{I} \setminus \mathcal{J}} \frac{s_j^i}{2^b} = \frac{1}{2^b} \cdot \prod_{j \in \mathcal{I} \cup \mathcal{J}} s_j^i.$$

Thus,

$$(12) = \frac{\prod_{j \in \mathcal{I} \cup \mathcal{J}} \frac{s_j^i}{2^b} - \prod_{j \in \mathcal{I}} \frac{s_j^i}{2^b} \cdot \prod_{j \in \mathcal{J}} \frac{s_j^i}{2^b}}{\sqrt{\mathbf{Var}\left[I_{y_1}\right] \cdot \mathbf{Var}\left[J_{y_1}\right]}}$$

Inserting into Eq. (11),

$$\mathbb{E}\left[s_{\mathcal{I}}^i \cdot s_{\mathcal{J}}^i\right] = \frac{\left(\prod_{j \in \mathcal{I} \cup \mathcal{J}} \frac{s_j^i}{2^b}\right) - \left(\prod_{j \in \mathcal{I}} \frac{s_j^i}{2^b} \cdot \prod_{j \in \mathcal{J}} \frac{s_j^i}{2^b}\right)}{\sqrt{\mathbf{Var}\left[I_{y_1}\right] \cdot \mathbf{Var}\left[J_{y_1}\right]}} \cdot \sqrt{2^b \cdot 2^b \cdot \mathbf{Var}\left[I_{y_1}\right] \cdot \mathbf{Var}\left[J_{y_1}\right]} +$$

$$\left(\sum_{y_1 \in \mathbb{F}_2^b} \mathbb{E}\left[I_{y_1}\right]\right) \cdot \left(\sum_{y_1 \in \mathbb{F}_2^b} \mathbb{E}\left[J_{y_1}\right]\right)$$

$$= 2^b \cdot \underbrace{\left(\prod_{j \in \mathcal{I} \cup \mathcal{J}} \frac{s_j^i}{2^b}\right)}_{p_{\mathcal{I} \cup \mathcal{J}}} - 2^b \cdot \underbrace{\left(\prod_{j \in \mathcal{I}} \frac{s_j^i}{2^b}\right)}_{p_{\mathcal{I}}} \cdot \underbrace{\left(\prod_{j \in \mathcal{J}} \frac{s_j^i}{2^b}\right)}_{p_{\mathcal{J}}}$$

$$+ 2^b \cdot \underbrace{\left(\prod_{j \in \mathcal{I}} \frac{s_j^i}{2^b}\right)}_{p_{\mathcal{I}}} \cdot 2^b \cdot \underbrace{\left(\prod_{j \in \mathcal{J}} \frac{s_j^i}{2^b}\right)}_{p_{\mathcal{J}}}$$

substituting $n = 2^b$ and inserting into Equation produces

$$(10) = \mathbf{Cov}\left[s_{\mathcal{I}}^i, s_{\mathcal{J}}^i\right] = \mathbb{E}\left[s_{\mathcal{I}}^i \cdot s_{\mathcal{J}}^i\right] - \mathbb{E}\left[s_{\mathcal{I}}^i\right] \cdot \mathbb{E}\left[s_{\mathcal{J}}^i\right]$$
$$= n \cdot p_{\mathcal{I}\cup\mathcal{J}} - n \cdot p_{\mathcal{I}} \cdot p_{\mathcal{J}} + n \cdot p_{\mathcal{I}} \cdot n \cdot p_{\mathcal{J}} - n \cdot p_{\mathcal{I}} \cdot n \cdot p_{\mathcal{J}}$$
$$= n \cdot p_{\mathcal{I}\cup\mathcal{J}} - n \cdot p_{\mathcal{I}} \cdot p_{\mathcal{J}},$$

which gives our claim in Lemma 4.

We show that we are allowed to apply Lemma 4. Since the permutations are independent from each other and the values are sampled independently at random, we can say that each value in $\mathcal{S}_u$, $\mathcal{S}_v$, $\mathcal{S}_w$ is chosen independently from the others. The size of all three lists is $n = 2^b$; moreover, we can instantiate the probabilities p_j, for $j \in [r]$ as $p_j = \frac{s_j^i}{2^b}$. In our case, this means

$$\mathbb{E}\left[s_{\mathcal{I}}^i \cdot s_{\mathcal{J}}^i\right] = 2^{2b} \cdot \prod_{i \in \mathcal{I}} p_j \cdot \prod_{j \in \mathcal{J}} p_j + \mathbf{Cov}\left[s_{\mathcal{I}}^i, s_{\mathcal{J}}^i\right], \quad \text{where}$$

$$\mathbf{Cov}\left[s_{\mathcal{I}}^i, s_{\mathcal{J}}^i\right] = 2^b \cdot \prod_{i \in \mathcal{I}\cup\mathcal{J}} p_i - 2^b \cdot \prod_{i \in \mathcal{I}} p_i \cdot \prod_{j \in \mathcal{J}} p_j.$$

For example, let $\mathcal{I} = \{1, 2\}$ and $\mathcal{J} = \{1, 3, 4\}$. Then,

$$\mathbf{Cov}\left[s_{1,2}^i, s_{1,3,4}^i\right] = 2^b \cdot \left(\frac{s_1^i s_2^i s_3^i s_4^i}{2^{4b}} - \frac{(s_1^i)^2 s_2^i s_3^i s_4^i}{2^{5b}}\right).$$

Decomposing c^i. Given the covariance, we can rewrite c^i. We define $\mathcal{C}_{t,r}$ for the set of t-out-of-r element combinations, e.g. $\mathcal{C}_{2,3} = \{(1,2), (1,3), (2,3)\}$.

$$c^i = \sum_{t_1=2}^{r} \sum_{t_2=2}^{r} (-1)^{t_1+t_2} \cdot c_{t_1,t_2,r}^i, \quad \text{where } c_{t_1,t_2,r}^i = \sum_{\mathcal{I}\in\mathcal{C}_{t_1,r}} \sum_{\mathcal{J}\in\mathcal{C}_{t_2,r}} \mathbf{Cov}[s_{\mathcal{I}}^{i,w_{\mathcal{I}}}, s_{\mathcal{J}}^{i,w_{\mathcal{J}}}].$$
$$\tag{13}$$

Lemma 4 yields

$$c_{t_1,t_2,r}^i = \sum_{\mathcal{I}\in\mathcal{C}_{t_1,r}} \sum_{\mathcal{J}\in\mathcal{C}_{t_2,r}} 2^b \cdot (p_{\mathcal{I}\cup\mathcal{J}} - p_{\mathcal{I}} p_{\mathcal{J}}) \tag{14}$$

$$= 2^b \cdot \underbrace{\left(\sum_{\mathcal{I}\in\mathcal{C}_{t_1,r}} \sum_{\mathcal{J}\in\mathcal{C}_{t_2,r}} \prod_{j\in\mathcal{I}\cup\mathcal{J}} p_j\right)}_{\overline{c}_{t_1,t_2,r}^i} - 2^b \cdot \underbrace{\left(\sum_{\mathcal{I}\in\mathcal{C}_{t_1,r}} \sum_{\mathcal{J}\in\mathcal{C}_{t_2,r}} \prod_{i\in\mathcal{I}} p_i \prod_{j\in\mathcal{J}} p_j\right)}_{\underline{c}_{t_1,t_2,r}^i}.$$
$$\tag{15}$$

Later, we will consider the case that $p_1 = p_2 = \cdots = p_r = p$. Then, we can write $c_{t_1,t_2,r}^i$ as

$$2^b \cdot \left(\overline{c}_{t_1,t_2,r}^i - \underline{c}_{t_1,t_2,r}^i\right) = 2^b \cdot \left(\sum_{j=0}^{u} \left(\overline{k}_{t_1,t_2,r,j}^i \cdot p^{\overline{\ell}_{t_1,t_2,r,j}^i}\right) - \underline{k}_{t_1,t_2,r}^i \cdot p^{\underline{\ell}_{t_1,t_2,r}^i}\right)$$

with $u =^{\mathrm{def}} \min(r - t_2, t_1)$ and j denotes the number of elements in $\mathcal{I}$ that are not contained in $\mathcal{J}$. Thus, we can reduce the task to that of finding the multiples

$$\overline{k}^i_{t_1,t_2,r,j} = |\{(\mathcal{I}, \mathcal{J}) \in \mathcal{C}_{t_1,r}, \times \mathcal{C}_{t_2,r} : |\mathcal{I} \cup \mathcal{J}| = t_2 + j\}| \quad \text{and} \quad \overline{\ell}^i_{t_1,t_2,r,j} = |\mathcal{I} \cup \mathcal{J}|. \tag{16}$$

and

$$\underline{k}^i_{t_1,t_2,r} = |\{(\mathcal{I}, \mathcal{J}) \in \mathcal{C}_{t_1,r}, \times \mathcal{C}_{t_2,r}\}| = |\mathcal{C}_{t_1,r}| \cdot |\mathcal{C}_{t_2,r}| = \binom{r}{t_1} \cdot \binom{r}{t_2} \quad \text{and} \tag{17}$$

$$\underline{\ell}^i_{t_1,t_2,r} = |\mathcal{I}| + |\mathcal{J}| = t_1 + t_2. \tag{18}$$

The exponent $\overline{\ell}^i_{t_1,t_2,r,j}$ is derived from the size of the union set $\mathcal{I} \cup \mathcal{J}$ when j elements of $\mathcal{I}$ are not in $\mathcal{J}$. Thus $\overline{\ell}^i_{t_1,t_2,r,j} = \max(t_1, t_2) + j$ for all $j \in [0..u]$ where $u =^{\mathrm{def}} \min(r - t_2, t_1)$. It remains to determine $\overline{k}^i_{t_1,t_2,r,j}$. For this purpose, we can use the simple combinatorial Lemma 5.

Lemma 5. Let t_1, t_2, r, j be fixed integers with $t_1 \le t_2 \le r$ and $j \in [t_2..r]$. Let $\mathcal{I}, \mathcal{J} \subseteq [r]$ be non-identical subsets of $[r]$ with $|\mathcal{I}| = t_1$ and $|\mathcal{J}| = t_2$. Then, the number of combinations of distributing $\mathcal{I}$ and $\mathcal{J}$ so that

$$|\{(\mathcal{I}, \mathcal{J}) \in \mathcal{C}_{t_1,r}, \times \mathcal{C}_{t_2,r} : |\mathcal{I} \cup \mathcal{J}| = t_2 + j\}| = \binom{r}{t_2} \cdot \binom{t_2}{t_1 - j} \cdot \binom{r - t_2}{j}.$$

Proof. W.l.o.g., we had fixed that $|\mathcal{I}| \le |\mathcal{J}|$ and therefore $t_1 \le t_2$. There are $\binom{r}{t_2}$ sets $\mathcal{J}$ among r elements. We defined that j elements of $\mathcal{I}$ are not in $\mathcal{J}$. For a fixed $\mathcal{J}$ and fixed j, there are $\binom{t_2}{t_1-j}$ combinations of the $t_1 - j$ values in $\mathcal{I} \cap \mathcal{J}$ and $\binom{r-t_2}{j}$ combinations of distributing j values from $\mathcal{I} \setminus \mathcal{J}$ outside of $\mathcal{J}$. □

We can rewrite Lemma 5 as Lemma 6, which will serve useful.

Lemma 6. Let t_1, t_2, r, ℓ be fixed integers with $t_1, t_2 \le r$. Let $\mathcal{I}, \mathcal{J} \subseteq [r]$ such that $|\mathcal{I}| = t_1$, $|\mathcal{J}| = t_2$, and $j = \ell - t_1$. Then, the number of combinations of distributing $\mathcal{I}$ and $\mathcal{J}$ so that $|\mathcal{I} \cup \mathcal{J}| = \ell$ is

$$|\{(\mathcal{I}, \mathcal{J}) \in \mathcal{C}_{t_1,r}, \times \mathcal{C}_{t_2,r} : |\mathcal{I} \cup \mathcal{J}| = \ell\}| = \binom{r}{t_1}\binom{t_1}{t_1 + t_2 - \ell}\binom{r - t_1}{\ell - t_1}(-1)^{t_1 + t_2}.$$

Proof. There are $\binom{r}{t_1}$ sets $\mathcal{I}$ among r elements. The overlap, i.e., the number of shared elements in the intersection $|\mathcal{I} \cap \mathcal{J}| = t_1 + t_2 - \ell$. Among the t_1 elements of $\mathcal{I}$, there are $\binom{t_1}{t_1+t_2-\ell}$ combinations what elements of $\mathcal{I}$ and $\mathcal{J}$ could be in the intersection. Then, the remaining $\ell - t_1$ elements in $\mathcal{J} \setminus \mathcal{I}$ can be distributed by $\binom{r-t_1}{\ell-t_1}$ combinations over the remaining $r - t_1$ elements not in $\mathcal{I}$. □

Upper Bounding c^i for General r. We aim at having a simplified upper bound for c^i for general r. The terms in c^i consist of multiples of powers of p from exponents 2 to $2r$. Now, we can find non-negative integer coefficients k^i_j, for all $j \in [2..r]$, so that

$$c^i = k^i_2 \cdot p^2 + k^i_3 \cdot p^3 + \sum_{j=2}^{r} \left(k^i_{2j} \cdot p^{2j}\right) . \tag{19}$$

We show that there the indices $j \in [2..2r]$ are the only potential positive coefficients k^i_j. For $k^i_\ell \cdot p^\ell$ with $k_\ell < 2$, there must exist $\overline{\ell}^i_{t_1,t_2,r,j} < 2$ or $\underline{\ell}^i_{t_1,t_2,r} < 2$ for some $t_1, t_2 \in [2..r]$ and $j \leq r$. Though, our sets always have $|\mathcal{I}|, |\mathcal{J}| \in [2..r]$. Hence,

$$\overline{\ell}^i_{t_1,t_2,r,j} = |\mathcal{I} \cup \mathcal{J}| \implies \overline{\ell}^i_{t_1,t_2,r,j} \in [2..2r] \quad \text{and}$$
$$\underline{\ell}^i_{t_1,t_2,r} = |\mathcal{I}| + |\mathcal{J}| \implies \underline{\ell}^i_{t_1,t_2,r} \in [4..2r] .$$

Thus, $k^i_\ell = 0$ for all $\ell \notin [2..2r]$. We want to reduce the bound to the terms with the few lowest exponents and show that we can upper bound the tail. In particular, we want a bound so that we can reduce Eq. (19) to

$$c^i \leq 2^b \cdot (k^i_2 \cdot p^2 + k^i_3 \cdot p^3).$$

We show the following lemma. Later, we also show that $p \leq 1/3r$ always holds.

Lemma 7. Let $r \geq 2$ be integer. For all even $\ell = 2j$ for some $j \in [2..r-1]$,

$$\frac{|k^i_{\ell+1}|}{|k^i_\ell|} \leq 3r, \quad k^i_{\ell+1} \geq 0, \quad k^i_\ell \leq 0 \quad \text{and} \quad k^i_{2r} \leq 0 .$$

We defer the proof of Lemma 7 to the full version of the paper [7]. Combined with our assumption that $p \leq 1/3r$, it follows for all $\ell = 2j$ for some $j \in [2..r-1]$, that

$$k^i_{\ell+1} \cdot p^{\ell+1} \leq \left(3r \cdot k^i_\ell\right) \cdot \left(p^\ell \cdot \frac{1}{3r}\right) = k^i_\ell \cdot p^\ell ,$$

and therefore

$$c^i = 2^b \cdot (k^i_2 p^2 + k^i_3 p^3 + \sum_{j=2}^{r-1} \underbrace{\left(-k^i_{2j} p^{2j} + k^i_{2j+1} p^{2j+1}\right)}_{\leq 0} - k^i_{2r} p^{2r}) \leq 2^b \cdot (k^i_2 p^2 + k^i_3 p^3) .$$

The factors k_2^i and k_3^i result from only few terms in c^i. In particular, they stem from $c_{2,2,r}^i$, $c_{2,3,r}^i = c_{3,2,r}^i$, and $c_{3,3,r}^i$. Given $r \geq 3$, they result from

$$k_2^i = \overline{k}_{2,2,r,0}^i = \binom{r}{2}\binom{2}{2}\binom{2}{0} = \binom{r}{2}$$

$$k_3^i = \overline{k}_{2,2,r,1}^i - \overline{k}_{2,3,r,0}^i - \overline{k}_{3,2,r,0}^i + \overline{k}_{3,3,r,0}^i = \binom{r}{3}.$$

Note that the statement also holds for $r = 2$, where $k_2^i = 1$, $k_4^i = 1$, and $k_j^i = 0$ for all positive integers $j \notin \{2, 4\}$. We obtain

$$c^i \leq 2^b \cdot \left(\binom{r}{2} \cdot p^2 + \binom{r}{3} \cdot p^3 \right). \tag{20}$$

Equal Probabilities p_i. It remains to show that $p_1 = \cdots = p_r$. The values of the a most significant bits of the permutation outputs, $\mathbf{V}_j^i = V_j^1, \ldots, V_j^i$, for all $j \in [r]$, are sampled uniformly and independently at random, also in the modified real world $\mathcal{O}_{\mathrm{real}}$ since we replace their sampling with that from a truncated permutation. Thus, every V_j^i has probability 2^{-a} to be equal to a specific a-bit value. Therefore

$$\mathbb{E}_{\mathbf{V}^{i-1}}[s_1^i] = \cdots = \mathbb{E}_{\mathbf{V}^{i-1}}[s_r^i] = \frac{i-1}{2^a}.$$

Thus, for all $j \in [r]$, we can use

$$p_j = \mathbb{E}\left[\frac{s_j^i}{2^b}\right] = \frac{\mathbb{E}[s_j^i]}{2^b} = \frac{i-1}{2^n}.$$

We have to show that the expectations of the quantities $s_1^i, \ldots, s_r^i$ are independent. We can adopt the argument from [10] here: it holds since they stem from pairwise independent permutations and hence

$$\mathbb{E}_{\mathbf{V}^{i-1}}[s_2^i | s_1^i] = \mathbb{E}_{\mathbf{V}^{i-1}}[s_2^i]$$

and similar statements can be derived for all other combinations. We can use

$$\mathbb{E}_{\mathbf{V}^{i-1}}[s_1^i s_2^i] = \mathbb{E}_{\mathbf{V}^{i-1}}[s_1^i] \cdot \mathbb{E}_{\mathbf{V}^{i-1}}[s_2^i]$$

and the other product combinations can be decomposed similarly.

Finalizing with the χ^2 Approach. We have that

$$\mathbb{E}\left[\left(\Pr_{\mathcal{O}_{\mathrm{real}}}[W^i = W | \mathbf{W}^{i-1}] - \Pr_{\mathcal{O}_{\mathrm{ideal}}}[W^i = W | \mathbf{W}^{i-1}]\right)^2\right] \leq \left(\frac{4}{3}\right)^{2r} \cdot \frac{1}{2^{2rb}} \cdot \mathbf{Var}[\overline{n}^i].$$

Using the χ^2 approach and inserting $\text{Pr}_{\mathcal{O}_{\text{ideal}}}[W^i = W|\mathbf{W}^{i-1}] = 2^{-(r-1)b}$, we obtain

$$(|\text{Pr}[\mathcal{O}_{\text{real}} = \tau] - \text{Pr}[\mathcal{O}_{\text{ideal}} = \tau]|)^2$$

$$\leq \frac{1}{2}\sum_{i=1}^{q}\mathbb{E}_{\mathcal{O}_{\text{real}}}[\chi^2(\mathbf{W}^{i-1})]$$

$$\leq \frac{1}{2}\sum_{i=1}^{q}\sum_{W\in(\mathbb{F}_2^b)^{r-1}}\mathbb{E}_{\mathcal{O}_{\text{real}}}\left[\frac{(\text{Pr}_{\mathcal{O}_{\text{real}}}[W^i = W|\mathbf{W}^{i-1}] - \text{Pr}_{\mathcal{O}_{\text{ideal}}}[W^i = W|\mathbf{W}^{i-1}])^2}{\text{Pr}_{\mathcal{O}_{\text{ideal}}}[W^i = W|\mathbf{W}^{i-1}]}\right]$$

$$\leq \frac{1}{2}\cdot 2^{(r-1)b}\cdot\sum_{i=1}^{q}\sum_{W\in(\mathbb{F}_2^b)^{r-1}}\mathbb{E}\left[\left(\Pr_{\mathcal{O}_{\text{real}}}[W^i = W|\mathbf{W}^{i-1}] - \Pr_{\mathcal{O}_{\text{ideal}}}[W^i = W|\mathbf{W}^{i-1}]\right)^2\right]$$

$$\leq \frac{1}{2}\cdot 2^{(r-1)b}\cdot\sum_{i=1}^{q}\sum_{W\in(\mathbb{F}_2^b)^{r-1}}\left(\left(\frac{4}{3}\right)^{2r}\cdot\frac{1}{2^{2rb}}\cdot c^i\right)$$

$$\leq \frac{1}{2^{2b+1}}\cdot\left(\frac{4}{3}\right)^{2r}\cdot\sum_{i=1}^{q}c^i. \tag{21}$$

From Eq. (20)

$$c^i \leq 2^b\left(\binom{r}{2}p^2 + \binom{r}{3}p^3\right)$$

and $p = (i-1)/2^n$, we obtain that

$$(21) = \sqrt{\frac{1}{2^{2b+1}}\cdot\left(\frac{4}{3}\right)^{2r}\cdot\sum_{i=1}^{q}2^b\cdot\left(\binom{r}{2}\frac{(i-1)^2}{2^{2n}} + \binom{r}{3}\frac{(i-1)^3}{2^{3n}}\right)}$$

$$= \sqrt{\frac{1}{2^{2b+1}}\cdot\left(\frac{4}{3}\right)^{2r}\cdot\frac{1}{2^a}\cdot\sum_{i=1}^{q}\left(\binom{r}{2}\frac{(i-1)^2}{2^n} + \binom{r}{3}\frac{(i-1)^3}{2^{2n}}\right)}$$

$$\leq \sqrt{\frac{1}{2^{2n-a}}\cdot\left(\frac{4}{3}\right)^{2r}\cdot\frac{1}{2}\cdot\left(\frac{r^2q^3}{2^n} + \frac{r^3q^4}{2^{2n}}\right)}$$

$$\leq \left(\frac{4}{3}\right)^{r}\cdot\frac{1}{2}\cdot\sqrt{\frac{r^2q^3}{2^{3n-a}} + \frac{r^3q^4}{2^{4n-a}}}$$

$$\leq \left(\frac{4}{3}\right)^{r}\cdot\left(\frac{rq}{2^{n-a/3}}\right)^{3/2},$$

where we used $q \leq 2^n/3r$ to upper bound

$$\frac{r^2q^3}{2^{3n-a}} + \frac{r^3q^4}{2^{4n-a}} \leq \frac{2r^3q^3}{2^{3n-a}}.$$

This yields the bound in Theorem 3.   □

We can obtain tighter constant factors for concrete values of r. We give the results for $r = 3, 4$ in Corollary 2 to aid the reader.

Corollary 2. Let a, b, q be positive integers and $\tau = (\mathbf{V}, \mathbf{W})$ be a good transcript such that $C_{\mathbf{V},j}(i) < 2^{b-2}$ holds for all $i \in [q]$ and $j \in [r]$ and $q \leq 2^n/9$. Then, it holds

$$\mathbb{E}_\tau[\Pr[\mathcal{O}_2 = \tau] - \Pr[\mathcal{O}_{\text{ideal}} = \tau]] \leq \begin{cases} 4 \cdot \left(\frac{q}{2^{n-a/3}}\right)^{3/2} & \text{for } r = 3 \\ 8 \cdot \left(\frac{q}{2^{n-a/3}}\right)^{3/2} & \text{for } r = 4 \,. \end{cases}$$

6 Conclusion

We have shown that XTH, the XORP-like extension of STH achieves a level of $O(n - a/2 - 1.5\log(r))$-bit PRF security. This is similar to the logarithmic loss in r of XORP compared to the sum of permutations, providing a trade-off between releasing more bits from each permutation call and from summation. Note that this work has considered the version with independent permutations. Future work can investigate variants that feed domain-separated inputs into the same permutation.

References

1. Bellare, M., Impagliazzo, R.: A tool for obtaining tighter security analyses of pseudorandom function based constructions, with applications to prp to prf conversion. Cryptology ePrint Archive, Report 1999/024 (1999). http://eprint.iacr.org/1999/024

2. Bellare, M., Krovetz, T., Rogaway, P.: Luby-Rackoff backwards: increasing security by making block ciphers non-invertible. In: Nyberg, K. (ed.) EUROCRYPT 1998. LNCS, vol. 1403, pp. 266–280. Springer, Heidelberg (1998). https://doi.org/10.1007/BFb0054132

3. Bhattacharya, S., Nandi, M.: Full indifferentiable security of the xor of two or more random permutations using the \chi˜2 Method. In: Nielsen, J.B., Rijmen, V. (eds.) EUROCRYPT 2018. LNCS, vol. 10820, pp. 387–412. Springer, Cham (2018). https://doi.org/10.1007/978-3-319-78381-9_15

4. Dai, W., Hoang, V.T., Tessaro, S.: Information-theoretic indistinguishability via the chi-squared method. In: Katz, J., Shacham, H. (eds.) CRYPTO 2017. LNCS, vol. 10403, pp. 497–523. Springer, Cham (2017). https://doi.org/10.1007/978-3-319-63697-9_17

5. Datta, N., Dutta, A., Nandi, M., Paul, G.: Double-block hash-then-sum: a paradigm for constructing BBB secure PRF. IACR Trans. Symm. Cryptol. **2018**(3), 36–92 (2018)

6. Dinur, I.: Tight indistinguishability bounds for the xor of independent random permutations by fourier analysis. In: Joye, M., Leander, G. (eds.) EUROCRYPT I, vol. 14651, pp. 33–62. Springer, Heidelberg (2024). https://doi.org/10.1007/978-3-031-58716-0_2

7. Dutta, A., List, E.: Proving the security of the extended summation-truncation hybrid. Cryptology ePrint Archive, Paper 2024/1723 (2024)

8. Dutta, A., Nandi, M., Saha, A.: Proof of mirror theory for $\xi_{\max} = 2$. IEEE Trans. Inf. Theory **68**(9), 6218–6232 (2022)

9. Gilboa, S., Gueron, S.: The advantage of truncated permutations. Disc. Appl. Math. **294**, 214–223 (2021)
10. Gunsing, A., Mennink, B.: The summation-truncation hybrid: reusing discarded bits for free. In: Micciancio, D., Ristenpart, T. (eds.) CRYPTO 2020. LNCS, vol. 12170, pp. 187–217. Springer, Cham (2020). https://doi.org/10.1007/978-3-030-56784-2_7
11. Hall, C., Wagner, D., Kelsey, J., Schneier, B.: Building PRFs from PRPs. In: Krawczyk, H. (ed.) CRYPTO 1998. LNCS, vol. 1462, pp. 370–389. Springer, Heidelberg (1998). https://doi.org/10.1007/BFb0055742
12. Iwata, T.: New blockcipher modes of operation with beyond the birthday bound security. In: Robshaw, M. (ed.) FSE 2006. LNCS, vol. 4047, pp. 310–327. Springer, Heidelberg (2006). https://doi.org/10.1007/11799313_20
13. Iwata, T., Mennink, B., Vizár, D.: CENC is optimally secure. IACR Cryptology ePrint Archive **2016**, 1087 (2016)
14. Iwata, T., Minematsu, K.: Stronger security variants of GCM-SIV. IACR Trans. Symm. Cryptol. **2016**(1), 134–157 (2016)
15. Jean, J., Nikolic, I., Peyrin, T., Seurin, Y.: The deoxys AEAD family. J. Cryptol. **34**(3), 31 (2021)
16. Lucks, S.: The Sum of PRPs is a secure PRF. In: Preneel, B. (ed.) EUROCRYPT 2000. LNCS, vol. 1807, pp. 470–484. Springer, Heidelberg (2000). https://doi.org/10.1007/3-540-45539-6_34
17. Naito, Y.: Blockcipher-based MACs: beyond the birthday bound without message length. In: Takagi, T., Peyrin, T. (eds.) ASIACRYPT 2017. LNCS, vol. 10626, pp. 446–470. Springer, Cham (2017). https://doi.org/10.1007/978-3-319-70700-6_16
18. Yasuda, K.: A new variant of PMAC: beyond the birthday bound. In: Rogaway, P. (ed.) CRYPTO 2011. LNCS, vol. 6841, pp. 596–609. Springer, Heidelberg (2011). https://doi.org/10.1007/978-3-642-22792-9_34
19. Zhang, L., Wu, W., Sui, H., Wang, P.: 3kf9: enhancing 3GPP-MAC beyond the birthday bound. In: Wang, X., Sako, K. (eds.) ASIACRYPT 2012. LNCS, vol. 7658, pp. 296–312. Springer, Heidelberg (2012). https://doi.org/10.1007/978-3-642-34961-4_19

Constructing WAPB Boolean Functions From the Direct Sum of WAPB Boolean Functions

Deepak Kumar Dalai[1]([✉]) [iD] and Krishna Mallick[2]

[1] School of Mathematical Sciences, National Institute of Science Education and Research, An OCC of Homi Bhabha National Institute, Bhubaneswar 752050, Odisha, India
deeepak@niser.ac.in
[2] School of Computer Sciences, National Institute of Science Education and Research, An OCC of Homi Bhabha National Institute, Bhubaneswar 752050, Odisha, India
krishna.mallick@niser.ac.in

Abstract. A Boolean function with good cryptographic properties over a set of vectors with constant Hamming weight is significant for stream ciphers like FLIP. Efficiency is another critical criterion for designing a stream cipher. The nonlinear filter function used in the cipher FLIP is a direct sum of three Boolean functions to be efficient with a high number of state variables. In our work, we have studied the construction of WAPB and WPB Boolean functions by the direct sum of two WAPB/WPB Boolean functions. A general result in this direction is presented. We have presented some cases when the direct sum results a WAPB/WPB Boolean function. We have introduced several constructions of WAPB/WPB Boolean functions in this context. Some results on the direct sum of WAPB/WPB Boolean functions presented by Carlet et al. in IACR Trans. Symmetric Cryptology 2017(3) and Zhu et al. in Discrete Applied Mathematics 2022(314) are direct consequences of the obtained results.

Keywords: Boolean function · Weightwise almost perfectly balanced (WAPB) · Weightwise perfectly balanced (WPB) · Direct sum · Nonlinearity

1 Introduction

An n-variable Boolean function f is a mapping from the n-dimensional vector space $\mathbb{F}_2^n$ to $\mathbb{F}_2$, where $\mathbb{F}_2$ is a finite field with two elements $\{0, 1\}$. Depending upon the underlying algebraic structure, the '+' symbol is used for the addition operation in both $\mathbb{F}_2$ and $\mathbb{R}$ (set of real numbers). Boolean functions are crucial in constructing nonlinear components in symmetric ciphers. In stream ciphers, Boolean functions are employed as filter functions to generate a pseudo-random sequence. All the cryptographic criteria have been established to analyze

S. Mukhopadhyay and P. Stănică (Eds.): INDOCRYPT 2024, LNCS 15495, pp. 188–209, 2025.
https://doi.org/10.1007/978-3-031-80308-6_9

and construct Boolean functions for use in ciphers that can withstand various attacks. The cryptographic criteria for a filter function are generally defined over the entire vector space $\mathbb{F}_2^n$. The study of the Boolean functions over a restricted domain became interesting after the appearance of the FLIP cipher in 2016 [15]. The main idea of proposing the FLIP cipher is to combine a symmetric cipher with homomorphic encryption to improve the efficiency of computations through cloud-based services. The new stream cipher design involves using a bit permutation generator to permute the inputs before entering them into the filter function for every updating process. Therefore, the entire setup is known as a filter permutator. As a result, the Hamming weight (i.e., the number of non-zero coordinates) of the inputs to the filter function remains the same as the Hamming weight of the initial vector. This restriction of the inputs significantly changes the viewpoint toward its security analysis. Further, the filter function is a direct sum of three Boolean functions, namely, an L-type, a Q-type, and a T-type function. As a result, this cipher is very efficient, although the filter function has 395 variables. In this article, we have studied direct sum Boolean functions, which are balanced in the domains of constant weight.

The motivation to construct Boolean functions in the FLIP frame of reference arises. An initial cryptographic study of Boolean function in a restricted domain is introduced by Carlet et al. in [3]. Boolean functions used as filter functions in stream cipher are distributed uniformly over $\mathbb{F}_2^n$ and are called balanced Boolean functions. In the FLIP cipher, the Hamming weight of the key register is $\frac{n}{2}$, for n is even. Therefore, the keystream generated by FLIP should look like a random sequence over the set of vectors of Hamming weight $\frac{n}{2}$ for n is even, or we can say the filter function in FLIP should be balanced over the set $E_{n,\frac{n}{2}} = \{x \in \mathbb{F}_2^n | \mathtt{wt}(x) = \frac{n}{2}\}$. The Boolean functions balanced over the subsets of $\mathbb{F}_2^n$ containing vectors with constant Hamming weight are said to be weightwise perfectly balanced (WPB). In order for WPB Boolean functions to exist, n must be of the form 2^l for $l \in \mathbb{N}$. In [3], the author extended the concept of WPB Boolean function for any $n \in \mathbb{N}$ and named these functions as weightwise almost perfectly balanced (WAPB) Boolean functions by allowing these functions to be almost balanced over the restricted domain, depending on the cardinality. The WPB and WAPB functions with good cryptographic criteria over the restricted domains are essential in the FLIP frame of reference. The class of WPB Boolean functions are balanced WAPB Boolean functions when n is of the form 2^l. Several cryptographic criteria of a Boolean function over a restricted domain are studied in [3]. An upper bound on the nonlinearity and algebraic immunity of a Boolean function on restricted inputs is presented in [3]. The nonlinearity bound is further improved in [18]. There are several proposed methods for constructing WAPB and WPB (see [3–7,9–12,16,17,20–25] etc.) are available in the literature. Following are some important constructions and studies on WAPB/WPB Boolean functions.

1. The first WPB Boolean function constructions were introduced by Carlet et al. in [3] in 2017. A construction is based on a sum of four Boolean functions (see Proposition 2 and Corollary 1). The author also presented a recursive

construction for weightwise almost perfectly balanced (WAPB) Boolean functions in the same paper. Upper bounds on the weightwise nonlinearities and weightwise algebraic immunity are presented in this paper (see Lemma 1).

2. Tang and Liu [21] proposed a construction of a class WAPB Boolean functions for an even number of variables, which satisfy optimal algebraic immunity.

3. Liu and Mesnager [12] presented a class of WPB Boolean functions that are 2-rotation symmetric, which have high weightwise nonlinearities and nonlinearity. This technique is generalized in [5] to get WAPB Boolean functions on any number of variables.

4. Several constructions of WPB and WAPB Boolean functions are presented in [4,16] by modifying the support of linear and quadratic functions.

5. In [25], Zhu and Su presented a method of constructing a WAPB Boolean function for an arbitrary number of variables using the direct sum of known WPB Boolean functions(see Proposition 3).

6. In [6–9], Gini and Méaux have proved several results on WAPB/WPB Boolean functions. The authors have discussed the Hamming weight of linear functions restricted to the set of vectors with constant Hamming weights that Krawtchouk polynomials can express. Furthermore, an upper and lower bound on the nonlinearity of $f \in \mathcal{W}_n$, where $\mathcal{W}_n$ is the set of all WPB Boolean functions, have been studied in [8].

Recently, such functions with high nonlinearity have been searched using genetic programming (GP) and genetic algorithm (GA) in [13]. However, these functions may not be suitable for cryptographic implementation due to their structureless representation. Therefore, analyzing the cryptographic properties like nonlinearity, algebraic immunity, and efficiency of a WAPB Boolean function is vital from the perspective of ciphers like FLIP. Indeed, the upper bounds for nonlinearity and weightwise nonlinearity are not tight for such Boolean functions. Moreover, it is also significant to figure out the algebraic structure of these WAPB Boolean functions such that these functions with better trade-offs on cryptographic properties can be constructed from the known Boolean functions in lower dimensions.

1.1   Our Contribution

Although many constructions of WAPB (specially, WPB) Boolean functions are available, achieving such a function with good cryptographic properties like nonlinearity, algebraic immunity, and efficiency is still tricky. We can find many constructions of WPB Boolean functions are available in the literature due to its simplicity (i.e., for any positive integer of the form $n = 2^l$). Therefore, construction of WAPB Boolean functions (i.e., for any positive integer n) needs more attention.

In FLIP, the filter function is a direct sum of three functions to achieve the efficiency criteria. Motivated by this cipher, we have studied to construct WAPB Boolean functions from the direct sum of two WAPB Boolean functions in this paper. Such classes of functions are efficient, have simple algebraic structures, and are easy to cryptanalyse. Further, each part of the direct sum are also WAPB functions.

We have presented a general result when the direct sum of the WAPB Boolean functions forms a WAPB Boolean function. We have studied some different cases when the direct sum results in a WAPB (and WPB) Boolean function. Our results provide more straightforward proof of existing direct sum results in [3, 25]. As WPB functions are a special class of WAPB Boolean functions, our construction of WAPB Boolean functions also includes WPB Boolean functions when the number of variables is of the form $n = 2^l$.

1.2   Organisation

The research objectives and our contributions are already outlined. The remaining part of the paper is organized as follows:

i. In Sect. 2, we precisely define all the required definitions and notations of Boolean functions and their cryptographic properties. Furthermore, the definitions of WAPB and WPB Boolean functions and some previous constructions of these Boolean functions are also discussed.

ii. Sect. 3 presents an elaborate study on forming a WAPB/WPB Boolean function using the direct sum of two WAPB Boolean functions. We explore various cases where the direct sum obtains a WAPB Boolean function. Additionally, we provide some constructions of classes of WAPB Boolean functions and analyse its cryptographic properties. Furthermore, we improved the weightwise algebraic immunity bound given in [3].

iii. The paper is concluded with future scope in Sect. 4.

2   Preliminaries

Let $\mathbb{F}_2 = \{0, 1\}$ be a finite field with addition '+' and multiplication '·'. The multiplication $x \cdot y$ is written as xy for $x, y \in \mathbb{F}_2$. $\mathbb{F}_2^n$ is the n-dimensional vector space over $\mathbb{F}_2$. We denote $[i, j] = \{i, i+1, \ldots, j\}$ for two integers i, j with $i \leq j$. An n-variable Boolean function is a mapping from $\mathbb{F}_2^n$ to $\mathbb{F}_2$ and $\mathcal{B}_n$ denotes the set of all n-variable Boolean functions. For any $v = (v_1, v_2, \ldots, v_n) \in \mathbb{F}_2^n$, the Hamming weight of v is defined as $\mathtt{wt}(v) = |\{i \in [1, n] : v_i = 1\}|$. The support of a Boolean function $f \in \mathcal{B}_n$ is $\mathtt{sup}(f) = \{v \in \mathbb{F}_2^n : f(v) = 1\}$ and Hamming weight of f is $\mathtt{wt}(f) = |\mathtt{sup}(f)|$. The truth table representation of a Boolean function $f \in \mathcal{B}_n$ is a 2^n-dimensional vector representation, i.e., $f = (f(0, 0, \ldots, 0), f(0, 0, \ldots, 1), \ldots, f(1, 1, \ldots, 1))$. The algebraic normal form (ANF) representation is defined as $f(x) = \sum_{u \in \mathbb{F}_2^n} a_u x^u$, where $a_u \in \mathbb{F}_2$ and $x^u = x_1^{u_1} x_2^{u_2} \cdots x_n^{u_n}$ for $x = (x_1, x_2, \ldots, x_n)$ and $u = (u_1, u_2, \ldots, u_n)$. The algebraic degree of a Boolean function $f \in \mathcal{B}_n$ is defined as $\deg(f) = \max\{\mathtt{wt}(u) : u \in \mathbb{F}_2^n, a_u \neq 0\}$. Any $f \in \mathcal{B}_n$ with $\deg(f) \leq 1$ is said to be an affine Boolean function, and the set of all affine Boolean functions in $\mathcal{B}_n$ is denoted by $\mathcal{A}_n$.

Denote $E_{n,k} = \{v \in \mathbb{F}_2^n : \mathtt{wt}(v) = k\}$ for every $k \in [0, n]$. The support and Hamming weight of f restricted to $E_{n,k}$ are denoted as $\mathtt{sup}_k(f) = \{v \in E_{n,k} :$

$f(v) = 1\}$ and $\mathtt{wt}_k(f) = |\mathtt{sup}_k(f)|$, respectively. The Hamming distance between two functions $f, g \in \mathcal{B}_n$ is given as $\mathtt{d}(f, g) = |\{v \in \mathbb{F}_2^n : f(v) \neq g(v)\}| = \mathtt{wt}(f+g)$ and the Hamming distance between two functions $f, g \in \mathcal{B}_n$ restricted to $E_{n,k}$ is given as $\mathtt{d}_k(f, g) = |\{v \in E_{n,k} : f(v) \neq g(v)\}| = \mathtt{wt}_k(f + g)$.

A Boolean function $f \in \mathcal{B}_n$ is balanced, if $\mathtt{wt}(f) = 2^{n-1}$. The nonlinearity of $f \in \mathcal{B}_n$ denoted as $\mathtt{nl}(f)$, is the minimum Hamming distance of f to any affine function. That is, $\mathtt{nl}(f) = \min_{g \in \mathcal{A}_n} \mathtt{d}(f, g)$. Similarly, all these cryptographic criteria are also defined for the n-variable Boolean function when the inputs are restricted to $E_{n,k}$ as follows.

Definition 1. *A Boolean function $f \in \mathcal{B}_n$ is said to be weightwise almost perfectly balanced (WAPB) if for all $k \in [0, n]$,*

$$\mathtt{wt}_k(f) = \begin{cases} \frac{\binom{n}{k}}{2} & \text{if } \binom{n}{k} \text{ is even,} \\ \frac{\binom{n}{k} \pm 1}{2} & \text{if } \binom{n}{k} \text{ is odd.} \end{cases}$$

For $k \in [0, n]$, $\delta_k^f \in \{-1, 0, 1\}$ is defined as $\delta_k^f = 2\mathtt{wt}_k(f) - \binom{n}{k}$. That is, $\mathtt{wt}_k(f) = \frac{1}{2}\left[\binom{n}{k} + \delta_k^f\right]$.

Hence, for any WAPB $f \in \mathcal{B}_n$, $\quad \delta_k^f = \begin{cases} 0 & \text{if } \binom{n}{k} \text{ is even,} \\ -1 & \text{if } \mathtt{wt}_k(f) < \frac{\binom{n}{k}}{2}, \\ 1 & \text{if } \mathtt{wt}_k(f) > \frac{\binom{n}{k}}{2}. \end{cases}$

Definition 2. *For $x = (x_1, x_2, \ldots, x_n), y = (y_1, y_2, \ldots, y_n) \in \mathbb{F}_2^n$, we say y covers x (i.e., $x \preceq y$), if $x_i \leq y_i, \forall i \in [1, n]$ i.e., $y_i = 1$ if $x_i = 1, \forall i \in [1, n]$.*

Given a positive integer n, denote $e(n) = \{e_1, e_2, \ldots, e_w\} \subseteq \mathbb{N} \cup \{0\}$ if $n = 2^{e_1} + 2^{e_2} + \cdots + 2^{e_w}$. Therefore, for two positive integers m, n, we denote $m \preceq n$ if $e(m) \subseteq e(n)$. Further, for a set of non-negative integers T (i.e., $T \subseteq \mathbb{N} \cup \{0\}$), we define (abusing the notation of power set) $2^T = \sum_{t \in T} 2^t$.

Proposition 1 (Lucas' Theorem). *For two nonnegative integers n and k,*

$$\binom{n}{k} = \begin{cases} 1 \,(mod\ 2) & \text{if } k \preceq n \\ 0 \,(mod\ 2) & \text{if } k \npreceq n. \end{cases}$$

It is straightforward from Proposition 1 that $\binom{n}{k}$ is even for all $k \in [1, n-1]$ iff $n = 2^m$ for a nonnegative integer m. Then we have the following special class of WAPB Boolean functions when $n = 2^m$ for a nonnegative integer m.

Definition 3. *A balanced Boolean function $f \in \mathcal{B}_n$ is said to be weightwise perfectly balanced (WPB) if the restriction of f to $E_{n,k}$, is balanced for all $k \in [1, n-1]$, i.e., $\binom{n}{k}$ is even and $\mathtt{wt}_k(f) = \frac{\binom{n}{k}}{2}$ for all $k \in [1, n-1]$.*

Therefore, if $f \in \mathcal{B}_n$ is an WPB Boolean function, then $n = 2^m$ for a nonnegative integer m with $\delta_k^f = 0$ for all $k \in [1, n-1]$ and $\delta_0^f = -\delta_n^f$ (i.e., $f(0,0,\ldots,0) \neq f(1,1,\ldots,1)$). Hence, if $n = 2^m$ then there are $2 \prod_{k=1}^{n-1} \binom{\binom{n}{k}}{\binom{n}{k}/2}$ WPB Boolean functions, else there is none.

Definition 4. *Let* $f \in \mathcal{B}_n$ *be a Boolean function and* $E \subseteq \mathbb{F}_2^n$. *The Walsh transform of* f *at* $a \in \mathbb{F}_2^n$ *is defined as* $W_f(a) = \sum_{x \in \mathbb{F}_2^n} (-1)^{f(x)+a \cdot x}$. *The Walsh transform of* f *restricted to* E *at* $a \in \mathbb{F}_2^n$ *is defined as* $W_{f,E}(a) = \sum_{x \in E} (-1)^{f(x)+a \cdot x}$. *If* $E = E_{n,k}$, *we denote* $W_{f,E}(a)$ *as* $W_{f,k}(a)$.

Therefore, $W_{f,k}(0) = \sum_{x \in E_{n,k}} (-1)^{f(x)} = \binom{n}{k} - 2\mathtt{wt}_k(f) = -\delta_k^f$.

The Boolean functions with good cryptographic criteria over $E_{n,k}$ are significant for the stream ciphers like FLIP. The nonlinearity and algebraic immunity of a Boolean function over a restricted domain is introduced in [3].

Definition 5 (Weightwise nonlinearity). *The nonlinearity of* $f \in \mathcal{B}_n$ *over* $E_{n,k}$, *denoted as* $\mathtt{nl}_k(f)$, *is the Hamming distance of* f *to the set of all affine functions* $\mathcal{A}_n$ *when evaluated over* $E_{n,k}$. *That is,* $\mathtt{nl}_k(f) = \min_{g \in \mathcal{A}_n} d_k(f, g) = \min_{g \in \mathcal{A}_n} \mathtt{wt}_k(f + g)$.

The following identity and upper bound on the nonlinearity of a Boolean function over $E_{n,k}$ can be derived. The upper bound is further improved by Mesnager et al. in [18].

Lemma 1. *[3] If* $f \in B_n$ *then for* $k \in [0, n]$,

$$\mathtt{nl}_k(f) = \frac{|E_{n,k}|}{2} - \frac{1}{2} \max_{a \in \mathbb{F}_2^n} \left| \sum_{x \in E_{n,k}} (-1)^{f(x)+a.x} \right| \ and$$

$$\mathtt{nl}_k(f) \leq \frac{1}{2} \left[|E_{n,k}| - \sqrt{|E_{n,k}|} \right] = \frac{1}{2} \left[\binom{n}{k} - \sqrt{\binom{n}{k}} \right].$$

Definition 6. *Given* $f \in \mathcal{B}_n$, *a nonzero* $g \in \mathcal{B}_n$ *is called an annihilator of* f *if* $fg = 0$, *i.e.,* $f(x)g(x) = 0$ *for all* $x \in \mathbb{F}_2^n$. *For* $E \subseteq \mathbb{F}_2^n$, *a function* $g \in \mathcal{B}_n$ *is called an annihilator of* f *over* E *if* $g(x) \neq 0$ *for some* $x \in E$ *and* $f(x)g(x) = 0$ *for all* $x \in E$. *The set of all annihilators of* $f \in \mathcal{B}_n$ *is denoted by* $Ann(f)$ *and the set of all annihilators of* f *over* E *is denoted by* $Ann_E(f)$. *The algebraic immunity of* $f \in \mathcal{B}_n$ *is defined as*

$$\mathtt{AI}(f) = \min\{\deg(g) : g \in Ann(f) \cup Ann(1 + f)\}.$$

For $E \subseteq \mathbb{F}_2^n$, *the algebraic immunity of* f *over* E *is defined by*

$$\mathtt{AI}_E(f) = \min\{\deg(g) : g \in Ann_E(f) \cup Ann_E(1 + f)\}.$$

For $E = E_{n,k}$, *we denote* $Ann_E(f)$ *and* $\mathtt{AI}_E(f)$ *as* $Ann_k(f)$ *and* $\mathtt{AI}_k(f)$, *respectively.*

Note 1. For $f \in \mathcal{B}_n$ and $E \subseteq \mathbb{F}_2^n$, if $g \in Ann_E(f)$ then there exists $x \in E$ such that $g(x) \neq 0$. This implies that an annihilator of f is not necessarily an annihilator of f on E. That is, $Ann(f) \not\subseteq Ann_E(f)$ and hence $\mathtt{AI}_E(f) \not\leq \mathtt{AI}(f)$ for $f \in \mathcal{B}_n$ and $E \subseteq \mathbb{F}_2^n$.

For cryptographic use, especially in lightweight ciphers, the Boolean functions need to have good cryptographic properties along with a simple way of expression for fast computation, inexpensive implementation, and energy efficiency. The first construction for the class of WAPB and WPB Boolean functions has been introduced in [3] using the direct sum of Boolean functions. The direct sum of $f \in \mathcal{B}_m$ and $g \in \mathcal{B}_n$ forms a Boolean function $h \in \mathcal{B}_{m+n}$ such that $h(x, y) = f(x) + g(y)$, where $x \in \mathbb{F}_2^m, y \in \mathbb{F}_2^n$. We will study the direct sum of two WAPB Boolean functions in detail in Sect. 3. The following propositions present some classes of WPB and WAPB Boolean functions by modifying the direct sum of two Boolean functions.

Proposition 2. *[3] Let $f, f', g, g' \in \mathcal{B}_n$ where f, f', g are WPB Boolean functions. Then the function $h \in \mathcal{B}_{2n}$ defined by*

$$h(x, y) = f(x) + \prod_{i=1}^{n} x_i + g(y) + (f(x) + f'(x))g'(y), \quad where \; x, y \in \mathbb{F}_2^n,$$

is a WPB Boolean function.

If we consider $f' = f$, then a simple construction of WPB Boolean function arises (see Corollary 1) and this result also a particular case of our result. The following WAPB construction presented in [25] is based on the direct sum of WPB Boolean functions of n_is variables where n_is are distinct powers of 2.

Proposition 3. *[25] Let $n = n_1 + n_2 + \cdots + n_p$ for n_i being the power of 2 for $1 \leq i \leq p$ and $0 < n_1 < n_2 < \cdots < n_p$. Let $f_{n_i} \in \mathcal{B}_{n_i}$ be WPB with $f_{n_i}(0, 0, \ldots, 0) = 0, f_{n_i}(1, 1, \ldots, 1) = 1$ for $1 \leq i \leq p$. Then $h \in \mathcal{B}_n$ defined as*

$$h_n(x_1, x_2, \ldots, x_n) = f_{n_1}(x_1, x_2, \ldots, x_{n_1}) + f_{n_2}(x_{n_1+1}, x_{n_1+2}, \ldots, x_{n_1+n_2}) + \cdots$$
$$+ f_{n_p}(x_{n-n_p+1}, x_{n-n_p+2}, \ldots, x_n)$$

is a WAPB Boolean function.

3   On the Direct Sum of WAPB Boolean Functions

In this section we have studied on the formation of WAPB Boolean functions from the direct sum of two WAPB Boolean functions.

Theorem 1. *Let $f \in \mathcal{B}_m$, $g \in \mathcal{B}_n$ be two WAPB Boolean functions. Let $h \in \mathcal{B}_{m+n}$ be defined as $h(x, y) = f(x) + g(y)$ for $x \in \mathbb{F}_2^m$ and $y \in \mathbb{F}_2^n$. Then*

$$\mathtt{wt}_k(h) = \frac{\binom{m+n}{k} - \sum_{i=0}^{k} \delta_i^f \delta_{k-i}^g}{2} \; for \; k \in [0, m+n].$$

Proof. As $f \in \mathcal{B}_m, g \in \mathcal{B}_n$ are WAPB Boolean functions, we have from Definition 1 that

$$\mathtt{wt}_k(f) = \frac{\binom{m}{k} + \delta_k^f}{2} \text{ for } k \in [0, m+n] \text{ where } \delta_i^f = 0 \text{ for } i \in [m+1, m+n] \text{ and}$$

$$\mathtt{wt}_k(g) = \frac{\binom{n}{k} + \delta_k^g}{2} \text{ for } k \in [0, m+n] \text{ where } \delta_i^g = 0 \text{ for } i \in [n+1, m+n].$$

Here, we could extend k (in $\mathtt{wt}_k(f)$ and δ_k^f) till $m+n$, as $\binom{m}{k} = 0$ for $k > m$ and $\binom{n}{k} = 0$ for $k > n$.

As $h(x, y) = f(x) + g(y)$, $h(x, y) = 1$ if and only if (i) $f(x) = 1$ and $g(y) = 0$ or, (ii) $f(x) = 0$ and $g(y) = 1$. Hence, the weight of $h(x, y) = f(x) + g(y)$ in the restricted domain of $E_{n,k}$ is

$$\mathtt{wt}_k(h) = \sum_{i=0}^{k} [\mathtt{wt}_i(f)\mathtt{wt}_{k-i}(1+g) + \mathtt{wt}_i(1+f)\mathtt{wt}_{k-i}(g)]$$

$$= \sum_{i=0}^{k} \left[\left(\frac{\binom{m}{i} + \delta_i^f}{2} \right) \left(\binom{n}{k-i} - \frac{\binom{n}{k-i} + \delta_{k-i}^g}{2} \right) \right.$$

$$\left. + \left(\binom{m}{i} - \frac{\binom{m}{i} + \delta_i^f}{2} \right) \left(\frac{\binom{n}{k-i} + \delta_{k-i}^g}{2} \right) \right]$$

$$= \frac{1}{4} \sum_{i=0}^{k} \left[\left(\binom{m}{i} + \delta_i^f \right) \left(\binom{n}{k-i} - \delta_{k-i}^g \right) \right.$$

$$\left. + \left(\binom{m}{i} - \delta_i^f \right) \left(\binom{n}{k-i} + \delta_{k-i}^g \right) \right]$$

$$= \frac{1}{4} \sum_{i=0}^{k} \left[2\binom{m}{i}\binom{n}{k-i} - 2\delta_i^f \delta_{k-i}^g \right]$$

$$= \frac{1}{2} \binom{m+n}{k} - \frac{1}{2} \sum_{i=0}^{k} \delta_i^f \delta_{k-i}^g = \frac{\binom{m+n}{k} - \sum_{i=0}^{k} \delta_i^f \delta_{k-i}^g}{2}.$$

$\square$

The proof of Theorem 1 can also be done using Piling-up lemma [14]. Here, the function h (in Theorem 1) is a WAPB if and only if the chosen functions f, g satisfies $\delta_k^h = -\sum_{i=0}^{k} \delta_i^f \delta_{k-i}^g \in \{-1, 0, 1\}$.

Example 1. Let $f \in \mathcal{B}_5$ and $g \in \mathcal{B}_4$ be two WAPB Boolean functions and $h \in \mathcal{B}_9$ be defined as $h(x, y) = f(x) + g(y)$ for $x \in \mathbb{F}_2^5, y \in \mathbb{F}_2^4$. Then $\delta_2^f = \delta_3^f = 0$ and $\delta_0^f, \delta_1^f, \delta_4^f, \delta_5^f \in \{-1, 1\}$. Further, $\delta_1^g = \delta_2^g = \delta_3^g = 0$ and $\delta_0^g, \delta_4^g \in \{-1, 1\}$. Here h is a WAPB iff $\delta_k^h = -\sum_{i=0}^{k} \delta_i^f \delta_{k-i}^g \in \{-1, 0, 1\}$ for $k \in [0, 9]$ with $\delta_k^h = 0$ for $k \not\le 9$ i.e., for $k = 2, 3, 4, 5, 6, 7$. Then the above conditions form the equations $\delta_0^h = -\delta_0^f \delta_0^g \implies \delta_0^f \delta_0^g = \pm 1;$

$$\delta_1^h = -\delta_0^f \delta_1^g - \delta_1^f \delta_0^g \implies \delta_1^f \delta_0^g = \pm 1;$$
$$\delta_2^h = -\sum_{i=0}^{2} \delta_i^f \delta_{2-i}^g \implies 0 = 0;$$
$$\delta_3^h = -\sum_{i=0}^{3} \delta_i^f \delta_{3-i}^g \implies 0 = 0;$$
$$\delta_4^h = -\sum_{i=0}^{4} \delta_i^f \delta_{4-i}^g \implies \delta_0^f \delta_4^g + \delta_4^f \delta_0^g = 0;$$
$$\delta_5^h = -\sum_{i=0}^{5} \delta_i^f \delta_{5-i}^g \implies \delta_1^f \delta_4^g + \delta_5^f \delta_0^g = 0;$$
$$\delta_6^h = -\sum_{i=0}^{6} \delta_i^f \delta_{6-i}^g \implies 0 = 0;$$
$$\delta_7^h = -\sum_{i=0}^{7} \delta_i^f \delta_{7-i}^g \implies 0 = 0;$$
$$\delta_8^h = -\sum_{i=0}^{8} \delta_i^f \delta_{8-i}^g \implies \delta_4^f \delta_4^g = \pm 1;$$
$$\delta_9^h = -\sum_{i=0}^{9} \delta_i^f \delta_{9-i}^g \implies \delta_5^f \delta_4^g = \pm 1.$$

If we consider all f, g and h are balanced, then $\delta_0^f + \delta_1^f + \delta_4^f + \delta_5^f = \delta_0^g + \delta_4^g = \delta_0^h + \delta_1^h + \delta_8^h + \delta_9^h = 0$. Using all the above equations, we have simplified conditions $\delta_0^g = -\delta_4^g = \pm 1$ and $\delta_0^f = -\delta_1^f = \delta_4^f = -\delta_5^f = \pm 1$. If we have $f(0) = g(0) = 0$ i.e., $\delta_0^f = \delta_0^g = -1$. Then the balanced functions $f \in \mathcal{B}_5, g \in \mathcal{B}_4$ satisfying $\delta_0^f = \delta_4^f = -1, \delta_1^f = \delta_5^f = 1$; and $\delta_0^g = -1, \delta_4^g = 1$ result a balanced WAPB $h(x, y) = f(x) + g(y) \in \mathcal{B}_9$. □

The terms $\delta_k^h, k \in [0, m + n]$ are the products of the convolutions of the sequences $\{\delta_i^f : i \in [0, m]\}$ and $\{\delta_j^g : j \in [0, n]\}$. Therefore, h is a WAPB if and only if the following matrix multiplication satisfies

$$\Delta^f \delta^g = \delta^h, \text{ that is, } \begin{bmatrix} \delta_0^f & 0 & 0 & \cdots & 0 \\ \delta_1^f & \delta_0^f & 0 & \cdots & 0 \\ \delta_2^f & \delta_1^f & \delta_0^f & \cdots & 0 \\ & & \vdots & & \\ 0 & 0 & \cdots & \delta_m^f & \delta_{m-1}^f \\ 0 & 0 & \cdots & 0 & \delta_m^f \end{bmatrix} \begin{bmatrix} \delta_0^g \\ \delta_1^g \\ \vdots \\ \delta_n^g \end{bmatrix} = \begin{bmatrix} \delta_0^h \\ \delta_1^h \\ \vdots \\ \delta_{m+n}^h \end{bmatrix}. \quad (1)$$

Here, the matrices Δ^f, δ^g and δ^h are of order $(m + n + 1) \times (n + 1), (n + 1) \times 1$ and $(m + n + 1) \times 1$ respectively. This multiplication can also be written as

$$\Delta^g \delta^f = \delta^h, \text{ that is, } \begin{bmatrix} \delta_0^g & 0 & 0 & \cdots & 0 \\ \delta_1^g & \delta_0^g & 0 & \cdots & 0 \\ \delta_2^g & \delta_1^g & \delta_0^g & \cdots & 0 \\ & & \vdots & & \\ 0 & 0 & \cdots & \delta_n^g & \delta_{n-1}^g \\ 0 & 0 & \cdots & 0 & \delta_n^g \end{bmatrix} \begin{bmatrix} \delta_0^f \\ \delta_1^f \\ \vdots \\ \delta_m^f \end{bmatrix} = \begin{bmatrix} \delta_0^h \\ \delta_1^h \\ \vdots \\ \delta_{m+n}^h \end{bmatrix}. \quad (2)$$

The matrices Δ^g and δ^f are of order $(m + n + 1) \times (m + 1)$ and $(m + 1) \times 1$ respectively. Here, $\delta_i^f, \delta_j^g, \delta_k^h \in \{-1, 0, 1\}$ with $\delta_i^f = 0$ if $i \nleq m$, $\delta_j^g = 0$ if $j \nleq n$ and $\delta_k^h = 0$ if $k \nleq m + n$. Now we will study some different cases of m and n.

Theorem 2. *Let m and n be positive integers such that $e(m) \cap e(n) = \emptyset$. Let $f \in \mathcal{B}_m$ and $g \in \mathcal{B}_n$ be two WAPB Boolean functions. Then $h \in \mathcal{B}_{m+n}$ defined* as $h(x, y) = f(x) + g(y)$ *for* $x \in \mathbb{F}_2^m, y \in \mathbb{F}_2^n$ *is a WAPB Boolean function with*

$$\delta_k^h = \begin{cases} 0 & \text{if } e(k) \not\subseteq e(m) \cup e(n) = e(m+n) \text{ i.e., } k \npreceq m+n \\ -\delta_s^f \delta_{k-s}^g & \text{if } e(k) \subseteq e(m) \cup e(n) = e(m+n) \text{ i.e., } k \preceq m+n \end{cases}$$

where $e(s) = e(k) \cap e(m)$ *(i.e.,* $s = k\&m$ *where* $\&$ *is the bitwise AND operation).*

Proof. Since $f \in \mathcal{B}_m$ and $g \in \mathcal{B}_n$ are WAPB Boolean functions, from Lucas' Theorem we have

$$\delta_i^f = \begin{cases} 0 & \text{if } e(i) \not\subseteq e(m) \\ \pm 1 & \text{if } e(i) \subseteq e(m) \end{cases} \quad \text{and} \quad \delta_j^g = \begin{cases} 0 & \text{if } e(j) \not\subseteq e(n) \\ \pm 1 & \text{if } e(j) \subseteq e(n). \end{cases} \quad (3)$$

Given that $e(m) \cap e(n) = \emptyset$. That implies, $e(m+n) = e(m) \cup e(n)$. As $h(x, y) = f(x) + g(y)$, for $k \in [0, m+n]$, substituting δ_i^f, δ_j^g from Eq. 3 in Theorem 1, we have

$$\text{wt}_k(h) = \frac{1}{2}\left[\binom{m+n}{k} - \sum_{i \in [0,k]} \delta_i^f \delta_{k-i}^g\right] = \frac{1}{2}\left[\binom{m+n}{k} - \sum_{\substack{i \in [0,k] \\ e(i) \subseteq e(m) \\ e(k-i) \subseteq e(n)}} \delta_i^f \delta_{k-i}^g\right].$$

Consider $i \in [0, k]$ and $e(i) = \{a_1, a_2, \ldots, a_p\} \subseteq e(m)$. Further consider $e(k-i) = \{b_1, b_2, \ldots, b_q\} \subseteq e(n)$. Then $e(i) \cap e(k-i) = \emptyset$ and $e(k) = e(i) \cup e(k-i) \subseteq e(m) \cup e(n)$. Hence, we have

$$\text{wt}_k(h) = \begin{cases} \frac{1}{2}\binom{m+n}{k} & \text{if } e(k) \not\subseteq e(m) \cup e(n) \\ \frac{1}{2}\left[\binom{m+n}{k} - \sum_{\substack{i \in [0,k] \\ e(i) \subseteq e(m), e(k-i) \subseteq e(n)}} \delta_i^f \delta_{k-i}^g\right] & \text{if } e(k) \subseteq e(m) \cup e(n). \end{cases}$$

Moreover from the above argument, we have that if $k = 2^{a_1} + 2^{a_2} + \cdots + 2^{a_p} + 2^{b_1} + 2^{b_2} + \cdots + 2^{b_q}$, then the only case when $e(i) \subseteq e(m), e(k-i) \subseteq e(n)$ is $i = 2^{a_1} + 2^{a_2} + \cdots + 2^{a_p}$ and $k - i = 2^{b_1} + 2^{b_2} + \cdots + 2^{b_q}$ i.e., $e(i) = e(k) \cap e(m)$. Hence,

$$\text{wt}_k(h) = \begin{cases} \frac{1}{2}\binom{m+n}{k} & \text{if } e(k) \not\subseteq e(m) \cup e(n) \\ \frac{1}{2}\left[\binom{m+n}{k} - \delta_i^f \delta_{k-i}^g\right] \text{ where } e(i) = e(k) \cap e(m) & \text{if } e(k) \subseteq e(m) \cup e(n). \end{cases}$$

$\square$

Example 2. Consider two WAPB Boolean functions $f \in \mathcal{B}_{10}$ and $g \in \mathcal{B}_4$. Then $\delta_i^f = 0$ for $i \in \{1, 3, 4, 5, 6, 7, 9\}$ and $\delta_0^f, \delta_2^f, \delta_8^f, \delta_{10}^f \in \{-1, 1\}$. Similarly, $\delta_j^g = 0$ for $j \in \{1, 2, 3\}$ and $\delta_0^g, \delta_4^g \in \{-1, 1\}$. Here two sets $e(10) = \{3, 1\}$ and $e(4) = \{2\}$ are disjoint. Then $h \in \mathcal{B}_{14}$ such that $h(x, y) = f(x) + g(y)$ is a WAPB with $\delta_k^h = 0$ for $k \in \{1, 3, 5, 7, 9, 11, 13\}$ and $\delta_0^h = -\delta_0^f \delta_0^g, \delta_2^h = -\delta_2^f \delta_0^g, \delta_4^h = -\delta_0^f \delta_4^g, \delta_6^h = -\delta_2^f \delta_6^g, \delta_8^h = -\delta_8^f \delta_0^g, \delta_{10}^h = -\delta_{10}^f \delta_0^g, \delta_{12}^h = -\delta_8^f \delta_4^g, \delta_{14}^h = -\delta_{10}^f \delta_4^g$.

$\square$

The result on the direct sum of WPB Boolean functions in [25, Theorem 3] (which is stated in Proposition 3) is a direct consequence of the Theorem 2. The theorem is stated and proved using our method and notation as follows.

Theorem 3. *Let n be a positive integer with $e(n) = \{a_1, a_2, \ldots, a_p\}$ with $0 \leq a_1 < a_2 < \cdots < a_p$. Denote $n_i = 2^{a_i}$ for $i \in [1,p]$. Let $f_{n_i} \in \mathcal{B}_{n_i}$ be WPB Boolean functions with $f_{n_i}(0,0,\ldots,0) = 0, f_{n_i}(1,1,\ldots,1) = 1$ for $1 \leq i \leq p$. Then $h \in \mathcal{B}_n$ defined as*

$$h_n(x_1, x_2, \ldots, x_n) = f_{n_1}(x_1, x_2, \ldots, x_{n_1}) + f_{n_2}(x_{n_1+1}, x_{n_1+2}, \ldots, x_{n_1+n_2}) + \cdots$$
$$+ f_{n_p}(x_{n-n_p+1}, x_{n-n_p+2}, \ldots, x_n)$$

is a WAPB, with

$$\delta_k^{h_n} = \begin{cases} 0 & \text{if } e(k) \not\subseteq e(n) \\ -(-1)^{|e(k)|} = (-1)^{\mathbf{wt}(k)+1} & \text{if } e(k) \subseteq e(n), \end{cases} \quad (4)$$

for $k \in [0,n]$.

Proof. It is given that $\delta_0^{f_{n_i}} = -1$ and $\delta_{n_i}^{f_{n_i}} = 1$. We will prove it using induction on $p = |e(n)|$. If $|e(n)| = 1$, it is already a WAPB and satisfying Eq. 4. Assume that it is true if $|e(n)| \leq p - 1$. Let $m = n_1 + n_2 + \cdots + n_{p-1}$ i.e., $e(m) = \{a_1, a_2, \ldots, a_{p-1}\}$. As per the assumption

$$h_m(x_1, \ldots, x_m) = f_{n_1}(x_1, \ldots, x_{n_1}) + \cdots + f_{n_{p-1}}(x_{m-n_{p-1}+1}, \ldots, x_m)$$

is a WAPB Boolean function with

$$\delta_k^{h_m} = \begin{cases} 0 & \text{if } e(k) \not\subseteq e(m) \\ -(-1)^{|e(k)|} & \text{if } e(k) \subseteq e(m), \end{cases}$$

for $k \in [0,m]$. Let $n = m + n_p$ where $n_p = 2^{a_p}$ and $n_p > m$. Further, let $h_n(x_1, \ldots, x_n) = h_m(x_1, \ldots, x_m) + f_{n_p}(x_{m+1}, \ldots, x_n)$. Here, $\delta_0^{f_{n_p}} = -1$ and $\delta_{n_p}^{f_{n_p}} = 1$. As $n_p = 2^{a_p}$ and $n_p > m$, $e(n_p) = \{a_p\}$ and $e(m) \cap e(n_p) = \emptyset$. Hence, for $k \in [0,n]$ (using Theorem 2),

$$\delta_k^{h_n} = \begin{cases} 0 & \text{if } e(k) \not\subseteq e(n) \\ -\delta_s^{f_{n_p}} \delta_{k-s}^{h_m} \text{ where } e(s) = e(k) \cap e(n_p) & \text{if } e(k) \subseteq e(n). \end{cases}$$

$$= \begin{cases} 0 & \text{if } e(k) \not\subseteq e(n) \\ -\delta_0^{f_{n_p}} \delta_k^{h_m} & \text{if } e(k) \subseteq e(n) \text{ and } e(n_p) \cap e(k) = \emptyset \\ -\delta_{n_p}^{f_{n_p}} \delta_{k-n_p}^{h_m} & \text{if } e(k) \subseteq e(n) \text{ and } e(n_p) \cap e(k) = \{a_p\} \end{cases}$$

$$= \begin{cases} 0 & \text{if } e(k) \not\subseteq e(n) \\ -(-1)(-(-1)^{|e(k)|}) & \text{if } e(k) \subseteq e(n) \text{ and } e(n_p) \cap e(k) = \emptyset \\ -(1)(-(-1)^{|e(k)|-1}) & \text{if } e(k) \subseteq e(n) \text{ and } e(n_p) \cap e(k) = \{a_p\} \end{cases}$$

$$= \begin{cases} 0 & \text{if } e(k) \not\subseteq e(n) \\ -(-1)^{|e(k)|} & \text{if } e(k) \subseteq e(n). \end{cases}$$

$\square$

Now we will study the possibility of formation of a WPB Boolean function by the direct sum of two WAPB Boolean functions. Let m and n be two nonnegative integers such that $e(m) = \{a_1, a_2, \ldots, a_p\}$ with $0 \le a_1 < a_2 < \cdots < a_p$ and $e(n) = \{b_1, b_2, \ldots, b_q\}$ with $0 \le b_1 < b_2 < \cdots < b_q$. If $m + n = 2^l$ for a nonnegative integer l then $e(m) \cap e(n) = \{a_1\}$ (i.e., $a_1 = b_1$), and $e(m) \cup e(n) = \{a_1, a_1 + 1, a_1 + 2, \ldots, l - 1\}$. For example, if $m = 76$ and $n = 52$ then $e(m) = \{2, 3, 6\}, e(n) = \{2, 4, 5\}$ and $m + n = 128 = 2^7$. Here, $e(m) \cap e(n) = \{2\}$ and $e(m) \cup e(n) = \{2, 3, 4, 5, 6\}$.

Theorem 4. *Let m and n be two positive integers such that $m + n = 2^l$ for a nonnegative integer l. Let $f \in \mathcal{B}_m$ and $g \in \mathcal{B}_n$ be two WAPB Boolean functions. Then $h \in \mathcal{B}_{m+n}$ defined as $h(x, y) = f(x) + g(y)$ for $x \in \mathbb{F}_2^m, y \in \mathbb{F}_2^n$ be a WPB Boolean function if there is a $c \in \{-1, 1\}$ such that*

$$\frac{\delta_0^f}{\delta_m^f} = -\frac{\delta_0^g}{\delta_n^g};$$

$$\frac{\delta_{2^{T_1 \setminus \{a_1\}}}^f}{\delta_{2^{T_1}}^f} = c \text{ for every } T_1 \subseteq e(m) \text{ with } a_1 \in T_1;$$

$$\frac{\delta_{2^{T_2 \setminus \{a_1\}}}^g}{\delta_{2^{T_2}}^g} = -c \text{ for every } T_2 \subseteq e(n) \text{ with } a_1 \in T_2;$$

$$\frac{\delta_{2^{T_1}}^f}{\delta_{2^{S_1}}^f} = -\frac{\delta_{2^{S_2}}^g}{\delta_{2^{T_2}}^g}.$$

where $S_1 = (T_1 \setminus \{s\}) \cup (e(m) \cap \{a_1, a_1 + 1, \ldots, s - 1\})$ and $S_2 = (T_2 \setminus \{s\}) \cup (e(n) \cap \{a_1, a_1 + 1, \ldots, s - 1\})$ with s be the smallest integer in $e(k)$.

Proof. Let $e(m) = \{a_1, a_2, \ldots, a_p\}$ with $0 \le a_1 < a_2 < \cdots < a_p$ and $e(n) = \{b_1, b_2, \ldots, b_q\}$ with $0 \le b_1 < b_2 < \cdots < b_q$. Since $m+n = 2^l$, $e(m) \cap e(n) = \{a_1\}$ (i.e., $a_1 = b_1$), and $e(m) \cup e(n) = \{a_1, a_1 + 1, a_1 + 2, \ldots, l - 1\}$.

Since $f \in \mathcal{B}_m$ and $g \in \mathcal{B}_n$ are two WAPB Boolean functions, for any $T \subseteq \mathbb{N} \cup \{0\}$ we have

$$\delta_{2^T}^f = \begin{cases} \pm 1 & \text{if } T \subseteq e(m) \\ 0 & \text{if } T \not\subseteq e(m), \end{cases} \quad \text{and} \quad \delta_{2^T}^g = \begin{cases} \pm 1 & \text{if } T \subseteq e(n) \\ 0 & \text{if } T \not\subseteq e(n). \end{cases}$$

Here, $\delta_0^h = \delta_0^f \delta_0^g \in \{-1, 1\}$ and $\delta_{m+n}^h = \delta_m^f \delta_n^g \in \{-1, 1\}$. For a balanced function h, we have

$$\delta_0^f \delta_0^g = -\delta_m^f \delta_n^g \implies \frac{\delta_0^f}{\delta_m^f} = -\frac{\delta_0^g}{\delta_n^g}. \tag{5}$$

For all other cases i.e., $k \in [0, m+n-1]$, we need to find conditions for $\delta_k^h = -\sum_{i=0}^k \delta_i^f \delta_{k-i}^g = 0$. We use $m = 76$ and $n = 52$ for the demonstration purpose during the following steps. We consider three different cases for $k \in [1, m+n-1]$ as follows.

Case I (When $e(k) \not\subseteq e(m) \cup e(n) = \{a_1, a_1 + 1, a_1 + 2, \ldots, l-1\}$): Then $e(k)$ contains an integer $d \in [0, a_1 - 1]$.

If $e(k)$ contains an integer $d \in [0, a_1 - 1]$ then we claim that for every $i \in [0, k]$, atleast one of the $\delta_i^f, \delta_{k-i}^g$ is 0. Let there be an $i \in [0, k]$ such that $\delta_i^f \neq 0$ i.e., $e(i) \subseteq e(m)$. Since $d < a_1$, $e(k - i)$ contains d i.e., $e(k - i) \not\subseteq e(n)$ and that implies $\delta_{k-i}^g = 0$. Hence, $\sum_{i=0}^{k} \delta_i^f \delta_{k-i}^g = 0$. No condition is formed in this case.

For example, if we take $k = 5$, then $e(k) = \{0, 2\} \not\subseteq e(m) \cup e(n) = \{2, 3, 4, 5, 6\}$ and $d = 0$. Here, $\sum_{i=0}^{5} \delta_i^f \delta_{5-i}^g = \delta_0^f \delta_5^g + \delta_1^f \delta_4^g + \delta_2^f \delta_3^g + \delta_4^f \delta_1^g + \delta_5^f \delta_0^g = 0$ as $\delta_5^g = \delta_1^f = \delta_3^g = \delta_1^g = \delta_5^f = 0$.

Case II (When $e(k) \subseteq e(m) \cup e(n)$ and $a_1 \in e(k)$): Let $T = e(k), T_1 = T \cap e(m)$ and $T_2 = T \cap e(n)$. Here, $T_1 \cap T_2 = \{a_1\}$ and $T_1 \cup T_2 = T$. Now we will find $i \in [0, k]$ such that both $\delta_i^f, \delta_{k-i}^g$ nonzero. Let $\delta_i^f \neq 0$ and $\delta_{k-i}^g \neq 0$ for some $i \in [0, k]$, i.e., $e(i) \subseteq e(m)$ and $e(k - i) \subseteq e(n)$.

Let $a_1 \notin e(i)$. Then $a_1 \in e(k - i)$ as a_1 is smallest integer in $e(m) \cup e(n)$ and $a_1 \in e(k)$. Thus, $e(i)$ and $e(k-i)$ are disjoint. That implies, $e(i) \cup e(k-i) = e(k)$ Hence, $e(k - i) = T_2$ and $e(i) = T_1 \setminus \{a_1\} = T \setminus T_2$. Similarly, if $a_1 \in e(i)$, $e(i) = T_1$ and $e(k - i) = T_2 \setminus \{a_1\}$. Hence, in this case, $-\delta_k^h = \sum_{i=0}^{k} \delta_i^f \delta_{k-i}^g = \delta_{2^{T \setminus T_2}}^f \delta_{2^{T_2}}^g + \delta_{2^{T_1}}^f \delta_{2^{T \setminus T_1}}^g \in \{-2, 0, 2\}$.

Therefore, $\delta_k^h = 0$ if $\delta_{2^{T \setminus T_2}}^f \delta_{2^{T_2}}^g + \delta_{2^{T_1}}^f \delta_{2^{T \setminus T_1}}^g = 0$ i.e., if $\dfrac{\delta_{2^{T \setminus T_2}}^f}{\delta_{2^{T_1}}^f} = -\dfrac{\delta_{2^{T \setminus T_1}}^g}{\delta_{2^{T_2}}^g}$. Hence,

$$\delta_k^h = 0 \text{ if } \frac{\delta_{2^{T_1 \setminus \{a_1\}}}^f}{\delta_{2^{T_1}}^f} = -\frac{\delta_{2^{T_2 \setminus \{a_1\}}}^g}{\delta_{2^{T_2}}^g}. \tag{6}$$

For example, consider $k = 44$ i.e., $e(k) = \{2, 3, 5\}$, with $m = 76$ and $n = 52$. Here $e(k) \subseteq e(m) \cup e(n)$ and $a_1 = 2 \in e(k)$. Then $T_1 = \{2, 3\}$ and $T_2 = \{2, 5\}$. Hence, $-\delta_{44}^h = \sum_{i=0}^{44} \delta_i^f \delta_{44-i}^g = \delta_{2^3}^f \delta_{2^2+2^5}^g + \delta_{2^2+2^3}^f \delta_{2^5}^g = \delta_8^f \delta_{36}^g + \delta_{12}^f \delta_{32}^g \in \{-2, 0, 2\}$. To impose $\delta_{44}^h = 0$, the WAPB functions f, g need to satisfy $\dfrac{\delta_8^f}{\delta_{12}^f} = -\dfrac{\delta_{32}^g}{\delta_{36}^g}$.

Hence, for a fixed $T_2 \subseteq e(n)$ containing a_1, the condition in Eq. 6 is satisfied for every $T_1 \subseteq e(m)$ containing a_1. Thus, $\dfrac{\delta_{2^{T_1 \setminus \{a_1\}}}^f}{\delta_{2^{T_1}}^f} = c \in \{-1, 1\}$ is constant for every $T_1 \subseteq e(m)$ containing a_1. Similarly, for a fixed $T_1 \subseteq e(m)$ containing a_1, the condition in Eq. 6 is satisfied for every $T_2 \subseteq e(n)$ containing a_1. Thus, $\dfrac{\delta_{2^{T_2 \setminus \{a_1\}}}^g}{\delta_{2^{T_2}}^g} = -c \in \{-1, 1\}$, is constant for every $T_2 \subseteq e(n)$ containing a_1. That is, for a $c \in \{1, -1\}$,

$$\frac{\delta_{2^{T_1 \setminus \{a_1\}}}^f}{\delta_{2^{T_1}}^f} = c \text{ for every } T_1 \subseteq e(m) \text{ with } a_1 \in T_1;$$

$$\frac{\delta_{2^{T_2 \setminus \{a_1\}}}^g}{\delta_{2^{T_2}}^g} = -c \text{ for every } T_2 \subseteq e(n) \text{ with } a_1 \in T_2. \tag{7}$$

In the above example, $\frac{\delta_0^f}{\delta_4^f} = \frac{\delta_8^f}{\delta_{12}^f} = \frac{\delta_{64}^f}{\delta_{68}^f} = \frac{\delta_{72}^f}{\delta_{76}^f} = c$ and $\frac{\delta_0^g}{\delta_4^g} = \frac{\delta_{16}^g}{\delta_{20}^g} = \frac{\delta_{32}^g}{\delta_{36}^g} = \frac{\delta_{48}^g}{\delta_{52}^g} = -c$.

Case III (When $e(k) \subseteq e(m) \cup e(n)$ and $a_1 \notin e(k)$): Let denote $T = e(k)$, $T_1 = T \cap e(m)$ and $T_2 = T \cap e(n)$. Here, $T_1 \cap T_2 = \emptyset$ and $T_1 \cup T_2 = T$. Now we will find $i \in [0, k]$ such that both $\delta_i^f, \delta_{k-i}^g$ are nonzero. Let $\delta_i^f \neq 0$ and $\delta_{k-i}^g \neq 0$ for some $i \in [0, k]$. That is, $e(i) \subseteq e(m)$ and $e(k-i) \subseteq e(n)$.

If $a_1 \notin e(i)$ then $e(i) \cup e(k-i) = e(k)$ as $e(i)$ and $e(k-i)$ are disjoint. Hence, $e(i) = T_1$ and $e(k-i) = T_2$ i.e., $i = 2^{T_1}$ and $k-i = 2^{T_2}$. In the example, if we take $k = 24$ i.e., $e(k) = \{3, 4\}$, we have $i = 2^{T_1} = 2^3 = 8$ and $k - i = 2^{T_2} = 2^4 = 16$ where $a_1 = 2 \notin e(i)$. Then $\delta_8^f \delta_{16}^g \neq 0$.

Further, let $a_1 \in e(i)$. Since a_1 is the smallest integer in $e(m) \cup e(n)$, $a_1 \in e(k-i)$. Let s be the smallest integer in $e(k)$. As $k = i + (k-i)$, $\{a_1, a_1 + 1, \ldots, s-1\} \subseteq e(i) \cup e(k-i)$ and $s \notin e(i) \cup e(k-i)$. Let denote $S_1 = (T_1 \setminus \{s\}) \cup (e(m) \cap \{a_1, a_1 + 1, \ldots, s-1\})$ and $S_2 = (T_2 \setminus \{s\}) \cup (e(n) \cap \{a_1, \ldots, s-1\})$. Hence $i = 2^{S_1}$ and $k - i = 2^{S_2}$

In the example, let take $k = 24$ i.e., $e(k) = \{3, 4\}$. Here, $e(k) \subseteq e(76) \cup e(52)$ and $a_1 = 2 \notin e(k)$. Further, $T_1 = \{3\}, T_2 = \{4\}$. Then the smallest integer in $e(k)$ is $s = 3$. Hence $S_1 = (T_1 \setminus \{3\}) \cup (e(76) \cap \{2\}) = \{2\}$ and $S_2 = (T_2 \setminus \{3\}) \cup (e(52) \cap \{2\}) = \{2, 4\}$. Hence, for $i = 2^{S_1} = 2^2 = 4$ and $k - i = 2^{S_2} = 2^2 + 2^4 = 20$, $\delta_4^f \delta_{20}^g \neq 0$. Now combining above two cases for the example $k = 24$, we have $\sum_{i=0}^{k} \delta_i^f \delta_{k-i}^g = \delta_{2^{T_1}}^f \delta_{2^{T_2}}^g + \delta_{2^{S_1}}^f \delta_{2^{S_2}}^g = \delta_8^f \delta_{16}^g + \delta_4^f \delta_{20}^g$.

Hence combining two cases, when $e(k) \subseteq e(m) \cup e(n)$ and $a_1 \notin e(k)$, we have $-\delta_k^h = \sum_{i=0}^{k} \delta_i^f \delta_{k-i}^g = \delta_{2^{T_1}}^f \delta_{2^{T_2}}^g + \delta_{2^{S_1}}^f \delta_{2^{S_2}}^g \in \{-2, 0, 2\}$. Therefore, $\delta_k^h = 0$ if $\delta_{2^{T_1}}^f \delta_{2^{T_2}}^g + \delta_{2^{S_1}}^f \delta_{2^{S_2}}^g = 0$. Hence,

$$\sum_{i=0}^{k} \delta_i^f \delta_{k-i}^g = 0 \implies \frac{\delta_{2^{T_1}}^f}{\delta_{2^{S_1}}^f} = -\frac{\delta_{2^{S_2}}^g}{\delta_{2^{T_2}}^g}.$$

$\square$

Example 3. Consider $m = 12$ and $n = 20$. Then $m + n = 32 = 2^5$ with $e(m) = \{2, 3\}$ and $e(n) = \{2, 4\}$. Here, $e(m) \cup e(n) = \{2, 3, 4\}$ and $a_1 = 2$. Let $f \in \mathcal{B}_{12}$ and $g \in \mathcal{B}_{20}$ be WAPB Boolean functions.

The first condition in Theorem 4 results that $\frac{\delta_0^f}{\delta_{12}^f} = -\frac{\delta_0^g}{\delta_{20}^g}$.

The second and third conditions in Theorem 4 result, for a $c \in \{1, -1\}$, that

$$\frac{\delta_{2^\emptyset}^f}{\delta_{2^{\{2\}}}^f} = \frac{\delta_{2^{\{3\}}}^f}{\delta_{2^{\{2,3\}}}^f} = \frac{\delta_0^f}{\delta_4^f} = \frac{\delta_8^f}{\delta_{12}^f} = c, \quad \text{and} \quad \frac{\delta_{2^\emptyset}^g}{\delta_{2^{\{2\}}}^g} = \frac{\delta_{2^{\{4\}}}^g}{\delta_{2^{\{2,4\}}}^g} = \frac{\delta_0^g}{\delta_4^g} = \frac{\delta_{16}^g}{\delta_{20}^g} = -c.$$

From the fourth condition in Theorem 4, we have
$\frac{\delta_8^f}{\delta_4^f} = -\frac{\delta_4^g}{\delta_0^g}$ (when $e(k) = \{3\}$); $\frac{\delta_8^f}{\delta_4^f} = -\frac{\delta_{20}^g}{\delta_{16}^g}$ (when $e(k) = \{3, 4\}$); $\frac{\delta_0^f}{\delta_{12}^f} = -\frac{\delta_4^g}{\delta_{16}^g}$
(when $e(k) = \{4\}$). Now combining all equations, we have $\frac{\delta_0^f}{\delta_{12}^f} = -\frac{\delta_0^g}{\delta_{20}^g} = -\frac{\delta_4^g}{\delta_{16}^g}$

and $\frac{\delta_0^f}{\delta_4^f} = \frac{\delta_8^f}{\delta_{12}^f} = \frac{\delta_4^f}{\delta_8^f} = -\frac{\delta_0^g}{\delta_4^g} = -\frac{\delta_{16}^g}{\delta_{20}^g}$. If we consider f such that $\delta_0^f = x$ and

$\delta_{12}^f = y$ for some $x, y \in \{1, -1\}$ then $\frac{x}{\delta_4^f} = \frac{\delta_8^f}{y} = \frac{\delta_4^f}{\delta_8^f} = c \implies c = xy$. Hence $\delta_8^f = x$ and $\delta_4^f = y$. Further, if we consider $\delta_0^g = z \in \{1, -1\}$ then $\delta_{16}^g = z$ and $\delta_4^g = \delta_{20}^g = -xyz$. Hence, for WAPB f, g satisfying $\delta_0^f = \delta_8^f = x; \delta_4^f = \delta_{12}^f = y;$ $\delta_0^g = \delta_{16}^g = z$ and $\delta_4^g = \delta_{20}^g = -xyz$ for $x, y, z \in \{1, -1\}$ result a WPB Boolean function $h \in \mathcal{B}_{32}$. □

Now we have a consequence of the possibility of existence of WPB function due to the direct sum in two WAPB Boolean functions i.e., when $m = n = 2^{l-1}$.

Lemma 2. *Let $n = 2^{l-1}$ be a positive integer and $f, g \in \mathcal{B}_n$ be two WAPB Boolean functions. Then $h \in \mathcal{B}_{2n}$ defined as $h(x, y) = f(x) + g(y)$ for $x, y \in \mathbb{F}_2^n$ be a WPB Boolean function if $\frac{\delta_0^f}{\delta_n^f} = -\frac{\delta_0^g}{\delta_n^g}$.*

Proof. Since $f, g \in \mathcal{B}_n$ are WAPB and n is a power of 2, $\delta_0^f, \delta_n^f, \delta_0^g, \delta_n^g \in \{1, -1\}$ and $\delta_i^f = \delta_i^g = 0$ for $i \in [1, n-1]$. Hence, from Theorem 1, we have

$$
\mathtt{wt}_k(h) = \begin{cases}
\frac{1}{2}\binom{2n}{k} & \text{if } k \neq 0, n, 2n, \\
\frac{1}{2}\left[\binom{2n}{k} - \delta_0^f \delta_0^g\right] & \text{if } k = 0, \\
\frac{1}{2}\left[\binom{2n}{k} - (\delta_0^f \delta_n^g + \delta_n^f \delta_0^g)\right] & \text{if } k = n, \\
\frac{1}{2}\left[\binom{2n}{k} - \delta_n^f \delta_n^g\right] & \text{if } k = 2n.
\end{cases}
$$

Here, for $k = 0$ and $2n$, the value of $\delta_0^f \delta_0^g, \delta_n^f \delta_n^g$ are already in $\{1, -1\}$. For $k = n$, $\delta_0^f \delta_n^g + \delta_n^f \delta_0^g \in \{-2, 0, 2\}$. For h being a WAPB, $\delta_0^f \delta_n^g + \delta_n^f \delta_0^g = 0$ i.e., $\frac{\delta_0^f}{\delta_n^f} = -\frac{\delta_0^g}{\delta_n^g}$. Further if $\frac{\delta_0^f}{\delta_n^f} = -\frac{\delta_0^g}{\delta_n^g}$, we have $\frac{\delta_0^f \delta_0^g}{\delta_n^f \delta_n^g} = -\frac{(\delta_0^g)^2}{(\delta_n^g)^2} = -1 \implies \delta_0^f \delta_0^g = -\delta_n^f \delta_n^g \implies \delta_0^h = -\delta_{2n}^h$. Hence h is balanced and results h is a WPB Boolean function.

Alternative Proof: From Eq. 1, the convolutions of the sequences $\{\delta_i^f : i \in [0, m]\}$ and $\{\delta_j^g : j \in [0, n]\}$ produces the sequence $\{\delta_0^f \delta_0^g, 0, 0, \ldots, 0, \delta_n^f \delta_n^g\}$. Hence, the direct sum of f and g, i.e., $h \in \mathcal{B}_{2n}$ such that $h_{2n}(x, y) = f_n(x) + g_n(y)$, is a WAPB. Further, h is a WPB if $\delta_0^f \delta_0^g + \delta_n^f \delta_n^g = 0$ i.e., $\frac{\delta_0^f}{\delta_n^f} = -\frac{\delta_n^g}{\delta_0^g} = -\frac{\delta_0^g}{\delta_n^g}$. □

In this above case, h will be a WPB Boolean function if one of f and g is balanced (i.e., WPB) and other one is not balanced (i.e., a WAPB but not WPB). However, we can construct a WPB Boolean function h from two WPB Boolean functions f and g as in the following corollary which is presented in [3]
.

Corollary 1. *[3] Let $f, g \in \mathcal{B}_n$ be two WPB Boolean functions where $n = 2^l$. Then $h \in \mathcal{B}_{2n}$ defined as $h(x, y) = f(x) + g(y) + y_1 y_2 \cdots y_n$ for $x, y = (y_1, y_2, \ldots, y_n) \in \mathbb{F}_2^n$ be a WPB Boolean function.*

Proof. Here $f, g \in \mathcal{B}_n$ are two WPB Boolean functions. Then $\hat{g}(x_1, \ldots, x_n) = g(x_1, \ldots, x_n) + x_1 x_2 \cdots x_n$ is a unbalanced WAPB Boolean function. Then $\frac{\delta_0^f}{\delta_n^f} =$

-1 and $\frac{\delta_0^{\hat{g}}}{\delta_n^{\hat{g}}} = 1$. Hence, from Lemma 2, we have $h(x,y) = f(x) + \hat{g}(y)$ for $x, y \in \mathbb{F}_2^n$ is a WPB Boolean function with $\delta_0^h = -\delta_0^f \delta_0^{\hat{g}} = -\delta_0^f \delta_0^g$ and $\delta_{2n}^h = -\delta_n^f \delta_n^{\hat{g}} = \delta_n^f \delta_n^g$. □

We will study some cryptographic properties of a class WPB functions recursively generated using Lemma 2 or, Corollary 1.

Theorem 5. *For $n = 2^l, l \geq 1$, let $f_n \in \mathcal{B}_n$ be defined recursively as*

$$f_n(x_1, \ldots, x_n) = f_{\frac{n}{2}}(x_1, \ldots, x_{\frac{n}{2}}) + f_{\frac{n}{2}}(x_{\frac{n}{2}+1}, \ldots, x_n) + \prod_{i=\frac{n}{2}+1}^{n} x_i, \text{ for } l \geq 2 \text{ and}$$

$f_2(x_1, x_2) = x_2$. *Then*

1. f_n *is WPB.*
2. $f_n(x_1, \ldots, x_n) = \sum_{2^1|i} x_i + \sum_{2^2|i} x_{i-1}x_i + \sum_{2^3|i} x_{i-3}x_{i-2}x_{i-1}x_i + \cdots + x_{\frac{n}{2}+1} \cdots x_n.$
3. $\mathtt{nl}(f_n) = 2^{n-1} - \frac{1}{2}(3^{\frac{n}{2}} - 1).$
4. $\mathtt{AI}(f_n) \leq 1 + \frac{n}{4}.$

Proof. 1. Here, f_2 is WPB and using Corollary 1, it can be proved inductively that f_n for $n = 2^l, l \geq 2$ is WPB.

2. The ANF of f_n can inductively be proved with the base case $f_2 \in \mathcal{B}_2$ such that $f_2(x_1, x_2) = x_2$. Assume that

$$f_m(x_1, \ldots, x_m) = \sum_{2^1|i} x_i + \sum_{2^2|i} x_{i-1}x_i + \sum_{2^3|i} x_{i-3}x_{i-2}x_{i-1}x_i + \cdots + x_{\frac{m}{2}+1} \cdots x_m.$$

Then $f_{2m}(x_1, \ldots, x_{2m}) = f_m(x_1, \ldots, x_m) + f_m(x_{m+1}, \ldots, x_{2m}) + \prod_{i=m+1}^{2m} x_i.$

As m is a power of 2, the result follows.

3. For $n = 2^l, l \geq 1$, let $l_n(x_1, \ldots, x_n) = \sum_{2|i} x_i$ and $g_n(x_1, \ldots, x_n) = f_n + l_n = g_{\frac{n}{2}}(x_1, \ldots, x_{\frac{n}{2}}) + g_{\frac{n}{2}}(x_{\frac{n}{2}+1}, \ldots, x_n) + x_{\frac{n}{2}+1} \cdots x_n$. That is, l_n and g_n are linear and nonlinear terms of f_n respectively. Further, denote $g'_n(x_1, x_2, \ldots, x_n) = g_n(x_1, x_2, \ldots, x_n) + x_1 x_2 \cdots x_n$. That is, $g_n(x_1, \ldots, x_n) = g_{\frac{n}{2}}(x_1, \ldots, x_{\frac{n}{2}}) + g'_{\frac{n}{2}}(x_{\frac{n}{2}+1}, \ldots, x_n)$. Hence, $\mathtt{wt}(g'_n) = \mathtt{wt}(g_n + x_1 x_2 \cdots x_n) = \mathtt{wt}(g_n) - 1$
We will prove that l_n is a nearest linear function from f_n and for that reason we will find $\mathtt{wt}(g_n)$. We can check that $\mathtt{wt}(g_2) = 0$ and $\mathtt{wt}(g_4) = \mathtt{wt}(x_3 x_4) = 4$. From the ANF of $g_n, n \geq 4$, we can compute $g_n(1, 1, \ldots, 1) = (\sum_{2^2|i} 1 + \sum_{2^3|i} 1 + \cdots + 1) \mod 2 = 1$ as every summations but the last one contains even number of 1s. Therefore, $\mathtt{wt}(g_n)$ satisfies the recursion

$$\mathtt{wt}(g_n) = \mathtt{wt}(g_{\frac{n}{2}})\mathtt{wt}(\overline{g'_{\frac{n}{2}}}) + \mathtt{wt}(\overline{g_{\frac{n}{2}}})\mathtt{wt}(g'_{\frac{n}{2}}) \text{ for } n = 2^l, l > 2.$$

Denote the sequence $\mathtt{wt}(g_n) = w_l$ for $n = 2^l$. Then $\mathtt{wt}(\overline{g_n}) = 2^n - w_l, \mathtt{wt}(g_n') = w_l - 1, \mathtt{wt}(\overline{g_n'}) = 2^n - w_l + 1$. Hence, we have

$$w_{l+1} = w_l(2^{2^l} - w_l + 1) + (2^{2^l} - w_l)(w_l - 1) = -2w_l^2 + 2(2^{2^l} + 1)w_l - 2^{2^l}$$

$$\implies -2w_{l+1} = 4w_l^2 - 4(2^{2^l} + 1)w_l + 2^{2^l+1}$$

$$= (2w_l)^2 - 2(2^{2^l} + 1)2w_l + (2^{2^l} + 1)^2 - (2^{2^l} + 1)^2 + 2^{2^l+1}$$

$$= (2^{2^l} + 1 - 2w_l)^2 - (2^{2^{l+1}} + 1)$$

$$\implies 2^{2^{l+1}} + 1 - 2w_{l+1} = (2^{2^l} + 1 - 2w_l)^2$$

$$\implies z_{l+1} = z_l^2 \text{ where } z_l = 2^{2^l} + 1 - 2w_l.$$

Here $z_2 = 2^{2^2} + 1 - 2w_2 = 2^4 + 1 - 8 = 9$. Using the back recursion, $z_1 = 3$. Then solving the nonlinear recursion [1], $z_l = z_{l-1}^2, l \geq 2$ and $z_1 = 3$, we have $z_l = 3^{2^{l-1}}$. Hence, $\mathtt{wt}(g_n) = w_l = \frac{2^{2^l}+1-z_l}{2} = \frac{2^{2^l}+1-3^{2^{l-1}}}{2} = 2^{2^l-1} - \frac{1}{2}(3^{2^{l-1}} - 1) = 2^{n-1} - \frac{1}{2}(3^{\frac{n}{2}} - 1)$. As $g_n(x_1, \ldots, x_n) = f_n(x_1, \ldots, x_n) + \sum_{i|n} x_i$, $\mathtt{nl}(f_n) \leq \mathtt{wt}(g_n) = 2^{n-1} - \frac{1}{2}(3^{\frac{n}{2}} - 1)$ for $n \geq 4$.

We can check that $l_4(x_1, \ldots, x_4) = x_2 + x_4$ is a nearest linear function with distance $2^3 - \frac{1}{2}(3^2 - 1) = 4$. Let assume that $\mathtt{nl}(f_n) = 2^{n-1} - \frac{1}{2}(3^{\frac{n}{2}} - 1)$ with a nearest linear function l_n. Then $\mathtt{nl}(f_n + x_1 x_2 \cdots x_n) = 2^{n-1} - \frac{1}{2}(3^{\frac{n}{2}} - 1) - 1$ with a nearest linear function l_n. As f_{2n} is a direct sum of f_n and $f_n + x_1 x_2 \cdots x_n$, from Proposition 4, $\mathtt{nl}(f_{2n}) = 2^n[\mathtt{nl}(f_n) + \mathtt{nl}(f_n + x_1 x_2 \cdots x_n)] - 2\mathtt{nl}(f_n)\mathtt{nl}(f_n + x_1 x_2 \cdots x_n) = 2^n[2^{n-1} - \frac{1}{2}(3^{\frac{n}{2}} - 1) + 2^{n-1} - \frac{1}{2}(3^{\frac{n}{2}} - 1) - 1] - 2(2^{n-1} - \frac{1}{2}(3^{\frac{n}{2}} - 1))(2^{n-1} - \frac{1}{2}(3^{\frac{n}{2}} - 1) - 1) = 2^n(2^n - 3^{\frac{n}{2}}) - \frac{1}{2}(2^n - 3^{\frac{n}{2}} + 1)(2^n - 3^{\frac{n}{2}} + 1) = 2^{2n-1} - \frac{1}{2}(3^n - 1)$.

4. We can verify from the ANF of f_n that $(1 + \sum_{2|i} x_i)(1 + x_4)(1 + x_8) \cdots (1 + x_n) = (1 + \sum_{2|i} x_i) \prod_{4|i}(1 + x_i)$ is an annihilator of f_n. Hence, $\mathtt{AI} \leq 1 + \frac{n}{4}$. □

The following lemma presents to get an n variable WPB Boolean function from the concatenation of two $n - 1$ variable WAPB Boolean functions.

Lemma 3. *Let* $n = 2^l$ *and* $f, g \in \mathcal{B}_{n-1}$ *are WAPB Boolean functions with* $\delta_i^f = -\delta_{i-1}^f, \delta_i^f = \delta_i^g$ *for* $i \in [1, n]$. *Then* $h \in \mathcal{B}_n$ *defined as* $h(x_1, x_2, \ldots, x_n) = x_n f(x_1, x_2, \ldots, x_{n-1}) + (1 + x_n)g(x_1, x_2, \ldots, x_{n-1})$ *is WPB.*

Proof. Here, $\mathtt{wt}_k(h) = \mathtt{wt}_k(f) + \mathtt{wt}_{k-1}(g)$ for $k \in [1, n-1]$, $\mathtt{wt}_0(h) = \mathtt{wt}_0(f)$ and $\mathtt{wt}_n(h) = \mathtt{wt}_{n-1}(g)$.

Case $k = 0, n$: $\mathtt{wt}_0(h) = \mathtt{wt}_0(f)$ and $\mathtt{wt}_n(h) = \mathtt{wt}_{n-1}(g)$. Hence, $\delta_0^h = \delta_0^f \in \{-1, 1\}$ and $\delta_n^h = \delta_{n-1}^g \in \{-1, 1\}$.

Case $k \in [1, n-1]$: $\mathtt{wt}_k(h) = \mathtt{wt}_k(f) + \mathtt{wt}_{k-1}(g) = \frac{1}{2}\left[\binom{n-1}{k} + \delta_k^f\right] + \frac{1}{2}\left[\binom{n-1}{k-1} + \delta_{k-1}^g\right] = \frac{1}{2}\left[\binom{n}{k} + \delta_k^f + \delta_{k-1}^g\right] = \frac{1}{2}\left[\binom{n}{k} + \delta_k^f + \delta_{k-1}^f\right] = \frac{1}{2}\binom{n}{k}$. Hence, $\delta_k^h = 0$ for $k \in [1, n]$ and that implies h is a WPB Boolean function on n-variables. □

Now we are defining a class of WAPB Boolean functions which will help us to generate WPB Boolean functions.

Definition 7. *An WAPB Boolean function $f \in \mathcal{B}_n$ satisfying $\delta_i^f = -\delta_{i-1}^f$ for $i \in [1, n]$ (i.e., $\delta_i^f = (-1)^i \delta_0^f$, for $i \in [0, n]$) is defined as an alternating WAPB (AWAPB) Boolean function.*

From Lucas Theorem (Proposition 1), it can be checked that an n-variable AWAPB Boolean function exists if and only if n is a positive integer of the form $2^l - 1$. We present a construction of a AWAPB Boolean function on $2^l - 1$ variable from two WAPB functions in the following lemma.

Lemma 4. *Let $n = 2^l$ be a positive integer. Let $f \in \mathcal{B}_{n-1}$ AWAPB Boolean function and $g \in \mathcal{B}_n$ be an unbalanced WAPB Boolean function. Then the direct sum $h \in \mathcal{B}_{2n-1}$ defined as $h(x, y) = f(x) + g(y)$ for $x \in \mathbb{F}_2^{n-1}, y \in \mathbb{F}_2^n$ is a AWAPB Boolean function if g is not balanced i.e., $\delta_0^g = \delta_n^g$.*

Proof. Since $f \in \mathcal{B}_{n-1}$ is AWAPB and $g \in \mathcal{B}_n$ is unbalanced WAPB, $\delta_i^f = (-1)^i \delta_0^f$ for $i \in [1, n-1]$, $\delta_i^g = 0$ for $i \in [1, n-1]$ and $\delta_0^g = \delta_n^g$. As $e(n) \cap e(n-1) = \emptyset$, for $k \in [0, 2n-1]$, we have (from Theorem 2)

$$\delta_k^h = \begin{cases} 0 & \text{if } e(k) \not\subseteq e(2n-1) \\ -\delta_s^f \delta_{k-s}^g \text{ where } e(s) = e(k) \cap e(n-1) & \text{if } e(k) \subseteq e(2n-1) \end{cases}$$

$$= \begin{cases} 0 & \text{if } e(k) \not\subseteq e(2n-1) \\ -\delta_k^f \delta_0^g = (-1)^{k+1} \delta_0^f \delta_0^g & \text{if } k < n \text{ and } e(k) \subseteq e(2n-1) \\ -\delta_{k-n}^f \delta_n^g = (-1)^{k-n+1} \delta_0^f \delta_n^g & \text{if } k \geq n \text{ and } e(k) \subseteq e(2n-1). \end{cases}$$

As $e(k) \subseteq e(2n-1) = e(2^{l+1} - 1)$ for every $k \in [0, 2n-1]$, the case $\delta_k^h = 0$ will never occur. Further, as n is an even integer for $n > 1$, $(-1)^{k-n+1} = (-1)^{k+1}$ for $k \in [n, 2n-1]$. Hence,

$$\delta_k^h = \begin{cases} 0 & \text{if } e(k) \not\subseteq e(2n-1), \text{which will never occur,} \\ (-1)^{k+1} \delta_0^f \delta_0^g & \text{if } k < n \text{ and } e(k) \subseteq e(2n-1), \\ (-1)^{k+1} \delta_0^f \delta_n^g & \text{if } k \geq n \text{ and } e(k) \subseteq e(2n-1). \end{cases}$$

Hence, if $\delta_n^g = \delta_0^g$ (i.e., g is not balanced), $\delta_k^h = (-1)^{k+1} \delta_0^f \delta_0^g$ which implies that h is a AWAPB with $\delta_0^h = -\delta_0^f \delta_0^g$. $\qquad\square$

If g is WPB, then it can be made as a unbalanced WPB by adding the term $\prod_{i=1}^n y_i$ with g and hence we have the following corollary.

Corollary 2. *Let $n = 2^l$ be a positive integer. Let $f \in \mathcal{B}_{n-1}$ be an AWAPB Boolean function and $g \in \mathcal{B}_n$ be a WPB Boolean function. Then $h \in \mathcal{B}_{2n-1}$ defined as $h(x, y) = f(x) + g(y) + \prod_{i=1}^n y_i$ for $x \in \mathbb{F}_2^{n-1}, y \in \mathbb{F}_2^n$ is a AWAPB Boolean function.*

Using Lemma 3 and Corollary 2, we can recursively generate AWAPB and WPB Boolean functions as illustrated in the following example.

Example 4. Consider $f_1 \in \mathcal{B}_1$ such that $f_1(x_1) = x_1$. Here f_1 is an AWAPB Boolean function with $\delta_0^f = -1$ and $\delta_1^f = 1$. Considering $f = g = f_1$ in Lemma 3, we have $f_2(x_1, x_2) = x_2 x_1 + (1 + x_2)x_1 = x_1$ is a WPB in $\mathcal{B}_2$. Now considering, $f = f_1, g = f_2$ in Corollary 2, we have $f_3(x_1, x_2, x_3) = x_1 + x_2 + x_2 x_3$ is AWAPB in $\mathcal{B}_3$.

Further, instead of using the same function f_3 for f and g in Lemma 3, we consider $f(x) = f_3(x)$ and $g(x) = f_3(Ax)$ where A is a permutation matrix which permutes the coordinates of the input vector x. As a result g is too an AWAPB Boolean function with $\delta_i^f = \delta_i^g$ for $i \in [0, n]$. Here, consider $g(x_1, x_2, x_3) = x_1 + x_3 + x_2 x_3$. Then $f_4(x_1, x_2, x_3, x_4) = x_4 f(x_1, x_2, x_3) + (1 + x_4)g(x_1, x_2, x_3) = x_1 + x_2 + x_2 x_3 + x_2 x_4 + x_3 x_4$ is WPB.

Then considering, $f = f_3, g = f_4$ in Corollary 2, we have $f_7(x_1, \ldots, x_7) = f_3(x_1, x_2, x_3) + f_4(x_4, x_5, x_6, x_7) + x_4 x_5 x_6 x_7$ is AWAPB in $\mathcal{B}_7$. Similarly, using Lemma 3 and Corollary 2 alternatively, we can generate WPB and AWAPB Boolean functions on higher number of variables. □

3.1   Cryptographic Properties of the Direct Sum

Now we will study and collect some existing results on cryptographic properties of direct sum of Boolean functions. We consider $f \in \mathcal{B}_m, g \in \mathcal{B}_n$ and $h \in \mathcal{B}_{m+n}$ defined as $h(x, y) = f(x) + g(y)$ for $x \in \mathbb{F}_2^m$ and $y \in \mathbb{F}_2^n$ in the following results. We also consider $a = (a', a'') \in \mathbb{F}_2^{m+n}$ where $a' \in \mathbb{F}_2^m$ and $a'' \in \mathbb{F}_2^n$.

Proposition 4. *[19]*

1. *The Walsh transform of h given as $W_h(a) = W_f(a')W_g(a'')$.*
2. *The nonlinearity of h is given as $\mathrm{nl}(h) = 2^n \mathrm{nl}(f) + 2^m \mathrm{nl}(g) - 2\mathrm{nl}(f)\mathrm{nl}(g)$.*

Proposition 5. *[3] For $k \in [0, m + n]$,*

1. *the Walsh transform of h over $E_{m+n,k}$ is given as*

$$W_{h,k}(a) = \sum_{i=0}^{k} W_{f,i}(a')W_{g,k-i}(a'').$$

2. *a bound on nonlinearity over $E_{m+n,k}$ is given as*

$$\mathrm{nl}_k(f) \geq \sum_{i=0}^{k} \left(\binom{m}{i} \mathrm{nl}_{k-i}(g) + \binom{n}{k-i} \mathrm{nl}_i(f) - 2\mathrm{nl}_i(f)\mathrm{nl}_{k-i}(g) \right).$$

Proposition 6.

1. *[2] $\max(\mathrm{AI}(f), \mathrm{AI}(g)) \leq \mathrm{AI}(h) \leq \min\{\max\{\deg(f), \deg(g)\}, \mathrm{AI}(f) + \mathrm{AI}(g)\}$.*
2. *[3] For $1 \leq k \leq \min\{m, n\}, \mathrm{AI}_k(h) \geq \min_{0 \leq j \leq k} \{\max\{\mathrm{AI}_j(f), \mathrm{AI}_{k-j}(g)\}\}$.*

The Item 2 of Proposition 6 can be generalized as follows.

Theorem 6. *For $0 \leq k \leq m+n$ and $m \leq n$,*

$$\min_{\max\{0,k-n\} \leq j \leq \min\{m,k\}} \{\max\{\mathtt{AI}_j(f), \mathtt{AI}_{k-j}(g)\}\} \leq \mathtt{AI}_k(h) \leq \deg(h).$$

Proof. Let A_k be an annihilator of h over $E_{m+n,k}$. Then $A_k(x,y)h(x,y) = 0$ for all $(x,y) \in E_{m+n,k}$ and there exists an $(x_0,y_0) \in E_{m+n,k}$ such that $A_k(x_0,y_0) = 1$. Let $\mathtt{wt}(x_0) = j$ (i.e., $x_0 \in E_{m,j}$) for some $\max\{0, k-n\} \leq j \leq \min\{m,k\}$. As $A_k(x_0,y_0) = 1$, $h(x_0,y_0) = f(x_0) + g(y_0) = 0$ i.e., either $f(x_0) = g(y_0) = 0$ or, $f(x_0) = g(y_0) = 1$. Now fixing x_0, we have $A_k(x_0,y)(f(x_0) + g(y)) = 0$ for all $y \in E_{n,k-j}$. That implies, $A_k(x_0,y) = A_{k,x_0}(y)$ is an annihilator of g or, $1+g$ over $E_{n,k-j}$ (as $A_k(x_0,y_0) = 1$). Similarly, fixing y_0, we have $A_k(x,y_0)(f(x)+g(y_0)) = 0$ for all $x \in E_{n,j}$. That implies, $A_{k,y_0}(x)$ is an annihilator of f or, $1 + f$ over $E_{n,j}$. As a summary, if A_k is an annihilator of h over $E_{m+n,k}$, then there is a $j \in [\max\{0, k-m\}, \min\{m,k\}]$ such that A_{k,y_0} is an annihilator of f or, $1 + f$ over $E_{n,j}$ and A_{k,x_0} is an annihilator of g or, $1 + g$ over $E_{n,k-j}$. Hence, $\deg A_k \geq \mathtt{AI}_k(h) \geq \min_{\max\{0,k-n\} \leq j \leq \min\{m,k\}} \{\max\{\mathtt{AI}_j(f), \mathtt{AI}_{k-j}(g)\}\}$.

$1 + h$ is an annihilator of h but not necessarily $1 + h$ is an annihilator of h in the domain $E_{n,k}$ if $1 + h$ is 0 in the domain $E_{n,k}$. In our case, h is a WPB and hence is a balance function in each domain $E_{n,k}$ for $k \in [1, n-1]$. That implies, $1 + h$ is not 0 in each domain $E_{n,k}$ for $k \in [1, n-1]$ and hence $1 + h$ is an annihilator of h in the domain $E_{n,k}$ and $\mathtt{AI}_k(h) \leq \deg(h)$. $\square$

4   Conclusion

We have studied the direct sum of two WAPB/WPB Boolean functions and conditioned when it obtains a new WAPB/WPB Boolean function. Some of the results presented in [3, 25] are consequences of our results. We have presented some constructions of WAPB/WPB Boolean functions in this direction. There is still an open problem to study the direct sum $h(x,y) = f(x)+g(y)$ in Theorem 2 when $e(m) \cap e(n) \neq \emptyset$.

References

1. Aho, A.V., Sloane, N.J.A.: Some doubly exponential sequences. Fibonacci Q. **11**, 429–437 (1973)
2. Braeken, A., Preneel, B.: On the algebraic immunity of symmetric Boolean functions. In: Maitra, S., Veni Madhavan, C.E., Venkatesan, R. (eds.) INDOCRYPT 2005. LNCS, vol. 3797, pp. 35–48. Springer, Heidelberg (2005). https://doi.org/10.1007/11596219_4
3. Carlet, C., Méaux, P., Rotella, Y.: Boolean functions with restricted input and their robustness; application to the FLIP cipher. IACR Trans. Symmetric Cryptol. **2017**(3), 192–227 (2017)
4. Dalai, D.K., Mallick, K.: A class of weightwise almost perfectly balanced Boolean functions. Adv. Math. Commun. **18**(2), 480–504 (2024)

5. Dalai, D.K., Mallick, K., Méaux, P.: A class of weightwise almost perfectly balanced Boolean functions. In: 9th International Workshop on Boolean Functions and their Applications-BFA (2024). https://boolean.w.uib.no/bfa-2024-accepted-abstracts/
6. Gini, A., Méaux, P.: On the weightwise nonlinearity of weightwise perfectly balanced functions. Discret. Appl. Math. **322**, 320–341 (2022)
7. Gini, A., Méaux, P.: Weightwise almost perfectly balanced functions: secondary constructions for all n and better weightwise nonlinearities. In: Isobe, T., Sarkar, S. (eds.) Progress in Cryptology - INDOCRYPT 2022. LNCS, vol. 13774, pp. 492–514. Springer, Cham (2022). https://doi.org/10.1007/978-3-031-22912-1_22
8. Gini, A., Méaux, P.: On the algebraic immunity of weightwise perfectly balanced functions. IACR Cryptol. ePrint Arch., p. 495 (2023). https://eprint.iacr.org/2023/495
9. Gini, A., Méaux, P.: Weightwise perfectly balanced functions and nonlinearity. In: El Hajji, S., Mesnager, S., Souidi, E.M. (eds.) Codes, Cryptology and Information Security - C2SI 2023. LNCS, vol. 13874, pp. 338–359. Springer, Cham (2023). https://doi.org/10.1007/978-3-031-33017-9_21
10. Guo, X., Su, S.: Construction of weightwise almost perfectly balanced Boolean functions on an arbitrary number of variables. Discret. Appl. Math. **307**, 102–114 (2022)
11. Li, J., Su, S.: Construction of weightwise perfectly balanced Boolean functions with high weightwise nonlinearity. Discret. Appl. Math. **279**, 218–227 (2020)
12. Liu, J., Mesnager, S.: Weightwise perfectly balanced functions with high weightwise nonlinearity profile. Des. Codes Crypt. **87**(8), 1797–1813 (2019)
13. Mariot, L., Picek, S., Jakobovic, D., Djurasevic, M., Leporati, A.: Evolutionary construction of perfectly balanced Boolean functions. In: IEEE Congress on Evolutionary Computation, CEC 2022, pp. 1–8. IEEE (2022)
14. Matsui, M.: Linear cryptanalysis method for DES cipher. In: Helleseth, T. (eds.) Advances in Cryptology - EUROCRYPT 1993. LNCS, vol. 765, pp. 386–397. Springer, Cham (1993). https://doi.org/10.1007/3-540-48285-7_33
15. Méaux, P., Journault, A., Standaert, F.-X., Carlet, C.: Towards stream ciphers for efficient FHE with low-noise ciphertexts. In: Fischlin, M., Coron, J.-S. (eds.) EUROCRYPT 2016. LNCS, vol. 9665, pp. 311–343. Springer, Heidelberg (2016). https://doi.org/10.1007/978-3-662-49890-3_13
16. Mesnager, S., Su, S.: On constructions of weightwise perfectly balanced Boolean functions. Cryptogr. Commun. **13**(6), 951–979 (2021). https://doi.org/10.1007/s12095-021-00481-3
17. Mesnager, S., Su, S., Li, J.: On concrete constructions of weightwise perfectly balanced functions with optimal algebraic immunity and high weightwise nonlinearity. In: The 6th International Workshop on Boolean Functions and Applications (2021). https://boolean.w.uib.no/files/2021/08/BFA_2021_abstract_9.pdf
18. Mesnager, S., Zhou, Z., Ding, C.: On the nonlinearity of Boolean functions with restricted input. Cryptogr. Commun. **11**(1), 63–76 (2019)
19. Seberry, J., Zhang, X.-M., Zheng, Y.: Relationships among nonlinearity criteria. In: De Santis, A. (ed.) EUROCRYPT 1994. LNCS, vol. 950, pp. 376–388. Springer, Heidelberg (1995). https://doi.org/10.1007/BFb0053452
20. Su, S.: The lower bound of the weightwise nonlinearity profile of a class of weightwise perfectly balanced functions. Discret. Appl. Math. **297**, 60–70 (2021)
21. Tang, D., Liu, J.: A family of weightwise (almost) perfectly balanced Boolean functions with optimal algebraic immunity. Cryptogr. Commun. **11**(6), 1185–1197 (2019)

22. Zhang, R., Su, S.: A new construction of weightwise perfectly balanced Boolean functions. Adv. Math. Commun. **17**(4), 757–770 (2023)
23. Zhao, Q., Jia, Y., Zheng, D., Qin, B.: A new construction of weightwise perfectly balanced functions with high weightwise nonlinearity. Mathematics **11**(5), 1193 (2023)
24. Zhao, Q., Li, M., Chen, Z., Qin, B., Zheng, D.: A unified construction of weightwise perfectly balanced Boolean functions. Discret. Appl. Math. **337**, 190–201 (2023)
25. Zhu, L., Su, S.: A systematic method of constructing weightwise almost perfectly balanced Boolean functions on an arbitrary number of variables. Discret. Appl. Math. **314**, 181–190 (2022)

Cryptographic Constructions

Multi-key Fully-Homomorphic Aggregate MAC for Arithmetic Circuits

Suvasree Biswas[✉] and Arkady Yerukhimovich

George Washington University, Washington, USA
{suvasree,arkady}@gwu.edu

Abstract. Homomorphic message authenticators allow a user to perform computation on previously authenticated data producing a tag σ that can be used to verify the authenticity of the computation. We extend this notion to consider a multi-party setting where we wish to produce a tag that allows verifying (possibly different) computations on all party's data at once. Moreover, the size of this tag should not grow as a function of the number of parties or the complexity of the computations. We construct the first aggregate homomorphic MAC scheme that achieves such aggregation of homomorphic tags. Moreover, the final aggregate tag consists of only a single group element. Our construction supports aggregation of computations that can be expressed by bounded-depth arithmetic circuits and is secure in the random oracle model based on the hardness of the Computational Co-Diffie-Hellman problem over an asymmetric bilinear map.

Keywords: homomorphic authenticators · aggregate MAC · verifiable computation

1 Introduction

Consider a scenario where a scientist wishes to collect aggregate measurements about some scientific phenomena, for example average or standard deviation of ocean temperature around the world. She can deploy sensors to collect data in different regions and, to minimize communication, have the sensors upload their readings to the cloud. Moreover, since the scientist is only interested in aggregate statistics, she does not need to view all of the uploaded measurements, just the resulting aggregate statistics. So, instead of having the cloud store all the collected data, she can have it compute the aggregates and only review the results.

However, this requires that the scientist rely on the cloud to compute the aggregate statistics. A malicious cloud can change the original readings or add error into the computations to produce faulty results. Thus, the scientist needs to check that the aggregate statistics are correct for *all* the sensors even without seeing the original sensor readings. Moreover, to reduce communication and storage, we want the "proof" of correctness to be small—of size not growing

S. Mukhopadhyay and P. Stănică (Eds.): INDOCRYPT 2024, LNCS 15495, pp. 213–233, 2025.
https://doi.org/10.1007/978-3-031-80308-6_10

with either the complexity of the computed statistics or the number of sensors. Moreover, even if the cloud is able to learn the secret keys of some of the sensors, we want the authenticity of the computation on the remaining sensors' data to be preserved.

To address this challenge we turn to the use of homomorphic authenticators [2,9,12,18,21,24,26]. Homomorphic authenticators allow a user to use their secret key to authenticate an input x. Anybody can then use an associated (public) evaluation key to evaluate a program $\mathcal{P}$ on the authenticated input producing the result along with a tag authenticating the output. Finally, a party with knowledge of the user's verification key—a public key in the case of signatures, and a secret key in the case of message authentication codes—can verify that the final output was computed correctly on the originally authenticated input. In our previously described application this would amount to the scientist provisioning secret keys to each sensor, the sensor authenticating their measurement and sending it to the cloud, who evaluates the program $\mathcal{P}$ allowing the scientist to verify the output.

Originally, homomorphic authenticators focused on the case of a single user authenticating their input, but more recent work [2,18] has considered the case of homomorphic authenticators over multiple users. These allowed authenticating computation that took input from multiple users. The first work to consider this setting by Fiore et al. [18] restricted the computation to low-depth circuits. A very recent breakthrough work by Anthoine et al. [2] overcame this limitation, but the resulting authenticators grow either with the number of users or the depth of the computation.

Our goal is different from these works. First, while we aim to support multiple users with different authentication keys so that compromising one user's key does not compromise the other users, we only consider computations that work over the inputs of a single user—i.e., we only wish to compute over readings for each sensor, not across sensors. However, to minimize storage and communication we insist that the size of the final authenticator be independent of the number of users whose computation is authenticated and the depth and size of the computed circuits. Specifically, it will consist of only a single group element. Thus our goal is both weaker and stronger than the prior work supporting a more restrictive class of computations, but enforcing a much stricter length bound on the authenticator.

In order to compress the authenticators to achieve this length bound, we turn to another well-studied tool in the cryptographic literature—aggregate signatures [4,10,25,27]. Aggregate signatures allow the compression of multiple signatures into a single short signature that does not grow with the number of input signatures. However, they do not provide any homomorphic functionality and the original inputs are necessary to verify the signature.

We build on these two tools to develop a primitive we call a HA-MAC which combines the benefits of both of these primitives allowing verification of computation over the sensors' inputs while only requiring an authenticator consisting of only one group element.

1.1   Our Construction

Our Homomorphic Aggregate MAC (HA-MAC) combines the properties of homomorphic authenticators and aggregate authenticator to achieve functionality and security that are a good fit for the outsourced measurement use-case described previously. We briefly describe these in what follows before sketching the intuition behind our construction.

Functionality. A HA-MAC is a symmetric-key primitive where each of the P users is given a secret key sk_i. Using this secret key each user can authenticate inputs of his choice. An evaluator who does not know the secret key, but has a corresponding evaluation key ek_i can evaluate (possibly different) programs over each user's inputs and aggregate the results of any subset of the users into a short aggregate MAC. Finally, a verifying party who knows all of the secret keys of the included users can use this MAC to verify that the evaluator correctly computed all of the aggregated values. Importantly, the verifier can do this even without knowing the original values authenticated by the users.

Security. Our HA-MAC also provides strong security guarantees. The MAC guarantees the integrity of an honest user's input and computation even if all of the remaining users collude with the evaluator meaning that users have nothing to fear from participating. Moreover, security is achieved even against an adversary who is additionally allowed to ask for an unbounded number of authentications, as well as MAC verifications and aggregate MAC verifications with tags including the honest user. We prove our HA-MAC secure in the random oracle model under the standard co-CDH assumption over pairing-friendly elliptic curves.

Specifically, we prove the following theorem:

Theorem 1 (Informal). *For any pairing friendly elliptic curve E with a bilinear map $e : \mathcal{G}_1 \times \mathcal{G}_2 \to \mathcal{G}_T$ and efficiently computable isomorphism $\psi : \mathcal{G}_2 \to \mathcal{G}_1$ such that the co-Computational Diffie-Hellman (co-CDH) problem is hard, the HA-MAC construction Θ given in Fig. 2 is a secure HA-MAC in the random oracle model.*

1.2   Overview of Our Techniques

We begin with the construction of a single-user homomorphic message authentication code due to Catalano and Fiore [12]. In their construction[1], all authentication tags correspond to polynomials in $\mathbb{Z}_p[x]$ represented as a vector of their coefficients. To authenticate an input $m \in \mathbb{Z}_p$ with a label[2] $\tau \in \{0,1\}^\lambda$ the initial authentication tag is a degree-1 polynomial $y \in \mathbb{Z}_p[x]$ such that $y(0) = m$ and $y(x) = \mathrm{PRF}_K(\tau)$ for a hidden point $x \in \mathbb{Z}_p$ and a PRF key K. Looking forward, x and K will be part of a user's secret key.

---

[1] In what follows we abuse notation to use symbol x both for indeterminate and a scalar point.

[2] Labels are used to identify the input wires in an arithmetic circuit describing a homomorphic computation.

For any depth-D arithmetic circuit f with n input labels, the homomorphic tag authenticating the output of f is a degree-D polynomial $y_f \in \mathbb{Z}_p[x]$ constructed as follows from the input polynomials y_i. Going through the circuit f gate-by-gate, for every addition gate, we add the input polynomials (by adding the coefficients). For every multiplication gate, we multiply the input polynomials (by computing a convolution of their coefficients). It is easy to see that these operations are naturally homomorphic with respect to the evaluation of the polynomials at any evaluation point. Specifically, we observe that $y_f(0) = f(y_1(0), \ldots, y_n(0)) = f(m_1, \ldots, m_n)$ where $y_1, \ldots, y_n$ are the original input tags on inputs $m_1, \ldots, m_n$ and $y_f(x) = f(y_1(x), \ldots, y_n(x)) = f(\mathrm{PRF}_K(\tau_1), \ldots, \mathrm{PRF}_K(\tau_n))$ where x is the hidden point used in constructing the initial authentication tags. Observe that if the hidden evaluation point x is random and unknown to an adversary, then to check the validity of a claimed Mac y', it is sufficient to check whether these two equalities hold. Specifically, on an input polynomial y' if $y'(0) = y_f(0)$ and $y'(x) = y_f(x)$ then by the Schwartz-Zippel lemma, with overwhelming probability, $y' = y_f$ is a valid tag. Note, however, that when computed in this way the degree of the output polynomial y_f is D and thus $D + 1$ coefficients are necessary to represent this polynomial resulting in a tag that grows linearly with the depth of the circuit f.

To avoid this increase in the size of the tags, Catalano and Fiore instead encode and evaluate these polynomials in the exponent in a multiplicative group $\mathcal{G}$. More concretely, for some program $\mathcal{P} = (f, (\tau_i)_{\forall i \in [n]})$, let $(y_i)_{\forall i \in D}$ be the coefficients of the polynomial y_f computed as before. Succinct tag Λ is computed as $\Lambda = \prod_{i=1}^{D}(u^{x^i})^{y_i}$ where the values u^{x^i} for $i \in [D]$ are part of a user's public evaluation key and u is a generator of the group $\mathcal{G}$. Intuitively, Λ is computed by taking an inner product between the coefficient vector of y with powers of x excluding the degree-0 coefficient. This corresponds to evaluating $y(x) - y(0)$ in the exponent, so $\Lambda = u^{y(x)-y(0)}$. Now, on input a tuple $(m, \mathcal{P}, \Lambda)$, we can perform the Schwartz-Zippel test in the exponent by checking that $u^m \cdot \Lambda = u^\rho$ where $\rho = y(x)$, the evaluation of the polynomial at the hidden point x. For purposes of the security proof and in order to allow further functionality instead of performing the above check directly—as done by Catalano and Fiore—we instead perform this check using a bilinear pairing to map the values into a target group $\mathcal{G}_T$. Specifically, for groups $(\mathcal{G}_1, \mathcal{G}_2)$ with a bilinear map $e : \mathcal{G}_1 \times \mathcal{G}_2 \rightarrow \mathcal{G}_T$, Λ is computed in $\mathcal{G}_1$. We then choose a random element $w = \mathcal{H}_2(m) \in \mathcal{G}_2$ (for a hash function $\mathcal{H}_2$ modeled as a random oracle) and use the bilinear map to compute $e(\Lambda, w)$, $e(u^m, w)$, and $e(u^\rho, w)$ and use these values to verify that $e(u^m, w) \cdot e(\Lambda, w) = e(u^\rho, w)$.

Armed with this single-user homomorphic Mac construction, we now recall our goal of an aggregate homomorphic Mac. Specifically, we wish to aggregate the Macs of a subset U of users to authenticate computations performed by each of the users. An immediate way to realize this aggregation is via the trivial approach of concatenating the homomorphic Macs of all the users in U. Specifically, each user $l \in U$ has a secret key sk_l and evaluation key ek_l, where ek_l contains l's generator u_l for the previously described Mac. Then, an aggregate Mac for U

can just consist of $(\Lambda_1, \ldots, \Lambda_{|U|})$ and can be verified by checking that for all $l \in U$ with $w_l = \mathcal{H}_2(m_l)$,

$$e(\Lambda_l, w_l) = e(u_l, w_l^{\rho}) \cdot e(u_l, w_l^{m_l})^{-1} \qquad (1)$$

However, this trivial aggregation fails to meet our goal of an aggregate Mac that does not grow with the number of parties in the set U. Thus, to achieve this goal, we modify the aggregation procedure to combine all of the users' Macs into a single group element. Intuitively, this is done by taking a random linear combination of the users' Macs in the exponent. To ensure that the verifier can recompute the linear combination for verification, but at the same time an adversary cannot predict the random weights that will be used in the linear combination we again turn to the hash function $\mathcal{H}_2$. Specifically, for any subset of users $U \subseteq P$, for each user in U the aggregate algorithm computes $e(\Lambda_l, w_l)$ as before and then it multiplies all of these into one group element in $\mathcal{G}_T$. Using $e(g_1, g_2)$ as the generator in $\mathcal{G}_T$, correct aggregation under evaluation keys $(\mathsf{ek}_l)_{\forall l \in U} \leftarrow (u_l^{x_{l_0}^k})_{\forall k \in D}$ evaluates to the following in the exponent:

$$\Sigma_{\forall l \in U} \, \gamma_l b_l \left(m_l + \sum_{k=1}^{D} y_{l,k} x_{l_0}^k \right) = \Sigma_{\forall l \in U} \, \gamma_l b_l f_l(r_{\tau_{l,1}}, \ldots, r_{\tau_{l,n}})$$

where γ_l and b_l are exponents such that $g_2^{b_l} \leftarrow w_l$, $g_1^{\gamma_l} \leftarrow u_l$ for some $\gamma_l \leftarrow_{\$} \mathbb{Z}_p$. Note that b_l and γ_l are both secret to the adversary since b_l is the discrete log of the hash output w_l and γ_l is the discrete log of the random generator u_l.

Verification of aggregate MAC σ^* involves checking whether σ^* is equal to $e(g_1, g_2)^{\Sigma_{\forall l \in U} \, \gamma_l b_l (f_l(r_{\tau_{l,1}}, \ldots, r_{\tau_{l,n}}) - m_l)}$. Since the adversary does not know γ_l or b_l, he cannot produce a forged Mac σ^* that passes this verification.

2   Related Work

Homomorphic Authenticators: Since their introduction by Ronald Rivest [34], homomorphic authenticators have been widely investigated (e.g. [2,9,12,15,18, 21,24,26]) both in the public-key (i.e., homomorphic signatures) and private-key (i.e., homomorphic Mac) variants. Of these, we follow the line of work on fully-homomorphic authenticators initiated by Gennaro and Wichs [21] and follow-on work [2,12,18].

In particular, our starting point is the work of Catalano and Fiore [12] who showed how to build practical fully-homomorphic Macs from the ℓ-Diffie-Hellman Inversion assumption. However, this work is restricted to the single-key setting whereas we are aiming to support authentications under multiple keys. Follow-on work by Fiore et al. [18] did achieve support for multi-key homomorphic Macs, however they were limited to only low-degree computations. Very recently, Anthoine et al. [2] showed how to lift this limitation achieving homomorphic Macs for arbitrary computation across inputs authenticated by multiple clients. However, their resulting Mac, while sublinear in the number of clients and the size of the computation, still required that the number of group elements grows with the number of parties. We, on the other hand, only aim for a weaker primitive allowing homomorphic computation only on inputs from each party separately followed by an aggregation of the results. But, this allows us to achieve a final Mac consisting of only a single group element.

Aggregate Signatures and MACs: Another line of work closely related to ours is aggregate signatures and Macs (e.g. [4, 10, 25, 27, 35, 36]). These primitives focus on aggregating multiple signatures or Macs into a single (very) short authenticator. However, they do not consider computation on these authenticators. Our construction directly builds on the work of Boneh et al. [10] and thus inherits their need of a random oracle. While later work [35] showed how to avoid this random oracle assumption, this was at the cost of very strong computational assumptions. We see constructing a HA-MAC scheme secure in the standard model as an interesting open question.

Verifiable Computation: Homomorphic aggregate MACs are also closely related to the more general notion of verifiable computation [3, 6, 13, 17, 19, 23, 29, 31, 33]. Generally, this line of work differs from homomorphic Macs in that verifiable computation often requires interaction whereas homomorphic Macs are non-interactive. Moreover, homomorphic Macs allow any party with knowledge of the public evaluation keys to verifiably compute functions on authenticated inputs with no additional communication. On the other hand, verifiable computation is heavily focused on reducing the amount of computation necessary to verify while we focus primarily on reducing communication. We refer readers to [21] for a more in-depth discussion.

Succinct Non-interactive Arguments of Knowledge: Another way to achieve a homomorphic aggregate Mac is via the use of SNARKs [8, 20, 30, 32]. While SNARKs do provide an alternative solution to ours, they are generally more expensive—especially for the prover, and necessarily rely on non-standard assumptions [22].

3   Preliminaries

3.1   Notation

We let $\lambda \in \mathbb{N}$ denote a security parameter. A function $\nu \colon \mathbb{N} \to \mathbb{N}$ is *negligible* if $\nu(\lambda) = O(\lambda^{-c})$ for every constant $c > 0$ and a function p is polynomial if $p(\lambda) = O(\lambda^c)$ for some constant $c > 0$. For a finite set S, we use $x \leftarrow_\$ S$ to denote sampling a uniformly random element from S. When analyzing the concrete security of a primitive, we say that a primitive is (t, ϵ)-secure if any adversary running for time *at most* t succeeds in breaking the scheme with probability *at most* ϵ. We use $\mathbb{F}$ to refer to a finite field, bold variables (e.g., $\mathbf{r}$) to represent vectors, and PRF to denote a pseudorandom function. We refer the reader to the text by Katz and Lindell [28] for the formal definition.

3.2 Arithmetic Circuits

Definition 1. *Arithmetic Circuit [1, 12, 16]* *An arithmetic circuit over a field* $\mathbb{F}$ *and a set of variables* $X = (\tau_i)_{\forall i \in [n]}$ *is a directed acyclic graph representing polynomial computation. We say that a polynomial* $f \in \mathbb{F}[\tau_1, \dots, \tau_n]$, *over a field* $\mathbb{F}$ *is computable by a circuit of size* s *and depth* D *if there exists a directed acyclic graph with* s *nodes and depth* D *such that its leaf nodes are labelled by variables or field constants, internal nodes are labelled with* $+$ *and* $\times$, *and* f *is the polynomial computed at the root node.*

In this paper, we consider evaluations of arithmetic circuits over field elements and over polynomials. We briefly describe these below:

- Computing on field elements: We interpret f as a bounded depth arithmetic circuit over $\mathbb{F}_p$. Meaning f on input $\mathbf{x} \in \mathbb{F}_p^n$ outputs a scalar in $\mathbb{F}_p$. Here the evaluation takes place gate by gate where each gate is either $+$ or $\times$ mod p.
- Computing on polynomials: We interpret f as a D degree polynomial that f computes. Meaning f on input n degree-1 polynomials ($\{y_i\}_{\forall i \in [n]}$ outputs a degree-D polynomial y^* over $\mathbb{F}_p$. We represent polynomials in coefficient notation, i.e., each $y_i = (y_i^0, y_i^1)$ and y^* is a vector of $D + 1$ coefficients. Here the evaluation is performed gate by gate evaluation where multiplication gates correspond to polynomial multiplication (corresponding to a convolution of the coefficient vectors) and addition gates correspond to polynomial additions (corresponding to component-wise addition of the coefficient vectors.)

3.3 Labeled Programs

We recall the definition of labeled programs [12, 21]. A labelled program is a tuple $\mathcal{P} = (f, \tau_1, \dots, \tau_n)$, where $f : \mathbb{F}^n \to \mathbb{F}$ is a function represented by an arithmetic circuit and $\tau_1, \dots, \tau_n$ are binary strings that are used as labels to identify the inputs of this function.

All data is authenticated with a label $\sigma \leftarrow Auth(sk, (\tau, m))$ which is used to identify a particular input to the circuit and to indicate which input of the circuit it corresponds to. Specifically, given a labeled program $\mathcal{P} = (f, \tau_1, \dots, \tau_n)$ and a set of tags $(\sigma_1, \dots, \sigma_n)$ that authenticate message m_i under label τ_i, homomorphic evaluation allows anyone to evaluate the program $\mathcal{P}$ on messages m_i as long as the labels τ_i match those specified in the program.

We denote the identity program associated with label τ as $\mathcal{I}\tau = (g_{id}, \tau)$, where g_{id} signifies the canonical identity function, i.e., $g_{id}(x) = x$ for all inputs x, and $\tau \in \{0, 1\}^*$ denotes an input label.

We recall the definition of well defined program [12].

Definition 2. *A labeled program* $\mathcal{P} = \big(f, (\tau_i)_{\forall i \in [n]}\big)$ *is considered **well-defined** over a table* $\mathcal{T}$ *if one of the following conditions holds:*

- *Either, for every label* $\tau_i \in \mathcal{P}$, $\mathcal{T}$ *contains an entry* $(\tau_i, m_i, \cdot)$ *for some message* m_i *previously authenticated under label* τ_i. *Or,*

– *if there exists an index $i \in [n]$ such that the tuple $(\tau_i, \cdot, \cdot)$ is absent from $\mathcal{T}$ then the output $f(\{m_j\}_{(\tau_j, m_j, \cdot) \in \mathcal{T}} \cup \{\tilde{m}_j\}_{(\tau'_j, \cdot, \cdot) \notin \mathcal{T}})$ remains constant over all possible choices of $\tilde{m}_j$ from $\mathcal{M}$.*
In this case the output of $f(\{m_j\}_{(\tau_j, m_j, \cdot) \in \mathcal{T}} \cup \{\tilde{m}_j\}_{(\tau'_j, \cdot, \cdot) \notin \mathcal{T}})$ remains the same over all possible choices for $\tilde{m}_j$ from $\mathcal{M}$. I.e., the output of f is fixed given the already authenticated inputs.

We also recall proposition 1 from [12] below for completeness.

Proposition 1. *[12] Let $\lambda, n \in \mathbb{N}$ and let $\mathcal{F}$ be the class of arithmetic circuits $f : \mathbb{F}^n \to \mathbb{F}$ over a finite field $\mathbb{F}$ of order p and such that the degree of f is at most d, with $\frac{d}{p} < \frac{1}{2}$. Then, there exists a probabilistic algorithm that given f st $f \in \mathcal{F}$ decides whether $\exists y \in \mathbb{F}$ such that $f(\mathbf{u}) = y$, $\forall \mathbf{u} \in \mathbb{F}^n$ (i.e., if f is constant) and is correct with probability at least $(1 - 2^{-\lambda})$.*

Proof. The algorithm begins by sampling uniformly at random $\lambda + 1$ tuples $\{\mathbf{u}_i\}_{i=0}^{\lambda}$ from $\mathbb{F}^n$. It then checks if $f(\mathbf{u}_0) = \cdots = f(\mathbf{u}_\lambda)$. If this condition holds, it concludes *constant*; otherwise, it concludes *non-constant*. If f is constant, the algorithm is correct with certainty. In the case where f is not constant, the probability of the algorithm being wrong is essentially the probability that $f(\mathbf{u}_0) = \cdots = f(\mathbf{u}_\lambda)$ over all possible choices of $\mathbf{u}_i$'s $i \in [0, \lambda]$. This probability can be bounded above by $(\Pr_{\mathbf{u}_i \leftarrow \mathbb{F}^n}[f(\mathbf{u}_i) = y_0 \mid y_0 = f(\mathbf{u}_0)])^\lambda \leq \left(\frac{d}{p}\right)^\lambda \leq 2^{-\lambda}$, where the upper bound by $\frac{d}{p}$ follows from the Schwartz-Zippel Lemma [14,37].

3.4   Computational Assumptions

We now recall the necessary definitions of a bilinear maps and associated computational hardness assumptions.

Let $\mathcal{G}_1$ and $\mathcal{G}_2$ are two (multiplicative) cyclic groups of prime order p. g_1 is a generator of $\mathcal{G}_1$ and g_2 is a generator of $\mathcal{G}_2$. ψ is an isomorphism from $\mathcal{G}_2$ to $\mathcal{G}_1$, with $\psi(g_2) = g_1$. e is a bilinear map e : $\mathcal{G}_1 \times \mathcal{G}_2 \to \mathcal{G}_T$ such that $e(g_1, g_2) \neq 1$. For our paper, we consider subgroups $\mathcal{G}_1, \mathcal{G}_2$ as $\mathcal{G}_1 \subseteq E(\mathbb{F}_q)$ and $\mathcal{G}_2 \subseteq E(\mathbb{F}_{q^l})$ where E is the elliptic curve over the respective finite field. Two groups $(\mathcal{G}_1, \mathcal{G}_2)$ are a bilinear group if the group action on either can be computed in one time unit, the map ψ from $\mathcal{G}_2$ to $\mathcal{G}_1$ can be computed in one unit time, a bilinear map e exists and e is computable in one unit time.

COMPUTATIONAL CO-DIFFIE HELLMAN PROBLEM. The Computational co-Diffie-Hellman problem(co-CDH) over groups $\mathcal{G}_1$ and $\mathcal{G}_2$ is defined as the problem of computing $h^a \in \mathcal{G}_1$ given $g_2, g_2^a \in \mathcal{G}_2$ and $h \in \mathcal{G}_1$.

Next we define the advantage of any PPT algorithm $\mathcal{A}$ in solving the Computational co-Diffie-Hellman problem in groups $\mathcal{G}_1$ and $\mathcal{G}_2$ as

$$\mathbf{Adv}_{\mathcal{A}}^{co-CDH} \stackrel{\text{def}}{=} \Pr[\mathcal{A}(g_2, g_2^a, h) = h^a | a \leftarrow_{\$} \mathbb{Z}_p, h \leftarrow_{\$} \mathcal{G}_1] \tag{2}$$

where the probability is taken over the uniform random choice of a from $\mathbb{Z}_p$ and h from $\mathcal{G}_1$ and over the coin tosses of $\mathcal{A}$.

Definition 3. co-CDH [10, 11] Co-CDH is (t', ϵ')-hard over $(\mathcal{G}_1, \mathcal{G}_2)$ if for any adversary running in time at most t', $\mathbf{Adv}_{\mathcal{A}}^{co-CDH} < \epsilon'$.

OBSERVATION: We note the following useful property that holds for any tuple (g_2, g_2^a, h, h^b) where $g_2 \in \mathcal{G}_2$, and $h \in \mathcal{G}_1$

$$a = b \bmod p \iff \mathrm{e}(h, g_2^a) = \mathrm{e}(h^b, g_2) \tag{3}$$

4 Multi-key Fully-Homomorphic Aggregate MAC (HA-MAC) for Bounded-Depth Arithmetic Circuits

We now describe our main primitive, a multi-key homomorphic MAC that allows authenticating arbitrary bounded-depth computation on each user's inputs and aggregating the resulting MACs into a single short tag that can be used to verify all users' computations. We now proceed to describe the functionality and security of this primitive.

4.1 Functionality

In the description below, we abuse notation to use P both as the set of parties and the number of parties (i.e., $|P|$). A multi-key fully-homomorphic aggregate MAC consists of the following six algorithms:

- KeyGen$(1^\lambda, 1^n, 1^D, 1^P)$: On input the security parameter $\lambda \in \mathbb{N}$, input size n, depth bound D, and number of users P, KeyGen outputs a secret key and a public evaluation key (sk, ek) for all P parties.
- Auth(sk_l, τ, m) : On input the secret key sk_l of user $l \in P$ and a message $m \in \mathcal{M}$ with input label $\tau \in \{0,1\}^\lambda$, Auth outputs a message authentication code σ.
- Eval$(\mathsf{ek}_l, f, \sigma_1, \ldots, \sigma_n)$: On input the evaluation key of party $l \in P$, a function $f : \mathcal{M}^n \to \mathcal{M}$ and input message authentication codes $\sigma_1, \ldots, \sigma_n$, Eval outputs a message authentication code Λ. Note that since the parties' ek's are public any party is able to run Eval if it has valid input MACs.
- Ver$(\mathsf{sk}_l, m, \mathcal{P}, \Lambda)$: On input secret key sk_l for some $l \in P$, a program $\mathcal{P} = (f, \tau_1, \ldots, \tau_n)$, a message m and a message authentication code Λ, Ver outputs 0/1 to indicate whether Λ is a valid tag authenticating that m is a valid output of computation of program $\mathcal{P}$.
- Aggregate$((\Lambda_l, m_l)_{\forall l \in U})$: On input $|U|$ tuples of message and message authentication codes for some set of parties $U \subseteq P$, Aggregate outputs an HA-MAC σ^*.
- AggVer$((m_l, \mathsf{sk}_l, \mathcal{P}_l)_{\forall l \in U}, \sigma^*)$: On input $|U|$ tuples of (message, secret key, program) for some subset of parties $U \subseteq P$ and HA-MAC σ^*, AggVer outputs a 0/1 indicating whether σ^* is a valid HA-MAC on messages $m_1, \ldots, m_{|U|}$ resulting from evaluating the programs $\mathcal{P}_1, \ldots, \mathcal{P}_{|U|}$ for each of the $|U|$ parties.

We now define three notions of correctness that a HA-MAC scheme must satisfy.

AUTHENTICATION CORRECTNESS:
For any message $m \in \mathcal{M}$, any label $\tau \in \{0,1\}^\lambda$, for all $(\mathsf{sk}, \mathsf{ek}) \leftarrow_\$ \mathsf{KeyGen}(1^\lambda, 1^n, 1^D, 1^P)$ and any tag $\sigma \leftarrow_\$ \mathsf{Auth}(\mathsf{sk}, \tau, m)$ we require that

$$Pr[\mathsf{Ver}(\mathsf{sk}, m, \mathcal{I}_\tau, \sigma) = 1] = 1$$

where $\mathcal{I}_\tau$ is a special identity program that on input m outputs m and the probability is taken over the random coins of KeyGen.

EVALUATION CORRECTNESS:
For any $(\mathsf{sk}, \mathsf{ek}) \leftarrow_\$ \mathsf{KeyGen}(1^\lambda, 1^n, 1^D, 1^P)$, a function $f : \mathcal{M}^n \to \mathcal{M}$, any set of message tag tuple $\{(m_i, \sigma_i)\}_{i=1}^n$, if for all $i \in [n]$ $\mathsf{Ver}(\mathsf{sk}, m_i, \mathcal{I}_{\tau_i}, \sigma_i) = 1$ then for $m^* \leftarrow f((m_i)_{i \in n})$ and $\Lambda \leftarrow \mathsf{Eval}(\mathsf{ek}, f, \sigma_1, \ldots, \sigma_n)$ we require that

$$\mathsf{Ver}(\mathsf{sk}, m^*, \mathcal{P}, \Lambda) = 1$$

AGGREGATION CORRECTNESS:
For any set of parties $U \subseteq P$, given $|U|$ tuples $(m_l, \Lambda_l, \mathcal{P}_l)_{\forall l \in U}$ such that $\mathsf{Ver}(\mathsf{sk}_l, m_l, \mathcal{P}_l, \Lambda_l) = 1$ for all $l \in U$ then,

$$Pr[\mathsf{AggVer}((\mathsf{sk}_l, m_l, \mathcal{P}_l)_{\forall l \in U}, \sigma^*) = 1] = 1$$

where $\sigma^* \leftarrow \mathsf{Aggregate}((m_l, \Lambda_l)_{\forall l \in U})$ and the probability is taken over the random coins of KeyGen.

4.2 Security

We now turn to defining security of an HA-MAC scheme. Roughly, we want a strong definition of security where an adversary can corrupt, and hence learn the secret keys of, all the parties except one designated *challenge* party. Without loss of generality we refer to this party as p_1.

   We allow the adversary to see authentications under secret key of p_1 on messages and labels of its choice and he can thus evaluate programs of his choice on these messages and to aggregate the resulting tags with arbitrary computations over inputs from all other parties. Additionally, we allow the adversary to ask verification and aggregate verfication queries on initial and aggregate MACs that include p_1 respectively.

   Informally, security of our HA-MAC requires that the adversary not be able to forge original or aggregate MACs that include p_1's input. Of course, the adversary can run evaluation on the MACs that he has received for p_1. Thus, what we require is that $\mathcal{A}$ not be able to produce a valid MAC or aggregate MAC that authenticates some program $\mathcal{P}'$ for p_1 such that $\mathcal{P}'$ cannot be computed from the inputs and labels on which $\mathcal{A}$ has requested tags. This is akin to the usual chosen message security of a MAC.

   This definition guarantees that even if all but one parties are controlled by the adversary any inputs and computations provided by an honest party will be validly authenticated. So, the verifier can be certain that the honest party's values are correctly computed given the final aggregate MAC.

Formally, we define existential unforgeability against a chosen message attack for a homomorphic aggregate MAC using the game $\mathbf{G}_{\Theta,\mathcal{A}}^{\text{HA-UF-CMA}}(1^\lambda, 1^n, 1^D, 1^P)$ between a challenger and an adversary $\mathcal{A}$ given in Fig. 1.

$\mathbf{G}_{\Theta,\mathcal{A}}^{\text{HA-UF-CMA}}(1^\lambda, 1^n, 1^D, 1^P)$

INITIALIZE$(1^\lambda, 1^n, 1^D, 1^P)$:
1 $\mathcal{T} \leftarrow \emptyset$
2 $(\mathsf{sk}_l, \mathsf{ek}_l)_{\forall l \in P} \leftarrow\!\!\$\ \mathsf{KeyGen}(1^\lambda, 1^n, 1^D, 1^P)$
3 Return $\mathsf{ek}_1, (\mathsf{ek}_l, \mathsf{sk}_l)_{\forall l \in [2,P]}$

AUTHO(τ, m):
4 if $(\tau, m, \cdot) \in \mathcal{T}$, $\sigma \leftarrow \mathcal{T}(\tau, m, \cdot)$
5 if $(\tau, m, \cdot) \notin \mathcal{T}$, $\sigma \leftarrow \mathsf{Auth}_{\mathsf{sk}_1}(\tau, m)$; $\mathcal{T} = \mathcal{T} \cup (\tau, m, \sigma)$
6 if $(\tau, \cdot, \cdot) \in \mathcal{T}$, ignore
7 Return σ

VERO$(m, \mathcal{P}, \sigma)$:
8 Return $\mathsf{Ver}_{\mathsf{sk}_1}(m, \mathcal{P}, \sigma)$

AVERO$((m'_j, \mathcal{P}'_j)_{\forall j \in U}, (\mathsf{ek}'_j)_{\forall j \in U \setminus \{1\}}, \sigma')$: // $U \subseteq P, (\mathsf{ek}'_j)_{\forall j \in U \setminus \{1\}} \subseteq (\mathsf{ek}_j)_{\forall j \in P}$
9 Return $\mathsf{AggVer}((m'_j, \mathcal{P}'_j)_{\forall j \in U}, (\mathsf{sk}_1, \mathsf{sk}_j)_{\forall j \in U \setminus \{1\}}, \sigma')$

FINALIZE(IN):
10 if $IN = ((m^*_j, \mathcal{P}^*_j)_{\forall j \in U}, (\mathsf{ek}^*_j)_{\forall j \in U \setminus \{1\}}, \sigma^*)$
11   Return $\mathsf{Check}_A((m^*_j, \mathcal{P}^*_j)_{\forall j \in U}, (\mathsf{ek}^*_j)_{\forall j \in U \setminus \{1\}}, \sigma^*))$
12 else if $IN = (m^*, \mathcal{P}^*, \sigma^*)$
13   Return $\mathsf{Check}_V(m^*, \mathcal{P}^*, \sigma^*)$

Fig. 1. Game defining required security game for Θ

OUTPUT. We say that $\mathcal{A}$ wins, i.e. that the output of $\mathbf{G}_{\Theta,\mathcal{A}}^{\text{HA-UF-CMA}} = 1$ iff FINALIZE returns 1. This happens if either $\mathsf{Check}_A = 1$ or $\mathsf{Check}_V = 1$.

We now describe the Check_A and Check_V algorithms that we use to define the notion of a forgery. These algorithms capture what is meant by a valid aggregate forgery and a valid forgery respectively. As mentioned previously an adversary can easily produce valid MACs and aggregate MACs on inputs for which he has already seen valid MACs, thus these algorithms check whether a claimed forgery was really for something new. To define these formally, we turn to the notion of well-defined program with respect to a table $\mathcal{T}$ as defined in Sect. 3.3.

ALGORITHM Check_A. For any input $((m^*_j, \mathcal{P}^*_j)_{\forall j \in U}, (\mathsf{ek}^*_j)_{\forall j \in U \setminus \{1\}}, \sigma^*)$,
$\mathsf{Check}_A\left((m^*_j, \mathcal{P}^*_j)_{\forall j \in U}, (\mathsf{ek}^*_j)_{\forall j \in U \setminus \{1\}}, \sigma^*\right) = 1$ iff

- $\mathsf{AggVer}((m^*_j, \mathcal{P}^*_j)_{\forall j \in U}, (\mathsf{sk}_j)_{\forall j \in U \setminus \{1\}}, \sigma^*) = 1$ and
- $m^*_1 \notin \mathcal{T}$, meaning the forgery is non-trivial, and

– one of the below two events happens:
  • (Type 1 Forgery) $\mathcal{P}_1^*$ is *not* well defined with respect to $\mathcal{T}$. This implies that the adversary has not seen MACs on all the inputs needed to compute $\mathcal{P}_1^*$.
  • (Type 2 Forgery) $\mathcal{P}_1^*$ is well defined with respect to $\mathcal{T}$ but $m_1^* \neq f_1^*\left((m_j)_{(\tau_{1,j}^*, m_j) \in \mathcal{T}}\right)$. This captures the case when m_1^* is not the correct output of the labeled program $\mathcal{P}_1^*$ when executed on previously authenticated messages $(m_j)_{\forall j \in [n]}$.

ALGORITHM Check$_V$. For any input $(m^*, \mathcal{P}^*, \sigma^*)$, Check$_V(m^*, \mathcal{P}^*, \sigma^*) = 1$ iff

– Ver$_{sk_1}(m^*, \mathcal{P}^*, \sigma^*) = 1$ and
– $m^* \notin \mathcal{T}$, meaning the forgery is non-trivial and
– one of the below two happens
  • (Type 1 Forgery) $\mathcal{P}^*$ is *not* well defined with respect to $\mathcal{T}$
  • (Type 2 Forgery) $\mathcal{P}^*$ is well defined with respect to $\mathcal{T}$ but $m^* \neq f^*\left((m_j)_{(\tau_{1,j}^*, m_j) \in \mathcal{T}}\right)$. This attests that m^* is not the correct output of the labeleld program $\mathcal{P}^*$ when executed on previously authenticated messages $(m_j)_{\forall j \in [n]}$.

DEFINITION. We say that a multi-key fully homomorphic aggregate MAC (HA-MAC) scheme is secure if for any PPT adversary $\mathcal{A}$ there exists a negligible function ν such that

$$\mathbf{Adv}_{\Theta, \mathcal{A}}^{\mathrm{HA}}(1^\lambda, 1^n, 1^D, 1^P) = \Pr[\mathbf{G}_{\Theta, \mathcal{A}}^{\mathrm{HA\text{-}UF\text{-}CMA}}(1^\lambda, 1^n, 1^D, 1^P) \Rightarrow 1] \leq \nu(\lambda) \quad (4)$$

where the probability is over the random coins of $\mathcal{A}$ and KeyGen

Understanding $\mathbf{G}_{\Theta, \mathcal{A}}^{\mathrm{HA\text{-}UF\text{-}CMA}}$: The game begins with the INITIALIZE procedure in which the challenger creates $P\ (sk, ek)$ pairs. He gives all of these except the challenge party's sk_1 to $\mathcal{A}$. This allows $\mathcal{A}$ to produce arbitrary MACs for any party other than p_1.

Additionally, $\mathcal{A}$ can make queries to AUTHO to get tags under sk_1 for (m, τ) message label pairs of his choice. The challenger records all such queries in a table $\mathcal{T}$ and answers consistently on repeated queries or if multiple values are requested with the same label.

$\mathcal{A}$ can also make queries to VERO and AVERO on any messages, programs, and MACs. In particular, for the AVERO queries $\mathcal{A}$ can choose any subset of parties to aggregate as long as they include p_1. Computing aggregate MACs without p_1 is trivial since $\mathcal{A}$ has all the other sk's.

Finally, $\mathcal{A}$ outputs in the FINALIZE phase either an attempted forgery of the original MAC or an attempted aggregate forgery. He wins if the produced forgery couldn't be computed by simply running Eval on MACs he has already seen (i.e., the ones contained in the table $\mathcal{T}$). The Check$_A$ and Check$_V$ procedures perform this check for AVERO and VERO respectively. Thus, the adversary wins if he is able to produce a valid MAC or aggregate MAC that cannot be computed from the table $\mathcal{T}$.

5   Construction of Multi Key Fully Homomorphic Aggregate MAC for Arithmetic Circuits

We are now ready to present the details of our HA-MAC constructions. Our construction largely follows the construction of Catalano and Fiore [12] with the major differences, that we highlight, to enable aggregation. The details of the construction are given in Fig. 2.

Setup: As described in Sect. 3.4, we assume the existence of groups $\mathcal{G}_1$ and $\mathcal{G}_2$ of order p with generators g_1 and g_2 respectively. With an efficiently computable isomorphism ψ from $\mathcal{G}_2$ to $\mathcal{G}_1$ such that $\psi(g_2) = g_1$ and a bilinear map $e : \mathcal{G}_1 \times \mathcal{G}_2 \to \mathcal{G}_T$ such that $e(g_1, g_2) \neq 1$ and the co-CDH problem in $\mathcal{G}_1$ and $\mathcal{G}_2$ is hard. We assume that all of these are public parameters that are available to all of the subsequent algorithms. We additionally assume the existence of a hash function $\mathcal{H}_2 : \{0,1\}^* \to \mathcal{G}_2$ which is modelled as random oracle [5].

KeyGen: For each party $l \in P$, we choose a public generator $u_l = g_1^{\gamma_l}$ for a random $\gamma_l \leftarrow_\$ \mathbb{Z}_p$. Additionally, we choose a random evaluation point $x_{l_0} \leftarrow_\$ \mathbb{Z}_p$ and a PRF key K_l. The secret key sk_l consists of (x_{l_0}, K_l, γ_l) while the public evaluation key ek_l consists of encodings of powers of x_{l_0} in the exponent. That is ek_l consists of $u_l, u_l^{x_{l_0}}, u_l^{x_{l_0}^2}, \ldots, u_l^{x_{l_0}^D}$ up to the specified depth bound D. We note that our construction differs from [12] in that we include the generator u_l in the public evaluation key, whereas their construction required keeping this value secret.

Auth: To authenticate a message m with label τ under user p_l's key, the tag σ is a degree-1 univariate polynomial y such that $y(0) = m$ and $y(x_{l_0}) = r_\tau$ where x_{l_0} is the secret evaluation point in sk_l and $r_\tau = PRF_k(\tau)$. We represent σ as a vector of the coefficients of y: (y_0, y_1).

Eval: To compute the homomorphic tag for an evaluation of a depth-D arithmetic circuit f and input tags $\sigma_1, \ldots, \sigma_n$ we follow the procedure described in Sect. 3.2 to evaluate f over the degree-1 polynomials in the σ's. That is, for every addition gate, we add the input polynomials and for every multiplication gate we do a polynomial multiplication corresponding to performing a convolution on the coefficients.

At the end of this computation, we get a degree-D polynomial y such that $y(0) = f(m_1, m_2, \ldots, m_n)$ and $y(x_0) = f(r_{\tau_1}, r_{\tau_2}, \ldots, r_{\tau_n})$[3]. We let y_k for $k \in \{0, \ldots, D\}$ be the coefficients of y.

Now to compute the homomorphic tag, we essentially evaluate $y(x_0) - y(0)$ in the exponent. Recall that ek contains $h_0 = u, h_1 = u^{x_0}, h_2 = u^{x_0^2}, \ldots, h_D = u^{x_0^D}$. Now, note that $y(x_0) = y_0 + y_1 x_0 + y_2 x_0^2 + \cdots + y_D x_0^D$ and $y(0) = y_0$. So, to evaluate $y(x_0) - y(0)$ in the exponent we compute the product

$$h_1^{y_1} \cdot h_2^{y_2} \cdots h_D^{y_D} = u^{y(x_0) - y(0)}.$$

---

[3] In what follows, for ease of presentation we drop the subscript l from all terms since we are only considering the case of player p_l.

Ver: On input the homomorphic tag Λ, program $\mathcal{P}$, and message m, we first compute $\rho = f(r_{\tau_1}, r_{\tau_2}, \ldots, r_{\tau_n}) = y(x_0)$ where f is the function from $\mathcal{P}$ and $r_{\tau_i} = PRF_K(\tau_i)$ and y is the polynomial computed in Eval. Now recall that for a valid tag $\Lambda = u^{y(x_0)-y(0)}$ where $y(0) = m$, the homomorphic output. So, to verify the homomorphic tag Λ, we just need to check whether $\Lambda \cdot u^m = u^\rho$. The case for verifying an original (i.e., non-homomorphic) tag is similar.

For reasons needed in the proof we deviate from [12]. Meaning, instead we do this verification in the target group $\mathcal{G}_T$ by pairing all terms in this equation with $w \leftarrow \mathcal{H}_2(m)$. Roughly, this is necessary to allow the security reduction to program in the challenge from its challenger. A summary of the proof is given in Sect. 6.

Aggregate: To aggregate homomorphic tags from a set of U users, we take a weighted sum of the tags in the exponent. Specifically, for each message m_i to aggregate, we compute $w_i \leftarrow \mathcal{H}_2(m_i)$. Viewing w_i as $g_2^{b_i}$ for random exponent b_i, we multiply Λ_i by b_i in the exponent by computing $e(\Lambda_i, w_i)$. Finally we multiply the resulting elements in $\mathcal{G}_T$ together to compute a random linear combination of the component tags.

AggVer: On input an aggregate mac on any subset of U users, its corresponding messages and evaluation keys, AggVer algorithm uses its secret keys for the users in set U to recompute the aggregate MAC and checks whether it is equal to the input aggregate MAC. It accepts iff the equality holds. Note that since ρ is computed by evaluating f at random points, aggregate verification does not require the verifier to know the original inputs, only the outputs of the homomorphic computations.

5.1 Correctness

We now prove that the construction in Fig. 2 satisfies the necessary correctness properties of HA-MAC.

AUTHENTICATION CORRECTNESS. For any $l \in P$, recall that $\rho_l \leftarrow y_0 + y_1 \cdot x_{l_0}$. Let $\gamma_l \leftarrow^{\$} \mathbb{Z}_p, b_l \leftarrow^{\$} \mathbb{Z}_p$ such that $u_l \leftarrow g_1^{\gamma_l}$ and $w_l \leftarrow g_2^{b_l}$. Therefore we get that:

$$e(u_l^{x_{l_0}y_1}, w_l) \cdot e(u_l, w_l^{y_0}) = e(g_1, g_2)^{\gamma_l \cdot b_l(y_0 + x_{l_0} \cdot y_1)} = e(u_l, w_l)^{\rho_l} \tag{5}$$

EVALUATION CORRECTNESS. For any $l \in P$, recall that $\rho_l \leftarrow m_l + \Sigma_{k=1}^{D} y_{l,k} \cdot x_{l_0}^k$. Therefore we get that

$$e(\Lambda_l, w_l) \cdot e(u_l, w_l)^{m_l} = e(g_1^{\gamma_l \cdot \Sigma_{k=1}^{D} y_{l,k} \cdot x_{l_0}^k}, g_2^{b_l}) \cdot e(g_1^{\gamma_l}, g_2^{b_l})^{m_l} = e(u_l, w_l)^{\rho_l} \tag{6}$$

AGGREGATION CORRECTNESS. For any subset of users U such that $U \subseteq P$ then for all l in U, let $b_l \leftarrow^{\$} \mathbb{Z}_p$, $\gamma_l \leftarrow^{\$} \mathbb{Z}_p$, hash of message m_l is $w_l \leftarrow g_2^{b_l}$ and generator $u_l \leftarrow g_1^{\gamma_l}$ then we get that:

$$\sigma^* \cdot (\Pi_{\forall l \in U} \, e(u_l, w_l^{m_l}))$$
$$= e(g_1, g_2)^{\Sigma_{\forall l \in U} (\gamma_l \cdot b_l \cdot (\Sigma_{k=1}^{D} y_{l,k} \cdot x_{l_0}^k))} \cdot e(g_1, g_2)^{\Sigma_{\forall l \in U} (\gamma_l \cdot b_l \cdot m_l)}$$
$$= e(g_1, g_2)^{\Sigma_{\forall l \in U} (\gamma_l \cdot b_l \cdot ((\Sigma_{k=1}^{D} y_{l,k} \cdot x_{l_0}^k) + m_l))} = \Pi_{\forall l \in U} \, e(u_l, w_l)^{\rho_l} \tag{7}$$

KeyGen($1^\lambda, 1^n, 1^D, 1^P$):

1. $p \leftarrow_\$ \mathsf{PrimeGen}(1^\lambda)$
2. $\forall l \in [P]$:
3. $\quad x_{l_0}, \gamma_l \leftarrow_\$ \mathbb{Z}_p^2, K_l \leftarrow_\$ \mathcal{K}, u_l \leftarrow g_1^{\gamma_l}$
4. $\quad \forall k \in [0, D], h_{l,k} \leftarrow u_l^{x_{l_0}^k}$
5. $\quad \mathsf{sk}_l \leftarrow (x_{l_0}, K_l, \gamma_l)$
6. $\quad \mathsf{ek}_l \leftarrow (h_{l,0}, h_{l,1}, h_{l,2}, \ldots h_{l,D})$
7. Return $(\mathsf{ek}_l, \mathsf{sk}_l)_{\forall l \in [P]}$

Auth($\mathsf{sk}, (\tau, m)$):

8. $(x_0, K, \gamma) \leftarrow \mathsf{sk}$
9. $r_\tau \leftarrow PRF_K(\tau)$
10. $y_0 \leftarrow m; y_1 \leftarrow \frac{r_\tau - m}{x_0} \mod p$
11. Return $\sigma \leftarrow (y_0, y_1)$

Ver($\mathsf{sk}, m, \mathcal{P}, \Lambda$):

12. $(f, \tau_1, \ldots, \tau_n) \leftarrow \mathcal{P}$
13. $(x_0, K, \gamma) \leftarrow \mathsf{sk}, w \leftarrow \mathcal{H}_2(m), u \leftarrow g_1^\gamma$
14. $\mathbf{r} \leftarrow PRF_K(\tau_i)_{\forall i \in [n]}, \rho \leftarrow f(\mathbf{r})$
15. If $\Lambda = (y_0, y_1)$ then
    return $e(u^{x_0 \cdot y_1}, w) \cdot e(u, w^{y_0}) \stackrel{?}{=} e(u, w^\rho)$
16. Else Return $e(u, w^m) \cdot e(\Lambda, w) \stackrel{?}{=} e(u, w^\rho)$

Eval($\mathsf{ek}, f, (\sigma_i)_{\forall i \in [n]}$):

17. $(y_k)_{k \in [0, D]} \leftarrow f((\sigma_i)_{\forall i \in [n]})$
18. $\Lambda \leftarrow \Pi_{k=1}^D h_k^{y_k}$
19. Return Λ

Aggregate($(\Lambda_l, m_l)_{\forall l \in U}$):

20. $w_l \leftarrow \mathcal{H}_2(m_l)$
21. $\sigma^* \leftarrow \Pi_{\forall l \in U} e(\Lambda_l, w_l)$
22. Return σ^*

AggVer($(\mathsf{sk}_l, \mathcal{P}_l, m_l)_{\forall l \in U}, \sigma^*$):

23. $\forall l \in U : (f_l, \tau_{l,i})_{\forall i \in [n]} \leftarrow \mathcal{P}_l, (x_{l_0}, K_l, \gamma_l) \leftarrow \mathsf{sk}_l$
24. $\quad w_l \leftarrow \mathcal{H}_2(m_l), u_l \leftarrow g_1^{\gamma_l}$
25. $\quad \mathbf{r}_l \leftarrow PRF_{K_l}(\tau_{l,i})_{\forall l \in U, \forall i \in [n]}, \rho_l \leftarrow f_l(\mathbf{r}_l)$
26. Return $\Pi_{\forall l \in U} e(u_l, w_l^{\rho_l}) \stackrel{?}{=} \sigma^* \cdot \Pi_{\forall l \in U} e(u_l, w_l^{m_l})$

Fig. 2. The construction of HA-MAC Θ for bounded depth arithmetic circuits

5.2 Performance and Security

Finally, our HA-MAC scheme achieves the following performance and security:
Efficiency: We note that the size our homomorphic tag and aggregate tag is *succinct* consisting of just one field element independent of the number of parties or complexity of the computed functions. Moreover, the original inputs into the homomorphic computation are not needed to verify the final aggregate signature so they do not need to be communicated. We observe that the evaluation key in our construction grows linearly in the depth of the homomorphic evaluation (as in prior work [12]). Eliminating this dependence on the depth of the function is an interesting open question.
Security: We say that a P-user HA-MAC scheme is $(t, Q_H, Q_A, Q_V, Q_{AV}, P, \epsilon)$-secure in the random oracle model if for any PPT adversary $\mathcal{A}(t, Q_H, Q_A, Q_V, Q_{AV}, P, \epsilon)$ in the $\mathbf{G}_{\Theta, \mathcal{A}}^{\text{HA-UF-CMA}}$ game (Fig. 1)

- $\mathcal{A}$ runs in time *at most* t,
- $\mathcal{A}$ makes *at most* Q_H queries to the hash function,

- $\mathcal{A}$ makes *at most* Q_A queries to the authenticate oracle,
- $\mathcal{A}$ makes *at most* Q_V queries to the verification oracle, and
- $\mathcal{A}$ makes *at most* Q_{AV} queries to the aggregate verification oracle then
- $\Pr[\mathbf{G}_{\Theta,\mathcal{A}}^{\text{HA-UF-CMA}} = 1] \leq \epsilon$

where the probability is over the random coins of $\mathcal{A}$ and KeyGen.

We can now formally state the security of our construction.

Theorem 2. *If co-CDH is (t', ϵ') hard over groups $(\mathcal{G}_1, \mathcal{G}_2)$ and PRF is ϵ'' secure then HA-MAC scheme as defined in Fig. 2 is $(t, Q_H, Q_A, Q_V, Q_{AV}, P, \epsilon)$-secure in the random oracle model for all t, ϵ satisfying*

$$\epsilon < \frac{Q}{2^\lambda} + \frac{DQ}{p - DQ} + \frac{DQ^2}{p} + e^{\frac{3}{p}} PQ \left(e^{\frac{1}{Q}} + 1 \right) \epsilon' + \epsilon''$$

and

$$t > t' - c(s + P + Q + PQ + sQ + sPQ)$$

where Q is the maximum of Q_H, Q_A, Q_V, Q_{AV}, c is the maximum time to compute any group operation, s is the size of the depth D function f as in definition 1 and P is the total number of users.

We sketch the intuition behind the proof of this theorem in Sect. 6. The full proof of security can be found in the full version [7].

6 Proof of Security

To prove the security of our construction, we define a series of games. We briefly describe these games and the intuition behind the proof here, the full proof can be found in the full version.

Game 0: Our starting point is the real-world security game $\mathbf{G}_{\Theta,\mathcal{A}}^{\text{HA-UF-CMA}}$ (Fig. 1). Game $\mathbf{G}_{\Theta,\mathcal{A}}^{0}$ is the same as the real game except for two changes. In every VERO query $(m, \mathcal{P}, \Lambda)$, the challenger uses the probabilistic test of Catalano and Fiore [12](recalled in Proposition 1) to test whether $\mathcal{P}$ is well defined with respect to table $\mathcal{T}$. Similarly, we also use this same test in every AVERO query to check whether $\mathcal{P}'_1$ (player p_1's program) is well defined. Since the test in Proposition 1 is correct with all but negligible probability, this only introduces a negligible difference between the games.

Game 1: Game $\mathbf{G}_{\Theta,\mathcal{A}}^{1}$ is the same as $\mathbf{G}_{\Theta,\mathcal{A}}^{0}$ except that the PRF in the Auth oracle is replaced by a trully random function(TRF) $\mathcal{R} : \{0,1\}^* \to \mathbb{Z}_p$. Clearly this change is undetectable by the security of the PRF.

Game 2: Game $\mathbf{G}_{\Theta,\mathcal{A}}^{2}$ is the same as $\mathbf{G}_{\Theta,\mathcal{A}}^{1}$ except the challenger adjusts how it responds to VERO and AVERO queries.

- VERO Queries: For any VERO query $(m, \mathcal{P}, \Lambda)$ such that $\mathcal{P}$ is not well defined in $\mathcal{T}$, the challenger returns reject. For any well-defined $\mathcal{P}$, the challenger acts exactly as in $\mathbf{G}^1$.
  Thus, the only difference between $\mathbf{G}^2$ and $\mathbf{G}^1$ occurs if a Mac for a not well-defined program is accepted in $\mathbf{G}^1$. We argue that for any adversary $\mathcal{A}$ such a *bad* query only occurs with negligible probability. Roughly, making such a query requires finding the correct value of ρ. However, finding ρ requires guessing r_τ for some τ not in $\mathcal{T}$. Since $r_\tau = \mathcal{R}(\tau)$ for a random function $\mathcal{R}$, this can only happen negligibly often.
- AVERO Queries: Similarly, $\mathbf{G}^1$ and $\mathbf{G}^2$ differ if $\mathcal{A}$ can find an aggregate Mac σ' such that the circuit for player p_1 is not well-defined relative to $\mathcal{T}$, but is still accepted in $\mathbf{G}^1$. Here to prove that the probability of $\mathcal{A}$ finding such a σ is negligible, we extract p_1's component of the aggregate (using the sk's of the other parties, which the challenger knows), and argue as above that he must have predicted a value r_τ.

Game 3: Game $\mathbf{G}^3_{\Theta,\mathcal{A}}$ is the same as $\mathbf{G}^2_{\Theta,\mathcal{A}}$ except the following change in answering Auth queries. If the random value for some tag τ queried in Auth has previously been used to answer a VERO or AVERO query, then just resample a new, independent value $r_\tau = \mathcal{R}(\tau)$. Since $\mathcal{A}$ can only learn a polynomial number of possible r_τ points, the probability that one of these is sampled is negligible. If this doesn't happen this game is the same as the previous one.

Game 4: Game $\mathbf{G}^4_{\Theta,\mathcal{A}}$ differs from $\mathbf{G}^3_{\Theta,\mathcal{A}}$ in how the challenger answers VERO queries for programs that are well-defined relative to $\mathcal{T}$. Specifically, for a query $(m, \mathcal{P}, \Lambda)^4$, for every input tag τ_i in $\mathcal{P}$, the challenger finds the corresponding input Mac σ_i and message m_i in $\mathcal{T}$. (Since $\mathcal{P}$ is well-defined relative to $\mathcal{T}$, only wires that have no impact on the output may not have an entry in $\mathcal{T}$, in which case the challenger can just choose a random tag.) Next the challenger uses these tags σ_i to recompute the homomorphic Mac Λ^{***} using Eval and checks if it is equal to the claimed Mac Λ queried by the adversary. The challenger returns accept if and only if they are equal.

This differs from $\mathbf{G}^3$ in that it eliminates the possibility of Type-2 forgeries in VERO queries. In a Type-2 forgery, the adversary somehow produces a VERO query for a well-defined program where the message used for one of the inputs is not equal to the one in $\mathcal{T}$. In this case the recomputed Λ^{***} will not be correct, but the query will still accept in $\mathbf{G}^3$.

We prove that an adversary can only make such a query with non-negligible probability. To do so, we introduce one more change into $\mathbf{G}^4$ in that we change how the challenger answers hash evaluation queries for $\mathcal{H}_2$. Specifically, instead of just returning a random element in $\mathcal{G}_2$, the challenger now samples a random exponent b and returns g_2^b as the random element in $\mathcal{G}_2$ while storing b. This

---

[4] $\mathbf{G}^4$ also handles the case of verification of initial Macs in addition to homomorphically computed Macs, but we omit this case from our discussion here to simplify presentation.

knowledge of the discrete log b, allows the challenger to detect when an adversary makes such a Type-2 forgery query.

Finally, we argue that such a query is negligibly likely by showing a reduction from an adversary making such a query to solving the co-CDH problem in $(\mathcal{G}_1, \mathcal{G}_2)$. To do so, intuitively the challenger roughly does the following. The challenger randomly picks one of the VERO queries and programs the random oracle to return the co-CDH challenge g_2^a as the output of the corresponding $\mathcal{H}_2$ query. We can show that by embedding the other part of the co-CDH challenge (the value $h \in \mathcal{G}_1$) in the evaluation key ek_1, a Type-2 forgery query must allow extraction of the value h^a, thus solving co-CDH.

Game 5: Game $\mathbf{G}^5_{\Theta, \mathcal{A}}$ makes a similar change to the one in $\mathbf{G}^4$, but to the AVERO queries. Here, again, the goal is to eliminate Type-2 forgeries in AVERO queries. The added challenge here is that we have to deal with the fact that some (in fact, all but one) of the macs included in an aggregate mac come from malicious parties. We show that the challenger is able to extract the claimed homomorphic Mac for party p_1, and can then compare this to the Mac recomputed from the table $\mathcal{T}$, accepting if and only if they match. Effectively, the challenger can recompute the random linear combination of the adversaries' macs by using $\mathcal{H}_2$ to recompute the weights (in the exponent) for this sum.

With this check in place, the only way that $\mathbf{G}^5$ and $\mathbf{G}^4$ differ is if the adversary queries an aggregate Mac that contains a Type-2 forgery for party 1. However, by a reduction similar to the one described in the previous game, we can show that any adversary that can make such a query must also solve the co-CDH problem on $(\mathcal{G}_1, \mathcal{G}_2)$.

Unforgeability: To conclude the proof of unforgeability, we observe that $\Pr[\mathbf{G}^5_{\Theta, \mathcal{A}}] = 0$ because all verification queries and aggregate verification queries for both Type-1 and Type-2 forgeries are answered with 0. Thus, there is no opportunity for any adversary to win in Game 5, and so for any adversary $\mathcal{A}$ its advantage in $\mathbf{G}^5$ is 0.

7   Conclusion

In this paper, we introduced the concept of a multi-key fully-homomorphic aggregate MAC (HA-MAC) for arithmetic circuits. This primitive enables an untrusted server to produce a short certificate to prove that he has performed correct (disjoint) computations on multiple users' data. The size of this proof is independent of the number of users or the complexity of the performed computations. We give a construction of this primitive based on the co-CDH assumption in the random oracle model.

Our paper leaves open a number of interesting problems for further study. Two immediate questions are removing the reliance on the random oracle and improving the computational efficiency of verification. While the final Mac in our construction is succinct—only 1 field element—verification still requires the verifier to evaluate all functions on random inputs. Very recent work by Anthoine et al. [2] showed how to amortize verification costs for multiple verifications of the same computation in a similar setting. It would be interesting to apply similar techniques to amortize verification costs for our HA-MAC. Further possible improvements to our construction include allowing verification given only an aggregate of the homomorphic outputs rather than requiring all of the outputs to verify an aggregate Mac and eliminating the bounded-degree requirement of our assumption. I.e., can we construct a scheme where the size of the keys required does not grow with the depth of the homomorphic evaluations supported.

Acknowledgements. We would like to thank Adam O'Neill, Ojaswi Acharya, Weiqi Feng for valuable discussions that led to the problem studied here. Arkady Yerukhimovich is supported in part by NSF grants CNS-1955620 and CNS-2144798 (CAREER).

References

1. Agrawal, M., Saptharishil, R.: Classifying polynomials and identity testing. In: Current Trends in Science. Platinum Jubilee Special. Indian Academy of Sciences (2009)
2. Anthoine, G., Balbás, D., Fiore, D.: Fully-succinct multi-key homomorphic signatures from standard assumptions. In: Annual International Cryptology Conference, pp. 317–351. Springer (2024)
3. Applebaum, B., Ishai, Y., Kushilevitz, E.: From secrecy to soundness: efficient verification via secure computation. In: Abramsky, S., Gavoille, C., Kirchner, C., Meyer auf der Heide, F., Spirakis, P.G. (eds.) ICALP 2010. LNCS, vol. 6198, pp. 152–163. Springer, Heidelberg (2010). https://doi.org/10.1007/978-3-642-14165-2_14
4. Bellare, M., Namprempre, C., Neven, G.: Unrestricted aggregate signatures. In: Arge, L., Cachin, C., Jurdziński, T., Tarlecki, A. (eds.) ICALP 2007. LNCS, vol. 4596, pp. 411–422. Springer, Heidelberg (2007). https://doi.org/10.1007/978-3-540-73420-8_37
5. Bellare, M., Rogaway, P.: The exact security of digital signatures-how to sign with RSA and Rabin. In: International Conference on the Theory and Applications of Cryptographic Techniques, pp. 399–416. Springer (1996)
6. Benabbas, S., Gennaro, R., Vahlis, Y.: Verifiable delegation of computation over large datasets. In: Rogaway, P. (ed.) CRYPTO 2011. LNCS, vol. 6841, pp. 111–131. Springer, Heidelberg (2011). https://doi.org/10.1007/978-3-642-22792-9_7
7. Biswas, S., Yerukhimovich, A.: Multi-key fully-homomorphic aggregate MAC for arithmetic circuits. Cryptology ePrint Archive, Paper 2024/1499 (2024). https://eprint.iacr.org/2024/1499
8. Bitansky, N., Canetti, R., Chiesa, A., Tromer, E.: From extractable collision resistance to succinct non-interactive arguments of knowledge, and back again. In: Goldwasser, S. (ed.) ITCS 2012: 3rd Innovations in Theoretical Computer Science, pp. 326–349. Association for Computing Machinery, Cambridge (2012). https://doi.org/10.1145/2090236.2090263

9. Boneh, D., Freeman, D.M.: Homomorphic signatures for polynomial functions. In: Paterson, K.G. (ed.) EUROCRYPT 2011. LNCS, vol. 6632, pp. 149–168. Springer, Heidelberg (2011). https://doi.org/10.1007/978-3-642-20465-4_10

10. Boneh, D., Gentry, C., Lynn, B., Shacham, H.: Aggregate and verifiably encrypted signatures from bilinear maps. In: Biham, E. (ed.) EUROCRYPT 2003. LNCS, vol. 2656, pp. 416–432. Springer, Heidelberg (2003). https://doi.org/10.1007/3-540-39200-9_26

11. Boneh, D., Lynn, B., Shacham, H.: Short signatures from the Weil pairing. In: International Conference on the Theory and Application of Cryptology and Information Security, pp. 514–532. Springer (2001)

12. Catalano, D., Fiore, D.: Practical homomorphic MACs for arithmetic circuits. In: Johansson, T., Nguyen, P.Q. (eds.) EUROCRYPT 2013. LNCS, vol. 7881, pp. 336–352. Springer, Heidelberg (2013). https://doi.org/10.1007/978-3-642-38348-9_21

13. Chung, K.-M., Kalai, Y., Vadhan, S.: Improved delegation of computation using fully homomorphic encryption. In: Rabin, T. (ed.) CRYPTO 2010. LNCS, vol. 6223, pp. 483–501. Springer, Heidelberg (2010). https://doi.org/10.1007/978-3-642-14623-7_26

14. DeMillo, R.A., Lipton, R.J.: A probabilistic remark on algebraic program testing. Inf. Process. Lett. **7**(4), 193–195 (1978)

15. Desmedt, Y.: Computer security by redefining what a computer is. In: Michael, J.B., Ashby, V., Meadows, C. (eds.) Proceedings on the 1992–1993 Workshop on New Security Paradigms, 22–24 September 1992; and 3–5 August 1993, Little Compton, pp. 160–166. ACM (1993). https://doi.org/10.1145/283751.283834

16. Dutta, P., Dwivedi, P., Saxena, N.: Demystifying the border of depth-3 algebraic circuits. In: 2021 IEEE 62nd Annual Symposium on Foundations of Computer Science (FOCS), pp. 92–103. IEEE (2022)

17. Fiore, D., Gennaro, R.: Publicly verifiable delegation of large polynomials and matrix computations, with applications. In: Proceedings of the 2012 ACM Conference on Computer and Communications Security, pp. 501–512 (2012)

18. Fiore, D., Mitrokotsa, A., Nizzardo, L., Pagnin, E.: Multi-key homomorphic authenticators. In: Cheon, J.H., Takagi, T. (eds.) ASIACRYPT 2016. LNCS, vol. 10032, pp. 499–530. Springer, Heidelberg (2016). https://doi.org/10.1007/978-3-662-53890-6_17

19. Gennaro, R., Gentry, C., Parno, B.: Non-interactive verifiable computing: outsourcing computation to untrusted workers. In: Rabin, T. (ed.) CRYPTO 2010. LNCS, vol. 6223, pp. 465–482. Springer, Heidelberg (2010). https://doi.org/10.1007/978-3-642-14623-7_25

20. Gennaro, R., Gentry, C., Parno, B., Raykova, M.: Quadratic span programs and succinct NIZKs without PCPs. In: Johansson, T., Nguyen, P.Q. (eds.) EUROCRYPT 2013. LNCS, vol. 7881, pp. 626–645. Springer, Heidelberg (2013). https://doi.org/10.1007/978-3-642-38348-9_37

21. Gennaro, R., Wichs, D.: Fully homomorphic message authenticators. In: International Conference on the Theory and Application of Cryptology and Information Security, pp. 301–320. Springer (2013)

22. Gentry, C., Wichs, D.: Separating succinct non-interactive arguments from all falsifiable assumptions. In: Proceedings of the Forty-Third Annual ACM Symposium on Theory of Computing, pp. 99–108 (2011)

23. Goldwasser, S., Kalai, Y.T., Rothblum, G.N.: Delegating computation: interactive proofs for muggles. In: Ladner, R.E., Dwork, C. (eds.) 40th Annual ACM Symposium on Theory of Computing, pp. 113–122. ACM Press, Victoria (2008). https://doi.org/10.1145/1374376.1374396

24. Gorbunov, S., Vaikuntanathan, V., Wichs, D.: Leveled fully homomorphic signatures from standard lattices. In: Servedio, R.A., Rubinfeld, R. (eds.) 47th Annual ACM Symposium on Theory of Computing, pp. 469–477. ACM Press, Portland (2015). https://doi.org/10.1145/2746539.2746576
25. Gorbunov, S., Wee, H.: Digital signatures for consensus. Cryptology ePrint Archive (2019)
26. Johnson, R., Molnar, D., Song, D., Wagner, D.: Homomorphic signature schemes. In: Preneel, B. (ed.) CT-RSA 2002. LNCS, vol. 2271, pp. 244–262. Springer, Heidelberg (2002). https://doi.org/10.1007/3-540-45760-7_17
27. Katz, J., Lindell, A.Y.: Aggregate message authentication codes. In: Cryptographers' Track at the RSA Conference, pp. 155–169. Springer (2008)
28. Katz, J., Lindell, Y.: Introduction to Modern Cryptography: Principles and Protocols. Chapman and Hall/CRC (2007)
29. Kilian, J.: A note on efficient zero-knowledge proofs and arguments (extended abstract). In: 24th Annual ACM Symposium on Theory of Computing, pp. 723–732. ACM Press, Victoria (1992). https://doi.org/10.1145/129712.129782
30. Lai, R.W., Tai, R.K., Wong, H.W., Chow, S.S.: Multi-key homomorphic signatures unforgeable under insider corruption. In: International Conference on the Theory and Application of Cryptology and Information Security, pp. 465–492. Springer (2018)
31. Micali, S.: CS proofs (extended abstracts). In: 35th Annual Symposium on Foundations of Computer Science, pp. 436–453. IEEE Computer Society Press, Santa Fe (1994). https://doi.org/10.1109/SFCS.1994.365746
32. Parno, B., Howell, J., Gentry, C., Raykova, M.: Pinocchio: nearly practical verifiable computation. Commun. ACM **59**(2), 103–112 (2016)
33. Parno, B., Raykova, M., Vaikuntanathan, V.: How to delegate and verify in public: verifiable computation from attribute-based encryption. In: Cramer, R. (ed.) TCC 2012. LNCS, vol. 7194, pp. 422–439. Springer, Heidelberg (2012). https://doi.org/10.1007/978-3-642-28914-9_24
34. Rivest, R.: Two new signature schemes, 2001. In: Cambridge Seminar (2001)
35. Rückert, M., Schröder, D.: Aggregate and verifiably encrypted signatures from multilinear maps without random oracles. In: International Conference on Information Security and Assurance, pp. 750–759. Springer (2009)
36. Waters, B., Wu, D.J.: Batch arguments for np and more from standard bilinear group assumptions. In: Annual International Cryptology Conference, pp. 433–463. Springer (2022)
37. Zippel, R.: Probabilistic algorithms for sparse polynomials. In: International Symposium on Symbolic and Algebraic Manipulation, pp. 216–226. Springer (1979)

Leakage-Resilient Key-Dependent Message Secure Encryption Schemes

Dhairya Gupta, Mahesh Sreekumar Rajasree[✉], and Harihar Swaminathan

Indian Institute of Technology, Delhi, India
srmahesh1994@gmail.com

Abstract. We introduce a new security notion called *leakage-resilient key-dependent message (LR-KDM)* security, which combines both leakage-resilience (LR) and key-dependent message (KDM) security in a unified framework. Leakage-resilient cryptography protects against side-channel attacks that leak partial information about secret keys, while KDM security ensures encryption remains secure even when messages are functions of the secret key. Although both notions have been studied extensively, their combination has not been explored. In this paper,

1. We demonstrate that there exist encryption schemes that are LR secure and KDM secure but do not achieve LR-KDM security.
2. We propose a construction for LR-KDM encryption based on a novel primitive called *leakage-resilient homomorphic* hash proof system, which could be of independent interest.
3. We prove that the encryption scheme proposed by Brakerski *et al.* (EUROCRYPT 2018) satisfies LR-KDM security.
4. We show that the KDM amplification techniques proposed by Waters and Wichs (CRYPTO 2023) and Applebaum (EUROCRYPT 2011) can be adapted in the LR-KDM setting.

Keywords: leakage · key-dependent message · encryption

1 Introduction

Cryptographic security has traditionally relied on the assumption that secret keys remain hidden from adversaries and that the messages being encrypted are independent of the keys. However, with the advancement of side-channel attacks, these assumptions are often unrealistic in practice. Side-channel attacks exploit physical leakage, such as power consumption [33], electromagnetic radiation [1], or timing information [34], to extract information about the secret keys. This has motivated the development of leakage-resilient (LR) cryptography, where the goal is to ensure security even when some information about the secret key is leaked.

Canetti *et al.* [16] and Dodis, Sahai and Smith [20] were among the first to study leakage-resilient encryption scheme. However, their results were limited to leakage that consisted of subsets of bits of the stored secret, rather than more

S. Mukhopadhyay and P. Stănică (Eds.): INDOCRYPT 2024, LNCS 15495, pp. 234–257, 2025.
https://doi.org/10.1007/978-3-031-80308-6_11

general functions of it. The works of Dziembowski [21] and Di Crescenzo, Lipton and Walfish [17] focused on arbitrary leakage functions where adversary can only obtain a fixed number of bits. Akavia, Goldwasser and Vaikuntanathan [2] considered arbitrary leakage in which the amount of leakage is not an absolute value but rather is expressed as a function of the secret-key size. This was followed by several works [8,14,19,22–24,27,28,38,39] that prioritised the leakage-rate, i.e., the ratio of the bits that is allowed to be leaked to the size of the secret key. Brakerski et al. [12] and Dodis et al. [18] expanded this model by considering continual leakage models, which allow adversaries to repeatedly learn information about the secret key in a bounded manner over time. This opened the door for more sophisticated constructions, such as the continual-leakage resilient encryption schemes proposed by Brakerski et al. [12] and Dodis et al. [18], and the works by Lewko, Lewko and Waters [36], which allowed for encryption in the face of continuous side-channel attacks. Recent efforts in the quantum setting include [3,15].

While leakage-resilient cryptography addresses the issue of key leakage, another major challenge arises when adversaries can influence the messages being encrypted, particularly when these messages are dependent on the secret key itself. This scenario is captured by the notion of key-dependent message (KDM) security. KDM security ensures that even if the adversary can encrypt messages that are functions of the secret key (for example, encrypting the key itself), they gain no useful information about the key or the underlying plaintexts.

The first formalization of KDM security was introduced by Black, Rogaway, and Shrimpton [9] in the context of public-key encryption, where the challenge was to provide security against adversaries who can choose messages dependent on the secret key. The difficulty of this problem stems from the fact that most encryption schemes break down when such correlations exist between the key and the message. Boneh, Halevi, Hamburg, and Ostrovsky [10] developed the first KDM-secure public-key encryption technique using the decisional Diffie-Hellman (DDH) assumption for affine functions. Their study revealed that KDM security may be achieved using ordinary cryptographic assumptions.

Following this, several constructions were introduced to achieve KDM security for various classes of functions. For instance, Applebaum et al. [5] explored KDM security based on the Learning with Errors (LWE) assumption, while the work of Barak, Haitner, Hofheinz and Ishai [7] achieved KDM security for bounded sized circuit under DDH assumption. There are several works in this area [4,6,29–32,35,37,40,41]. The importance of KDM security cannot be overstated, especially in scenarios where circular encryption or key-dependent operations are unavoidable, such as in encrypted file systems and certain network protocols.

To the best of our knowledge, no earlier study has systematically combined LR and KDM security into a single cryptographic scheme, despite various designs that achieve both independently [10,11,13,26]. In this work, we introduce leakage-resilient key-dependent message security and begin the investigation into this security model.

1.1 Our Results

We outline the various results in this paper.

- **Separation Results:** We show that there exists a PKE scheme which is leakage-resilient and secure against KDM attacks separately, but is not LR-KDM secure.
- **LR-KDM Secure Schemes from Standard Assumptions:** We formally define leakage-resilient version of *homomorphic hash proof system* and use it to construct a 1-ciphertext LR-KDM scheme. We also show that the construction given by Brakerski *et al.* [13] that is leakage-resilient and key-dependent message secure is also LR-KDM secure for affine functions with leakage-rate 1.
- **LR-KDM Amplification:** We show that the amplification given by Waters and Wichs [40] and Applebaum [4] can also be adapted in the LR-KDM setting to give a construction for LR-KDM secure public key encryption scheme for the class of circuits assuming the existence of LR-KDM SKE for the class for projection functions. Therefore, amplifying the LR-KDM schemes from batch encryption gives LR-KDM secure for bounded-sized circuit functions with leakage-rate 1 from the hardness of CDH, LWE, LPN.

1.2 Related Works

We highlight a few prominent works that have given constructions that are separately leakage-resilient and key-dependent message secure. Naor and Segev [38] observed that the circular-KDM secure scheme of Boneh *et al.* [10] is already leakage-resilient under the DDH assumption. Brakerski and Goldwasser [11] demonstrated how to construct schemes with $1 - o(1)$ leakage-rate and affine-KDM security based on the Quadratic Residuosity (QR) and Decisional Composite Residuosity (DCR). Hajiabadi, Kapron and Srinivasan [26] developed $1 - o(1)$ leakage-rate and projective-KDM secure schemes using homomorphic hash proof systems. Brakerski *et al.* [13] used a novel tool called batch encryption to design $1 - o(1)$ leakage-rate and affine-KDM secure schemes based on DDH, Learning with Parity (LPN), and other standard assumptions.

1.3 Open Problem

As this is the first work to address leakage-resilience and key-dependent message (KDM) security simultaneously, it opens up several directions for further research. We outline a few key open problems:

1. **Multi-Key LR-KDM Security:** In this paper, we focus on the *single-key* version of LR-KDM security, where the security game involves a single pair of public-secret keys. A natural extension is the *multi-key* version, where the adversary interacts with multiple pairs of public-secret keys $(\mathsf{pk}_i, \mathsf{sk}_i)_{i \in [k]}$. In the challenge phase, the adversary could select functions from the class

$f : \mathcal{SK}^k \rightarrow \mathcal{M}$ and receive either an encryption of $\mathbf{0}$ or $f(\{\mathsf{sk}_i\}_i)$. In the leakage phase, two variants are possible: the adversary could send either a single leakage function $h : \mathcal{SK}^k \rightarrow \{0,1\}^\ell$ and receive $h(\{\mathsf{sk}_i\}_i)$, or k separate leakage functions $h_i : \mathcal{SK} \rightarrow \{0,1\}^\ell$ and receive $\{h_j(\mathsf{sk}_i)_i\}_j$. These variations define different levels of security in the multi-key LR-KDM setting, which remain unexplored.

2. **LR-KDM Security under Chosen-Ciphertext Attacks (CCA):** Another interesting direction is to investigate LR-KDM security in the presence of chosen-ciphertext attacks (CCA). While there has been significant work on constructing CCA-LR secure and CCA-KDM secure encryption schemes, as well as transformations from CPA-secure to CCA-secure KDM schemes, the landscape of CCA-secure LR-KDM encryption remains largely unexplored. This offers rich opportunities for exploration and further development in this area.

2   Technical Overview

In this section, we present an overview of the results presented in this paper.

2.1   Separation Results

We begin by demonstrating that the notion of leakage-resilient key-dependent message (LR-KDM) security is inherently *non-trivial*. Specifically, an encryption scheme that is independently leakage-resilient (LR) and key-dependent message (KDM) secure does not immediately imply that it is LR-KDM secure. Whether or not such a separation between these security notions must exist is not immediately clear, and it requires careful analysis to establish.

To illustrate this separation, we construct a symmetric key encryption (SKE) scheme that is independently LR secure and KDM secure but fails to achieve LR-KDM security. Let SKE' represent an SKE scheme that is both LR secure and circular-KDM secure. Circular-KDM security means that the scheme remains secure even if the ciphertext encrypts the secret key itself. From this base scheme, we construct a new SKE scheme SKE by combining SKE' with a pseudorandom function (PRF).

In the construction of SKE, the secret key sk consists of two components: k, which is the key for the PRF, and $\mathsf{ske.sk}$, which is a randomly generated key for the encryption scheme SKE'. The encryption for a message m proceeds as follows: first, it checks if m is equal to $\mathsf{ske.sk}$. If so, it computes $c_0 = \mathsf{PRF.Eval}(k, 1)$; otherwise, it computes $c_0 = \mathsf{PRF.Eval}(k, 0)$. It then generates $c_1 \leftarrow \mathsf{SKE.Enc}(\mathsf{ske.sk}, m)$ and outputs the ciphertext $\mathsf{ct} = (c_0, c_1)$. Decryption is straightforward: the scheme returns $\mathsf{SKE.Dec}(\mathsf{ske.sk}, c_1)$, while the first component c_0 is ignored.

We first argue that SKE retains leakage resilience. Suppose there is an adversary $\mathcal{A}$ capable of breaking the LR security of SKE. We can then construct another adversary $\mathcal{B}$ that breaks the LR security of the underlying scheme SKE'. Adversary $\mathcal{B}$ generates the PRF key k independently and can simulate the LR security game for SKE'. For any leakage function $f(k, \mathsf{ske.sk})$ that $\mathcal{A}$ expects, $\mathcal{B}$ hardcodes the value of k into f and queries its own challenger to receive the appropriate leakage. When $\mathcal{A}$ sends messages m_0 and m_1, $\mathcal{B}$ forwards them to its challenger and receives the challenge ciphertext c_1^*. It then computes $c_0 = \mathsf{PRF.Eval}(k, \mathbf{0})$ and sends the ciphertext $\mathsf{ct}^* = (c_0, c_1^*)$ to $\mathcal{A}$. Since, with very high probability, neither m_0 nor m_1 equals $\mathsf{ske.sk}$, the ciphertext ct^* is correctly formed. Therefore, $\mathcal{B}$'s probability of winning is negligibly close to $\mathcal{A}$'s probability of success, ensuring that SKE is leakage-resilient.

Next, we argue that SKE is KDM secure with respect to the function $f(k, \mathsf{ske.sk}) = \mathsf{ske.sk}$, meaning that the scheme should remain secure when encrypting the secret key. The key observation here is that the second part of the KDM ciphertext, $c_1 \leftarrow \mathsf{SKE.Enc}(\mathsf{ske.sk}, f(k, \mathsf{ske.sk}))$, depends only on $\mathsf{ske.sk}$ and not on the PRF key k. By leveraging the security of the underlying PRF (see Sect. 3.1), we can replace the value of c_0 with a truly random value without the adversary detecting the change. It follows that any adversary $\mathcal{A}$ that breaks the KDM security of SKE using this modified ciphertext could be transformed into an adversary that breaks the KDM security of the base scheme SKE'.

Finally, we demonstrate that despite being both LR secure and KDM secure, the scheme SKE fails to achieve LR-KDM security. We parameterize the scheme so that the size of the PRF key k is smaller than the allowed leakage but still larger than the security parameter. In fact, using the GGM construction [25], we can construct PRFs where the size of the key is equivalent to the security parameter. The adversary can exploit this by leaking the entire PRF key k. Once the adversary receives the challenge ciphertext $\mathsf{ct}^* = (c_0, c_1)$, it checks whether $c_0 = \mathsf{PRF.Eval}(k, \mathbf{0})$. This allows the adversary to distinguish between an encryption of a zero message and an encryption of $f(\mathsf{ske.sk})$, effectively breaking the LR-KDM security. Further details on this separation are provided in Sect. 4.1.

2.2   LR-KDM Secure Schemes from Standard Assumptions

To construct a leakage-resilient key-dependent message (LR-KDM) secure public key encryption (PKE) scheme, we introduce the concept of a leakage-resilient homomorphic hash proof system (HPS). Wee [41] originally defined homomorphic hash proof systems and demonstrated their use in constructing KDM-secure public key encryption. By enhancing this primitive to be leakage-resilient, we aim to achieve 1-ciphertext LR-KDM security (i.e., the adversary obtains only a single challenge ciphertext) for public key encryption schemes.

An HPS is typically associated with a language $\mathcal{L}$ that satisfies the property of computational indistinguishability: it should be hard to distinguish whether a given string belongs to the language or not. In addition to this language, an HPS comprises three algorithms: setup, encapsulation, and decapsulation. The setup algorithm generates a public key pk and a secret key sk. The encapsulation algorithm takes as input the public key pk, an element $x \in \mathcal{L}$, and a corresponding witness w (demonstrating that $x \in \mathcal{L}$). It then outputs a string k. On the other hand, the decapsulation algorithm takes as input the secret key sk and a string x and produces the same string k. The correctness of the system ensures that if the encapsulation and decapsulation algorithms are run on the same input x, they will output the same value k.

Moreover, HPSs must satisfy the *smoothness* property, which guarantees that the decapsulation output is statistically close to a uniform distribution if the input string x does not belong to the language $\mathcal{L}$. Another crucial feature of homomorphic HPSs is that the underlying language supports a group structure, and the decapsulation algorithm exhibits homomorphic properties, meaning that it preserves group operations over the input.

Wee [41] provided a construction of a homomorphic HPS based on the d-Linear (d-Lin) assumption. Let G be a cyclic group of prime order with generator g. The language $\mathcal{L}$ consists of elements of the form $x = g^{r^\top P}$, where $P \leftarrow \mathbb{Z}_p^{d \times k}$ is a random matrix, and $r \leftarrow \mathbb{Z}_q^d$ is a vector that serves as the witness for x. The public key contains the term g^{Ps}, where $s \leftarrow \mathbb{Z}_q^k$, and the secret key is the vector s. On input a public key g^{Ps} and element along with its witness $((g^{r^\top P}), r)$, the encapsulation algorithm outputs $g^{r^\top Ps}$. The decapsulation algorithm outputs the same using the secret key s and element $g^{r^\top P}$. The homomorphic property of this scheme is evident from the group operations over G.

To extend this to leakage-resilient homomorphic hash proof systems, we require that the smoothness property holds even in the presence of partial leakage of the secret key sk. We can show that the above construction remains leakage-resilient by appealing to the Leakage-Hiding Lemma (LHL). This lemma states that for a randomly chosen matrix A, a random vector s, and some leakage function $f(s)$, the distribution of $(A, As, f(s))$ is statistically close to $(A, t, f(s))$, where t is uniformly random. This property ensures that the system's security degrades gracefully even in the presence of bounded leakage. For a more detailed discussion on leakage-resilient HPS and LR-KDM constructions from LR-HPS, refer to Sect. 5.

Following the definition of leakage-resilient homomorphic HPS, we turn our attention to encryption schemes. Specifically, we show that the encryption scheme proposed by Brakerski *et al.* [13], which leverages *batch encryption*, is both leakage-resilient and KDM secure and can be extended to satisfy LR-KDM security. The core idea of batch encryption is to encrypt multiple messages at once, which allows for efficient handling of key-dependent messages and leakage. Our proof that this scheme is LR-KDM secure closely follows the one provided by the original authors. We selected this particular scheme due to its simplicity, though other candidates for LR and KDM secure encryption could also have been explored. For more detailed information on the LR-KDM secure scheme from batch encryption, see Sect. 6.

2.3 LR-KDM Amplification

Waters and Wichs [40] introduced a novel construction of a key-dependent message (KDM) secure public key encryption (PKE) scheme for arbitrary circuits, using a combination of two core assumptions. Their approach begins with a public key encryption scheme that is semantically secure under chosen plaintext attacks (CPA-secure) and relies on the existence of a symmetric key encryption (SKE) scheme that is secure against circular-KDM attacks. Essentially, their construction leverages the circular-KDM security of the underlying symmetric encryption scheme to"amplify" the security from circular-KDM to circuit-KDM, thereby achieving KDM security for more complex classes of functions, specifically arbitrary circuits.

We build upon their work by introducing a similar amplification for LR-KDM security. In our approach, we replace the CPA-secure PKE scheme in the Waters-Wichs construction with a leakage-resilient PKE scheme. Simultaneously, we substitute the circular-KDM secure SKE assumption with a leakage-resilient circular-KDM secure SKE scheme. This adjustment allows us to achieve a leakage-resilient KDM secure PKE scheme that is secure for circuits, extending the same amplification technique introduced by Waters and Wichs to the leakage-resilience setting.

However, this amplification does not come without cost. One significant trade-off that arises in our construction is related to the amount of leakage that can be tolerated. Specifically, as we extend the security from circular-KDM to circuit-KDM, the amount of leakage that the system can handle becomes more constrained. For more details, see Sect. 7.

To address the disadvantage of leakage degradation, we apply Applebaum's amplification [4], which starts with a stronger assumption: projection-KDM secure PKE. As we show in Sect. 8, by starting with a projection-LR-KDM secure PKE scheme, we can amplify it to circuit-LR-KDM security without any loss in the leakage rate. This approach allows us to achieve strong security without the trade-off of reduced leakage tolerance.

3 Preliminaries

Throughout this paper, we will use λ to denote the security parameter and $\mathrm{negl}(\cdot)$ to denote a negligible function in the input. We will use the short-hand notation PPT for "probabilistic polynomial time". For any finite set X, $x \leftarrow X$ denotes the process of picking an element x from X uniformly at random. Similarly, for any distribution $\mathcal{D}$, $x \leftarrow \mathcal{D}$ denotes an element x drawn from the distribution $\mathcal{D}$. For any natural number $n \in \mathbb{N}$, $[n]$ denotes the set $\{1, 2, \ldots, n\}$. For any language $\mathcal{L} \subseteq \{0, 1\}^*$, we use the notation $\overline{\mathcal{L}}$ to denote the set of element not in the language $\mathcal{L}$.

3.1 Pseudorandom Functions

A family of pseudorandom functions PRF = (Setup, Eval) with key space $\{\mathcal{K}_\lambda\}_\lambda$, input space $\{\mathcal{X}_\lambda\}_\lambda$ and output space $\{\mathcal{Y}_\lambda\}_\lambda$ consists of the following algorithms.

- Setup(1^λ) : The key generation algorithm is a randomized algorithm that takes as input the security parameter 1^λ and outputs a key $k \in \mathcal{K}_\lambda$.
- Eval(k, x) : The evaluation algorithm is a deterministic algorithm that takes as input a key $k \in \mathcal{K}_\lambda$ and $x \in \mathcal{X}_\lambda$ and outputs $y \in \mathcal{Y}_\lambda$.

Definition 1. A PRF scheme PRF is secure if for all PPT adversary $\mathcal{A}$, there exists a negligible function $\text{negl}(\cdot)$ such that for all $\lambda \in \mathbb{N}$,

$$|\Pr[\mathcal{A}^{\text{Eval}(k,\cdot)}(1^\lambda) = 1 : k \leftarrow \text{Setup}(1^\lambda)] - \Pr[\mathcal{A}^{R(\cdot)}(1^\lambda) = 1 : R \leftarrow \mathcal{U}_\lambda]| \le \text{negl}(\lambda)$$

where $\mathcal{U}_\lambda$ is the set of all functions from $\mathcal{X}_\lambda$ to $\mathcal{Y}_\lambda$.

3.2 Public Key Encryption (PKE)

A public key encryption scheme PKE = (Setup, Enc, Dec) with message space $\mathcal{M}$ consists of the following PPT algorithms.

- Setup(1^λ) : The setup algorithm takes as input the security parameter λ and outputs a pair of public and secret key (pk, sk).
- Enc(pk, m) : The encryption algorithm is a randomized algorithm that takes as input a public key pk and message $m \in \mathcal{M}$ and outputs a ciphertext ct.
- Dec(sk, ct) : The decryption algorithm takes as input a secret key sk and a ciphertext ct and outputs either a message $m \in \mathcal{M}$ or $\perp$.

Correctness. For correctness, we require for all $\lambda \in \mathbb{N}$, (pk, sk) $\leftarrow$ Setup(1^λ) and $m \in \mathcal{M}$,
$$\Pr[\text{Dec}(\text{sk}, \text{Enc}(\text{pk}, m)) = m] = 1,$$

where the probability is over the random bits used in the encryption algorithm. *CPA Security.* A public key encryption scheme PKE is said to satisfy CPA security if no PPT adversary $\mathcal{A}$ can win the following security game with probability greater than $\frac{1}{2} + \text{negl}(\lambda)$.

- **Initialization Phase:** The challenger runs (pk, sk) $\leftarrow$ PKE.Setup(1^λ) and sends pk to $\mathcal{A}$. The challenger also samples a uniformly random bit b.
- **Challenge Phase:** The adversary sends a query (m_0, m_1). The challenger replies with ct $\leftarrow$ PKE.Enc(pk, m_b).
- **Response Phase:** $\mathcal{A}$ outputs a bit b' and wins if $b = b'$.

KDM Security. A public key encryption scheme PKE is said to be KDM secure over a class of functions $\mathcal{F} = \{\mathcal{F}_\lambda\}_\lambda$ if no PPT adversary $\mathcal{A}$ can win the following security game with probability greater than $\frac{1}{2} + \text{negl}(\lambda)$.

- **Initialization Phase:** The challenger runs (pk, sk) $\leftarrow$ PKE.Setup(1^λ) and sends pk to $\mathcal{A}$. The challenger also samples a uniformly random bit b.
- **Challenge Phase:** The adversary sends adaptive queries $f_1, , \cdots, f_Q$ with each $f_i \in \mathcal{F}_\lambda$. The challenger replies to the ith query with ct $\leftarrow$ PKE.Enc(pk, m_b) where $m_0 = \mathbf{0}$ and $m_1 = f_i(\text{sk})$.
- **Response Phase:** $\mathcal{A}$ outputs a bit b' and wins if $b = b'$.

We will work with the following class of functions for KDM security in this paper.

- **Circular:** A function is circular[1] if it is of the form $f_i(x_1, \ldots, x_n) = x_i$.
- **Projection:** A function is a projection function if each of its output bits depends on at most a single input bit.
- **Affine:** A function is an affine function if the function can be represented as $f(x) = Ax + b$ where A is a matrix and b is a vector.
- **Circuits of A-priori Bounded Size** s: A function in this class can be described by a circuit of size s.

LR Security. A public key encryption scheme PKE is said to be ℓ-LR secure if no PPT adversary $\mathcal{A}$ can win the following security game with probability greater than $\frac{1}{2} + \mathrm{negl}(\lambda)$.

- **Initialization Phase:** The challenger runs $(\mathsf{pk}, \mathsf{sk}) \leftarrow \mathsf{Setup}(1^\lambda)$ and sends pk to $\mathcal{A}$.
- **Leakage Phase:** $\mathcal{A}$ outputs a function $h : \mathcal{SK}_\lambda \rightarrow \{0,1\}^*$. The challenger sends the leakage $h(\mathsf{sk})$ to $\mathcal{A}$, given $|h(\mathsf{sk})| \leq \ell(\lambda)$.
- **Challenge Phase:** $\mathcal{A}$ sends (m_0, m_1) it to the challenger. The challenger randomly chooses $b \in \{0,1\}$ and computes a ciphertext $\mathsf{ct}^* = \mathsf{Enc}(\mathsf{pk}, m_b)$. It sends ct^* to $\mathcal{A}$.
- **Response Phase:** $\mathcal{A}$ receives ct^* and outputs b'. $\mathcal{A}$ wins the experiment if $b = b'$.

3.3 Garbling Scheme

A garbling scheme $\mathsf{GC} = (\mathsf{Garble}, \mathsf{Eval})$ for a class of circuits $\{\mathcal{C}_\lambda\}_\lambda$ consists of the following algorithms.

- $\mathsf{Garble}(1^\lambda, C)$: The garbling algorithm is a randomized algorithm that takes as input the security parameter 1^λ and a circuit $C \in \mathcal{C}_\lambda$ such that $C : \{0,1\}^n \rightarrow \{0,1\}^m$ and outputs a garbled circuit $\hat{C}$ and a set of labels $\{\mathsf{lab}_{i,b}\}_{i \in [n], b \in \{0,1\}}$.
- $\mathsf{Eval}(\hat{C}, \{\mathsf{lab}_i\}_{i \in [n]})$: The evaluation algorithm takes as input a garbled circuit $\hat{C}$ and a set of n labels $\{\mathsf{lab}_i\}_{i \in [n]}$ and outputs $y \in \{0,1\}^m$.

Correctness. For correctness of a garbling scheme GC for a class of circuits $\{\mathcal{C}_\lambda\}_\lambda$, we require that for all $\lambda \in \mathbb{N}, C \in \mathcal{C}_\lambda, x \in \{0,1\}^n$,

$$\mathsf{Eval}(\hat{C}, \{\mathsf{lab}_{i,x_i}\}_{i \in [n]}) = C(x)$$

where $(\hat{C}, \{\mathsf{lab}_{i,b}\}_{i \in [n], b \in \{0,1\}}) \leftarrow \mathsf{Garble}(1^\lambda, C)$.

Definition 2 (Simulation Security). A garbling scheme $\mathsf{GC} = (\mathsf{Garble}, \mathsf{Eval})$ for a class of circuits $\{\mathcal{C}_\lambda\}_\lambda$ is said to be secure if there exists a PPT algorithm Sim such that for all PPT adversaries $\mathcal{A} = (\mathcal{A}_1, \mathcal{A}_2)$, there exists a negligible function $\mathrm{negl}(\cdot)$ such that for all $\lambda \in \mathbb{N}$, the following holds.

---

[1] This definition is also used in [5].

$$\left| \Pr\left[\mathcal{A}_2(\hat{C}, \{\mathsf{lab}_{i,x_i}\}_{i\in[n]}, \mathsf{aux}) = 1 : \begin{matrix} (C, x, \mathsf{aux}) \leftarrow \mathcal{A}_1(1^\lambda), \\ (\hat{C}, \{\mathsf{lab}_{i,x_i}\}_{i\in[n]}) \leftarrow \mathsf{Sim}(1^\lambda, 1^n, 1^{|C|}, C(x)) \end{matrix} \right] \right.$$

$$\left. - \Pr\left[\mathcal{A}_2(\hat{C}, \{\mathsf{lab}_{i,x_i}\}_{i\in[n]}, \mathsf{aux}) = 1 : \begin{matrix} (C, x, \mathsf{aux}) \leftarrow \mathcal{A}_1(1^\lambda), \\ (\hat{C}, \{\mathsf{lab}_{i,b}\}_{i\in[n],b\in\{0,1\}}) \leftarrow \mathsf{Garble}(1^\lambda, C)) \end{matrix} \right] \right| \leq \frac{1}{2} + \mathrm{negl}(\lambda).$$

Using the above security definition, we have the following theorems.

Theorem 1. *Let* $\mathsf{GC} = (\mathsf{Garble}, \mathsf{Eval})$ *be a secure garbling scheme. Then for all circuits* C_1, C_2, *with the same sizes, input-output spaces and input* x *such that* $C_1(x) = C_2(x)$,

$$\left(\hat{C}_1, \{\mathsf{lab}_{i,x_i}\} \mid (\hat{C}_1, \{\mathsf{lab}_{i,b}\}) \leftarrow \mathsf{Garble}(1^\lambda, C_1) \right) \approx_c \left(\hat{C}_2, \{\mathsf{lab}_{i,x_i}\} \mid (\hat{C}_2, \{\mathsf{lab}_{i,b}\}) \leftarrow \mathsf{Garble}(1^\lambda, C_2) \right).$$

Theorem 2. *Let* $\mathsf{GC} = (\mathsf{Garble}, \mathsf{Eval})$ *be a secure garbling scheme. Then for all circuits* C_1, C_2, *with the same input-output spaces and size,*

$$\left(\hat{C}_1 \mid (\hat{C}_1, \{\mathsf{lab}_{i,b}\}) \leftarrow \mathsf{Garble}(1^\lambda, C_1) \right) \approx_c \left(\hat{C}_2 \mid (\hat{C}_2, \{\mathsf{lab}_{i,b}\}) \leftarrow \mathsf{Garble}(1^\lambda, C_2) \right).$$

3.4 Hash Proof System (HPS)

A hash proof system $\mathsf{HPS} = (\mathsf{Setup}, \mathsf{Encaps}, \mathsf{Decaps})$ associated with an efficiently sampleable language $\mathcal{L}$ consists of the following algorithms.

- $\mathsf{Setup}(1^\lambda)$: The setup algorithm takes as input the security parameter and outputs a pair of public and secret key $(\mathsf{pk}, \mathsf{sk})$.
- $\mathsf{Encaps}(\mathsf{pk}, (x, w))$: The encoding algorithm takes as input a public key pk and a pair of string and witness (x, w) from the language $\mathcal{L}$ and outputs a string $k \in \{0,1\}^\ell$.
- $\mathsf{Decaps}(\mathsf{sk}, x)$: The decoding algorithm takes as input a public key pk and string and outputs a string $k \in \{0,1\}^\ell$.

Correctness. For all $\lambda, (\mathsf{pk}, \mathsf{sk}) \leftarrow \mathsf{Setup}(1^\lambda)$ and $(x, w) \leftarrow \mathcal{L}$,

$$\mathsf{Encaps}(\mathsf{pk}, (x, w)) = \mathsf{Decaps}(\mathsf{sk}, \mathrm{x})$$

Language Indistinguishability. For large enought λ, the following distributions are computationally close – $\mathcal{L} \approx_c \mathcal{L} \cup \overline{\mathcal{L}}^2$.

Smoothness Property. For large enough λ, the following distributions are statistically close,

$$(\mathsf{pk}, x, \mathsf{Decaps}(\mathsf{sk}, x)) \approx_s (\mathsf{pk}, x, t)$$

where the distribution is over $(\mathsf{pk}, \mathsf{sk}) \leftarrow \mathsf{Setup}(1^\lambda)$, $x \leftarrow \mathcal{L} \cup \overline{\mathcal{L}}$ and $t \leftarrow \{0,1\}^\ell$.

Definition 3. A hash proof system is said to be homomorphic if the language $\mathcal{L} \cup \overline{\mathcal{L}}$ has a group structure with the operator $\odot$ and for all $\lambda \in \mathbb{N}, (\mathsf{pk}, \mathsf{sk}) \leftarrow \mathsf{Setup}(1^\lambda)$ and $x, y \in \mathcal{L} \cup \overline{\mathcal{L}}$, $\mathsf{Decaps}(\mathsf{sk}, x \odot y) = \mathsf{Decaps}(\mathsf{sk}, x) \oplus \mathsf{Decaps}(\mathsf{sk}, y)$.

---

[2] The above definition is from [41]. There are other works that define language indistinguishability as $\mathcal{L} \approx_c \overline{\mathcal{L}}$.

3.5 *d*-Linear Assumption (*d*-LIN)

Let $\mathbb{G}$ be a cyclic group or prime-order q and g be a generator of $\mathbb{G}$. We denote $\mathcal{G} = (\mathbb{G}, q, g)$ and $[x] = g^x$ where $x \in \mathbb{Z}_q$. The notation $[\cdot]$ can be naturally extended to vectors and matrices. We say that $\mathbb{G}$ satisfies d-Linear assumption if

$$(\mathcal{G}, [\mathbf{x}], [\mathbf{Ar}]) \approx_c (\mathcal{G}, [\mathbf{x}], [\mathbf{A'r'}])$$

where $\mathbf{x}, \mathbf{r'} \leftarrow \mathbb{Z}_q^{d+1}, \mathbf{r} \leftarrow \mathbb{Z}_q^d$, and $\mathbf{A} \in \mathbb{Z}_q^{(d+1) \times d}, \mathbf{A'} \in \mathbb{Z}_q^{(d+1) \times (d+1)}$ such that

$$\mathbf{A} = \begin{bmatrix} x_1 & 0 & 0 & \ldots & 0 \\ 0 & x_2 & 0 & \ldots & 0 \\ \vdots & & & & \\ 0 & 0 & 0 & \ldots & x_d \\ x_{d+1} & x_{d+1} & x_{d+1} & \ldots & x_{d+1} \end{bmatrix}, \mathbf{A'} = \begin{bmatrix} x_1 & 0 & 0 & \ldots & 0 & 0 \\ 0 & x_2 & 0 & \ldots & 0 & 0 \\ \vdots & & & & & \\ 0 & 0 & 0 & \ldots & x_d & 0 \\ 0 & 0 & 0 & \ldots & 0 & x_{d+1} \end{bmatrix}$$

3.6 Batch Encryption

A batch encryption scheme $\mathsf{BENC} = (\mathsf{Setup}, \mathsf{Gen}, \mathsf{Enc}, \mathsf{Dec})$ consists of the following algorithms

- $\mathsf{Setup}(1^\lambda)$: The setup algorithm takes as input the security parameter λ and outputs a common reference string crs.
- $\mathsf{Gen}(\mathsf{crs}, \mathsf{sk})$: The key generation algorithm takes as input a common reference string crs and a secret key sk. It outputs a public key pk.
- $\mathsf{Enc}(\mathsf{crs}, \mathsf{pk}, \mathbf{M})$: The encryption algorithm takes as input a common reference string crs, public key pk, matrix $\mathbf{M}$ and outputs a ciphertext ct.
- $\mathsf{Dec}(\mathsf{crs}, \mathsf{sk}, \mathsf{ct})$: The decryption algorithm takes as input a common reference string crs, secret key sk, ciphertext ct and outputs a message m.

Correctness. For all $\mathsf{crs} \leftarrow \mathsf{Setup}(1^\lambda), \mathsf{sk}, \mathsf{pk} \leftarrow \mathsf{Gen}(\mathsf{crs}, \mathsf{sk}), \mathbf{M}, \mathsf{ct} \leftarrow \mathsf{Enc}(\mathsf{crs}, \mathsf{pk}, \mathbf{M})$ and $m' = \mathsf{Dec}(\mathsf{crs}, \mathsf{sk}, \mathsf{ct})$, we must have $m'_i = \mathbf{M}_{i, \mathsf{sk}_i}$.

Security. A batch encryption scheme BENC is secure if no PPT adversary $\mathcal{A}$ can win the following security game with probability greater than $\frac{1}{2} + \mathsf{negl}(\lambda)$.

- **Initialization Phase:** The adversary takes 1^λ as input and sends sk to the challenger. The challenger runs $\mathsf{crs} \leftarrow \mathsf{Setup}(1^\lambda)$ and sends crs to $\mathcal{A}$. The challenger also samples a uniformly random bit b.
- **Challenge Phase:** The adversary sends a query $(\mathbf{M}^{(0)}, \mathbf{M}^{(1)})$ such that $\mathbf{M}^{(0)}_{i, \mathsf{sk}_i} = \mathbf{M}^{(1)}_{i, \mathsf{sk}_i}$ for all $i \in [n]$. The challenger computes $\mathsf{pk} = \mathsf{Gen}(\mathsf{crs}, \mathsf{sk})$ and replies with $\mathsf{ct} \leftarrow \mathsf{Enc}(\mathsf{crs}, \mathsf{pk}, \mathbf{M}^{(b)})$.
- **Response Phase:** $\mathcal{A}$ outputs a bit b' and wins if $b = b'$.

Theorem 3. *Assuming the hardness of $X \in \{\mathsf{CDH}, \mathsf{LWE}, \mathsf{LPN}\}$, there exists secure batch encryption schemes.*

4   Leakage-Resilient Key-Dependent Message Secure Encryption Scheme: Definition

In this section, we proceed to formally define leakage-resilient key-dependent secure encryption scheme. Let $\mathsf{PKE} = (\mathsf{Setup}, \mathsf{Enc}, \mathsf{Dec})$ be a public key encryption scheme with secret key space $\{\mathcal{SK}_\lambda\}_\lambda$ and message space $\{\mathcal{M}_\lambda\}_\lambda$. Consider the following experiment with an adversary $\mathcal{A}$ where $\mathcal{F} = \{\mathcal{F}_\lambda\}_\lambda$ such that $\mathcal{F}_\lambda$ is a class of functions $f : \mathcal{SK}_\lambda \to \mathcal{M}_\lambda$.

- **Initialization Phase:** The challenger runs $(\mathsf{pk}, \mathsf{sk}) \leftarrow \mathsf{Setup}(1^\lambda)$ and sends pk to $\mathcal{A}$. The challenger also samples a uniformly random bit b.
- **Leakage Phase:** $\mathcal{A}$ outputs a function $h : \mathcal{SK}_\lambda \to \{0,1\}^*$. The challenger sends the leakage $h(\mathsf{sk})$ to $\mathcal{A}$.
- **Challenge Phase:** The adversary sends adaptive queries $f_1, , \cdots, f_Q$ with each $f_i \in \mathcal{F}_\lambda$. The challenger replies to the ith query with $\mathsf{ct}_i \leftarrow \mathsf{PKE.Enc}(\mathsf{pk}, m_b)$ where $m_0 = \mathbf{0}$ and $m_1 = f_i(\mathsf{sk})$.
- **Response Phase:** $\mathcal{A}$ receives ct^* and outputs b'. $\mathcal{A}$ wins the experiment if $b = b'$.

Definition 4. A PKE scheme is said to be $(\ell, \mathcal{F})$-LR-KDM secure if for all PPT adversary $\mathcal{A}$, there exists a negligible function $\mathrm{negl}(\cdot)$ such that for all $\lambda \in \mathbb{N}$, $\Pr[\mathcal{A} \text{ wins in the above experiment}] \leq \frac{1}{2} + \mathrm{negl}(\lambda)$, provided $|h(\mathsf{sk})| \leq \ell(\lambda)$.

Definition 5. A PKE scheme is said to be 1-ciphertext $(\ell, \mathcal{F})$-LR-KDM secure if for all PPT adversary $\mathcal{A}$, there exists a negligible function $\mathrm{negl}(\cdot)$ such that for all $\lambda \in \mathbb{N}$, $\Pr[\mathcal{A} \text{ wins in the above experiment}] \leq \frac{1}{2} + \mathrm{negl}(\lambda)$, provided $|h(\mathsf{sk})| \leq \ell(\lambda)$ and $\mathcal{A}$ queries a single f in the challenge phase.

4.1   Non-triviality of Leakage-Resilient Key-Dependent Message Security

We will show that there exists a scheme that is ℓ-leakage-resilient and $\mathcal{F}$-KDM secure, but is not $(\ell, \mathcal{F})$-LR-KDM secure. Let $\mathsf{SKE}' = (\mathsf{SKE.Setup}, \mathsf{SKE.Enc}, \mathsf{SKE.Dec})$ be a SKE scheme that is ℓ-leakage-resilient and circular-KDM secure where $\ell(\lambda) \geq \lambda^3$. Let $\mathsf{PRF} = (\mathsf{PRF.Setup}, \mathsf{PRF.Eval})$ be a secure PRF that has key size, input size and output size equal to the security parameter λ. Consider a SKE scheme SKE with the following description.

- $\mathsf{Setup}(1^\lambda)$: The setup algorithm takes as input the security parameter λ. It generates $k \leftarrow \mathsf{PRF.Setup}(1^\lambda)$ and $\mathsf{ske.sk} \leftarrow \mathsf{SKE.Setup}(1^\lambda)$. It outputs $\mathsf{sk} := (k, \mathsf{ske.sk})$.
- $\mathsf{Enc}(\mathsf{sk}, m)$: The encryption algorithm takes as input a secret key $\mathsf{sk} = (k, \mathsf{ske.sk})$ and a message m. If $m \neq \mathsf{ske.sk}$, it sets $c_0 := \mathsf{PRF.Eval}(k, \mathbf{0})$. Otherwise, it sets $c_0 = \mathsf{PRF.Eval}(k, \mathbf{1})$. It computes $c_1 \leftarrow \mathsf{SKE.Enc}(\mathsf{ske.sk}, m)$ and outputs $\mathsf{ct} = (c_0, c_1)$.

---

[3] Clearly, the size of the secret key, $|\mathsf{ske.sk}|$ must also be greater than λ.

- Dec(sk, ct) : The decryption algorithm takes as input a secret key sk = $(k, \text{ske.sk})$ and a ciphertext ct = (c_0, c_1). It returns SKE.Dec(ske.sk, c_1).

The correctness of the above scheme is straight-forward. We will now show that SKE is ℓ-leakage-resilient and f-KDM secure separately, where f will be defined later.

Claim 1. Assuming that SKE' is ℓ-leakage-resilient, then SKE is ℓ-leakage-resilient secure.

Proof. Assume the contrary that there exists an adversary $\mathcal{A}$ that wins the leakage-resilient game for SKE scheme with non-negligible advantage. We will construct an adversary $\mathcal{B}$ that wins the leakage-resilient game for SKE'.

- **Initialization Phase:** The challenger runs ske.sk $\leftarrow$ SKE.Setup(1^λ). $\mathcal{B}$ generates $k \leftarrow$ PRF.Setup(1^λ).
- **Leakage Phase:** $\mathcal{A}$ sends a function $h(\cdot, \cdot)$ to $\mathcal{B}$. $\mathcal{B}$ sends the function $h'(\cdot) = h(k, \cdot)$ to the challenger. The challenger sends the leakage $h'(\text{ske.sk}) = h(k, \text{ske.sk})$ to $\mathcal{B}$ who relays it to $\mathcal{A}$.
- **Challenge Phase:** $\mathcal{A}$ sends m_0, m_1 to $\mathcal{B}$ who relays it to the challenger. The challenger randomly chooses $b \in \{0, 1\}$ and computes a ciphertext $c_1^* =$ SKE.Enc(ske.sk, m_0) if $b = 0$, else it computes $c_1^* =$ SKE.Enc(ske.sk, m_1). It sends c_1^* to $\mathcal{B}$. $\mathcal{B}$ generates $c_0 =$ PRF.Eval($k, \mathbf{0}$) and sends ct* := (c_0, c_1^*) to $\mathcal{A}$.
- **Response Phase:** $\mathcal{A}$ receives ct* and returns b'. $\mathcal{B}$ outputs b'.

Observe that $\mathcal{B}$ has exactly simulated the leakage-resilient game for $\mathcal{A}$ except when ske.sk is either equal to m_0 or m_1. But, this happens with negligible probability[4]. So,

$$\Pr[\mathcal{B} \text{ wins the experiment}] = \Pr[\mathcal{A} \text{ wins the experiment}] - \text{negl}(\lambda).$$

This is a contradiction because SKE' is ℓ-leakage-resilient. Hence, SKE is ℓ-leakage-resilient secure.   □

Claim 2. Assuming that PRF is a secure PRF and SKE' is circular-KDM secure, then SKE is f-KDM secure where $f(x, y) = y$.

Proof. It is crucial to note that the output of the function f on input sk is only ske.sk. In other words, SKE.Enc(ske.sk, $f(\text{sk})$) = SKE.Enc(ske.sk, ske.sk), and this fact will be used in our proof. Let us first consider a hybrid experiment for the f-KDM game (refer Sect. 3.2) for the above scheme where the challenger does not generate a PRF key. As a consequence, it generates a truly random value c_0 to compute the challenge ciphertext.

- **Initialization Phase:** The challenger runs ske.sk $\leftarrow$ SKE.Setup(1^λ).

---

[4] Otherwise, SKE' is not even CPA-secure.

– **Challenge Phase:** $\mathcal{A}$ sends the function f to the challenger[5]. The challenger randomly chooses $b \in \{0, 1\}$ and computes a ciphertext $c_1^* =$ SKE.Enc(ske.sk, $\mathbf{0}$) if $b = 0$, else it computes $c_1^* =$ SKE.Enc(ske.sk, ske.sk). It randomly generates c_0 and sends ct$^* := (c_0, c_1^*)$ to $\mathcal{A}$.

– **Response Phase:** $\mathcal{A}$ receives ct* and returns b'.

Observe that there is no PPT adversary $\mathcal{A}$ that can distinguish the original KDM game from this hybrid game because of the PRF security. This is because the PRF key k is only used to generate c_0 and isn't used in the generation of c_1. In other words, if there exists an adversary $\mathcal{A}$ that can distinguish these two games, then we can break the security of the underlying PRF scheme (see Sect. 3.1).

Now, assume the contrary that there exists an adversary $\mathcal{A}$ that wins in this hybrid game. We will construct an adversary $\mathcal{B}$ that wins the KDM game for SKE$'$.

– **Initialization Phase:** The challenger runs ske.sk $\leftarrow$ SKE.Setup(1^λ).
– **Challenge Phase:** $\mathcal{A}$ sends the function f to $\mathcal{B}$. $\mathcal{B}$ sends the function $f'(y) = y$ to the challenger. The challenger randomly chooses $b \in \{0, 1\}$ and computes a ciphertext $c_1^* =$ SKE.Enc(ske.sk, $\mathbf{0}$) if $b = 0$. Else, it computes $c_1^* =$ SKE.Enc(ske.sk, ske.sk). It sends c_1^* to $\mathcal{B}$. $\mathcal{B}$ randomly generates c_0 and sends ct$^* := (c_0, c_1^*)$ to $\mathcal{A}$.
– **Response Phase:** $\mathcal{A}$ receives ct* and returns b'. $\mathcal{B}$ outputs b'.

Observe that $\mathcal{B}$ has exactly simulated this hybrid game for $\mathcal{A}$. So,

$$\Pr[\mathcal{B} \text{ wins the experiment}] = \Pr[\mathcal{A} \text{ wins the experiment}]$$

This is a contradiction. Hence, SKE is f-KDM secure.   □

We will now show that SKE is not (ℓ, f)-LR-KDM secure.

Claim 3. SKE is not (ℓ, f)-LR-KDM secure.

Proof. We give the description of an adversary $\mathcal{A}$ that wins the LR-KDM security game for SKE.

– **Initialization Phase:** The challenger runs ske.sk $\leftarrow$ SKE.Setup(1^λ) and $k \leftarrow$ PRF.Setup(1^λ) and set sk $= (k, \text{ske.sk})$.
– **Leakage Phase:** $\mathcal{A}$ sends a function $h(\cdot, \cdot)$ such that $h(x, y) = x$. The challenger sends the leakage $h(k, \text{ske.sk}) = k$ to $\mathcal{A}$.
– **Challenge Phase:** $\mathcal{A}$ sends the function f to the challenger. The challenger randomly chooses $b \in \{0, 1\}$ and computes a ciphertext $c_1^* =$ SKE.Enc(ske.sk, $\mathbf{0}$) and $c_0^* =$ PRF.Eval($k, \mathbf{0}$) if $b = 0$, else it computes $c_1^* =$ SKE.Enc(ske.sk, ske.sk) and $c_0^* =$ PRF.Eval($k, \mathbf{1}$). It sends ct$^* := (c_0^*, c_1^*)$ to $\mathcal{A}$.

---

[5] Here, we consider the case where the security is against a single challenge ciphertext. It can be trivially extended to multi-challenge ciphertexts.

– **Response Phase:** $\mathcal{A}$ receives ct^*. It checks if and outputs 0 if $c_0^* = \mathsf{PRF.Eval}(k, \mathbf{0})$. Else, it outputs 1.

We first note that $h(\mathsf{sk}) = k$, therefore, $|h(\mathsf{sk})| = |k| = \lambda \leq \ell(\lambda)$. Therefore, $\mathcal{A}$ is a valid adversary that follows the required bounds in the experiment. Observe that $\mathcal{A}$ wins this game with high probability except when $\mathsf{PRF.Eval}(k, \mathbf{1}) = \mathsf{PRF.Eval}(k, \mathbf{0})$. But, this happens with negligible probability, otherwise a PPT adversary can distinguish PRF from a truly random function. This is because for a truly random function F whose outputs are of length λ, the probability of $F(0) = F(1)$ is negligible in λ. $\qquad\qquad\square$

5   LR-KDM from Hash Proof Systems

In this section, we present our LR-KDM construction from *leakage-resilient* HPS and show that by using LR-HPS in the construction given by Wee [41], we can achieve LR-KDM.

5.1   LR-KDM from Leakage-Resilient Homomorphic Hash Proof System

In this section, we start by formally defining a leakage-resilient homomorphic HPS.

Definition 6 (Leakage-Resilient Homomorphic Hash Proof System). A homomorphic hash proof system $\mathsf{HPS} = (\mathsf{Setup}, \mathsf{Encaps}, \mathsf{Decaps})$ is said to be ℓ-leakage-resilient if it satisfies the following leakage-resilient smoothness property - For large enough λ, any function f such that the size of the output of f is at most $\ell(\lambda)$,

$$(\mathsf{pk}, f(\mathsf{sk}), x, \mathsf{Decaps}(\mathsf{sk}, x)) \approx_s (\mathsf{pk}, f(\mathsf{sk}), x, t)$$

where the distribution is over $(\mathsf{pk}, \mathsf{sk}) \leftarrow \mathsf{Setup}(1^\lambda)$, $x \leftarrow \overline{\mathcal{L}}$ and t is chosen from uniform distribution.

Construction of Homomorphic LR-Hash Proof System We will present a construction for homomorphic ℓ-LR-HPS from d-LIN assumptions

Let $\mathbb{G}$ be a group of prime order q. Let $k \geq (d+2)\log q + \ell$. Sample $\mathbf{P} \leftarrow_R \mathbb{Z}_q^{d \times k}$ and output

$$\mathsf{pp} := (\mathbb{G}, q, g, g^{\mathbf{P}})$$

We define

$$\mathcal{L} := \left\{ g^{\mathbf{r}^\top \mathbf{P}} : \mathbf{r} \in \mathbb{Z}_q^d \right\} \text{ and } \overline{\mathcal{L}} := \left\{ g^{\mathbf{a}^\top} : \mathbf{a} \in \mathbb{Z}_q^k \right\} \setminus \mathcal{L}$$

– $\mathsf{Setup}(1^\lambda)$: The setup algorithm takes as input the security parameter 1^λ. It computes $\mathsf{pp} = (G, q, g, g^{\mathbf{P}})$ and randomly generates $\mathbf{s} \leftarrow \{0, 1\}^k$. It sets $\mathsf{pk} := (\mathsf{pp}, g^{\mathbf{Ps}})$ and $\mathsf{sk} := \mathbf{s}$.

- Encaps $\left(\mathsf{pk}, (x = g^{\mathbf{r}^\top \mathbf{P}}, w = \mathbf{r})\right)$: The encapsulation algorithm takes as input a public key $\mathsf{pk} := (\mathsf{pp}, g^{\mathbf{Ps}})$ and $x \in \mathcal{L}$ along with its witness w. Using $\mathbf{r} = w$, and $g^{\mathbf{Ps}}$, it outputs $\mathsf{Encaps}(\mathsf{pk}, (x, w)) := g^{\mathbf{r}^\top \mathbf{Ps}}$.
- Decaps $(\mathsf{sk}, x = g^{\mathbf{a}^\top})$: The decapsulation algorithm takes as input a secret key $\mathsf{sk} = \mathbf{s}$ and $x = g^{\mathbf{a}^\top}$. It computes $\mathsf{Decaps}(\mathsf{sk}, x) := g^{\mathbf{a}^\top \mathbf{s}}$

Theorem 4. *The above scheme is a valid Leakage-Resilient Hash Proof System*

Proof (Proof Sketch). Correctness holds by construction. The language indistinguishability follows from the d-LIN assumption[6] and homomorphism is also trivial. For leakage-resilient smoothness, we want to show that

$$(\mathsf{pk}, f(\mathsf{sk}), x, \mathsf{Decaps}(\mathsf{sk}, x)) \approx_c (\mathsf{pk}, f(\mathsf{sk}), x, t)$$

where t is chosen uniformly at random. More precisely,

$$((\mathbb{G}, q, g, g^{\mathbf{P}}, g^{\mathbf{Ps}}), f(\mathbf{s}), g^{\mathbf{a}^\top}, g^{\mathbf{a}^\top \mathbf{s}}) \approx_s ((\mathbb{G}, q, g, g^{\mathbf{P}}, g^{\mathbf{Ps}}), f(\mathbf{s}), g^{\mathbf{a}^\top}, g^{\mathbf{t}})$$

This reduces to

$$(\mathbf{P}, \mathbf{a}^\top, \mathbf{Ps}, \mathbf{a}^\top \mathbf{s}, f(\mathbf{s})) \approx_s (\mathbf{P}, \mathbf{a}^\top, \mathbf{Ps}, \mathbf{t}, f(\mathbf{s}))$$

which follows from LHL theorem because $|f(\mathbf{s})| \leq \ell$.

LR-KDM Secure Encryption from LR Hash Proof System In this section, we give a construction for Leakage-Resilient Key-Dependent Secure Encryption from LR-HPS.

Construction Let $\mathsf{HPS} = (\mathsf{HPS.Setup}, \mathsf{HPS.Encaps}, \mathsf{HPS.Decaps})$ be a ℓ-leakage-resilient homomorphic hash proof system. Consider the following public key encryption scheme $\mathsf{PKE} = (\mathsf{Setup}, \mathsf{Enc}, \mathsf{Dec})$:

- $\mathsf{Setup}(1^\lambda)$: The setup algorithm takes as input the security parameter 1^λ and runs $(\mathsf{hps.pk}, \mathsf{hps.sk}) \leftarrow \mathsf{HPS.Setup}(1^\lambda)$. It outputs $\mathsf{pk} := \mathsf{hps.pk}$ and $\mathsf{sk} := \mathsf{hps.sk}$.
- $\mathsf{Enc}(\mathsf{pk}, m)$: The encryption algorithm takes as input a public key $\mathsf{pk} = \mathsf{hps.pk}$ and a message m. It samples a string along with its witness $(x, w) \in \mathcal{R}_{\mathcal{L}}$ and outputs $\mathsf{ct} = (x, \mathsf{HPS.Encaps}(\mathsf{pk}, (x, w)) \oplus m)$.
- $\mathsf{Dec}(\mathsf{sk}, \mathsf{ct})$: The decryption algorithm takes as input a secret key $\mathsf{sk} = \mathsf{hps.sk}$ and a ciphertext $\mathsf{ct} = (c_0, c_1)$. It computes $m = \mathsf{HPS.Decaps}(\mathsf{sk}, c_0) \oplus c_1$ and outputs m.

Correctness. A ciphertext of a message m in the above scheme is $(c_0, c_1) = (x, k \oplus m)$ where $k \leftarrow \mathsf{HPS.Encaps}(\mathsf{pk}, (x, w))$. From the correctness of HPS, we have $k = \mathsf{HPS.Decaps}(\mathsf{sk}, x)$. Therefore, we get $\mathsf{HPS.Decaps}(\mathsf{sk}, c_0) \oplus c_1 = k \oplus (k \oplus m) = m$.

---

[6] For more details, see [38, Appendix A].

Theorem 5. *Assuming that* HPS *is an ℓ-leakage resilient homomorphic hash proof system, the above encryption scheme is a 1-ciphertext $(\ell, \mathcal{F})$-LR-KDM where $\mathcal{F} = \{\mathcal{F}_\lambda\}$*

$$\mathcal{F}_\lambda = \{f_{e,k}(\mathsf{sk}) = \mathsf{Decaps}(\mathsf{sk}, e) \oplus k\}_{e,k}$$

We defer the proof to the full-version.

Since, the LR-HPS scheme based on d-Lin presented in the previous section has the property that $\mathsf{Decaps}(\mathbf{s}, g^{\mathbf{a}^\top}) = g^{\mathbf{a}^\top \mathbf{s}} = [\mathbf{a}^\top \mathbf{s}]$ which is an linear function in $\mathbf{s}$ with respect to $[\cdot]$, we have the following theorem.

Theorem 6. *Assuming the hardness of d-LIN, there exists 1-ciphertext LR-KDM secure encryption scheme for the class of affine functions.*

6   LR-KDM from Batch Encryption

In this section, we show that the construction presented by Brakerski *et al.* [13] is LR-KDM secure over the class of affine functions over the bits of the secret key. Similar to the assumption made in [13, Section 7.2], we will also assume that the public key of the batch encryption scheme is of size λ.

Construction. Let $\mathsf{BENC} = (\mathsf{Batch.Setup}, \mathsf{Batch.Gen}, \mathsf{Batch.Enc}, \mathsf{Batch.Dec})$ be a batch encryption scheme with public key size λ. Consider the following public key encryption scheme $\mathsf{PKE} = (\mathsf{Setup}, \mathsf{Enc}, \mathsf{Dec})$:

- $\mathsf{Setup}(1^\lambda)$: The setup algorithm takes as input the security parameter λ. It generates $\mathsf{crs} \leftarrow \mathsf{Batch.Setup}(1^{\lambda+n})$ and samples a uniformly random $\mathsf{sk} \leftarrow \{0,1\}^n$. It computes $\mathsf{pk} = \mathsf{Batch.Gen}(\mathsf{crs}, \mathsf{sk})$ and outputs $(\mathsf{crs}, \mathsf{pk}, \mathsf{sk})$.
- $\mathsf{Enc}(\mathsf{crs}, \mathsf{pk}, m)$: The encryption algorithm takes as input crs, the public key pk and a message m. It constructs an XOR secret sharing $\{v_i\}_{i \in [n]}$ for m so that $m = v_1 \oplus \cdots \oplus v_n$. It constructs a $n \times 2$ matrix $\mathbf{M}$ such that $\mathbf{M}_{i,b} = v_i$ for $i \in [n], b \in \{0,1\}$. Finally, it outputs $\mathsf{ct} = \mathsf{Batch.Enc}(\mathsf{crs}, \mathsf{pk}, \mathbf{M})$.
- $\mathsf{Dec}(\mathsf{crs}, \mathsf{sk}, \mathsf{ct})$: The decryption algorithm takes as input crs, the secret key sk and a ciphertext ct. It computes $\{v_i'\} = \mathsf{Batch.Dec}(\mathsf{crs}, \mathsf{sk}, \mathsf{ct})$. It outputs $v_1' \oplus \cdots \oplus v_n'$.

Correctness. From the correctness of $\mathsf{Batch.Enc}$, we get

$$\mathsf{Dec}(\mathsf{crs}, \mathsf{sk}, \mathsf{Enc}(\mathsf{crs}, \mathsf{pk}, m)) = \bigoplus_{i=1}^{n} \mathsf{Batch.Dec}(\mathsf{crs}, \mathsf{sk}, \mathsf{Batch.Enc}(\mathsf{crs}, \mathsf{pk}, \mathbf{M}))_i$$

$$= \bigoplus_{i=1}^{n} \mathbf{M}_{i,\mathsf{sk}_i}$$

$$= v_1 \oplus \cdots \oplus v_n = m,$$

where we had computed $\mathbf{M}_{i,b} = v_i$ with $v_1 \oplus \cdots \oplus v_n = m$.

Theorem 7. *Assuming* BENC *is a secure batch encryption, for any* $n = \text{poly}(\lambda)$, *the above encryption scheme is LR-KDM secure over the class of affine functions over the bits of the secret key*

$$\mathcal{F}_\lambda = \left\{ f_a \mid f_a(\text{sk}) = a_{n+1} \oplus \left(\bigoplus_{i=1}^{n} a_i \cdot \text{sk}_i \right) \right\}$$

and with leakage at most $n - 2\lambda^7$.

Proof. We will show this using a sequence of hybrid arguments.
G_0: This is the original LR-KDM game.

– **Initialization Phase:**
  1. The challenger randomly generates $\text{crs} \leftarrow \text{Batch.Setup}(1^{\lambda+n})$ and $\text{sk} \leftarrow \{0,1\}^n$.
  2. It sets $\text{pk} = \text{Batch.Gen}(\text{crs}, \text{sk})$ and sends crs, pk to the adversary $\mathcal{A}$
– **Leakage Phase:**
  1. $\mathcal{A}$ sends a function h to the challenger.
  2. The challenger responds with $h(\text{sk})$.
– **Challenge Phase:**
  1. The challenger randomly picks $\beta \leftarrow \{0,1\}$.
  2. The adversary makes Q queries as follows:
     • $\mathcal{A}$ sends a function $f_a \in \mathcal{F}_\lambda$ to the challenger.
     • Let $m_0 = \mathbf{0}$ and $m_1 = f_a(\text{sk})$.
     • Construct the matrix $\mathbf{M}$ such that $\mathbf{M}_{i,b} = v_i$ for all $i \in [n], b \in \{0,1\}$ where $\bigoplus_{i=1}^{n} v_i = m_\beta$.
     • It sends $\text{Batch.Enc}(\text{crs}, \text{pk}, \mathbf{M})$ to $\mathcal{A}$.
– **Response Phase:** $\mathcal{A}$ outputs $\beta' \in \{0,1\}$.

G_1: The challenger modifies the matrix being encrypted. Observe that the knowledge of sk is not required while generating the matrix $\mathbf{M}$.

– **Challenge Phase:**
  1. The challenger randomly picks $\beta \leftarrow \{0,1\}$.
  2. The adversary makes Q queries as follows:
     • $\mathcal{A}$ sends a function $f_a \in \mathcal{F}_\lambda$ to the challenger.
     • Let $m_0 = \mathbf{0}$ and $m_1 = f_a(\text{sk})$.
     • If $\beta = 0$, then construct the matrix $\mathbf{M}$ as in the previous game. If $\beta = 1$, construct the matrix $\mathbf{M}$ such that $\mathbf{M}_{i,b} = ba_i \oplus v_i$ for $i < n$ and $\mathbf{M}_{n,b} = a_{n+1} \oplus ba_n \oplus v_n$ where $v_1 \oplus \cdots \oplus v_n = 0$.
     • It sends $\text{Batch.Enc}(\text{crs}, \text{pk}, \mathbf{M})$ to $\mathcal{A}$.
– **Response Phase:** $\mathcal{A}$ outputs $\beta' \in \{0,1\}$.

G_2: The challenger modifies the matrix being encrypted.

---

[7] By setting $n = \omega(\lambda)$, we get $1 - o(1)$ leakage-rate.

- **Challenge Phase:**
  1. The challenger randomly picks $\beta \leftarrow \{0, 1\}$.
  2. The adversary makes Q queries as follows:
     - $\mathcal{A}$ sends a function $f_a \in \mathcal{F}_\lambda$ to the challenger.
     - Let $m_0 = \mathbf{0}$ and $m_1 = f_a(\mathsf{sk})$.
     - The matrix $\mathbf{M}$ is sampled uniformly at random. .
     - It sends $\mathsf{Batch.Enc}(\mathsf{crs}, \mathsf{pk}, \mathbf{M})$ to $\mathcal{A}$.
- **Response Phase:** $\mathcal{A}$ outputs $\beta' \in \{0, 1\}$.

Analysis: Let $p_{\mathcal{A}, i}$ denote the probability that $\mathcal{A}$ wins in Game G_i. We will show that the success probability in each game is close to the previous one.

Lemma 1. *From the semantic security of* BENC, *for all* PPT *adversaries* $\mathcal{A}$, *there exists a negligible function* negl *such that* $|p_{\mathcal{A}, 1} - p_{\mathcal{A}, 0}| \leq \mathrm{negl}(\lambda)$.

Proof. We show the result by constructing $Q - 1$ hybrids between the games where in each hybrid we successively modify the matrix $\mathbf{M}$ for the ith query to be constructed as per game 1. Observe that for each successive hybrid, the distributions of $\{\mathbf{M}_{i,\mathsf{sk}_i}\}_{i \in [n]}$ are identical. Thus by the security of BENC, the hybrids are computationally indistinguishable. Summing over the hybrids and using the triangle inequality gives us our desired result. □

Lemma 2. *By the leftover hash lemma, for all* PPT *adversaries* $\mathcal{A}$, *there exists a negligible function* negl *such that* $|p_{\mathcal{A}, 2} - p_{\mathcal{A}, 1}| \leq \mathrm{negl}(\lambda)$.

Proof. We again split the two games with $Q-1$ hybrids in between where in each hybrid we successively modify the matrix $\mathbf{M}$ for the ith query to be constructed as per game 2. Consider the hybrids i and $i + 1$. We construct a hybrid $\mathcal{H}'$ between these two where the challenger constructs the matrix $\mathbf{M}$ uniformly at random subject to the condition that $\mathbf{M}_{1,\mathsf{sk}_1} \oplus \cdots \oplus \mathbf{M}_{n,\mathsf{sk}_n} = m_\beta$. Then the ith hybrid and $\mathcal{H}'$ are computationally indistinguishable by the security of BENC.

In $\mathcal{H}'$, the matrix $\mathbf{M}$ is sampled uniformly at random subject to the condition $\mathbf{M}_{1,\mathsf{sk}_1} \oplus \cdots \oplus \mathbf{M}_{n,\mathsf{sk}_n} = m_\beta$. Equivalently, every element of $\mathbf{M}$ except $\mathbf{M}_{n,0}$ is sampled uniformly at random and we set

$$\mathbf{M}_{n,0} = m_\beta \oplus \left(\bigoplus_{i=1}^{n-1} \mathbf{M}_{i,0} \right) \oplus \left(\bigoplus_{i=1}^{n} \mathsf{sk}_i \cdot (\mathbf{M}_{i,0} \oplus \mathbf{M}_{i,1}) \right).$$

By the leftover hash lemma, the distribution $\mathsf{sk} \mid \mathsf{crs}, \mathsf{pk}, h(\mathsf{sk})$ has average min-entropy at least $n - \lambda - (n - 2\lambda) = \lambda$. This is because crs is independent of sk and $|\mathsf{pk}| = \lambda, |h(\mathsf{sk})| \leq n - 2\lambda$. Thus the distribution of $\bigoplus_{i=1}^{n} \mathsf{sk}_i \cdot (\mathbf{M}_{i,0} \oplus \mathbf{M}_{i,1})$ is statistically indistinguishable from a uniformly random value independent of $\mathsf{crs}, \mathsf{pk}$ and $(\mathbf{M}_{i,0} \oplus \mathbf{M}_{i,1})_i$. Since the distribution of $\mathbf{M}$ used in the two games are statistically indistinguishable, $\mathcal{H}'$ and the $(i + 1)$th hybrid are also statistically indistinguishable. Summing over the hybrids and using the triangle inequality gives us our desired result. □

Using the above lemmas and the fact that $p_{\mathcal{A},2} = \frac{1}{2}$, for all PPT adversaries $\mathcal{A}$, there exists a negligible function $\mathrm{negl}(\cdot)$ such that for all $\lambda \in \mathbb{N}$, $p_{\mathcal{A},0} \leq \frac{1}{2} + \mathrm{negl}(\lambda)$.

7   LR-KDM SKE to LR-KDM PKE

In this section, we give a construction for LR-KDM secure PKE scheme for circuits from LR-KDM secure SKE for projection functions and LR secure PKE.
Our Construction. Let PKE = (PKE.Setup, PKE.Enc, PKE.Dec) be an LR-PKE scheme and (Garble, Eval) be a garbled circuit scheme. Let SKE = (SKE.Setup, SKE.Enc, SKE.Dec) be an LR-KDM SKE for projection functions with secret key size $\beta = \beta(\lambda)$. Consider the following public key encryption scheme:

- Setup(1^λ): The key generation algorithm takes as input the security parameter λ. For each $i \in [\beta]$ and $b \in \{0,1\}$, it samples $(\mathsf{pk}_{i,b}, \mathsf{sk}_{i,b}) \leftarrow \mathsf{PKE.Setup}(1^\lambda)$. It then generates $k \leftarrow \{0,1\}^\beta$. Finally, it sets

$$\overline{\mathsf{pk}} = \left\{\mathsf{pk}_{i,b}\right\}_{i\in[\beta],b\in\{0,1\}}, \ \overline{\mathsf{sk}} = \left(k, \{\mathsf{sk}_{i,k_i}\}_{i\in[\beta]}\right)$$

- Enc($\overline{\mathsf{pk}}, m$): The encryption algorithm takes as input the public key $\overline{\mathsf{pk}}$ and a message m. Let $\left(\widehat{C}, \{\mathsf{lab}_{i,b}\}_{i\in[\beta],b\in\{0,1\}}\right) \leftarrow \mathsf{Garble}(1^\lambda, C_m)$ where C_m is the circuit that outputs m on any input. For each $i \in [\beta]$ and $b \in \{0,1\}$, it samples $\mathsf{ct}_{i,b} \leftarrow \mathsf{PKE.Enc}(\mathsf{pk}_{i,b}, \mathsf{lab}_{i,b})$ and returns $\overline{\mathsf{ct}} = \left(\widehat{C}, \{\mathsf{ct}_{i,b}\}_{i\in[\beta],b\in\{0,1\}}\right)$.

- Dec($\overline{\mathsf{sk}}, \overline{\mathsf{ct}}$): The decryption algorithm takes as input the secret key $\overline{\mathsf{sk}}$ and a ciphertext $\overline{\mathsf{ct}}$. For each $i \in [\beta]$, it computes $\mathsf{lab}_{i,k_i} = \mathsf{PKE.Dec}(\mathsf{sk}_{i,k_i}, \mathsf{ct}_{i,k_i})$. It returns $m \leftarrow \mathsf{Eval}(\widehat{C}, \{\mathsf{lab}_{i,k_i}\})$.

Correctness. Let $\overline{\mathsf{ct}}$ be an encryption of m using the keys $\overline{\mathsf{pk}} = \left\{\mathsf{pk}_{i,b}\right\}_{i\in[\beta],b\in\{0,1\}}$ and let $\overline{\mathsf{sk}} = \left(k, \{\mathsf{sk}_{i,k_i}\}_{i\in[\beta]}\right)$. The decryption algorithm gets the right labels $\{\mathsf{lab}_{i,k_i}\}_{i\in[\beta]}$ used during encryption by the correctness of the underlying PKE scheme. By the correctness of the garbling scheme, we have $\mathsf{Eval}(\widehat{C}, \{\mathsf{lab}_{i,k_i}\}_{i\in[\beta]}) = C_m(k) = m$, as desired.
Leakage Rate. The leakage rate of PKE is $\frac{\ell(\lambda)}{|\mathsf{pke.sk}|}$, whereas the above scheme's leakage rate is $\frac{\ell(\lambda)}{\beta(|\mathsf{pke.sk}|+1)}$. There is a degradation factor of β in the leakage ratio.

Theorem 8. *Let* PKE = (PKE.Setup, PKE.Enc, PKE.Dec) *be a ℓ-LR PKE scheme,* SKE = (SKE.Setup, SKE.Enc, SKE.Dec) *be an $(\ell, \mathcal{P})$-LR-KDM secure SKE where $\mathcal{P}$ is the class of all projection functions and* (Garble, Eval) *a secure garbling scheme. Then the above scheme is a secure $(\ell, \mathcal{F})$-LR-KDM incompressible public key encryption scheme where $\mathcal{F} = \{\mathcal{F}_\lambda\}$ and $\mathcal{F}_\lambda$ is the class of bounded polynomial sized circuits.*

We defer the proof to the full-version.

8   Functionality Amplification for LR-KDM

In this section, we give a construction for LR-KDM secure PKE scheme for circuits from LR-KDM secure PKE for projection functions using garbling schemes.

Our Construction. Let PKE = (PKE.Setup, PKE.Enc, PKE.Dec) be an public key encryption scheme and (Garble, Eval) be a garbled circuit scheme. Consider the following public key encryption scheme:

- Setup(1^λ): The key generation algorithm takes as input the security parameter λ. It samples (pk, sk) $\leftarrow$ PKE.Setup(1^λ) and returns (pk, sk).
- Enc(pk, m): The encryption algorithm takes as input a public key pk and a message m. Let $\left(\widehat{C}, \{\mathsf{lab}_i\}_{i\in[|m|]}\right) \leftarrow$ Sim($1^\lambda, |m|, |C_m|, m$) where C_m is the circuit that outputs m on any input. It computes pke.ct $\leftarrow$ PKE.Enc(pk, $\{\mathsf{lab}_i\}_i$) and outputs ct $:= (\widehat{C}, \mathsf{pke.ct})$.
- Dec(sk, ct): The decryption algorithm takes as input the secret key sk and a ciphertext ct = (Garble$C$, pke.ct). It decrypts ct using sk to obtain $\{\mathsf{lab}_i\}_i$. It returns $m = \mathsf{Eval}(\widehat{C}, \{\mathsf{lab}_i\})$.

Correctness. Let ct be an encryption of m using the keys pk and let sk be the associated secret key. The decryption algorithm gets the right labels $\{\mathsf{lab}_i\}_{i\in[|m|]}$ used during encryption by the correctness of the underlying PKE scheme. By the correctness and security of the garbling scheme, we have $\mathsf{Eval}(\widehat{C}, \{\mathsf{lab}_i\}_{i\in[\beta]}) = m$, as desired.

Theorem 9. *Let* PKE = (PKE.Setup, PKE.Enc, PKE.Dec) *be a ℓ-LR-KDM PKE scheme for projection functions, and* (Garble, Eval) *a secure garbling scheme. Then the above scheme is a secure $(\ell, \mathcal{C})$-LR-KDM PKE scheme where $\mathcal{C}$ is the class of all bounded polynomial-sized circuits.*

We defer the proof to the full-version. By combining Theorem 3, Theorem 7 and Theorem 9, we obtain the following theorem.

Theorem 10. *Assuming the hardness of $X \in \{\mathsf{CDH}, \mathsf{LWE}, \mathsf{LPN}\}$, there exists a leakage-rate-1 LR-KDM secure scheme for the class of all bounded polynomial-sized circuits.*

References

1. Agrawal, D., Archambeault, B., Rao, J.R., Rohatgi, P.: The EM side-channel (s). In: Cryptographic Hardware and Embedded Systems-CHES 2002: 4th International Workshop Redwood Shores, 13–15 August 2002, Revised Papers 4, pp. 29–45. Springer (2003)
2. Akavia, A., Goldwasser, S., Vaikuntanathan, V.: Simultaneous hardcore bits and cryptography against memory attacks. In: Theory of Cryptography Conference, pp. 474–495. Springer (2009)
3. Ananth, P., Kaleoglu, F., Yuen, H.: Simultaneous Haar indistinguishability with applications to unclonable cryptography. arXiv preprint arXiv:2405.10274 (2024)

4. Applebaum, B.: Key-dependent message security: generic amplification and completeness. J. Cryptol. **27**(3), 429–451 (2014)
5. Applebaum, B., Cash, D., Peikert, C., Sahai, A.: Fast cryptographic primitives and circular-secure encryption based on hard learning problems. In: Halevi, S. (ed.) CRYPTO 2009. LNCS, vol. 5677, pp. 595–618. Springer, Heidelberg (2009). https://doi.org/10.1007/978-3-642-03356-8_35
6. Applebaum, B., Harnik, D., Ishai, Y.: Semantic security under related-key attacks and applications. Cryptology ePrint Archive (2010)
7. Barak, B., Haitner, I., Hofheinz, D., Ishai, Y.: Bounded key-dependent message security. In: Gilbert, H. (ed.) EUROCRYPT 2010. LNCS, vol. 6110, pp. 423–444. Springer, Heidelberg (2010). https://doi.org/10.1007/978-3-642-13190-5_22
8. Bhushan, K., Goyal, R., Koppula, V., Narayanan, V., Prabhakaran, M., Rajasree, M.S.: Leakage-Resilient Incompressible Cryptography: Constructions and Barriers. Springer-Verlag (2024)
9. Black, J., Rogaway, P., Shrimpton, T.: Encryption-scheme security in the presence of key-dependent messages. In: Selected Areas in Cryptography: 9th Annual International Workshop, SAC 2002 St. John's, Newfoundland, 15–16 August 2002, Revised Papers 9, pp. 62–75. Springer (2003)
10. Boneh, D., Halevi, S., Hamburg, M., Ostrovsky, R.: Circular-secure encryption from decision Diffie-Hellman. In: Wagner, D. (ed.) CRYPTO 2008. LNCS, vol. 5157, pp. 108–125. Springer, Heidelberg (2008). https://doi.org/10.1007/978-3-540-85174-5_7
11. Brakerski, Z., Goldwasser, S.: Circular and leakage resilient public-key encryption under subgroup indistinguishability. In: Rabin, T. (ed.) CRYPTO 2010. LNCS, vol. 6223, pp. 1–20. Springer, Heidelberg (2010). https://doi.org/10.1007/978-3-642-14623-7_1
12. Brakerski, Z., Kalai, Y.T., Katz, J., Vaikuntanathan, V.: Overcoming the hole in the bucket: public-key cryptography resilient to continual memory leakage. In: 2010 IEEE 51st Annual Symposium on Foundations of Computer Science, pp. 501–510. IEEE (2010)
13. Brakerski, Z., Lombardi, A., Segev, G., Vaikuntanathan, V.: Anonymous IBE, leakage resilience and circular security from new assumptions. In: Annual International Conference on the Theory and Applications of Cryptographic Techniques, pp. 535–564. Springer (2018)
14. Braverman, M., Hassidim, A., Kalai, Y.T.: Leaky pseudo-entropy functions. In: ICS, vol. 2011, p. 2 (2011)
15. Cakan, A., Goyal, V., Liu-Zhang, C.D., Ribeiro, J.: Unbounded leakage-resilience and intrusion-detection in a quantum world. Cryptology ePrint Archive (2023)
16. Canetti, R., Dodis, Y., Halevi, S., Kushilevitz, E., Sahai, A.: Exposure-resilient functions and all-or-nothing transforms. In: Preneel, B. (ed.) EUROCRYPT 2000. LNCS, vol. 1807, pp. 453–469. Springer, Heidelberg (2000). https://doi.org/10.1007/3-540-45539-6_33
17. Di Crescenzo, G., Lipton, R., Walfish, S.: Perfectly secure password protocols in the bounded retrieval model. In: Theory of Cryptography Conference, pp. 225–244. Springer (2006)
18. Dodis, Y., Haralambiev, K., López-Alt, A., Wichs, D.: Cryptography against continuous memory attacks. In: 2010 IEEE 51st Annual Symposium on Foundations of Computer Science, pp. 511–520. IEEE (2010)

19. Dodis, Y., Haralambiev, K., López-Alt, A., Wichs, D.: Efficient public-key cryptography in the presence of key leakage. In: Abe, M. (ed.) ASIACRYPT 2010. LNCS, vol. 6477, pp. 613–631. Springer, Heidelberg (2010). https://doi.org/10.1007/978-3-642-17373-8_35

20. Dodis, Y., Sahai, A., Smith, A.: On perfect and adaptive security in exposure-resilient cryptography. In: Pfitzmann, B. (ed.) EUROCRYPT 2001. LNCS, vol. 2045, pp. 301–324. Springer, Heidelberg (2001). https://doi.org/10.1007/3-540-44987-6_19

21. Dziembowski, S.: Intrusion-resilience via the bounded-storage model. In: Theory of Cryptography Conference, pp. 207–224. Springer (2006)

22. Faonio, A., Nielsen, J.B., Venturi, D.: Mind your coins: fully leakage-resilient signatures with graceful degradation. In: International Colloquium on Automata, Languages, and Programming, pp. 456–468. Springer (2015)

23. Faust, S., Kiltz, E., Pietrzak, K., Rothblum, G.N.: Leakage-resilient signatures. In: Micciancio, D. (ed.) TCC 2010. LNCS, vol. 5978, pp. 343–360. Springer, Heidelberg (2010). https://doi.org/10.1007/978-3-642-11799-2_21

24. Galindo, D., Vivek, S.: A leakage-resilient pairing-based variant of the Schnorr signature scheme. In: Cryptography and Coding: 14th IMA International Conference, IMACC 2013, Oxford, 17–19 December 2013. Proceedings 14, pp. 173–192. Springer (2013)

25. Goldreich, O., Goldwasser, S., Micali, S.: How to construct random functions. J. ACM 33(4), 792–807 (1986)

26. Hajiabadi, M., Kapron, B.M., Srinivasan, V.: On generic constructions of circularly-secure, leakage-resilient public-key encryption schemes. In: Public-Key Cryptography–PKC 2016, pp. 129–158. Springer (2016)

27. Han, S., Liu, S., Gu, D.: Almost tight multi-user security under adaptive corruptions and leakages in the standard model. In: Annual International Conference on the Theory and Applications of Cryptographic Techniques, pp. 132–162. Springer (2023)

28. Hazay, C., López-Alt, A., Wee, H., Wichs, D.: Leakage-resilient cryptography from minimal assumptions. J. Cryptol. 29(3), 514–551 (2016)

29. Kitagawa, F., Matsuda, T.: CPA-to-CCA transformation for KDM security. In: Hofheinz, D., Rosen, A. (eds.) TCC 2019. LNCS, vol. 11892, pp. 118–148. Springer, Cham (2019). https://doi.org/10.1007/978-3-030-36033-7_5

30. Kitagawa, F., Matsuda, T.: Circular Security Is Complete for KDM Security. In: Moriai, S., Wang, H. (eds.) ASIACRYPT 2020. LNCS, vol. 12491, pp. 253–285. Springer, Cham (2020). https://doi.org/10.1007/978-3-030-64837-4_9

31. Kitagawa, F., Matsuda, T., Tanaka, K.: Simple and efficient KDM-CCA secure public key encryption. In: Galbraith, S.D., Moriai, S. (eds.) ASIACRYPT 2019. LNCS, vol. 11923, pp. 97–127. Springer, Cham (2019). https://doi.org/10.1007/978-3-030-34618-8_4

32. Kitagawa, F., Matsuda, T., Tanaka, K.: CCA security and trapdoor functions via key-dependent-message security. J. Cryptol. 35(2), 9 (2022)

33. Kocher, P., Jaffe, J., Jun, B.: Differential power analysis. In: Wiener, M. (ed.) CRYPTO 1999. LNCS, vol. 1666, pp. 388–397. Springer, Heidelberg (1999). https://doi.org/10.1007/3-540-48405-1_25

34. Kocher, P.C.: Timing attacks on implementations of Diffie-Hellman, RSA, DSS, and Other Systems. In: Koblitz, N. (ed.) CRYPTO 1996. LNCS, vol. 1109, pp. 104–113. Springer, Heidelberg (1996). https://doi.org/10.1007/3-540-68697-5_9

35. Koppula, V., Kumar, A., Rajasree, M.S., Swaminathan, H.: Incompressible encryption beyond CPA/CCA security. Manuscript submitted (2024)

36. Lewko, A., Lewko, M., Waters, B.: How to leak on key updates. In: Proceedings of the Forty-Third Annual ACM Symposium on Theory of Computing, pp. 725–734 (2011)
37. Marcedone, A., Pass, R., Shelat, A.: Bounded KDM security from IO and OWF. In: International Conference on Security and Cryptography for Networks, pp. 571–586. Springer (2016)
38. Naor, M., Segev, G.: Public-key cryptosystems resilient to key leakage. In: Halevi, S. (ed.) CRYPTO 2009. LNCS, vol. 5677, pp. 18–35. Springer, Heidelberg (2009). https://doi.org/10.1007/978-3-642-03356-8_2
39. Nishimaki, R., Yamakawa, T.: Leakage-resilient identity-based encryption in bounded retrieval model with nearly optimal leakage-ratio. In: IACR International Workshop on Public Key Cryptography, pp. 466–495. Springer (2019)
40. Waters, B., Wichs, D.: Universal amplification of KDM security: from 1-key circular to multi-key KDM. In: Annual International Cryptology Conference, pp. 674–693. Springer (2023)
41. Wee, H.: KDM-security via homomorphic smooth projective hashing. In: Cheng, C.-M., Chung, K.-M., Persiano, G., Yang, B.-Y. (eds.) PKC 2016. LNCS, vol. 9615, pp. 159–179. Springer, Heidelberg (2016). https://doi.org/10.1007/978-3-662-49387-8_7

An Efficient Toolkit for Computing Third-Party Private Set Intersection

Kai Chen[1,2] , Yongqiang Li[1,2(✉)] , and Mingsheng Wang[1,2]

[1] Key Laboratory of Cyberspace Security Defense, Institute of Information Engineering, CAS, Beijing, China
{chenkai1621,liyongqiang,wangmingsheng}@iie.ac.cn
[2] School of Cyber Security, University of Chinese Academy of Sciences, Beijing, China

Abstract. Third-party private set intersection (PSI) allows two parties to compute the intersection of their private input sets without revealing any more information than the result to an inputless third party. In this work, we leverage homomorphic encryption and oblivious pseudorandom function techniques for the first time to design third-party PSI protocols. We present two highly efficient third-party PSI protocols characterized by linear communication and computational complexity, along with a requirement of only 2 communication rounds. These protocols significantly lower the computational workload compared to prior work. Furthermore, we extend our investigation to third-party PSI cardinality protocols. Our constructions to achieve the cardinality functionality attain linear communication and computational complexity. Finally, we implement our protocols in C++ and perform a comprehensive evaluation, an aspect previously unexplored in third-party PSI research. The results demonstrate that our OPRF-based third-party PSI can obtain a 4.6–13.78 times faster improvement over the HE-based third-party PSI with a single thread in LAN setting. Moreover, the results indicate that our OPRF-based third-party PSI will yield even greater improvements as the set size increases, compared to HE-based third-party PSI.

Keywords: Third-party PSI · Homomorphic encryption · Oblivious pseudorandom function

1 Introduction

Private set intersection (PSI) enables mutually distrustful parties to compute the intersection on their sets while ensuring that no more additional information about their private inputs is disclosed beyond what can be inferred from the intersection result [28,35]. Over the past decade, significant advancements have been made in the development of PSI, leading to considerable improvements in their efficiency and applicability. These protocols have gained notable attention and have been widely deployed in both theoretical research and practical implementations [6,33,41,42,44–46].

© The Author(s), under exclusive license to Springer Nature Switzerland AG 2025
S. Mukhopadhyay and P. Stănică (Eds.): INDOCRYPT 2024, LNCS 15495, pp. 258–281, 2025.
https://doi.org/10.1007/978-3-031-80308-6_12

Existing PSI protocols can be broadly classified based on several approaches. PSI was initially proposed from Diffie-Hellman-based oblivious pseudorandom functions (OPRFs), primarily due to the low communication cost associated with this method [28,35]. In subsequent years, several protocols were developed to enhance performance [4,12,29]. The development and refinement of Oblivious transfer (OT) extension further led to the emergence of a class of protocols that offered reduced computational costs, albeit with a trade-off of increased communication overhead compared to the DH-OPRF approaches [33,38,41,46,48]. Recently, a range of PSI and PSI-variant protocols arose from Homomorphic encryption (HE) in conjunction with oblivious polynomial evaluation and batching techniques [8,21,25]. Given the computational intensity of the HE technique, they are particularly advantageous in specific scenarios, such as unbalanced PSI and labeled PSI [7,10,27]. Additionally, circuit-based PSI protocols have demonstrated the potential to privately compute functions over the set intersection, although they typically require many communication rounds [5,44,45,49].

Third-Party PSI. Most recently, Yeo and Ying introduced a novel variant of PSI, named third-party private set intersection [52], which enables the private computation of the intersection of sets held by two distinct parties P_1 and P_2, with the result revealed only to an inputless third party P_3. The key challenge in efficiently achieving third-party PSI lies in the observation that P_3 possesses no information that can be utilized to constrain the elements potentially present in the intersection. Furthermore, P_3 must remain oblivious to the sets owned by P_1 and P_2, except for the intersection itself.

1.1   Motivation

Similar to the well-researched PSI, which has numerous applications including botnet detection [37], private contact discovery [30], online advertising [42], privacy-preserving ride-sharing [23], and contact tracing [51], third-party PSI also holds significant practical utility. It is particularly relevant in scenarios where the intersection result is disclosed solely to a third party for privacy reasons.

Consider the following scenario, where the regulatory authority intends to obtain relevant information from two organizations. A third-party PSI protocol effectively prevents the exposure of sensitive information between the participating parties while enabling the regulatory authority to achieve its objective. For instance, in the context of a pandemic-related disease outbreak, it is crucial for public health authorities to rapidly identify potential asymptomatic carriers. In this scenario, the public health authority functions as the third-party entity, while the premises that maintain visitor records, including time stamps, act as the participating entities. This arrangement allows the health regulatory authority to access a database of individuals who were present at specific locations during particular timeframes, thereby fulfilling its public health mandate while preserving individual privacy.

Table 1. Comparisons of asymptotic communication costs (in bits) and computation costs (in operations) of third-party PSI protocols within the semi-honest setting.

Protocol	Comm.	Comp.	Rounds
$[52]_1$	$O(n)$	$O(n)$	4
$[52]_2$	$O(n^{1.5+o(1)})$	$O(n^{2.5+o(1)})$	3
$[53]_1$	$O(n)$	$O(n^{1.5+o(1)})$	3
$[53]_2$	$O(n^{1+\delta})$	$O(n^{1+\epsilon})$	3
Ours$_{HE}$	$O(n)$	$O(n)$	2
Ours$_{OPRF}$	$O(n)$	$O(n)$	2

In this context, n denotes the size of each input set, $\delta > 0$ denotes the security parameter associated with the key agreement protocol, and $0 < \epsilon < 1$ is any positive constant [53]. For third-party PSI protocols based on OPRF, the communication rounds involving κ base OTs, where κ is the computational security parameter, are disregarded in this comparison. Subscripts 1 and 2 are used to distinguish between the first and second protocols, respectively.

1.2   Related Work and Challenges

It is important to note that existing PSI protocols designed for contact tracing operate within a different framework, typically involving two roles, a sender and a receiver, where conventional PSI protocols can be more directly applied. For example, in the scenarios described in [14,16,50,51], the receiver holds a set of identifiers while the sender processes a database of contact tokens from infected users, typically managed by the public health authority. The receiver can then assess their exposure to infected individuals by performing a PSI protocol with the server. The works of [14,16,50,51] are specifically tailed for unbalanced PSI, where the receiver's set is considerably smaller than that of the server. However, in our proposed use case, the scenario involves two senders and an inputless receiver, with the objective of identifying potential individuals at risk in the event of an outbreak at specific premises.

Other variants of PSI have been explored include server-aided setting [31,32,34] and multi-party PSI [1,5,25]. In the server-aided setting, the receiver processes the input, whereas in third-party PSI, the receiver is inputless. Moreover, the challenge becomes more complex when the participants with inputs are prohibited from gaining any information throughout the process. In the case of multi-party PSI, a potential solution involves assigning the third party the entire universe of possible input elements. However, this approach is clearly not ideal both theoretically and practically.

Recently, Yeo and Ying introduced innovative third-party PSI protocols, the first utilizing a commutative cipher, and the second employing a key agreement protocol [52]. While both protocols are communication-efficient, requiring minimal communication and a small number of communication rounds, they impose significant computational costs in practice. To address this, Yeo and Ying subsequently proposed two enhancements to the second protocol in [52] to mitigate the computational burden [53]. The first improvement involves modifying the protocol through polynomial interpolation, where the inputs are not transmitted directly but are instead used to interpolate a polynomial. This modification results in a reduction of computational cost by approximately a factor of n. The second improvement employs a hash function to distribute the parties' inputs into multiple buckets before applying the protocol independently to each bucket. It should be noted, as Yeo and Ying stated, that the security of their protocols is based on the underlying key agreement (KA) protocol. If KA is computationally secure, then the protocols are secure against computational adversaries. If KA is quantum-safe, the protocols are secure against quantum adversaries.

Table 1 summarizes the asymptotic complexities of the highest-performing PSI protocols in the semi-honest setting. Regarding both communication and computational complexity, the first protocol presented in [52] and our protocol are among the most efficient. However, as highlighted in [52], it imposes a significant computational burden in practice due to its underlying cryptographic assumptions, rendering it less efficient than our approach. In terms of round complexity, our protocols demonstrate superior efficiency. Detailed performance results and comparisons are presented in Sect. 5.

1.3   Our Contributions

In this work, we investigate novel techniques for designing both third-party PSI and third-party PSI cardinality protocols. We propose efficient and secure protocols in Sects. 3 and 4, and implement them, providing detailed comparisons in Sect. 5. Specifically, our main contributions are summarized as follows.

1. **New Protocols Based on Homomorphic Encryption.** We introduce a novel and secure construction based on homomorphic encryption, referred to as $\prod_{\text{PSI}}^{\text{third,HE}}$, for realizing PSI. While this protocol demonstrates efficiency for balanced sets, it can be extended to accommodate unbalanced scenarios where the input size of P_1 is significantly larger than that of P_2. Furthermore, we modify the $\prod_{\text{PSI}}^{\text{third,HE}}$ to develop an efficient and secure third-party PSI cardinality protocol, termed $\prod_{\text{PSI-card}}^{\text{third,HE}}$. This protocol is applicable to both balanced and unbalanced settings.

2. **New Protocols Based on Oblivious Pseudorandom Functions.** For the first time, we present a scalable and secure construction based on oblivious pseudorandom functions (OPRF), termed $\prod_{\text{PSI}}^{\text{third,OPRF}}$, for realizing third-party PSI. It is noteworthy that the protocols designed by Yeo and Ying [52, 53] encounter significant computational burdens during execution, especially for larger sets. Additionally, we extend our investigation to third-party PSI cardinality protocol, introducing an efficient and secure protocol, designated as $\prod_{\text{PSI-card}}^{\text{third,OPRF}}$.

3. **Implementation and Evaluation.** For the first time, we present the implementation of third-party private set intersection and intersection-cardinality

protocols. Our protocols are implemented in C++ and undergo a comprehensive evaluation in Sect. 5. The results demonstrate that our OPRF-based $\prod_{\text{PSI}}^{\text{third,OPRF}}$ is 4.6–13.78 times faster than the HE-based $\prod_{\text{PSI}}^{\text{third,HE}}$, and our OPRF-based $\prod_{\text{PSI-card}}^{\text{third,OPRF}}$ is 4.389–13.589 times faster than HE-based $\prod_{\text{PSI-card}}^{\text{third,HE}}$. Furthermore, the results indicate that our OPRF-based protocols exhibit increasingly significant performance improvements as the set size grows, compared to HE-based protocols.

2   Preliminaries and Notation

Notation. Let λ denote the statistical security parameter, κ the computational security parameter, and $[m]$ represent the set $\{1, 2, \ldots, m\}$. The set $X = \{x_1, \ldots, x_n\}$ has a cardinality $|X| = n$. Section 2.2 provides a detailed explanation of our notation related to homomorphic encryption schemes. For a participant P_i with a key pair (pk_i, sk_i) in a public key encryption scheme, we denote the encryption and decryption operations as Enc_{pk_i} and Dec_{sk_i}, respectively.

2.1   Security Model

Our PSI and PSI-card protocol involves three parties. We first review the definitions of third-party PSI and third-party PSI-card protocols as presented in [52,53], and adhere to the static semi-honest security model outlined in this work.

Parameters:
Set size for P_1 and P_2 is n.
Functionality:

- Wait for input $\mathbf{X} = \{x_1, \cdots, x_n\}$ from P_1.
- Wait for input $\mathbf{Y} = \{y_1, \cdots, y_n\}$ from P_2.
- Give output $\mathbf{X} \cap \mathbf{Y}$ to P_3.

Fig. 1. Ideal Functionality $\mathcal{F}_{\text{PSI}}^{\text{third}}$ for PSI

PSI and PSI-Card Functionality. We review the definition of third-party PSI protocol as established in [52] and present the corresponding ideal functionality, denoted as $\mathcal{F}_{\text{PSI}}^{\text{third}}$, illustrated in Fig. 1. Additionally, we review the definition of third-party PSI cardinality protocol as defined in [53] and present the corresponding ideal functionality, denoted as $\mathcal{F}_{\text{PSI-card}}^{\text{third}}$, depicted in Fig. 2.

Definition 1 (Third-party PSI protocol). *In a third-party PSI protocol, 2 parties P_1 and P_2 each holds a set with elements in $\{0, 1\}^*$, while a third-party P_3 has no input. At the end of the protocol, P_3 outputs the set intersection while P_1 and P_2 output $\perp$.*

Definition 2 (Third-party PSI Cardinality protocol). *In a third-party PSI cardinality protocol, 2 parties P_1 and P_2 each holds a set with elements in $\{0,1\}^*$, while a third-party P_3 has no input. At the end of the protocol, P_3 outputs the cardinality of the set intersection while P_1 and P_2 output $\perp$.*

Parameters:
  Set size for P_1 and P_2 is n.
Functionality:

- Wait for input $\mathbf{X} = \{x_1, \cdots, x_n\}$ from P_1.
- Wait for input $\mathbf{Y} = \{y_1, \cdots, y_n\}$ from P_2.
- Give output $|\mathbf{X} \cap \mathbf{Y}|$ to P_3.

Fig. 2. Ideal Functionality $\mathcal{F}_{\text{PSI-card}}^{\text{third}}$ for PSI-card

Static Semi-honest Security. The protocol involves three parties, denoted as P_1, P_2 and P_3. Let $f_i(X, Y, \perp)$ represent the output for P_i in the ideal functionality $\mathcal{F}$ and $f(X, Y, \perp) = (f_1(X, Y, \perp), f_2(X, Y, \perp), f_3(X, Y, \perp))$ denote the joint output. During an execution of protocol $\prod$ on inputs $(X, Y, \perp)$, the view of P_i, denoted as $\text{view}_i^{\prod}(X, Y, \perp)$, comprises the input X, Y or $\perp$, the contents of P_i's internal random tape, and the messages received throughout the execution. Similarly, the output of P_i, denoted as $\text{output}_i^{\prod}(X, Y, \perp)$, is derived from P_i's view during the execution of $\prod$ on inputs $(X, Y, \perp)$. The joint output of all parties $\text{output}^{\prod}(X, Y, \perp)$ is:

$$(\text{output}_1^{\prod}(X, Y, \perp), \text{output}_2^{\prod}(X, Y, \perp), \text{output}_3^{\prod}(X, Y, \perp)).$$

Definition 3. *A protocol $\prod$ securely computes $\mathcal{F}$ against static semi-honest adversaries if there exist probabilistic polynomial time (PPT) algorithms Sim_1, Sim_2 and Sim_3 such that:*

$$(\text{Sim}_1(X, f_1(X, Y, \perp)), f(X, Y, \perp)) \stackrel{c}{\equiv} (\text{view}_1^{\prod}(X, Y, \perp), \text{output}^{\prod}(X, Y, \perp)),$$

$$(\text{Sim}_2(Y, f_2(X, Y, \perp)), f(X, Y, \perp)) \stackrel{c}{\equiv} (\text{view}_2^{\prod}(X, Y, \perp), \text{output}^{\prod}(X, Y, \perp)),$$

$$(\text{Sim}_3(\perp, f_3(X, Y, \perp)), f(X, Y, \perp)) \stackrel{c}{\equiv} (\text{view}_3^{\prod}(X, Y, \perp), \text{output}^{\prod}(X, Y, \perp)).$$

2.2 Building Blocks

We provide a brief overview of the key cryptographic tools utilized, including Bloom filters, Homomorphic encryption, Oblivious pseudorandom functions, Cuckoo hashing, and Simple hashing.

Bloom Filters. Bloom filters, first introduced by Bloom in [2], are lightweight data structures designed to efficiently represent data sets and perform membership testing using only hash function evaluations. Let X denote the data set. A Bloom filter is initially represented by a bit vector of length B, with all bits set to 0. The Bloom filter employs k hash functions, denoted as $h_i : \{0, 1\}^* \to \{1, \ldots, B\}$ for $i \in \{1, \ldots, k\}$. To insert an element $x \in X$ into the Bloom filter, the hash functions $h_1(x), \ldots, h_k(x)$ are evaluated, and bits at the corresponding indices are set from 0 to 1. If a bit is already set to 1, it remains unchanged. These hash functions can subsequently be used by any participant to verify whether an element is potentially present in the Bloom filter.

Definition 4 (Represented elements). *Let BF be the Bloom filter. We say that an element s is represented in the BF, if we have that*

$$\{BF[h_i(s)] = 1\}_{i \in \{1, \ldots, k\}},$$

where $\{h_1(x), \ldots, h_k(x)\}$ are the hash functions used in conjunction with BF. We say that the set S is represented by BF, if we have that every element $s \in S$ is represented in BF [11].

Optimal Bloom Filter Parameters. It is important to note that Bloom filters are subject to the constraint of potentially yielding false positives during membership checks. Specifically, an element $y \notin X$ may appear to belong to X if all k hash outputs correspond to bits that have been set to 1. As demonstrated in [15], let $p = 1 - (1 - 1/B)^{kn}$ denote the probability that any given bit in the Bloom filter is set to 1. The upper bound on the false-positive probability is given by

$$\epsilon = p^k \times (1 + O(\frac{k}{p}\sqrt{\frac{\ln B - k \ln p}{B}})),$$

which is negligible in the number of hash functions k. In practice, given a set of size n and a desired small false-positive rate ϵ (e.g. 2^{-40}), one can select appropriate values for k and B to minimize storage when constructing a Bloom filter. As noted in [15], optimal performance is achieved when

$$k = \frac{B}{n} \ln 2, \text{ and } B \geq n \log_2 e \cdot \log_2 1/\epsilon, \tag{1}$$

where e denotes the base of the natural logarithm. To minimise B, we get the optimal value of k to be

$$k = \log_2 1/\epsilon. \tag{2}$$

Note that these parameters are consistently chosen as outlined in [15] in this paper. The proofs that these values are optimal can be found in [3,9].

Inverting and Encrypting Bloom Filters. In this paper, we employ a variant of the Bloom filter wherein each entry is inverted prior to encryption. Additionally, rather than representing each entry as a bit, we utilize 0 and 1 elements from the plaintext space of the given encryption scheme.

Definition 5 (Encrypted Bloom filters). *Let BF be the Bloom filter calculated for the set S using hash functions $h_1, \ldots, h_k$ and B entries. We say that the corresponding encrypted Bloom filter EBF has B entries where each entry is defined in the following way:*

$$EBF[i] = Enc_{pk}(BF[i])$$

for some public key pk. In the following, we define $EBF = \{C[1], \ldots, C[B]\}$. And for $s \in S$, we have $EBF[h_i(s)] = C_i$ for $i = \{1, \ldots, k\}$ where h_i is the ith hash function used in computing the original Bloom filter. In this case C_i is the ciphertext obtained by querying the ith hash function for EBF on s.

Definition 6 (Inverted Bloom filters). *Let BF be the Bloom filter calculated for the set S using hash functions $h_1, \ldots, h_k$ and B entries. We say that the corresponding inverted Bloom filter IBF has B entries where each entry is defined in the following way:*

$$IBF[i] = \begin{cases} 1 \ \text{if } BF[i] = 0, \\ 0 \ \text{otherwise.} \end{cases}$$

We denote $EIBF$ as an encrypted, inverted Bloom filter in the following section.

Homomorphic Encryption. Let (pk, sk) denote the key pair in a public key encryption scheme. Define $\tilde{x} = Enc_{pk}(x)$ and $\tilde{y} = Enc_{pk}(y)$. The encryption scheme is additively homomorphic if it satisfies the following properties:

- Let $+_H$ denote the homomorphic addition, then $Dec_{sk}(\tilde{x} +_H \tilde{y}) = x + y$.
- Let r denote a scalar, then $Dec_{sk}(\tilde{x} \cdot r) = x \cdot r$.

Additively homomorphic encryption is a special type of homomorphic encryption that enables homomorphic addition operations. Various techniques and algorithms can achieve additively homomorphic encryption [17, 22, 26], for simplicity, we focus on the Paillier cryptosystem [40].

Oblivious Pseudorandom Function. The Oblivious Pseudorandom Function (OPRF) is a widely used two-party protocol in which the Receiver selects a random PRF key k and the Sender takes x as input and obtains $F_k(x)$. Although numerous OPRF protocols exist [20, 24, 33, 43], we focus on the protocol presented in [33], which is based on Oblivious Transfer (OT) and primarily employs efficient symmetric cryptographic operations, with only a constant number of initial public key operations for base OTs. Due to its ability to efficiently generate a large number of OPRF instances, this protocol is commonly employed in Private Set Operation (PSO) protocols to enhance performance and may serve as the foundation for implementing multi-point OPRFs and batched OPRFs.

Cuckoo Hashing. Pagh and Rodler introduced the first cuckoo hashing scheme, where γ hash functions $h_1, \ldots, h_\gamma$ are employed to map n elements into $b = \epsilon n$ bins, along with an auxiliary stash [39]. The i-th bin is denoted as B_i. Cuckoo hashing scheme avoids collisions, ensuring that each bin contains only a single item, through the following approaches:

When inserting an item x, it will choose any empty bin of $B_{h_1(x)}, \ldots, B_{h_\gamma(x)}$. If all bins are occupied, a bin $B_{h_i(x)}$ is randomly chosen among the γ bins, and the prior item y in $B_{h_i(x)}$ (where $B_{h_j(y)} = B_{h_i(x)}$) is relocated to a new bin $B_{h_k(y)}$, where $k \neq j$. This process is repeated until no further evictions are necessary or the eviction threshold is reached (in this case, the last item is placed in the stash). According to the protocols in [13, 19] and empirical analysis in [46], the stash size can be reduced to 0 while achieving a hash failure probability of 2^{-40} by appropriately adjusting the values of γ and ϵ.

Simple Hashing. A simple hashing scheme uses γ hash functions $h_1, \ldots, h_\gamma : \{0,1\}^* \to [b]$ to map n items into b bins $B_1, \ldots, B_b$. Each item is inserted into bins $B_{h_1(x)}, \ldots, B_{h_\gamma(x)}$, regardless of whether these bins are empty. The maximum bin size ρ can be set to ensure that no bin will contain more than ρ items except with probability $2^{-\lambda}$ when hashing n items into b bins, as described by the following inequality [36, 47]:

$$\Pr[\exists \text{bin with } \geq \rho \text{items}] \leq b \left[\sum_{i=\rho}^{n} \binom{n}{i} \cdot \left(\frac{1}{b} \right)^i \cdot \left(1 - \frac{1}{b} \right)^{n-i} \right].$$

3 Private Set Intersection and Intersection-Cardinality Based on Homomorphic Encryption

In this section, we propose a novel third-party PSI protocol $\prod_{\text{PSI}}^{\text{third,HE}}$ and a new third-party PSI-cardinality protocol $\prod_{\text{PSI-card}}^{\text{third,HE}}$ by leveraging Homomorphic encryption and Bloom filter techniques.

3.1 Protocol $\prod_{\text{PSI}}^{\text{third}}$: Third-Party Private Set Intersection

In the subsequent section, we provide a brief overview of this protocol, followed by a detailed presentation in Fig. 3.

We assume that P_1's set is $X = \{x_1, \ldots, x_n\}$ and P_2's set is $Y = \{y_1, \ldots, y_n\}$. Then the protocol proceeds as follows. Firstly, P_3 chooses the parameters of Homomorphic encryption, including the public key pk_3 and the secret key sk_3, and sends pk_3 to P_1 and P_2. And P_1 chooses the parameters of Bloom Filter, including the number of bins B and k hash functions $h_1, \ldots, h_k$. Then P_1 calculates Bloom Filter representing X using the set of hash functions $h_1, \ldots, h_k$. Please refer to Sect. 2.2 for the details of Bloom Filter. After successfully inserting X into the Bloom Filter, P_1 sends parameters to P_2. Hereafter, we denote by

BF_1 the Bloom Filter representing X and $BF_1[i]$ the element in the i-th index of the bloom filter.

Secondly, P_1 inverts each entry in BF_1 to retrieve inverted Bloom filter IBF_1, and separately encrypts each element $\{IBF_1[i]\}_{i \in [B]}$ using pk_3. After this, P_1 obtains the encrypted, inverted Bloom filter $EIBF_{1,pk_3}$ and sends it to P_2.

Next, for each $y_i \in Y$, P_2 generates a pair (p_i, c_i) so that P_3 can test if $y_i \in X$ via (p_i, c_i). If yes, P_3 can obtain the item y_i. The details are shown below.

Note that for each item $y_i \in Y$, P_2 evaluates with the same k hash functions and retrieves $\{c_1^i, \ldots, c_k^i\}$ where $c_j^i = EIBF_1[h_j(y_i)]$. Subsequently, P_2 generates $c_i = c_1^i +_H \cdots +_H c_k^i$. Hence, to test if $y_i \in X$, we need only to check if $D_{sk_3}(c_i)$ equals to 0. If yes, we can obtain y_i through (p_i, c_i). If not, we cannot disclose any information about y_i. To do so, P_2 samples random value r_i and evaluates $p_i = r_i \cdot c_i +_H Enc_{pk_3}(y_i)$. At the end, P_2 sends (p_i, c_i) to P_3. Recall that if $y_i \in X$, then $D_{sk_3}(c_i) = 0$ and $p_i = r_i \cdot c_i +_H Enc_{pk_3}(y_i) = y_i$.

Finally, upon receiving (p_i, c_i), P_3 checks if the decryption of c_i is 0. If yes, P_3 decrypts p_i and obtains the intersection. Otherwise, P_3 obtains nothing.

Next, we first argue that the protocol $\prod_{PSI}^{third,HE}$ in Fig. 3 realizes the functionality $\mathcal{F}_{PSI}^{third}$ correctly, and then show it satisfies the security properties.

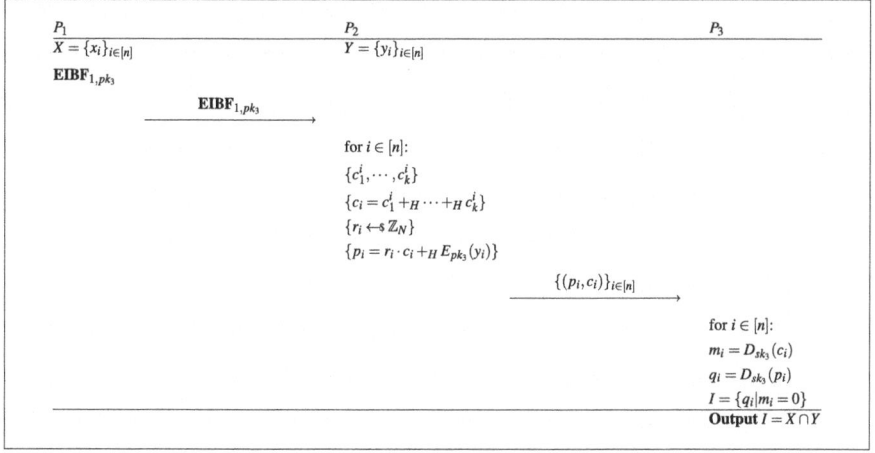

Fig. 3. Protocol $\prod_{PSI}^{third,HE}$ Based on Homomorphic Encryption

Correctness. P_3 obtains $\{(p_i, c_i)\}_{i \in [B]}$. For an item $y^* \in Y$, if $y^* \in X$, say $y^* = x_i$, then the bins of $EIBF_{1,pk_3}$ indexed by $\{h_j(x_i)\}_{j \in [k]}$ must be the encryption of 0. Hence, the encryption value c^* of y^* held by P_3 must be the encryption of 0. So in this case $m^* = D_{sk_3}(c^*) = 0$, P_3 decrypts p^* and obtains the intersection y_i. Otherwise (i.e., $y^* \notin X$), there exists at least one bin indexed by hash functions are the encryption of 1. Thus the decryption $m^* = D_{sk_3}(c^*) \in [k]$, P_3 decrypts p^*

and obtains $q^* = r^* \cdot m^* + y^*$. Since r^* is uniformly random in P_3's perspective, y^* will be randomly masked by P_2, and P_3 will obtain nothing about y^*.

We remark that the correctness error comes from the following collision. That is, for $y^* \notin X$ and all $j \in [k]$, $IBF_1[h_j(y^*)] = 0$ holds. To make the correctness hold with an overwhelming probability, we need to ensure the probability of collision happening is less than $2^{-\lambda}$.

Security. Now we proceed to show that $\prod_{PSI}^{third,HE}$ is secure against a semi-honest adversary corrupting a single party.

Theorem 1. $\prod_{PSI}^{third,HE}$ *is secure against a semi-honest* P_1.

Proof. This is clear as P_1 receives nothing from P_2 and P_3.   □

Theorem 2. $\prod_{PSI}^{third,HE}$ *is secure against a semi-honest* P_2.

Proof. The simulator receives $(|X|, Y, \mathcal{T}, \phi)$, where $\mathcal{T}$ is the entire message transcript that P_2 witnesses and ϕ denotes the empty output received. Specifically, in the perspective of P_2, $\mathcal{T}$ simply contains an encrypted, inverted Bloom filter $(EIBF_{1,pk_3})$ sent by P_1. Hence, the simulator is only tasked with constructing an $EIBF_{1,pk_3}$ that is indistinguishable from the one provided in the real execution.

From the knowledge of $(|X|, (h_1, \ldots, h_k))$, the simulator is capable of constructing an empty Bloom filter with the correct parameters and the same hash functions. The simulator then inverts each entry of the Bloom filter and encrypts it using the IND-CPA Homomorphic encryption scheme. Let $\mathcal{T}'$ denote the simulated transcript. Both $\mathcal{T}$ and $\mathcal{T}'$ just contain IND-CPA encrypted and inverted Bloom filters. It follows that any adversary capable of distinguishing between these transcripts would, by implication, be able to compromise the IND-CPA security of the homomorphic encryption scheme. This leads to the conclusion that the transcripts are indistinguishable.   □

Theorem 3. $\prod_{PSI}^{third,HE}$ *is secure against a semi-honest* P_3.

Proof. The simulator receives $(|X|, |Y|, \mathcal{T}, I)$, where $\mathcal{T}$ is the entire message transcript that P_3 witnesses and I denotes the intersection received. Specifically, for P_3, $\mathcal{T}$ comprises n pairs of encrypted values $\{(p_i, c_i)\}_i$. Hence, the simulator is tasked to generate a sequence $\{(p_i, c_i)\}_{i \in [n]}$ that is indistinguishable from the pairs observed during the real execution. The simulator proceeds with the following construction:

Let $L = |I|$ and $J = |Y| - L$. The simulator generates L pairs of encryptions (p_i', c_i') corresponding to the elements in I and 0. It then samples J pairs of random elements $(r_i' \in \mathbb{Z}_N$ and $m_i' \in [k])$, computes the corresponding encryptions, and subsequently shuffles the entire set of ciphertext pairs. Let $\mathcal{T}'$ be the simulated transcript containing these encrypted messages.

Since $\mathcal{T}'$ contains exactly L pairs of encryptions $\{(i, 0)\}_{i \in I}$, it is evident that the adversary will obtain the same intersection output as in the case of $\mathcal{T}'$. For the remaining J pairs, given that r_i is randomly sampled by P_2, $q_i = r_i \cdot m_i + y_i$ is uniformly distributed across the domain. Therefore, q_i is identically distributed to r_i', making it impossible for P_3 to distinguish between these two values.

Furthermore, in the real-world execution, the pairs (p_i, c_i) are encrypted by IND-CPA Homomorphic encryption scheme. Similarly, $\mathcal{T}'$ encrypts the pairs (p_i', c_i') using the same IND-CPA Homomorphic encryption scheme. Consequently, if an adversary is capable of distinguishing between $\mathcal{T}$ and $\mathcal{T}'$, it implies that they are able to break the IND-CPA security of the encryption scheme. $\square$

The third-party Private Set Intersection-Cardinality protocol is closely related to the third-party Private Set Intersection protocol. Consequently, we modify the $\prod_{\text{PSI}}^{\text{third,HE}}$ to derive the $\prod_{\text{PSI-card}}^{\text{third,HE}}$ as outlined below.

3.2 Protocol $\prod_{\text{PSI-card}}^{\text{third,HE}}$: Third-Party Private Set Intersection-Cardinality

Similar to $\prod_{\text{PSI}}^{\text{third,HE}}$, the core idea of designing $\prod_{\text{PSI-card}}^{\text{third,HE}}$ is to apply the Bloom filter. A brief description is given below and the details are shown in Fig. 4.

Similarly, P_1's set and P_2's set are assumed to be $X = \{x_1, \ldots, x_n\}$ and $Y = \{y_1, \ldots, y_n\}$, respectively. Then the protocol works as follows. Firstly, P_3 chooses the parameters of Homomorphic encryption, including the public key pk_3 and the secret key sk_3, and sends pk_3 to P_1 and P_2. And P_1 chooses the parameters of Bloom Filter, including the number of bins B and k hash functions $h_1, \ldots, h_k$. Then P_1 calculates Bloom Filter representing X using the set of hash functions $h_1, \ldots, h_k$. After successfully inserting X into the Bloom Filter, P_1 sends parameters to P_2. Hereafter, we denote by BF_1 the Bloom Filter representing X and $BF_1[i]$ the element in the i-th bin of the filter.

Secondly, P_1 inverts each entry in BF_1 to retrieve inverted Bloom filter IBF_1, and separately encrypts each element $\{IBF_1[i]\}_{i \in [B]}$ using pk_3. After this, P_1 obtains the encrypted, inverted Bloom filter $EIBF_{1,pk_3}$ and sends it to P_2.

Next, for each $y_i \in Y$, P_2 generates ciphertext c_i so that P_3 can test if $y_i \in X$ via c_i. If yes, P_3 increments the counter. The details are shown below.

Note that for each item $y_i \in Y$, P_2 evaluates with k hash functions and retrieves $\{c_1^i, \ldots, c_k^i\}$ where $c_j^i = EIBF_1[h_j(y_i)]$. Subsequently, P_2 generates $c_i = c_1^i +_H \cdots +_H c_k^i$. Hence, to test if $y_i \in X$, we need only to check if $D_{sk_3}(c_i)$ equals to 0. If yes, we can increment the counter. If not, we cannot disclose any information about y_i.

Finally, upon receiving c_i, P_3 checks if the decryption of c_i is 0. If yes, P_3 increments the counter and obtains the intersection cardinality. Otherwise, P_3 obtains nothing.

In what follows, we first show the correctness of the protocol $\prod_{\text{PSI-card}}^{\text{third,HE}}$ in Fig. 4 and then argue that it securely realizes the functionality $\mathcal{F}_{\text{PSI-card}}^{\text{third}}$.

Correctness. The analysis follows a similar approach to that of $\prod_{\text{PSI}}^{\text{third,HE}}$, and is therefore summarized briefly here. Upon obtaining $\{c_i\}_{i \in [n]}$, P_3 decrypts c^* for each item $y^* \in Y$. If $y^* \in X$, P_3 will obtain $m^* = D_{sk_3}(c^*) = 0$ and will increment the counter accordingly. Otherwise, it does nothing.

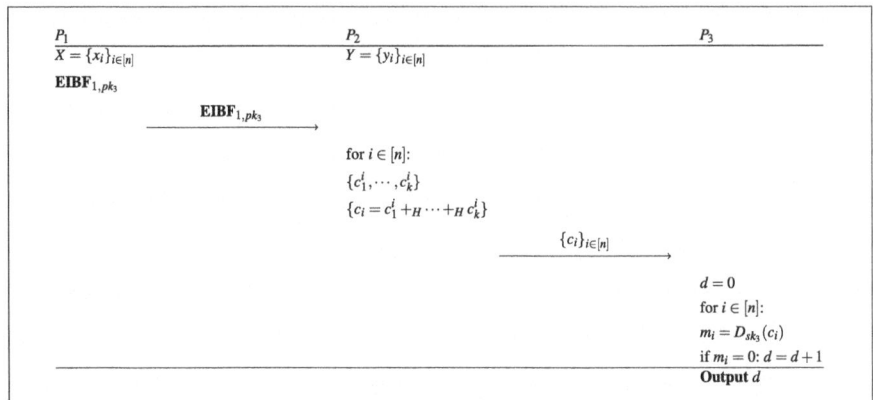

Fig. 4. Protocol $\prod_{\text{PSI-card}}^{\text{third,HE}}$ Based on Homomorphic Encryption

Security. Now we proceed to show that $\prod_{\text{PSI-card}}^{\text{third,HE}}$ is secure against a semi-honest adversary corrupting a single party.

Theorem 4. $\prod_{PSI-card}^{third,HE}$ is secure against a semi-honest P_1.

Proof. This is clear as P_1 receives nothing from P_2 and P_3.   □

Theorem 5. $\prod_{PSI-card}^{third,HE}$ is secure against a semi-honest P_2.

Proof. This condition is identical to that of $\prod_{\text{PSI}}^{\text{third,HE}}$. Given space constraints, we omit the detailed explanation.   □

Theorem 6. $\prod_{PSI-card}^{third,HE}$ is secure against a semi-honest P_3.

Proof. The simulator receives $(|X|, |Y|, \mathcal{T}, d)$, where $\mathcal{T}$ is the entire message transcript that P_3 witnesses and d denotes the intersection cardinality received. Specifically, for P_3, $\mathcal{T}$ contains n encryptions $\{c_i\}_i$. Hence, the simulator is tasked with constructing $\{c_i\}_{i \in [n]}$ that is indistinguishable from the encryptions provided in the real execution. Now the simulator constructs as follows:

Let $L = d$ and $J = |Y| - L$. The simulator constructs L encryptions c_i' of 0. Then it samples J random elements $m_i' \in [k]$, and calculates the encryptions. Subsequently, it shuffles the entire set of encryptions. Let $\mathcal{T}'$ be the simulated transcript containing these messages.

Since $\mathcal{T}'$ contains exact L encryptions 0, it is clear that the adversary learns the same cardinality output in the case of $\mathcal{T}'$. As for the rest of J encryptions, since m_i is generated by P_2 according to the hash functions and Bloom filter, and Bloom filter is randomly distributed in the perspective of P_3, hence, m_i is identically distributed to m_i', and thus P_3 cannot distinguish these two values.

Besides, in the real-world execution, the ciphertexts c_i are encrypted by IND-CPA Homomorphic encryption scheme. Since $\mathcal{T}'$ encrypts c_i' by IND-CPA Homomorphic encryption scheme either. We can show that any adversary is able to distinguish $\mathcal{T}$ and $\mathcal{T}'$ then they must break its IND-CPA security.   □

4   Private Set Intersection and Intersection-Cardinality Based on Oblivious Pseudorandom Function

In this section, we propose a scalable PSI protocol $\prod_{PSI}^{third,OPRF}$ and an efficient PSI-cardinality protocol $\prod_{PSI-card}^{third,OPRF}$ by leveraging Oblivious pseudorandom function (OPRF) and Cuckoo hashing techniques.

4.1   Protocol $\prod_{PSI}^{third}$: Third-Party Private Set Intersection

In the subsequent section, we provide a brief overview of this protocol, followed by a detailed presentation in Fig. 5.

We assume that P_1's set is $X = \{x_1, \ldots, x_n\}$ and P_2's set is $Y = \{y_1, \ldots, y_n\}$. Then the protocol proceeds as follows. Firstly, P_1 chooses the parameters of Cuckoo hashing without a stash, including the number of bins $b = \epsilon \cdot n$ and γ hash functions $h_1, \ldots, h_\gamma$. Then P_1 inserts X into this table and pads each empty bin with a dummy item d. Please refer to Sect. 2.2 for the details of Cuckoo hashing. After successfully inserting X into the Cuckoo hashing table, P_1 sends the parameters to P_2. Hereafter, we denote T_1 the Cuckoo hashing table filled with X, and $T_1[i]$ the item in the i-th bin of the table. Then P_2 inserts Y into a simple hash table with b bins by using the same γ hash functions and pads each empty bin with a dummy item e. Hereafter, we denote T_2 the simple hashing table filled with Y, and $T_2[i][j]$ the j-th item in the i-th bin of the table.

Secondly, P_1 and P_2 invoke $\mathcal{F}_{mpOPRF}$ with input T_1 and $\perp$, respectively. After this, P_1 obtains $\{F_{k_i}(T_1[i])\}_{i \in [b]}$, and P_2 obtains $\{k_i\}_{i \in [b]}$.

Next, for each $x_i \in T_1$, P_1 generates a value $T_3[i]$ so that P_3 can test if $x_i \in Y$ via $T_3[i]$. If yes, P_3 can obtain the item x_i. However, we observe that if P_1 simply encode x_i with $F_{k_i}(x_i)$, it is hard for P_3 to check whether the decoded message is in the intersection or a random value. To solve this problem, P_1 attaches x_i with a hashing value $h(x_i)$, the details of which are shown below.

Note that for each item $T_1[i]$, if there is an item $y \in Y$ equal to $T_1[i]$, then y must be inserted into $T_2[i]$ according to the property of the Cuckoo hashing and Simple hashing. Hence, to test if $T_1[i] \in Y$, we need only to check if $T_1[i] \in \{T_2[i][1], \ldots, T_2[i][|T_2[i]|]\}$. To do so, P_1 first attaches a label value $H(T_1[i])$ to $T_1[i]$ via hash function, and then encodes it with $F_{k_i}(T_1[i])$. At the end, P_1 sends the encoding value $T_3[i] = F_{k_i}(T_1[i]) \oplus (T_1[i]||h(T_1[i]))$ to P_3. At the same time, P_2 evaluates each item in $T_2[i]$ with the key k_i obtained in the $\mathcal{F}_{mpOPRF}$, and sends $T_4[i] = \{F_{k_i}(T_4[i][1]), \ldots, F_{k_i}(T_4[i][|T_4[i]|])\}$ to P_3.

Finally, upon on receiving $T_3[i]$ and $T_4[i]$, P_3 checks if there exists decoding $m_{i,j} = T_3[i] \oplus T_4[i][j]$ satisfying $m_{i,j} = z_i||h(z_i)$. If yes, P_3 adds z_i to intersection, otherwise obtains nothing. At last, P_3 outputs the intersection.

Next we first argue the protocol $\prod_{PSI}^{third,HE}$ in Fig. 5 realizes the functionality $\mathcal{F}_{PSI}^{third}$ correctly, and then show it satisfies the security properties.

Correctness. P_3 obtains the Cuckoo hash table T_3 from P_1, which is filled with the set X and dummy items d in the form $T_3[i] = F_{k_i}(T_1[i]) \oplus (T_1[i]||h(T_1[i]))$.

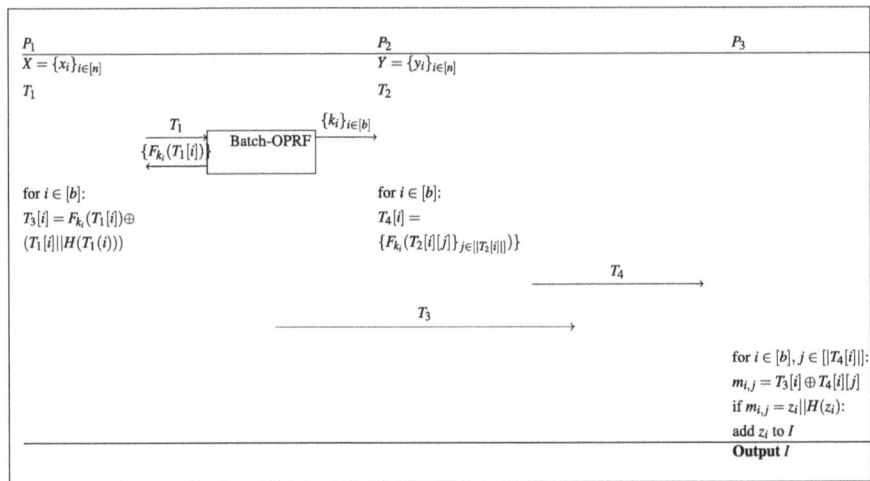

Fig. 5. Protocol $\prod_{\text{PSI}}^{\text{third,OPRF}}$ Based on Oblivious Pseudorandom Function

Additionally, P_3 acquires the Simple hash table T_4 from P_2, filled with the set Y and dummy items e in the form $T_4[i][j] = F_{k_i}(T_2[i][j])$. For an item $x^* \in X$, if $x^* \in Y$, say $x^* = y_i$, then y_i must be inserted into the corresponding bin of T_3 indexed by $h_j(x^*)$, where $j \in [\gamma]$. Without loss of generality, we assume that the index is $t = h_j(x^*) \in [b]$, then $T_3[t] \oplus T_4[t][j] = F_{k_t}(x^*) \oplus (x^*||h(x^*)) \oplus F_{k_t}(y_i) = x^*||h(x^*)$, which satisfies the form of $z_i||h(z_i)$. So in this case P_3 adds x^* into the intersection. Otherwise (i.e. $x^* \notin Y$), we have that $T_3[t] \oplus T_4[t][j] = F_{k_t}(x^*) \oplus (x^*||h(x^*)) \oplus F_{k_t}(T_4[t][j]) \neq x^*||h(x^*)$ with an overwhelming probability. Thus P_3 will obtain nothing about the intersection.

We remark that the correctness error comes from the following two types of collisions. Specifically, the first type occurs due to the Pseudorandom Function (PRF), where $F_k(x^*) = F_k(y^*)$ for some $x^* \neq y^*$. The second type of collision is caused by hashing, where $x_r = h(x_l)$ for some uniformly random $x_l||x_r$. To ensure correctness with overwhelming probability, the likelihood of a collision must be kept below $2^{-\lambda}$.

Security. Now we proceed to show that $\prod_{\text{PSI}}^{\text{third,OPRF}}$ is secure against a semi-honest adversary corrupting a single party.

Theorem 7. $\prod_{PSI}^{third,OPRF}$ *is secure against a semi-honest P_1.*

Proof. The simulator receives $(X, |Y|, \mathcal{T}, \phi)$, where $\mathcal{T}$ represents the entire message transcript observed by P_1, and ϕ denotes the empty output received. Specifically, in the perspective of P_1, $\mathcal{T}$ contains an array of pseudorandom values $\{F_{k_i}(T_1[i])\}_{i \in [b]}$ generated within $\mathcal{F}_{mpOPRF}$ functionality. Consequently, the simulator is tasked with constructing an array of strings that remains indistinguishable from the values generated during the real execution.

From knowledge of $(X, h_1, \ldots, h_\gamma)$, the simulator can construct a Cuckoo hash table using the correct parameters and identical hash functions. The simulator then samples a uniformly random value for each bin of Cuckoo hash table. Let $\mathcal{T}'$ denote the simulated transcript. It is trivial to show that any adversary who can distinguish between these transcripts can break the security of the multi-point Oblivious Pseudorandom Function scheme. □

Theorem 8. $\prod_{PSI}^{third, OPRF}$ *is secure against a semi-honest P_2.*

Proof. The simulator receives $(|X|, Y, \mathcal{T}, \phi)$, where $\mathcal{T}$ is the entire message transcript that P_2 witnesses and ϕ denotes the empty output received. Specifically, in the perspective of P_2, $\mathcal{T}$ contains an array of keys $\{k_i\}_{i\in[b]}$ generated in $\mathcal{F}_{mpOPRF}$. Hence, the simulator is only tasked with constructing an array of keys that is indistinguishable from the one provided in the real execution.

From knowledge of $(Y, h_1, \ldots, h_\gamma)$, the simulator is able to construct a Simple hash table using the correct parameters and the same hash functions. The simulator samples a uniformly random key for each bin of Simple hash table. Let $\mathcal{T}'$ denote the simulated transcript. It is trivial to show that any adversary who can distinguish between these transcripts can break the security of the multi-point Oblivious Pseudorandom Function scheme. □

Theorem 9. $\prod_{PSI}^{third, OPRF}$ *is secure against a semi-honest P_3.*

Proof. The simulator receives $(|X|, |Y|, \mathcal{T}, I)$, where $\mathcal{T}$ represents the entire message transcript that P_3 witnesses and I denotes the intersection received. From P_3's perspective, $\mathcal{T}$ contains encoding values T_3 originating from P_1 and an array of pseudorandom values T_4 generated by P_2. Hence, the simulator's task is to construct an array of encoding values and pseudorandom values that are those produced during the actual execution. The simulator proceeds with the construction as follows:

Let $L = |I|$ and $J = |X| - L$. Consider the set $\{K_i = |T_4[i]|\}_{i\in[b]}$. The simulation process begins by sampling L random values r'_i, and generating the corresponding encoding values $p'_i = r'_i \oplus (z_i || h(z_i))$, where $z_i \in I$. Subsequently, an empty set c'_i is initialized, and r'_i is inserted along with $K_i - 1$ additional random values. Next, the simulator samples J random values r'_j, and generates the encoding value $p'_j = r'_j$. An empty set c'_j is then created, into which K_j random values are inserted. Finally, it shuffles the entire set of $\{p'_i\}_{i\in[b]}$ and $\{c'_i\}_{i\in[b]}$, respectively. It is important to note that within each set c'_i, the elements $\{c'_i[j]\}_{j\in[K_i]}$ are shuffled. The resulting simulated transcript is denoted by $\mathcal{T}'$, which includes all these messages.

Given that $\mathcal{T}'$ contains exactly L pairs of encoded values $(p'_i, c'_i[j]) = (r'[i] \oplus (z_i || h(z_i)), r'[i])$, it follows that the adversary obtains the same intersection result in the case of $\mathcal{T}'$. For the remaining J bins, since $F_{k_i}(T_3[i])$ is generated pseudorandomly by $\mathcal{F}_{mpOPRF}$, $F_{k_i}(T_3[i]) \oplus (T_3[i] || h(T_3[i]))$ is uniformly distributed across the domain. Hence, $F_{k_i}(T_3[i]) \oplus (T_3[i] || h(T_3[i]))$ is indistinguishable from p'_i, making it impossible for P_3 to differentiate between these two values. Similarly, since the pseudorandom keys k_i are generated by $\mathcal{F}_{mpOPRF}$, $F_{k_i}(T_4[i][j])$ is uniformly distributed across the domain. Therefore, $F_{k_i}(T_4[i][j])$ is indistinguishable from $c_i'[j]$, and P_3 cannot differentiate between these two values.   □

The third-party Private Set Intersection-Cardinality protocol is closely related to the third-party Private Set Intersection protocol. Consequently, we modify the $\prod_{\mathrm{PSI}}^{\mathrm{third,OPRF}}$ to derive the $\prod_{\mathrm{PSI\text{-}card}}^{\mathrm{third,OPRF}}$ as outlined below.

4.2   Protocol $\prod_{\mathrm{PSI\text{-}card}}^{\mathrm{third,OPRF}}$: Third-Party Private Set Intersection-Cardinality

Similar to $\prod_{\mathrm{PSI}}^{\mathrm{third,OPRF}}$, the core idea of designing $\prod_{\mathrm{PSI\text{-}card}}^{\mathrm{third,OPRF}}$ is to apply the Oblivious pseudorandom function and Cuckoo hashing. A brief description is given below and the details are shown in Fig. 6.

Similarly, P_1's set and P_2's set are assumed to be $X = \{x_1, \ldots, x_n\}$ and $Y = \{y_1, \ldots, y_n\}$, respectively. Then the protocol works as follows. Firstly, P_1 chooses the parameters of Cuckoo hashing table without a stash, including the number of bins $b = \epsilon \cdot n$ and γ hash functions $h_1, \ldots, h_\gamma$. Then P_1 inserts X into this table and pads each empty bin with a dummy item d. After successfully inserting X into the Cuckoo hashing table, P_1 sends the parameters to P_2. Hereafter, we denote T_1 the Cuckoo hashing table filled with X, and $T_1[i]$ the item in the i-th bin of the table. Then P_2 inserts Y into a simple hash table with b bins by using the same γ hash functions and pads each empty bin with a dummy item e. Hereafter, we denote T_2 the simple hashing table filled with Y, and $T_2[i][j]$ the j-th item in the i-th bin of the table.

Secondly, P_1 and P_2 invoke $\mathcal{F}_{mpOPRF}$ with input T_1 and $\bot$, respectively. After this, P_1 obtains $\{F_{k_i}(T_1[i])\}_{i \in [b]}$, and P_2 obtains $\{k_i\}_{i \in [b]}$.

Next, for each $x_i \in T_1$, P_1 generates a value $T_3[i]$ so that P_3 can test if $x_i \in Y$ via $T_3[i]$. If yes, P_3 increments the counter. The details are shown below.

Note that for each item $T_1[i]$, if there is an item $y \in Y$ equal to $T_1[i]$, then y must be inserted into $T_2[i]$ according to the property of the Cuckoo hashing and Simple hashing. Hence, to test if $T_1[i] \in Y$, we need only to check if $T_1[i] \in \{T_2[i][1], \ldots, T_2[i][|T_2[i]|]\}$. To do so, P_1 generates $F_{k_i}(T_1[i])$ via $\mathcal{F}_{mpOPRF}$ and sends to P_3. At the same time, P_2 evaluates each item in $T_2[i]$ with the key k_i obtained in the $\mathcal{F}_{mpOPRF}$, and sends to P_3.

Finally, upon receiving $T_3[i]$ and $T_4[i]$, P_3 checks if $T_3[i] = T_4[i][j]$. If yes, P_3 increments the counter and obtains the intersection cardinality. Otherwise, P_3 obtains nothing.

In what follows, we first show the correctness of the protocol $\prod_{\mathrm{PSI\text{-}card}}^{\mathrm{third,OPRF}}$ in Fig. 6 and then argue that it securely realizes the functionality $\mathcal{F}_{\mathrm{PSI\text{-}card}}^{\mathrm{third}}$.

Correctness. The analysis follows a similar approach to that of $\prod_{\mathrm{PSI}}^{\mathrm{third,OPRF}}$, and is therefore summarized briefly here. Upon obtaining $\{T_4[i][j]\}_{i \in [b], j \in [|T_4[i]|]}$ and $\{T_3[i]\}_{i \in [b]}$, P_3 checks if $T_3[i] \in T_4[i]$. if yes, P_3 increments the counter. Otherwise, it does nothing.

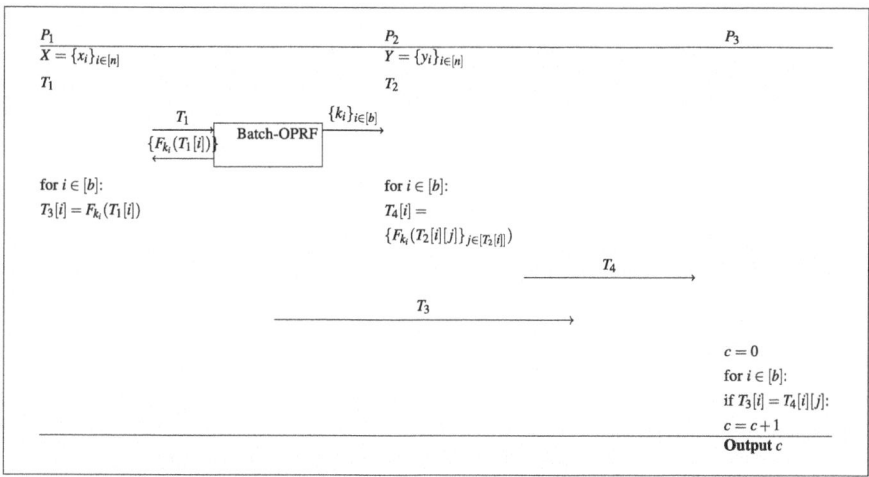

Fig. 6. Protocol $\prod_{\mathrm{PSI}}^{\mathrm{third,OPRF}}$ Based on Oblivious Pseudorandom Function

Security. Now we proceed to show that $\prod_{\mathrm{PSI}}^{\mathrm{third,OPRF}}$ is secure against a semi-honest adversary corrupting a single party.

Theorem 10. $\prod_{PSI\text{-}card}^{third,OPRF}$ *is secure against a semi-honest* P_1.

Proof. This condition is identical to that of $\prod_{\mathrm{PSI}}^{\mathrm{third,OPRF}}$. Given space constraints, we omit the detailed explanation. □

Theorem 11. $\prod_{PSI\text{-}card}^{third,OPRF}$ *is secure against a semi-honest* P_2.

Proof. This condition is identical to that of $\prod_{\mathrm{PSI}}^{\mathrm{third,OPRF}}$. Given space constraints, we omit the detailed explanation. □

Theorem 12. $\prod_{PSI\text{-}card}^{third,OPRF}$ *is secure against a semi-honest* P_3.

Proof. The simulator receives $(|X|, |Y|, \mathcal{T}, c)$, where $\mathcal{T}$ represents the entire message transcript that P_3 witnesses and c denotes the cardinality received. Specifically, From P_3's perspective, $\mathcal{T}$ contains b pseudorandom values T_3 from P_1 and T_4 from P_2. Hence, the simulator is tasked with constructing two arrays of pseudorandom values that are indistinguishable from those provided in the real execution. Now the simulator constructs as follows:

Let $L = c$, $J = |X| - L$, and $\{K_i = |T_4[i]|\}_{i \in [b]}$. The simulator begins by sampling L random values r_i', and generating the corresponding pseudorandom value $p_i' = r_i'$. Subsequently, an empty set d_i' is initialized, and r_i' is inserted along with additional $K_i - 1$ random values. Next, the simulator samples J random values r_j', and generates the encoding value $p_j' = r_j'$. An empty set d_j' is then created, into which K_j random values are inserted. Finally, it shuffles the entire set of $\{p_i'\}_{i \in [b]}$ and $\{d_i'\}_{i \in [b]}$, respectively. Note that within each set d_i', the elements $\{d_i'[j]\}_{j \in [K_i]}$ are shuffled. Let $\mathcal{T}'$ be the simulated transcript.

Since $\mathcal{T}'$ contains exactly L pairs of values $(p'_i, c'_i[j]) = (r'[i], r'[i])$, the adversary learns the same intersection cardinality in the case of $\mathcal{T}'$. For the remaining J bins, $F_{k_i}(T_3[i])$, generated by the $\mathcal{F}_{mpOPRF}$, is indistinguishable from p'_i, hence, P_3 cannot differentiate between these values. As the keys k_i are generated by $\mathcal{F}_{mpOPRF}$, $F_{k_i}(T_4[i][j])$ is uniformly distributed over the domain, making it indistinguishable from $c'_i[j]$. Thus, P_3 cannot distinguish these values. □

5   Performance Evaluation

In this section, we experimentally evaluate our PSI and PSI-card protocols $\Pi_{\text{PSI}}^{\text{third,HE}}$, $\Pi_{\text{PSI}}^{\text{third,OPRF}}$, $\Pi_{\text{PSI-card}}^{\text{third,HE}}$, and $\Pi_{\text{PSI-card}}^{\text{third,OPRF}}$. In Sect. 5.1, we describe the benchmarking environment. In Sect. 5.2, we compare our protocols in terms of single-threaded runtime, with the results detailed in Tables 2 and 3. The full implementation is available on GitHub: https://github.com/ckay0/thirdPSI.

5.1   Benchmarking Environment

We implement $\Pi_{\text{PSI}}^{\text{third,HE}}$, $\Pi_{\text{PSI}}^{\text{third,OPRF}}$, $\Pi_{\text{PSI-card}}^{\text{third,HE}}$ and $\Pi_{\text{PSI-card}}^{\text{third,OPRF}}$ in C++ and conducted our experiments on Intel@ Core^{TM} i7-10700 CPU @ 2.90 FHz × 16 GB of RAM. The protocols are evaluated in a LAN environment with 10 Gbps bandwidth and 0.02 ms RTT, emulated using the Linux tc command. The statistical security parameter was set to $\lambda = 40$.

Our HE-based PSI and PSI-card protocols utilize Bloom filters and the Paillier cryptosystem [40]. We provide our own implementation of the encrypted, inverted Bloom filter functionality, and employ the Paillier scheme using the library design in [18]. Our OPRF-based PSI and PSI-card protocols are constructed using OPRF and Cuckoo hashing table, with the OPRF implementation sourced from [6,33].

Parameters About Bloom Filter. For the experiments, we examine running times for sets sizes ranging from 2^{10} to 2^{20} elements, consistent with prior work. We select a false positive probability of $\epsilon = 2^{-40}$ and optimize the Bloom filter parameters as outlined in Sect. 2.2. For instance, with $k = 40$, the required Bloom filter size is $B = kn \log_2 e$ by Eq. (1) for sets of size n.

Parameters About Cuckoo Hashing. Since items in the stash of a Cuckoo hashing table must be compared with each item in the other set, rather than only those selected by the hash functions. Hence, it is crucial to minimize the stash size to 0. Empirical analysis in [46] demonstrates that increasing the number of hash functions significantly reduces the number of bins and the required stash size. Based on this, we implement Cuckoo hashing using 3 hash functions, which requires $1.27 \cdot n$ bins with a stash size of 0.

5.2 Performance Comparisons

Since [52,53] do not provide any implementation, we limit our discussion in Sect. 1.2 to theoretical comparisons, as illustrated in Table 1. This section presents a comparative analysis of the runtime performance between $\prod_{\text{PSI}}^{\text{third,HE}}$ and $\prod_{\text{PSI}}^{\text{third,OPRF}}$, with the results summarized in Table 2. Besides, we compare the runtime performance of $\prod_{\text{PSI-card}}^{\text{third,HE}}$ and $\prod_{\text{PSI-card}}^{\text{third,OPRF}}$, as detailed in Table 3.

Specifically, when compared to $\prod_{\text{PSI}}^{\text{third,HE}}$, our $\prod_{\text{PSI}}^{\text{third,OPRF}}$ demonstrates a significant performance enhancement, achieving a 4.6 times faster improvement for small set ($n = 2^{10}$) in LAN setting and a 13.78 times faster improvement for larger sets ($n = 2^{18}$). Notably, the improvement gains of our $\prod_{\text{PSI}}^{\text{third,OPRF}}$ increase as the set size expands. Similarly, the $\prod_{\text{PSI-card}}^{\text{third,OPRF}}$ exhibits superior performance, with a 4.389 times faster improvement for small sets ($n = 2^{10}$) in LAN setting and a 13.589 times faster improvement for larger sets ($n = 2^{18}$).

Table 2. Comparisons of total runtime (in seconds) and respective runtime (in seconds) between HE-based $\prod_{\text{PSI}}^{\text{third,HE}}$ and OPRF-based $\prod_{\text{PSI}}^{\text{third,OPRF}}$ with a single thread in LAN setting. Best results are marked in bold.

Protocol		2^{10}	2^{12}	2^{14}	2^{16}	2^{18}	2^{20}
$\prod_{\text{PSI}}^{\text{third,HE}}$	P_1	0.3471706	1.41192	5.623202	22.16618	254.714	–
	P_2	0.115061	0.460324	1.84586	7.28272	29.197153	–
	P_3	0.011339	0.036721	0.131207	0.519475	2.01733	–
	Sum	0.462232	1.87224	7.46906	29.4489	283.911	–
$\prod_{\text{PSI}}^{\text{third,OPRF}}$	P_1	0.080859	0.170165	0.532038	2.01944	8.08216	33.7933
	P_2	0.09861	0.210888	0.640433	2.77528	20.3685	85.8857
	P_3	0.000921	0.001833	0.006039	0.022515	0.08662	0.344163
	Sum	**0.099531**	**0.212721**	**0.646472**	**2.797795**	**20.45512**	**86.229863**

Table 3. Comparisons of total runtime (in seconds) and respective runtime (in seconds) between HE-based $\prod_{\text{PSI-card}}^{\text{third,HE}}$ and OPRF-based $\prod_{\text{PSI-card}}^{\text{third,OPRF}}$ with a single thread in LAN setting. Best results are marked in bold

Protocol		2^{10}	2^{12}	2^{14}	2^{16}	2^{18}	2^{20}
$\prod_{\text{PSI-card}}^{\text{third,HE}}$	P_1	0.342722	1.43299	5.66005	22.2455	254.925	–
	P_2	0.0964466	0.387186	1.613198	6.380176	25.572383	–
	P_3	0.0080178	0.0197356	0.0751962	0.2737504	1.0439745	–
	Sum	0.439169	1.82018	7.27325	28.6257	280.497	–
$\prod_{\text{PSI-card}}^{\text{third,OPRF}}$	P_1	0.080666	0.171368	0.53547	2.03603	8.12223	33.8245
	P_2	0.098735	0.210805	0.64066	2.77604	20.4886	84.9362
	P_3	0.00132	0.003163	0.012608	0.041021	0.153384	0.588081
	Sum	**0.100055**	**0.213968**	**0.653268**	**2.817061**	**20.641984**	**85.524281**

6   Conclusion

In this paper, we introduce novel methods for computing third-party PSI with linear complexities. Our theoretical analysis shows that our protocols outperform the current state-of-the-art [52,53]. Notably, Our third-party PSI and PSI-card protocols represent the first implementation of their kind. Experimental results confirm the practicality of our approach, establishing our designs as an efficient toolkit for third-party PSI computations.

Acknowledgement. The authors are very appreciate for the reviewers' valuable comments which are helpful for improving the presentation of the paper. The paper is supported by the National Natural Science Foundation of China under Grant 12371525, and also supported by the Strategic Priority Research Program of the Chinese Academy of Sciences under Grant XDB0690200.

References

1. Ben-Efraim, A., Nissenbaum, O., Omri, E., Paskin-Cherniavsky, A.: PSImple: practical multiparty maliciously-secure private set intersection. In: Proceedings of the 2022 ACM on Asia Conference on Computer and Communications Security, pp. 1098–1112 (2022)
2. Bloom, B.H.: Space/time trade-offs in hash coding with allowable errors. Commun. ACM **13**(7), 422–426 (1970)
3. Bose, P., et al.: On the false-positive rate of bloom filters. Inf. Process. Lett. **108**(4), 210–213 (2008)
4. Buddhavarapu, P., Knox, A., Mohassel, P., Sengupta, S., Taubeneck, E., Vlaskin, V.: Private matching for compute. Cryptology ePrint Archive (2020)
5. Chandran, N., Dasgupta, N., Gupta, D., Obbattu, S.L.B., Sekar, S., Shah, A.: Efficient linear multiparty psi and extensions to circuit/quorum psi. In: Proceedings of the 2021 ACM SIGSAC Conference on Computer and Communications Security, pp. 1182–1204 (2021)
6. Chase, M., Miao, P.: Private set intersection in the internet setting from lightweight oblivious PRF. In: Micciancio, D., Ristenpart, T. (eds.) CRYPTO 2020. LNCS, vol. 12172, pp. 34–63. Springer, Cham (2020). https://doi.org/10.1007/978-3-030-56877-1_2
7. Chen, H., Huang, Z., Laine, K., Rindal, P.: Labeled PSI from fully homomorphic encryption with malicious security. In: Proceedings of the 2018 ACM SIGSAC Conference on Computer and Communications Security, pp. 1223–1237 (2018)
8. Chen, H., Laine, K., Rindal, P.: Fast private set intersection from homomorphic encryption. In: Proceedings of the 2017 ACM SIGSAC Conference on Computer and Communications Security, pp. 1243–1255 (2017)
9. Christensen, K., Roginsky, A., Jimeno, M.: A new analysis of the false positive rate of a bloom filter. Inf. Process. Lett. **110**(21), 944–949 (2010)
10. Cong, K., et al.: Labeled psi from homomorphic encryption with reduced computation and communication. In: Proceedings of the 2021 ACM SIGSAC Conference on Computer and Communications Security, pp. 1135–1150 (2021)

11. Davidson, A., Cid, C.: An efficient toolkit for computing private set operations. In: Pieprzyk, J., Suriadi, S. (eds.) Information Security and Privacy: 22nd Australasian Conference, ACISP 2017, Auckland, New Zealand, 3–5 July 2017, Proceedings, Part II 22, pp. 261–278. Springer, Cham (2017). https://doi.org/10.1007/978-3-319-59870-3_15

12. De Cristofaro, E., Tsudik, G.: Practical private set intersection protocols with linear complexity. In: Sion, R. (eds.) International Conference on Financial Cryptography and Data Security, pp. 143–159. Springer, Cham (2010). https://doi.org/10.1007/978-3-642-14577-3_13

13. Devroye, L., Morin, P.: Cuckoo hashing: further analysis. Inf. Process. Lett. **86**(4), 215–219 (2003)

14. Dittmer, S., et al.: Function secret sharing for PSI-CA: with applications to private contact tracing. arXiv preprint arXiv:2012.13053 (2020)

15. Dong, C., Chen, L., Wen, Z.: When private set intersection meets big data: an efficient and scalable protocol. In: Proceedings of the 2013 ACM SIGSAC Conference on Computer & Communications Security, pp. 789–800 (2013)

16. Duong, T., Phan, D.H., Trieu, N.: Catalic: delegated PSI cardinality with applications to contact tracing. In: Moriai, S., Wang, H. (eds.) International Conference on the Theory and Application of Cryptology and Information Security, pp. 870–899. Springer, Cham (2020). https://doi.org/10.1007/978-3-030-64840-4_29

17. ElGamal, T.: A public key cryptosystem and a signature scheme based on discrete logarithms. IEEE Trans. Inf. Theory **31**(4), 469–472 (1985)

18. Flavio, B., et al.: pailliercryptolib (2023)

19. Fotakis, D., Pagh, R., Sanders, P., Spirakis, P.: Space efficient hash tables with worst case constant access time. Theory Comput. Syst. **38**(2), 229–248 (2005)

20. Freedman, M.J., Ishai, Y., Pinkas, B., Reingold, O.: Keyword search and oblivious pseudorandom functions. In: Kilian, J. (ed.) TCC 2005. LNCS, vol. 3378, pp. 303–324. Springer, Heidelberg (2005). https://doi.org/10.1007/978-3-540-30576-7_17

21. Freedman, M.J., Nissim, K., Pinkas, B.: Efficient private matching and set intersection. In: Cachin, C., Camenisch, J.L. (eds.) EUROCRYPT 2004. LNCS, vol. 3027, pp. 1–19. Springer, Heidelberg (2004). https://doi.org/10.1007/978-3-540-24676-3_1

22. Goldwasser, S., Micali, S.: Probabilistic encryption & how to play mental poker keeping secret all partial information. In: Providing Sound Foundations for Cryptography: on the Work of Shafi Goldwasser and Silvio Micali, pp. 173–201 (2019)

23. Hallgren, P., Orlandi, C., Sabelfeld, A.: PrivatePool: privacy-preserving ridesharing. In: 2017 IEEE 30th Computer Security Foundations Symposium (CSF), pp. 276–291. IEEE (2017)

24. Hazay, C., Lindell, Y.: Efficient protocols for set intersection and pattern matching with security against malicious and covert adversaries. In: Canetti, R. (eds.) Theory of Cryptography Conference, pp. 155–175. Springer, Cham (2008). https://doi.org/10.1007/978-3-540-78524-8_10

25. Hazay, C., Venkitasubramaniam, M.: Scalable multi-party private set-intersection. In: Fehr, S. (eds.) IACR International Workshop on Public Key Cryptography, pp. 175–203. Springer, Cham (2017). https://doi.org/10.1007/978-3-662-54365-8_8

26. Hoffstein, J., Pipher, J., Silverman, J.H.: NTRU: a ring-based public key cryptosystem. In: Buhler, J.P. (eds.) International Algorithmic Number Theory Symposium, pp. 267–288. Springer, Cham (1998). https://doi.org/10.1007/BFb0054868

27. Hu, J., Chen, J., Dai, W., Wang, H.: Fully homomorphic encryption-based protocols for enhanced private set intersection functionalities. Cryptology ePrint Archive (2023)

28. Huberman, B.A., Franklin, M., Hogg, T.: Enhancing privacy and trust in electronic communities. In: Proceedings of the 1st ACM Conference on Electronic Commerce, pp. 78–86 (1999)

29. Ion, M., et al.: On deploying secure computing: private intersection-sum-with-cardinality. In: 2020 IEEE European Symposium on Security and Privacy (EuroS&P), pp. 370–389. IEEE (2020)

30. Kales, D., Rechberger, C., Schneider, T., Senker, M., Weinert, C.: Mobile private contact discovery at scale. In: 28th USENIX Security Symposium (USENIX Security 2019), pp. 1447–1464 (2019)

31. Kamara, S., Mohassel, P., Raykova, M., Sadeghian, S.: Scaling private set intersection to billion-element sets. In: Christin, N., Safavi-Naini, R. (eds.) FC 2014. LNCS, vol. 8437, pp. 195–215. Springer, Heidelberg (2014). https://doi.org/10.1007/978-3-662-45472-5_13

32. Kerschbaum, F.: Outsourced private set intersection using homomorphic encryption. In: Proceedings of the 7th ACM Symposium on Information, Computer and Communications Security, pp. 85–86 (2012)

33. Kolesnikov, V., Kumaresan, R., Rosulek, M., Trieu, N.: Efficient batched oblivious PRF with applications to private set intersection. In: Proceedings of the 2016 ACM SIGSAC Conference on Computer and Communications Security, pp. 818–829 (2016)

34. Le, P.H., Ranellucci, S., Gordon, S.D.: Two-party private set intersection with an untrusted third party. In: Proceedings of the 2019 ACM SIGSAC Conference on Computer and Communications Security, pp. 2403–2420 (2019)

35. Meadows, C.: A more efficient cryptographic matchmaking protocol for use in the absence of a continuously available third party. In: 1986 IEEE Symposium on Security and Privacy, pp. 134–134. IEEE (1986)

36. Motwani, R., Raghavan, P.: Randomized algorithms. ACM Comput. Surv. (CSUR) **28**(1), 33–37 (1996)

37. Nagaraja, S., Mittal, P., Hong, C.Y., Caesar, M., Borisov, N.: {BotGrep}: finding {P2P} bots with structured graph analysis. In: 19th USENIX Security Symposium (USENIX Security 2010) (2010)

38. Orrù, M., Orsini, E., Scholl, P.: Actively secure 1-out-of-N OT extension with application to private set intersection. In: Handschuh, H. (ed.) CT-RSA 2017. LNCS, vol. 10159, pp. 381–396. Springer, Cham (2017). https://doi.org/10.1007/978-3-319-52153-4_22

39. Pagh, R., Rodler, F.F.: Cuckoo hashing. J. Algorithms **51**(2), 122–144 (2004)

40. Paillier, P.: Public-key cryptosystems based on composite degree residuosity classes. In: Stern, J. (ed.) EUROCRYPT 1999. LNCS, vol. 1592, pp. 223–238. Springer, Heidelberg (1999). https://doi.org/10.1007/3-540-48910-X_16

41. Pinkas, B., Rosulek, M., Trieu, N., Yanai, A.: SpOT-Light: lightweight private set intersection from sparse ot extension. In: Boldyreva, A., Micciancio, D. (eds.) CRYPTO 2019. LNCS, vol. 11694, pp. 401–431. Springer, Cham (2019). https://doi.org/10.1007/978-3-030-26954-8_13

42. Pinkas, B., Schneider, T., Segev, G., Zohner, M.: Phasing: private set intersection using permutation-based hashing. In: 24th USENIX Security Symposium (USENIX Security 2015), pp. 515–530 (2015)

43. Pinkas, B., Schneider, T., Smart, N.P., Williams, S.C.: Secure two-party computation is practical. In: Matsui, M. (ed.) ASIACRYPT 2009. LNCS, vol. 5912, pp. 250–267. Springer, Heidelberg (2009). https://doi.org/10.1007/978-3-642-10366-7_15

44. Pinkas, B., Schneider, T., Tkachenko, O., Yanai, A.: Efficient circuit-based PSI with linear communication. In: Ishai, Y., Rijmen, V. (eds.) EUROCRYPT 2019. LNCS, vol. 11478, pp. 122–153. Springer, Cham (2019). https://doi.org/10.1007/978-3-030-17659-4_5

45. Pinkas, B., Schneider, T., Weinert, C., Wieder, U.: Efficient circuit-based PSI via Cuckoo hashing. In: Nielsen, J.B., Rijmen, V. (eds.) EUROCRYPT 2018. LNCS, vol. 10822, pp. 125–157. Springer, Cham (2018). https://doi.org/10.1007/978-3-319-78372-7_5

46. Pinkas, B., Schneider, T., Zohner, M.: Scalable private set intersection based on OT extension. ACM Trans. Privacy Secur. (TOPS) **21**(2), 1–35 (2018)

47. Raab, M., Steger, A.: "Balls into Bins"—a simple and tight analysis. In: Luby, M., Rolim, J.D.P., Serna, M. (eds.) RANDOM 1998. LNCS, vol. 1518, pp. 159–170. Springer, Heidelberg (1998). https://doi.org/10.1007/3-540-49543-6_13

48. Rindal, P., Rosulek, M.: Malicious-secure private set intersection via dual execution. In: Proceedings of the 2017 ACM SIGSAC Conference on Computer and Communications Security, pp. 1229–1242 (2017)

49. Rindal, P., Schoppmann, P.: VOLE-PSI: fast OPRF and Circuit-PSI from Vector-OLE. In: Canteaut, A., Standaert, F.-X. (eds.) EUROCRYPT 2021. LNCS, vol. 12697, pp. 901–930. Springer, Cham (2021). https://doi.org/10.1007/978-3-030-77886-6_31

50. Trieu, N., Shehata, K., Saxena, P., Shokri, R., Song, D.: Epione: lightweight contact tracing with strong privacy. arXiv preprint arXiv:2004.13293 (2020)

51. Wu, M., Yuen, T.H.: Efficient unbalanced private set intersection cardinality and user-friendly privacy-preserving contact tracing. In: 32nd USENIX Security Symposium (USENIX Security 2023), pp. 283–300 (2023)

52. Yeo, F.Y., Ying, J.H.: Third-party private set intersection. In: 2023 IEEE International Symposium on Information Theory (ISIT), pp. 1633–1638. IEEE (2023)

53. Yeo, F.Y., Ying, J.H.: A near-linear quantum-safe third-party private set intersection protocol. Cryptology ePrint Archive (2024)

Revisiting Generic Conversion from Non-adaptive to Adaptively Secure IBS: Tightness and an Extension

Sanjit Chatterjee[1] and Tapas Pandit[2]([✉])

[1] Department of Computer Science and Automation, Indian Institute of Science Bangalore, Bangalore, India
`sanjit@iisc.ac.in`
[2] Plaksha University, Mohali, India
`tapas.pandit@plaksha.edu.in`

Abstract. Existential unforgeability under chosen-identity/message attack (EUF-ID-CMA) is the standard security model for identity-based signatures (IBS), where an attacker can adaptively choose message-identity pairs and identities respectively for the signing and key oracles. Recently, Pan and Wagner (PQC 2021) claimed to have realized tightly EUF-ID-CMA-secure IBS schemes from lattices using a two-stage approach: first, constructing a non-adaptively tightly secure IBS scheme from lattices, and then lifting this scheme to adaptive security (EUF-ID-CMA) using two generic approaches - one based on chameleon hashes in the standard model and the other on hash functions in the random oracle model (ROM).

In this paper, we critically analyze the adaptive unforgeability model used by Pan and Wagner, as well as the consequent security reduction of their generic conversions from non-adaptive to adaptive security, both in the standard model and in the ROM. We identify some subtle yet serious gaps in their approach. To bridge these gaps, we propose new security reductions for their generic conversions while arguing why the Pan-Wagner technique by itself is unlikely to yield a tight reduction. On the other hand, we show that their generic technique can be extended to application scenarios, such as IoT, where a user can register more than one signing devices under a single identity. The extended scheme has hardly any additional overhead and we argue its security in a suitably modified version of EUF-ID-CMA model.

Keywords: Identity-Based Signature · Adaptive security · Non-adaptive security · Generic construction · Tightness

1 Introduction

Identity-based cryptosystems, introduced by Shamir in 1984 [Sha84], simplify the certificate management process of traditional public-key systems

© The Author(s), under exclusive license to Springer Nature Switzerland AG 2025
S. Mukhopadhyay and P. Stănică (Eds.): INDOCRYPT 2024, LNCS 15495, pp. 282–304, 2025.
https://doi.org/10.1007/978-3-031-80308-6_13

[DH76, ElG84]. Since then, numerous proposals for identity-based signatures (IBS) [Hes02, Pat02, CC03] have been studied under various settings like RSA, discrete logarithm, bilinear pairing etc. More recently, in the post-quantum settings such as lattices and multivariate quadratic polynomials, several IBS constructions [Rüc10, CLND19, Luy19, CDP21, PW21] have been proposed.

Existential unforgeability under chosen-identity/message attack (EUF-ID-CMA) is the standard security model [BNN04, LPLL20] for identity-based signatures (IBS). An attacker in this model can adaptively choose query message-identity pairs and identities respectively for the sign and key oracles. In the literature, a couple of generic techniques to construct IBS from different primitives have been investigated. For instance, in one of the earliest and comprehensive treatment, [BNN04, BNN09] proposed a generic construction that uses ordinary digital signatures at two levels: one for generating public parameters and the master secret key, and the other for generating keys for individual identities. In their second generic construction [BNN04, BNN09] considered a standard identification scheme along with a trapdoor sampleable relation (TSR) followed by Fiat-Shamir transform [FS86]. However, these generic approaches lack tight security reductions.

Note that tightly secure cryptographic schemes offer better concrete security assurance than their non-tight counterparts with comparable parameters [CMS11, CKMS16]. So, some follow-up works focused on that aspect. For example, Zhang et al. proposed a generic construction of IBS [ZLGL19] using two digital signatures, one secure in the single-user setting and the other in the multi-user setting. They showed that their construction achieves tight security in a restricted version of EUF-ID-CMA model. More recently, Lee et al. [LPLL20] showed that the same construction also achieves tight security in the standard EUF-ID-CMA model without requiring any additional assumptions on the primitive signature schemes. It's worth noting that this construction is the same as the Bellare-Namprempre-Neven IBS construction [BNN09], except that the latter construction uses two instances of the same signature. This type of construction is called certificate-based IBS construction, which is inefficient compared to the other generic construction, as a signature under this IBS contains a certificate consisting of a public key and a signature.

A different approach has been pursued by Pan-Wagner in [PW21]. This work claims to realize tightly EUF-ID-CMA secure IBS schemes from lattices following a two-stage approach. They first constructed IBS scheme from lattices achieving tight reduction in a non-adaptive security model. They then lift such a scheme to adaptive security (EUF-ID-CMA) using two generic approaches: one based on chameleon hashes in the standard model, and the other based on hash functions in the random oracle model (ROM). The above Pan-Wagner result has recently been extended [SPMC23] in the quantum random oracle model (QROM).

Our Result. In this paper, we critically analyze the adaptive unforgeability model of [PW21] as well as the security reductions of their generic conversions from the non-adaptive to the adaptive security both in the standard model and ROM. We identify some crucial gaps in their approach and then propose new security reductions for their generic conversions. We also argue why the Pan-Wagner generic conversion technique is unlikely to yield a tight reduction. On the other hand, we show that their generic technique can be extended for application scenarios where a user can register more than one signing devices under a single identity. With very little modification, their proposal can realize tight security in a slightly restricted but realistic model. Our results are detailed as follows:

1. Section 3 revisits the claims of [PW21]. In Sect. 3.1, we identify some serious gap in the definition of adaptive unforgeability security used in [PW21]. To distinguish from the accepted notion of EUF-ID-CMA security (c.f., Definition 4), here we refer to it as EUF-ID-CMA-PW (c.f., Definition 5). Note that, in the IBS (c.f., Definition 3), a user with identity id generates signatures on (id, m) for different messages m using the same private key $\mathsf{sk}_{\mathsf{id}}$ and that environment is correctly captured in the EUF-ID-CMA model. In contrast, in the EUF-ID-CMA-PW model, for the same identity id, on input (id, m), the signing oracle $\mathcal{O}_{\mathsf{sign}}$ each time generates the signature using a *freshly generated* private key. Due to such significant deviation from the standard notion, the EUF-ID-CMA-PW security definition fails to accurately capture the actual protocol environment. In Sect. 3.2, we demonstrate gaps in the security reduction of the generic constructions in [PW21] in the standard EUF-ID-CMA model. This is done by considering two scenarios where the attacker (i) makes two (or more) signing queries on the same identity, and (ii) first makes a signing query followed by a key query on the same identity. In both cases, we show that the simulation does not follow the standard security game.

2. The above gaps are addressed in Sect. 4 through new security reductions for both the generic constructions of Pan-Wagner. We first establish a tight reduction of their constructions in a slightly restricted version of the standard EUF-ID-CMA model, where the identity sets for key queries and sign queries do not overlap. We then demonstrate how to generically lift this security to the standard EUF-ID-CMA model, with a tightness loss linear in the number of key queries. We also argue why the Pan-Wagner technique by itself is unlikely to yield tight security reduction in the EUF-ID-CMA model.

3. Finally, in Sect. 5, we extend the Pan-Wagner results to a setting where a user can register a polynomial number of signing devices under a single identity. For this purpose, their proposed constructions are modified slightly, which does not incur any additional cost in signature generation/verification or signature length. The only additional cost is to run the key generation algorithm of the underlying non-adaptive IBS scheme for each registered device. We show that the extended schemes are tightly secure in a restricted model which can be relaxed at the cost of a loss in tightness.

2   Preliminaries

The basic notations, the definition of chameleon hash function and its security along with the syntax and standard security model of identity-based signature are given in this section.

Notations. For a set S, the notation $x \xleftarrow{\text{U}} S$ denotes that x is drawn uniformly at random from S. For an algorithm A and its input x, the notation $y \leftarrow A(x)$ denotes that when A is run on x, it outputs y, where A may take its internal random coin r as additional input. The notation $y \leftarrow A(x;r)$ denotes that the random coin r is supplied as part of input from outside. For $n \in \mathbb{N}$, we define $[n] = \{x \in \mathbb{N} : x \leq n\}$.

Definition 1 (Chameleon Hash Function [PW21]). *An ϵ_{trap}-chameleon hash function is a triplet* $\text{CHF} = (\text{CHGen}, \text{CHash}, \text{CHColl})$, *where*

- $\text{CHGen}(\kappa)$ *takes as input a security parameter κ, and outputs a hash key* hk *and a trapdoor* td. *We assume that* hk *defines a message space* $\mathcal{M}$, *a randomness (or salt) space* SaltSp *and a hash value space* $\mathcal{D}$.
- $\text{CHash}(\text{hk}, m; r)$ *takes as input the hash key* hk, *a message m and a salt r from* SaltSp *(internally), and outputs a hash value from* $\mathcal{D}$.
- $\text{CHColl}(\text{hk}, \text{td}, m, r, m')$ *takes as input the hash key* hk, *the trapdoor* td, *two messages m, m' and a salt r, and outputs a salt $r' \in \text{SaltSp}$.*
- *For all* $(\text{hk}, \text{td}) \leftarrow \text{CHGen}(\kappa)$, *for all $m, m' \in \mathcal{M}$ the following two distribution will have statistical distance at most ϵ_{trap}:*
    1. $\{(r,h) : r \xleftarrow{\text{U}} \text{SaltSp}, h := \text{CHash}(\text{hk}, m; r)\}$
    2. $\{(r,h) : r' \xleftarrow{\text{U}} \text{SaltSp}, h := \text{CHash}(\text{hk}, m'; r'), r \leftarrow \text{CHColl}(\text{hk}, \text{td}, m', r', m)\}$

When ϵ_{trap} is negligible function in κ, we simply say CHF *is a chameleon hash function.*

Definition 2 (Collision Resistance). *A chameleon hash function* $\text{CHF} = (\text{CHGen}, \text{CHash}, \text{CHColl})$ *is collision resistant, if for all PPT algorithms $\mathcal{A}$, the advantage*

$$\text{Adv}_{\mathcal{A}}^{\text{Collision}}(\kappa) := \Pr\left[\text{Exp}_{\mathcal{A}}^{\text{Collision}}(\kappa) = 1\right]$$

in $\text{Exp}_{\mathcal{A}}^{\text{Collision}}(\kappa)$ *defined in Fig. 1 is a negligible function in κ.*

$\text{Exp}_{\mathcal{A}}^{\text{Collision}}(\kappa)$:
1: $(\text{hk}, \text{td}) \leftarrow \text{CHGen}(1^\kappa)$
2: $(m, r, m', r') \leftarrow \mathcal{A}(\text{hk})$
3: **if** $\text{CHash}(\text{hk}, m; r) = \text{CHash}(\text{hk}, m'; r')$ and $(m, r) \neq (m', r')$ **then**
4:     **return** 1
5: **end if**
6: **return** 0

Fig. 1. Experiment for collision resistant property of chameleon hash function.

We recall the standard definition and correctness of IBS as follows.

Definition 3 (IBS Scheme). *It consists of four PPT algorithms -* Setup, KeyGen, Sign *and* Ver.

- Setup*: It takes as input a security parameter κ, and outputs public parameters and master secret key pair* $(\mathsf{pp}, \mathsf{msk})$.
- KeyGen*: It takes as input public parameters* pp, *master secret key* msk *and an identity* $\mathsf{id} \in \mathcal{ID}$, *where* $\mathcal{ID}$ *is the identity space, and outputs a signing key* $\mathsf{sk_{id}}$.
- Sign*: It takes as input public parameters* pp, *a message* $m \in \mathcal{M}$, *where* $\mathcal{M}$ *is the message space, and a secret key* $\mathsf{sk_{id}}$, *and outputs a signature* σ.
- Ver*: It takes as input public parameters* pp, *an identity* id, *a message* m *and a signature* σ. *It outputs a value 1, if* σ *is a valid signature for* (id, m), *else it outputs 0.*

Correctness: *For all* $(\mathsf{pp}, \mathsf{msk}) \leftarrow \mathsf{Setup}(1^\kappa)$, *for all* $\mathsf{id} \in \mathcal{ID}$, $\mathsf{sk_{id}} \leftarrow \mathsf{KeyGen}(\mathsf{pp}, \mathsf{msk}, \mathsf{id})$ *and for all* $m \in \mathcal{M}$, *it is required that*

$$\mathsf{Ver}(\mathsf{pp}, \mathsf{id}, m, \mathsf{IBS.Sign}(\mathsf{pp}, m, \mathsf{sk_{id}})) = 1.$$

$\mathsf{Exp}_\mathcal{A}^{\text{EUF-ID-CMA}}(\kappa)$:	$\mathcal{O}_\text{key}(\mathsf{id})$:	$\mathcal{O}_\text{sign}(\mathsf{id}, m)$:
1: $\mathcal{Q}_\text{key} := \emptyset, \mathcal{Q}_\text{sign} := \emptyset, \mathcal{L}_\text{sk} := \emptyset$	1: **if** $\mathsf{id} \notin \mathcal{Q}_\text{key}$ **then**	1: **if** $(\mathsf{id}, m) \notin \mathcal{Q}_\text{sign}$ **then**
2: $(\mathsf{pp}, \mathsf{msk}) \leftarrow \mathsf{Setup}(1^\kappa)$	2: $\mathcal{Q}_\text{key} := \mathcal{Q}_\text{key} \cup \{\mathsf{id}\}$	2: $\mathcal{Q}_\text{sign} := \mathcal{Q}_\text{sign} \cup \{(\mathsf{id}, m)\}$
3: $(\mathsf{id}^*, m^*, \sigma^*) \leftarrow \mathcal{A}^{\{\mathcal{O}_\text{sign}, \mathcal{O}_\text{key}\}}(1^\kappa, \mathsf{pp})$	3: **end if**	3: **end if**
4: **if** $\mathsf{id}^* \in \mathcal{Q}_\text{key}$ or $(\mathsf{id}^*, m^*) \in \mathcal{Q}_\text{sign}$ **then**	4: **if** $(\mathsf{id}, \mathsf{sk_{id}}) \in \mathcal{L}_\text{sk}$ **then**	4: **if** $(\mathsf{id}, \mathsf{sk_{id}}) \notin \mathcal{L}_\text{sk}$ **then**
5: **return** 0	5: **return** $\mathsf{sk_{id}}$	5: $\mathsf{sk_{id}} \leftarrow \mathsf{KeyGen}(\mathsf{pp}, \mathsf{msk}, \mathsf{id})$
6: **end if**	6: **end if**	6: $\mathcal{L}_\text{sk} := \mathcal{L}_\text{sk} \cup \{(\mathsf{id}, \mathsf{sk_{id}})\}$
7: **return** $\mathsf{Ver}(\mathsf{pp}, \mathsf{id}^*, m^*, \sigma^*)$	7: $\mathsf{sk_{id}} \leftarrow \mathsf{KeyGen}(\mathsf{pp}, \mathsf{msk}, \mathsf{id})$	7: **end if**
	8: $\mathcal{L}_\text{sk} := \mathcal{L}_\text{sk} \cup \{(\mathsf{id}, \mathsf{sk_{id}})\}$	8: **return** $\mathsf{Sign}(\mathsf{pp}, m, \mathsf{sk_{id}})$
	9: **return** $\mathsf{sk_{id}}$	

Fig. 2. Experiment for EUF-ID-CMA of IBS according to [BNN04,LPLL20]

We recall here the EUF-ID-CMA security model from [BNN04] which is generally accepted as the standard definition of security for IBS and has been followed in numerous papers (see, for example, [LPLL20]).

Definition 4 (EUF-ID-CMA [BNN04,LPLL20]). *An IBS scheme is said to be EUF-ID-CMA secure, if for all PPT algorithms* $\mathcal{A}$, *the advantage*

$$\mathsf{Adv}_\mathcal{A}^{\text{EUF-ID-CMA}}(\kappa) := \Pr\left[\mathsf{Exp}_\mathcal{A}^{\text{EUF-ID-CMA}}(\kappa) = 1\right]$$

in $\mathsf{Exp}_\mathcal{A}^{\text{EUF-ID-CMA}}(\kappa)$ *defined in the left column of Fig. 2 is a negligible function in* κ, *where* $\mathcal{A}$ *is provided access to signature oracle* $\mathcal{O}_\text{sign}$ *(shown in the right column) and key-gen oracle* $\mathcal{O}_\text{key}$ *(shown in the middle column) at most polynomial number of times, and* $\mathcal{Q}_\text{key}$ *is the set of identities on which key-gen queries were made and* $\mathcal{Q}_\text{sign}$ *is the set of identity-message pairs on which signature queries were made.*

$\mathsf{Exp}_{\mathcal{A}}^{\text{EUF-ID-CMA-PW}}(\kappa):$	$\mathcal{O}_{\text{key}}(\text{id}):$
1: $\mathcal{Q}_{\text{key}} := \emptyset, \mathcal{Q}_{\text{sign}} := \emptyset$	1: $\mathcal{Q}_{\text{key}} := \mathcal{Q}_{\text{key}} \cup \{\text{id}\}$
2: $(\text{pp}, \text{msk}) \leftarrow \mathsf{Setup}(1^{\kappa})$	2: **return** $\mathsf{KeyGen}(\text{pp}, \text{msk}, \text{id})$
3: $(\text{id}^*, m^*, \sigma^*) \leftarrow \mathcal{A}^{\{\mathcal{O}_{\text{sign}}, \mathcal{O}_{\text{key}}\}}(1^{\kappa}, \text{pp})$	
4: **if** $\text{id}^* \in \mathcal{Q}_{\text{key}}$ or $(\text{id}^*, m^*) \in \mathcal{Q}_{\text{sign}}$ **then**	$\mathcal{O}_{\text{sign}}(\text{id}, m):$
5: **return** 0	1: $\mathcal{Q}_{\text{sign}} := \mathcal{Q}_{\text{sign}} \cup \{(\text{id}, m)\}$
6: **end if**	2: $\text{sk}_{\text{id}} \leftarrow \mathsf{KeyGen}(\text{pp}, \text{msk}, \text{id})$
7: **return** $\mathsf{Ver}(\text{pp}, \text{id}^*, m^*, \sigma^*)$	3: **return** $\mathsf{Sign}(\text{pp}, m, \text{sk}_{\text{id}})$

Fig. 3. Experiment for EUF-ID-CMA-PW of IBS according to [PW21]

3   Revisiting the Pan-Wagner Security Claim

3.1   Security Definition

In this section, we take a critical look at the Pan-Wagner generic results [PW21]. First we reproduce their adaptive security definition of IBS which is denoted here as EUF-ID-CMA-PW to distinguish from the standard notion given in Definition 4.

Definition 5 (EUF-ID-CMA-PW [PW21]). *An IBS scheme is said to be EUF-ID-CMA-PW secure, if for all PPT algorithms $\mathcal{A}$, the advantage*

$$\mathsf{Adv}_{\mathcal{A}}^{\text{EUF-ID-CMA-PW}}(\kappa) := \Pr\left[\mathsf{Exp}_{\mathcal{A}}^{\text{EUF-ID-CMA-PW}}(\kappa) = 1\right]$$

in $\mathsf{Exp}_{\mathcal{A}}^{\text{EUF-ID-CMA-PW}}(\kappa)$ defined in Fig. 3 is a negligible function in κ, where $\mathcal{A}$ is provided access to signature oracle $\mathcal{O}_{\text{sign}}$ and key-gen oracle $\mathcal{O}_{\text{key}}$ at most polynomial number of times, and $\mathcal{Q}_{\text{key}}$ is the set of identities on which key-gen queries were made and $\mathcal{Q}_{\text{sign}}$ is the set of identity-message pairs on which signature queries were made.

Pan and Wagner also defined a non-adaptive security model [PW21], denoted by EUF-naCMA. As the name suggests, it requires an attacker to declare the sets of identities and identity-message pairs respectively for key and sign oracles before observing the public parameters. It is formally reproduced as follows.

Definition 6 (EUF-naCMA [PW21]). *An IBS scheme is said to be EUF-naCMA secure, if for all PPT algorithms $\mathcal{A}$, the advantage*

$$\mathsf{Adv}_{\mathcal{A}}^{\text{EUF-naCMA}}(\kappa) := \Pr\left[\mathsf{Exp}_{\mathcal{A}}^{\text{EUF-naCMA}}(\kappa) = 1\right]$$

in $\mathsf{Exp}_{\mathcal{A}}^{\text{EUF-naCMA}}(\kappa)$ defined in Fig. 4 is a negligible function in κ.

$\mathsf{Exp}_{\mathcal{A}}^{\text{EUF-naCMA}}(\kappa)$:

1: $(\mathcal{Q}_{\text{key}}, \mathcal{Q}_{\text{sign}}) \leftarrow \mathcal{A}(1^{\kappa})$
2: $(\text{pp}, \text{msk}) \leftarrow \mathsf{Setup}(1^{\kappa})$
3: **for** $\text{id} \in \mathcal{Q}_{\text{key}}$ **do**
4:     $\text{sk}_{\text{id}} \leftarrow \mathsf{KeyGen}(\text{pp}, \text{msk}, \text{id})$
5:     $\mathcal{L}_{\text{sk}} := \mathcal{L}_{\text{sk}} \cup \{\text{sk}_{\text{id}}\}$
6: **end for**
7: **for** $(\text{id}, m) \in \mathcal{Q}_{\text{sign}}$ **do**
8:     $\sigma \leftarrow \mathsf{Sign}(\text{pp}, m, \text{sk}_{\text{id}})$
9:     $\mathcal{L}_{\text{sign}} := \mathcal{L}_{\text{sign}} \cup \{\sigma\}$
10: **end for**
11: $(\text{id}^*, m^*, \sigma^*) \leftarrow \mathcal{A}(\text{pp}, \mathcal{L}_{\text{sk}}, \mathcal{L}_{\text{sign}})$
12: **if** $\text{id}^* \in \mathcal{Q}_{\text{key}}$ **or** $(\text{id}^*, m^*) \in \mathcal{Q}_{\text{sign}}$ **then**
13:     **return** 0
14: **end if**
15: **return** $\mathsf{Ver}(\text{pp}, \text{id}, m, \sigma)$

Fig. 4. Experiment EUF-naCMA of identity-based signature according to [PW21]

Deviations from Standard Definition: Even though [PW21] presented Definition 5 as the notion of adaptive security for IBS, there are some subtle differences from the established notion of EUF-ID-CMA security for IBS, particularly in the key and sign queries. Please refer to lines 1–6 of $\mathcal{O}_{\text{sign}}$ and $\mathcal{O}_{\text{key}}$ in Fig. 2. Note that in the EUF-ID-CMA definition, a list $\mathcal{L}_{\text{sk}}$ consisting of $(\text{id}, \text{sk}_{\text{id}})$ is maintained and updated every time a secret key is generated for a particular identity. Furthermore, there are two *if* statements (respectively in line 1 and line 4) within the key and sign oracles in the standard EUF-ID-CMA definition.

In the IBS, key generation for any identity id is typically a one-time affair and the corresponding sk_{id} is then used by the owner of the corresponding id to generate polynomial many signatures. The security game naturally needs to simulate this protocol environment which is satisfied by Definition 4. In the corresponding security game (Fig. 2), the first time some identity id is queried to either of the two oracles, a secret key (sk_{id}) is generated for id and stored in $\mathcal{L}_{\text{sk}}$. This ensures that the *same* sk_{id} is used to generate all the signatures corresponding to that particular identity. Moreover, suppose that $\mathcal{A}$ makes a key query after some sign queries, then the private key returned for that identity is the one used to generate those signatures. However, these conditional statements are dropped from the definition of EUF-ID-CMA-PW (see Fig. 3). Thus, the corresponding security game will deviate significantly from the standard EUF-ID-CMA game and hence, the IBS protocol environment, unless key generation is a deterministic algorithm. We next elaborate on this gap in the concrete context of the security reductions given in [PW21].

3.2 Security Reductions

Pan and Wagner [PW21] proposed generic constructions of adaptively secure IBS from non-adaptively secure IBS, which are claimed to achieve tight reduction in both the random oracle and standard models. We now revisit the Pan-Wagner construction and its security reduction to demonstrate a gap in their argument in terms of the accepted notion of IBS security, i.e., Definition 4. This point is illustrated here in the context of their chameleon hash based construction in the standard model and a similar argument works for the construction in the random oracle model.

Let $\mathsf{IBS}' = (\mathsf{IBS}'.\mathsf{Setup}, \mathsf{IBS}'.\mathsf{KeyGen}, \mathsf{IBS}'.\mathsf{Sign}, \mathsf{IBS}'.\mathsf{Ver})$ be a non-adaptively secure primitive identity-based signature scheme with identity space $\mathcal{ID}'$ and message space $\mathcal{M}'$. Let $\mathsf{CHF} = (\mathsf{CHGen}, \mathsf{CHash}, \mathsf{CHColl})$ be a chameleon hash function. Let $\mathsf{IBS} = (\mathsf{IBS}.\mathsf{Setup}, \mathsf{IBS}.\mathsf{KeyGen}, \mathsf{IBS}.\mathsf{Sign}, \mathsf{IBS}.\mathsf{Ver})$ be the target IBS scheme constructed using chameleon hash functions and primitive IBS scheme IBS' as given in Fig. 5 with identity space $\mathcal{ID}$ and message space $\mathcal{M}$.[1]

$\mathsf{IBS}.\mathsf{Setup}(\kappa)$:	$\mathsf{IBS}.\mathsf{Sign}(\mathsf{pp}, m, \mathsf{sk}_{\mathsf{id}})$:
1: $(\mathsf{hk}, \mathsf{td}) \leftarrow \mathsf{CHGen}(\kappa)$	1: parse $\mathsf{sk}_{\mathsf{id}}$ as $\mathsf{sk}_{\mathsf{id}} = (\mathsf{sk}_{\mathsf{id}'}, r)$
2: $(\mathsf{pp}', \mathsf{msk}') \leftarrow \mathsf{IBS}'.\mathsf{Setup}(\kappa)$	2: $s \xleftarrow{\mathsf{U}} \mathsf{SaltSp}$
3: $\mathsf{pp} := (\mathsf{pp}', \mathsf{hk})$ and $\mathsf{msk} := \mathsf{msk}'$	3: $m' \leftarrow \mathsf{CHash}(\mathsf{hk}, m; s)$
4: **return** $(\mathsf{pp}, \mathsf{msk})$	4: $\sigma' \leftarrow \mathsf{IBS}'.\mathsf{Sign}(\mathsf{pp}', m', \mathsf{sk}_{\mathsf{id}'})$
	5: $\sigma := (\sigma', r, s)$
$\mathsf{IBS}.\mathsf{KeyGen}(\mathsf{pp}, \mathsf{msk}, \mathsf{id})$:	6: **return** σ
1: $r \xleftarrow{\mathsf{U}} \mathsf{SaltSp}$	
2: $\mathsf{id}' \leftarrow \mathsf{CHash}(\mathsf{hk}, \mathsf{id}; r)$	$\mathsf{IBS}.\mathsf{Ver}(\mathsf{pp}, \mathsf{id}, m, \sigma)$:
3: $\mathsf{sk}_{\mathsf{id}'} \leftarrow \mathsf{IBS}'.\mathsf{KeyGen}(\mathsf{pp}', \mathsf{msk}', \mathsf{id}')$	1: parse σ as $\sigma = (\sigma', r, s)$
4: $\mathsf{sk}_{\mathsf{id}} := (\mathsf{sk}_{\mathsf{id}'}, r)$	2: $\mathsf{id}' \leftarrow \mathsf{CHash}(\mathsf{hk}, \mathsf{id}; r)$
5: **return** $\mathsf{sk}_{\mathsf{id}}$	3: $m' \leftarrow \mathsf{CHash}(\mathsf{hk}, m; s)$
	4: **return** $\mathsf{IBS}'.\mathsf{Ver}(\mathsf{pp}', \mathsf{id}', m', \sigma')$

Fig. 5. Construction of EUF-ID-CMA-PW secure IBS from EUF-naCMA secure IBS' using chameleon hash

We briefly discuss how [PW21] reduces the adaptive security of IBS to the non-adaptive security of IBS' (see Sect. 3.1 of [PW21] for further details). Let $\mathcal{A}$ be an attacker and S be a simulator that plays the role of a challenger in the EUF-ID-CMA-PW security game of IBS (see Fig. 3). Using $\mathcal{A}$, S breaks the EUF-naCMA security (Definition 6) of IBS'. In the EUF-naCMA security game, at the very beginning S asks its challenger all the key and sign queries. Let $\mathcal{Q}'_{\mathsf{key}}$ and $\mathcal{Q}'_{\mathsf{sign}}$ be the lists of identities and message-identity pairs respectively

---

[1] In [PW21], the identity space and message space were denoted by $\mathcal{ID}$ and $\mathcal{M}$ respectively, for both IBS' and IBS whereas we use separate notations for the two schemes.

for which its challenger provides S the corresponding keys and signatures. S utilizes the chameleon hash key hk to prepare the lists $\mathcal{Q}'_{key}$ and $\mathcal{Q}'_{sign}$ so that it can later respond to $\mathcal{A}$ who adaptively queries for keys and signatures in the EUF-ID-CMA-PW game. In particular, S prepares the i-th entry of $\mathcal{Q}'_{sign}$ as $id'_i = \mathsf{CHash}(hk, 0; r'_i)$ and $m'_i = \mathsf{CHash}(hk, 0; s'_i)$, where r'_i and s'_i are the random salts chosen by S. Suppose σ'_i is the signature that S obtained from its challenger for (id'_i, m'_i). Later, when $\mathcal{A}$ makes the i-th signature query on some (id_i, m_i) of its choice, S utilizes the trapdoor td corresponding to the chameleon hash and calls the CHColl() algorithm. This way S *obtains* r_i and s_i such that $\mathsf{CHash}(hk, id_i; r_i) = id'$ and $\mathsf{CHash}(hk, m_i; s_i) = m'$ and returns (r_i, s_i, σ'_i) to $\mathcal{A}$. A similar approach is applied to prepare the list $\mathcal{Q}'_{key}$ and answer adaptive key queries of $\mathcal{A}$. Thus, S is able to map the i-th signature (resp. j-th key) query of $\mathcal{A}$ to the respective elements of $\mathcal{Q}'_{sign}$ and $\mathcal{Q}'_{key}$. Finally, when $\mathcal{A}$ returns a forgery, S uses the chameleon hash function to map that signature to a fresh identity-message pair corresponding to IBS′ yielding a tight reduction.

However, the proof given in [PW21] is not a *valid* reduction according to the standard EUF-ID-CMA security game (Definition 4). To see this consider the following two cases.

- Case I: $\mathcal{A}$ makes the i-th and j-th signature queries on the same identity as: (id, m_i) and (id, m_j). As per the protocol as well as the standard EUF-ID-CMA game, the same secret key sk_{id} has to be used to generate the two signatures. Hence, the same randomizer r must be included as part of the two returned signatures. However, in the reduction of [PW21] the randomizers will be different as the two queries are respectively mapped to the i-th and j-th elements of $\mathcal{Q}'_{sign}$. In particular, to generate the respective signatures, S first has to compute $r_1 = \mathsf{CHColl}(hk, td, 0, r_i, id)$ and $r_2 = \mathsf{CHColl}(hk, td, 0, r_j, id)$. Clearly, $r_1 \neq r_2$ due to the collision resistance property of the chameleon hash function.
- Case II: $\mathcal{A}$ first makes a signature query on some identity $\hat{id}$, followed by a key query on $\hat{id}$. The protocol definition of IBS as well as the EUF-ID-CMA game demand that the secret key returned against the key query for $\hat{id}$ must be the one used to respond to the previous sign query. This, in turn, implies that the r component in the two responses must be the same. However, as in the above case, the two queries yield different randomizers r, thus violating this condition in the reduction provided in [PW21].

We note that arguing in a similar line one can establish a gap in the security reduction of Pan-Wagner construction in the random oracle model. Thus, the main claim of [PW21] of realizing tightly secure IBS in EUF-ID-CMA model does not hold. It's worth noting that the results of [PW21] were recently extended in [SPMC23] to argue for tight security in the quantum random oracle model. Specifically, the authors of the latter paper claimed that the hash functions-based generic construction is EUF-ID-CMA secure in the QROM. However, similar to the results of Pan et al., this security claim also holds with respect to Definition 5 but not under the standard EUF-ID-CMA security model (c.f., Definition 4).

4   New Security Reduction

Given the gaps identified in the security proofs of [PW21], it's worth asking whether there exists a way to address the problem. We have both good and bad news on that front. We propose new security proofs of Pan-Wagner [PW21] generic constructions in the standard EUF-ID-CMA model and also argue why tight reduction is unlikely.

While it's possible to give a direct reduction, for ease of understanding as well as to maintain the generality, we prove security thorough two separate reductions. In the first step, we introduce a restriction on the adaptive adversary, $\mathcal{A}$. Let $\mathcal{Q}_{\mathsf{key}}$ (resp. $\mathcal{Q}_{\mathsf{id}}$) be the sets of identities on which $\mathcal{A}$ asks for key (resp. sign) queries, and let q_k (resp. q_s) denote the number of key (resp. sign) queries. As illustrated in Case II above, the Pan-Wagner reduction faces a problem whenever a sign query is followed by a key query on the same identity. So, first we consider a restricted version of the standard EUF-ID-CMA security model (c.f., Definition 4), denoted by WModel-I, by imposing the condition that $\mathcal{Q}_{\mathsf{key}} \cap \mathcal{Q}_{\mathsf{id}} = \emptyset$. We show in Theorem 1 that the Pan-Wagner construction in the random oracle model (reproduced in Fig. 6) admits a tight security reduction in WModel-I. Note that we still need to address the issue in Case I, namely, multiple signature queries on the same identity which is handled by setting $|\mathcal{Q}'_{\mathsf{sign}}| = q_s^2$, where q_s entries in $\mathcal{Q}'_{\mathsf{sign}}$ are associated with the same identity (and hence, the same r).

IBS.Setup(κ):

1: $(\mathsf{pp}', \mathsf{msk}') \leftarrow \mathsf{IBS}'.\mathsf{Setup}(\kappa)$
2: $\mathsf{pp} := \mathsf{pp}'$ and $\mathsf{msk} := \mathsf{msk}'$
3: **return** $(\mathsf{pp}, \mathsf{msk})$

IBS.KeyGen($\mathsf{pp}, \mathsf{msk}, \mathsf{id}$):

1: $r \xleftarrow{\mathsf{U}} \mathsf{SaltSp}$
2: $\mathsf{id}' \leftarrow \mathcal{H}_1(\mathsf{id}, r)$
3: $\mathsf{sk}_{\mathsf{id}'} \leftarrow \mathsf{IBS}'.\mathsf{KeyGen}(\mathsf{pp}', \mathsf{msk}', \mathsf{id}')$
4: $\mathsf{sk}_{\mathsf{id}} := (\mathsf{sk}_{\mathsf{id}'}, r)$
5: **return** $\mathsf{sk}_{\mathsf{id}}$

IBS.Sign($\mathsf{pp}, m, \mathsf{sk}_{\mathsf{id}}$):

1: parse $\mathsf{sk}_{\mathsf{id}}$ as $\mathsf{sk}_{\mathsf{id}} = (\mathsf{sk}_{\mathsf{id}'}, r)$
2: $s \xleftarrow{\mathsf{U}} \mathsf{SaltSp}$
3: $m' \leftarrow \mathcal{H}_2(m, s)$
4: $\sigma' \leftarrow \mathsf{IBS}'.\mathsf{Sign}(\mathsf{pp}', m', \mathsf{sk}_{\mathsf{id}'})$
5: $\sigma := (\sigma', r, s)$
6: **return** σ

IBS.Ver($\mathsf{pp}, \mathsf{id}, m, \sigma$):

1: parse σ as $\sigma = (\sigma', r, s)$
2: $\mathsf{id}' \leftarrow \mathcal{H}_1(\mathsf{id}, r)$
3: $m' \leftarrow \mathcal{H}_2(m, s)$
4: **return** $\mathsf{IBS}'.\mathsf{Ver}(\mathsf{pp}', \mathsf{id}', m', \sigma')$

Fig. 6. Construction of adaptively secure IBS from non-adaptively secure IBS', where $\mathcal{H}_1 : \mathcal{ID} \times \mathsf{SaltSp} \rightarrow \mathcal{ID}'$ and $\mathcal{H}_2 : \mathcal{M} \times \mathsf{SaltSp} \rightarrow \mathcal{M}'$ are hash functions and $\mathsf{SaltSp} = \{0, 1\}^\ell$.

Theorem 1. *Suppose there exists an adversary $\mathcal{A}$ who can break the WModel-I security of the target identity-based signature scheme IBS (see Fig. 6) in the random oracle model. Then, using $\mathcal{A}$ as a subroutine, we can create PPT algorithm*

S *for breaking the EUF-naCMA security (c.f., Definition 6) of the primitive scheme* IBS' *with the following relation of the respective advantages:*

$$\mathsf{Adv}_{\mathcal{A}}^{\mathsf{WModel\text{-}I}}(\kappa) \leq \mathsf{Adv}_{\mathsf{S}}^{\mathsf{EUF\text{-}naCMA}}(\kappa) + (q_k + 2q_s) \cdot \frac{q_h}{2^\ell} + \frac{q_k}{|\mathcal{ID}'|} + \frac{q_s^2}{|\mathcal{ID}'| \cdot |\mathcal{M}'|}$$

where q_k, q_s, and q_h are the numbers of key queries, sign queries and hash queries respectively.

Proof. We consider hybrid arguments through two different games: G_0 and G_1. If $\mathcal{A}$ is an attacker in game G_i, then its advantage in winning game G_i is denoted by $\mathsf{Adv}_{\mathcal{A}}^{G_i}(\kappa)$. We treat $\mathcal{H}_1$ and $\mathcal{H}_2$ as random oracles.

Let G_0 be defined as the security model WModel-I. Let G_1 be defined similarly to G_0, except that whenever the random oracle values, such as $\mathcal{H}_1(\mathsf{id}, r)$ and $\mathcal{H}_2(m, s)$, are already defined while answering key or sign queries, the game is aborted. Then, we have:

$$\left| \mathsf{Adv}_{\mathcal{A}}^{G_0}(\kappa) - \mathsf{Adv}_{\mathcal{A}}^{G_1}(\kappa) \right| \leq \Pr[\text{abort in } G_1].$$

Note that for each key query on id, abort will happen if the value $\mathcal{H}_1(\mathsf{id}, r)$ for a random salt r already appears in the list for $\mathcal{H}_1$. This will happen in a single key query with probability at most $q_h/2^\ell$. Similarly, the probability of an abort in a single sign query is at most $2 \cdot q_h/2^\ell$, where the factor of 2 in the numerator arises from two random oracle queries: one with the message and the other with the identity. Therefore, we have:

$$\left| \mathsf{Adv}_{\mathcal{A}}^{G_0}(\kappa) - \mathsf{Adv}_{\mathcal{A}}^{G_1}(\kappa) \right| \leq (q_k + 2q_s) \cdot q_h/2^\ell. \tag{1}$$

Let $\mathcal{A}$ be a WModel-I attacker of the target scheme IBS. We present a reduction of the EUF-naCMA security of IBS' from the restricted EUF-ID-CMA security of IBS in the context of environment G_1. To achieve this, we construct an algorithm (simulator), denoted by S, which breaks the EUF-naCMA security of the primitive scheme IBS' using $\mathcal{A}$ as a black-box.

As part of the EUF-naCMA security game, S must first provide two lists, $\mathcal{Q}'_{\mathsf{key}}$ and $\mathcal{Q}'_{\mathsf{sign}}$, containing identities and identity-message pairs respectively, to the challenger $\mathcal{C}$. It also maintains two additional lists, $\mathcal{L}'_{\mathsf{key}}$ and $\mathcal{L}'_{\mathsf{sign}}$, to store information such as $(i, \mathsf{id}'_i, r_i, -)$ and $(i, j, \mathsf{id}'_i, r_i, m'_j, s_j, -)$, where the final blank entries will be filled with the replied key $\mathsf{sk}_{\mathsf{id}'_i}$ and signature σ'_{ij}, respectively. These lists facilitate consistent responses to key and sign queries. The simulator S prepares these lists as follows:

List : $\mathcal{Q}'_{\mathsf{key}}$	List : $\mathcal{Q}'_{\mathsf{sign}}$
1: **for** each $i \in [q_k]$ **do**	1: **for** each $i \in [q_s]$ **do**
2: $(\mathsf{id}'_i, r_i) \xleftarrow{\mathsf{U}} \mathcal{ID}' \times \mathsf{SaltSp}$	2: $(\mathsf{id}'_i, r_i) \xleftarrow{\mathsf{U}} \mathcal{ID}' \times \mathsf{SaltSp}$
3: $\mathcal{Q}'_{\mathsf{key}} := \mathcal{Q}'_{\mathsf{key}} \cup \{\mathsf{id}'_i\}$	3: **for** each $j \in [q_s]$ **do**
4: $\mathcal{L}'_{\mathsf{key}} \qquad := \qquad \mathcal{L}'_{\mathsf{key}} \quad \cup$	4: $(m'_j, s_j) \xleftarrow{\mathsf{U}} \mathcal{M}' \times \mathsf{SaltSp}$
$\{(i, \mathsf{id}'_i, r_i, -)\}$	5: $\mathcal{Q}'_{\mathsf{sign}} := \mathcal{Q}'_{\mathsf{sign}} \cup \{(\mathsf{id}'_i, m'_j)\}.$
5: **end for**	6: $\mathcal{L}'_{\mathsf{sign}} \qquad := \qquad \mathcal{L}'_{\mathsf{sign}} \quad \cup$
	$\{(i, j, \mathsf{id}'_i, r_i, m'_j, s_j, -)\}$
	7: **end for**
	8: **end for**

The simulator S then sends $\mathcal{Q}'_{\mathsf{key}}$ and $\mathcal{Q}'_{\mathsf{sign}}$ to the challenger $\mathcal{C}$. After receiving the keys $\mathsf{sk}_{\mathsf{id}'_i}$ and signatures σ'_{ij} from $\mathcal{C}$, S updates $\mathcal{L}'_{\mathsf{key}}$ and $\mathcal{L}'_{\mathsf{sign}}$ accordingly. The entries in $\mathcal{L}'_{\mathsf{key}}$ now become $(i, \mathsf{id}'_i, r_i, \mathsf{sk}_{\mathsf{id}'_i})$ and those in $\mathcal{L}'_{\mathsf{sign}}$ become $(i, j, \mathsf{id}'_i, r_i, m'_j, s_j, \sigma'_{ij})$. Note that i and (i, j) involved in the tuples in $\mathcal{L}'_{\mathsf{key}}$ and $\mathcal{L}'_{\mathsf{sign}}$ work as indices while answering key and sign queries respectively. Furthermore, for any $i \in [q_s]$, there will be q_s tuples in $\mathcal{L}'_{\mathsf{sign}}$ with the same i as the first entry. Consequently, these tuples will contain the same id'_i and r_i.

Answering Queries: We now illustrate in Fig. 7 how S answers key, sign, and random oracle queries using only the resources $\mathcal{L}'_{\mathsf{key}}$ and $\mathcal{L}'_{\mathsf{sign}}$. Note that we employ several lists here, including $\mathcal{L}_{\mathsf{id}}$, $\mathcal{Q}_{\mathsf{key}}$, $\mathcal{Q}_{\mathsf{sign}}$, and $\mathcal{L}_{\mathsf{sk}}$, all initially empty. The latter three lists serve the same purpose as in the definition of WModel-I, while $\mathcal{L}_{\mathsf{id}}$ guarantees that the same salt component r is used in all signatures under the same identity. Additionally, we maintain two lists, $\mathcal{L}_{\mathcal{H}_1}$ and $\mathcal{L}_{\mathcal{H}_2}$, to store information related to $\mathcal{H}_1$ and $\mathcal{H}_2$ queries, respectively. These lists are updated while answering random oracle, key, and sign queries. We also use counters ctr, $\overline{\mathsf{ctr}}$, and ctr_i for each $i \in [q_s]$, initially set to zero, to accurately map the information in $\mathcal{L}'_{\mathsf{key}}$ and $\mathcal{L}'_{\mathsf{sign}}$.

Forgery: At the end, $\mathcal{A}$ produces the forgery $(\mathsf{id}^*, m^*, \sigma^*)$ with $\sigma^* = (\sigma'^*, r^*, s^*)$. The simulator S computes $\mathsf{id}'^* \leftarrow \mathcal{H}_1(\mathsf{id}^*, r^*)$ and $m'^* \leftarrow \mathcal{H}_2(m^*, s^*)$, and supplies the forgery $(\mathsf{id}'^*, m'^*, \sigma'^*)$ to $\mathcal{C}$.

Analysis: All the queries are answered perfectly thanks to the condition of WModel-I: $\mathcal{Q}_{\mathsf{key}} \cap \mathcal{Q}_{\mathsf{id}} = \emptyset$. Note that if $\mathsf{IBS}.\mathsf{Ver}(\mathsf{pp}, \mathsf{id}^*, m^*, \sigma^*) = 1$, then $\mathsf{IBS}'.\mathsf{Ver}(\mathsf{pp}', \mathsf{id}'^*, m'^*, \sigma'^*) = 1$. If $(\mathsf{id}^*, m^*, \sigma^*)$ is a valid forgery, then we have

$$\mathsf{id}^* \notin \mathcal{Q}_{\mathsf{key}} \tag{2}$$

$$(\mathsf{id}^*, m^*) \notin \mathcal{Q}_{\mathsf{sign}} \tag{3}$$

We have to show the following two conditions as required for a valid forgery against EUF-naCMA attack:

$$\mathsf{id}'^* \notin \mathcal{Q}'_{\mathsf{key}} \tag{4}$$

$$(\mathsf{id}'^*, m'^*) \notin \mathcal{Q}'_{\mathsf{sign}} \tag{5}$$

$\mathcal{O}_{\text{key}}(\text{id}):$

1: **if** id $\in \mathcal{Q}_{\text{key}}$ **then**
2:     **return** sk_{id} ▷ Available in the list $\mathcal{L}_{\text{sk}}$
3: **end if**
4: $\mathcal{Q}_{\text{key}} := \mathcal{Q}_{\text{key}} \cup \{\text{id}\}$
5: $\overline{\text{ctr}} := \overline{\text{ctr}} + 1$
6: set $i := \overline{\text{ctr}}$ and $\text{id}_i := \text{id}$
7: set $x_i := (\text{id}_i, r_i, \text{id}'_i)$ ▷ r_i and id'_i are available at the i-th entry of $\mathcal{L}'_{\text{key}}$
8: **if** $\exists (*, *, \text{id}'_i) \in \mathcal{L}_{\mathcal{H}_1}$ **then**
9:     $\boxed{\text{abort}}$ ▷ As defined in G_1
10: **end if**
11: $\mathcal{L}_{\mathcal{H}_1} := \mathcal{L}_{\mathcal{H}_1} \cup \{x_i\}$
12: set $\text{sk}_{\text{id}_i} := (\text{sk}_{\text{id}'_i}, r_i)$ ▷ Available at the i-th entry of $\mathcal{L}'_{\text{key}}$
13: $\mathcal{L}_{\text{sk}} := \mathcal{L}_{\text{sk}} \cup \{(\text{id}_i, \text{sk}_{\text{id}_i})\}$
14: **return** sk_{id_i}

$\mathcal{O}_{\mathcal{H}_1}(\text{id}, r)$

1: **if** $(\text{id}, r, \mathcal{H}_1(\text{id}, r)) \notin \mathcal{L}_{\mathcal{H}_1}$ **then**
2:     $\text{id}' \xleftarrow{\text{U}} \mathcal{ID}'$
3:     set $\mathcal{H}_1(\text{id}, r) := \text{id}'$
4:     $\mathcal{L}_{\mathcal{H}_1} := \mathcal{L}_{\mathcal{H}_1} \cup \{(\text{id}, r, \mathcal{H}_1(\text{id}, r)\}$
5: **end if**
6: **return** $\mathcal{H}_1(\text{id}, r)$

$\mathcal{O}_{\mathcal{H}_2}(m, s)$

1: **if** $(m, s, \mathcal{H}_2(m, s)) \notin \mathcal{L}_{\mathcal{H}_2}$ **then**
2:     $m' \xleftarrow{\text{U}} \mathcal{M}'$
3:     set $\mathcal{H}_2(m, s) := m'$
4:     $\mathcal{L}_{\mathcal{H}_2} := \mathcal{L}_{\mathcal{H}_2} \cup \{(m, s, \mathcal{H}_2(m, s))\}$
5: **end if**
6: **return** $\mathcal{H}_2(m, s)$

$\mathcal{O}_{\text{sign}}(\text{id}, m):$

1: **if** $(*, \text{id}) \notin \mathcal{L}_{\text{id}}$ **then**
2:     $\text{ctr} := \text{ctr} + 1$
3:     set $i := \text{ctr}$ and $\text{id}_i := \text{id}$
4:     set $x_i := (\text{id}_i, r_i, \text{id}'_i)$ ▷ r_i and id'_i are available at the i-th entry of $\mathcal{L}'_{\text{sign}}$
5:     **if** $\exists (*, *, \text{id}'_i) \in \mathcal{L}_{\mathcal{H}_1}$ **then**
6:         $\boxed{\text{abort}}$ ▷ As defined in G_1
7:     **end if**
8:     $\mathcal{L}_{\mathcal{H}_1} := \mathcal{L}_{\mathcal{H}_1} \cup \{x_i\}$
9:     $\mathcal{L}_{\text{id}} := \mathcal{L}_{\text{id}} \cup \{(i, \text{id}_i)\}$
10: **end if**
11: $\exists i \in [q_s]$ such that $(i, \text{id}) \in \mathcal{L}_{\text{id}}$
12: $\text{ctr}_i := \text{ctr}_i + 1$
13: set $j := \text{ctr}_i$ and $m_j := m$
14: set $y_j := (m_j, s_j, m'_j)$ ▷ s_j and m'_j are available at the (i, j)-th entry of $\mathcal{L}'_{\text{sign}}$
15: **if** $\exists (*, *, m'_j) \in \mathcal{L}_{\mathcal{H}_2}$ **then**
16:     $\boxed{\text{abort}}$ ▷ As defined in G_1
17: **end if**
18: $\mathcal{L}_{\mathcal{H}_2} := \mathcal{L}_{\mathcal{H}_2} \cup \{y_j\}$
19: **return** $\sigma_{ij} := (\sigma'_{ij}, r_i, s_j)$ ▷ σ'_{ij} is available at the (i, j)-th entry $\mathcal{L}'_{\text{sign}}$

Fig. 7. Illustration of handling key, sign and random oracle queries. Note that if a key query is made on an identity id for which a sign query has already been made, then ⊥ will be returned as the answer. The same treatment applies if the order of the queries is reversed.

It is easy to see that Eqs. 2 and 3 guarantee Eqs. 4 and 5 except with probability at most $q_k / |\mathcal{ID}'|$ and $q_s^2 / (|\mathcal{ID}'| \cdot |\mathcal{M}'|)$ respectively. In fact, given $\text{id}^* \notin \mathcal{Q}_{\text{key}}$, $\text{id}'^* \in \mathcal{Q}'_{\text{key}}$ could be possible if there exists $(\text{id}, r, \text{id}'^*) \in \mathcal{L}_{\mathcal{H}_1}$ such that $\mathcal{H}_1(\text{id}^*, r^*) = \text{id}'^*$, and thus, we have:

$$\Pr\left[\text{id}'^* \in \mathcal{Q}'_{\text{key}} : \text{id}^* \notin \mathcal{Q}_{\text{key}}\right] \leq \frac{q_k}{|\mathcal{ID}'|}.$$

Similarly, we have:

$$\Pr\left[(\mathsf{id'}^*, m'^*) \in \mathcal{Q}'_{\mathsf{sign}} : (\mathsf{id}^*, m^*) \notin \mathcal{Q}_{\mathsf{sign}}\right] \leq \frac{q_s^2}{|\mathcal{ID}'| \cdot |\mathcal{M}'|}.$$

Hence, we can write:

$$\mathsf{Adv}_{\mathcal{A}}^{G_1}(\kappa) \leq \mathsf{Adv}_{\mathsf{S}}^{\mathsf{EUF\text{-}naCMA}}(\kappa) + \frac{q_k}{|\mathcal{ID}'|} + \frac{q_s^2}{|\mathcal{ID}'| \cdot |\mathcal{M}'|}. \tag{6}$$

Therefore, using Eqs. 1 and 6, we can bound $\mathsf{Adv}_{\mathcal{A}}^{\mathsf{WModel\text{-}I}}(\kappa)$ as follows:

$$\mathsf{Adv}_{\mathcal{A}}^{\mathsf{WModel\text{-}I}}(\kappa) \stackrel{\text{By Defn}}{=} \mathsf{Adv}_{\mathcal{A}}^{G_0}(\kappa)$$

$$\stackrel{\text{Eqn 1}}{\leq} \mathsf{Adv}_{\mathcal{A}}^{G_1}(\kappa) + (q_k + 2q_s) \cdot \frac{q_h}{2^\ell}$$

$$\stackrel{\text{Eqn 6}}{\leq} \mathsf{Adv}_{\mathsf{S}}^{\mathsf{EUF\text{-}naCMA}}(\kappa) + \frac{q_k}{|\mathcal{ID}'|} + \frac{q_s^2}{|\mathcal{ID}'| \cdot |\mathcal{M}'|} + (q_k + 2q_s) \cdot \frac{q_h}{2^\ell}$$

$$= \mathsf{Adv}_{\mathsf{S}}^{\mathsf{EUF\text{-}naCMA}}(\kappa) + (q_k + 2q_s) \cdot \frac{q_h}{2^\ell} + \frac{q_k}{|\mathcal{ID}'|} + \frac{q_s^2}{|\mathcal{ID}'| \cdot |\mathcal{M}'|}.$$

This completes the proof.

An analogous tight reduction for the chameleon hash based construction (given in Fig. 5) can be easily established in the standard model. We omit the details.

Remark 1. In the above security reduction, as per WModel-I, the attacker is restricted by the condition that $\mathcal{Q}_{\mathsf{key}} \cap \mathcal{Q}_{\mathsf{id}} = \emptyset$. One can relax this slightly, as considered in [ZLGL19], by allowing $\mathcal{A}$ to make sign queries on (id, m) for any message m after making key query on id. In fact, such subsequent sign queries are effectively redundant and can be easily answered by running the signing algorithm using the secret key $\mathsf{sk}_{\mathsf{id}}$ that is already provided to $\mathcal{A}$.

We now argue that the restricted model, WModel-I, is polynomially equivalent to the EUF-ID-CMA security model (c.f., Definition 4). While the proof is completely generic and holds both in standard as well random oracle model, the crux lies in addressing scenario like Case II of previous section. Suppose $\mathcal{A}$ first makes a sign query on some (id, m) followed by a key query on the same identity id. For a proper simulation, the returned key $\mathsf{sk}_{\mathsf{id}}$ must be the one used to generate the signature σ returned earlier. To complicate the matter further, $\mathcal{A}$ is likely to make sign queries on certain identities but not a follow-up key query (in fact, $\mathcal{A}$ cannot corrupt the identity on which it will eventually produce the forgery). Hence, before answering a sign query on (id, m) for which no preceding key query on id exists, the simulator must somehow predict whether a key query on id will be made subsequently. If the prediction is incorrect, the simulator has to abort, as otherwise, it will fail to provide a proper simulation of the EUF-ID-CMA security game. This entails a degradation in the reduction which is at best proportional to either q_k or q_s (and at worst the number of random oracle queries in the ROM). Such degradation appears to be intrinsic to the very technique followed in [PW21] to realize adaptive security.

Theorem 2. *Suppose there exists an adversary $\mathcal{A}$ who can break the EUF-ID-CMA security (c.f., Definition 4) of an identity-based signature scheme* IBS. *Then, using $\mathcal{A}$ as a subroutine, we can create PPT algorithm* S *for breaking the* WModel-I *security of the same scheme* IBS *with the following relation of the respective advantages:*

$$\mathsf{Adv}_{\mathcal{A}}^{\mathrm{EUF\text{-}ID\text{-}CMA}}(\kappa) \leq q_k \cdot \mathsf{Adv}_{\mathsf{S}}^{\mathsf{WModel\text{-}I}}(\kappa),$$

where q_k is the number of key queries.

Proof. Let $\mathcal{A}$ be an attacker who can break the EUF-ID-CMA security (c.f., Definition 4) of the given scheme IBS. Let $\mathcal{C}$ be the challenger of the WModel-I security game. Let S be the challenger in the EUF-ID-CMA game who breaks the WModel-I security of IBS using $\mathcal{A}$ as a black box. The treatment is same for both the standard model and the random oracle model except that in ROM all random oracle queries of $\mathcal{A}$ are answered simply by relaying them to the challenger $\mathcal{C}$ for the WModel-I security game.

Note that the simulator S has access to oracles $\mathcal{O}_{\mathsf{key}}$ and $\mathcal{O}_{\mathsf{sign}}$ with the restriction that $\mathcal{Q}_{\mathsf{key}} \cap \mathcal{Q}_{\mathsf{id}} = \emptyset$. Suppose the first query that $\mathcal{A}$ makes on some identity id is a sign query. One simple strategy for S could be to first get $\mathsf{sk}_{\mathsf{id}}$ by making a key query to $\mathcal{O}_{\mathsf{key}}$ followed by running the sign algorithm using that key. However, the problem is that $\mathcal{A}$ will eventually produce a forgery on an identity for which sign queries are made but no follow-up key query. Hence, S cannot use that forgery if it has already queried $\mathcal{O}_{\mathsf{key}}$ on the corresponding identity. To resolve this issue, we adopt Coron's partitioning technique [Cor00]: The identity space $\mathcal{ID}$ is partitioned into two disjoint subsets, $\mathcal{ID}_\lambda$ and $\mathcal{ID} \setminus \mathcal{ID}_\lambda$, based on the outcome of a biased coin toss with a head probability of λ (the value of λ will be determined later). In fact, whenever an identity id $\in \mathcal{ID}$ is queried for the first time, toss the biased coin, and let η be the outcome of the toss (note that $\Pr[\eta = 1] = \lambda$). Then, assign id to $\mathcal{ID}_\lambda$ if $\eta = 1$; otherwise, assign it to $\mathcal{ID} \setminus \mathcal{ID}_\lambda$. The simulator S answers the key and sign queries of $\mathcal{A}$ as follows.

1. If a key query identity belongs to $\mathcal{ID}_\lambda$, then abort. Otherwise, answer the query by forwarding to the challenger $\mathcal{C}$ of the WModel-I security game.
2. For a signature query on (id, m), do the following:
   (a) If id $\in \mathcal{ID}_\lambda$, then answer by forwarding (id, m) to $\mathcal{C}$.
   (b) Else, get the key $\mathsf{sk}_{\mathsf{id}}$ first (by making key query to $\mathcal{C}$) and then generate signature using $\mathsf{sk}_{\mathsf{id}}$.

When $\mathcal{A}$ produces the forgery $(\mathsf{id}^*, m^*, \sigma^*)$, the simulator S checks whether $\mathsf{id}^* \notin \mathcal{ID}_\lambda$. If yes, then S aborts. Otherwise, S forwards the same to the challenger $\mathcal{C}$ of the WModel-I security game. When the game is not aborted, then for all key and sign queries, we have $\mathcal{Q}_{\mathsf{key}} \cap \mathcal{Q}_{\mathsf{id}} = \emptyset$ thanks to $\mathcal{Q}_{\mathsf{key}} \cap \mathcal{ID}_\lambda = \emptyset$. Therefore, the advantage in breaking the WModel-I security of IBS is given by:

$$\mathsf{Adv}_S^{\mathsf{WModel\text{-}I}}(\kappa) \geq (1 - \lambda)^{q_k} \cdot \lambda \cdot \mathsf{Adv}_\mathcal{A}^{\mathrm{EUF\text{-}ID\text{-}CMA}}(\kappa)$$

$$\approx \frac{1}{e \cdot q_k} \cdot \mathsf{Adv}_\mathcal{A}^{\mathrm{EUF\text{-}ID\text{-}CMA}}(\kappa)$$

$$\approx \frac{1}{q_k} \cdot \mathsf{Adv}_\mathcal{A}^{\mathrm{EUF\text{-}ID\text{-}CMA}}(\kappa),$$

where the quantity $(1 - \lambda)^{q_k} \cdot \lambda$ attains the maximum value of $1/(e \cdot q_k)$ when $\lambda = 1/(1 + q_k)$. This completes the proof.

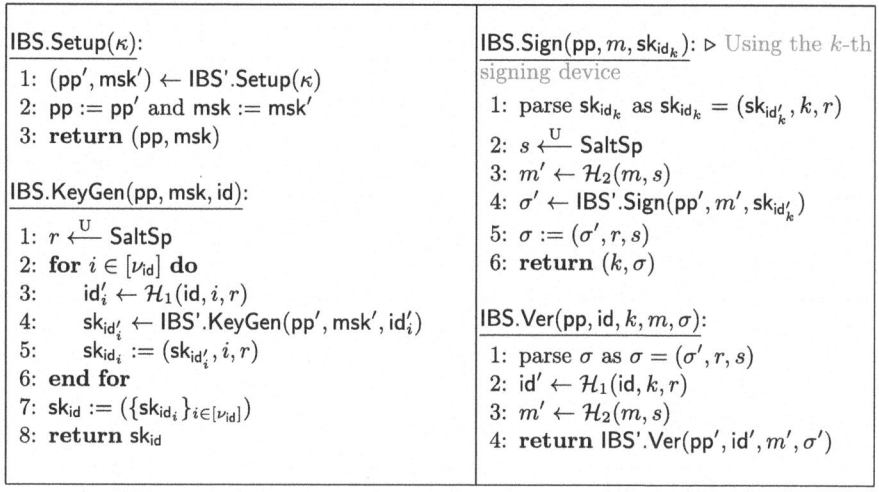

Fig. 8. A slightly modified construction of an adaptively secure IBS from a non-adaptively secure IBS', where $\mathcal{H}_1 : \mathcal{ID} \times [\nu] \times \mathsf{SaltSp} \to \mathcal{ID}'$ and $\mathcal{H}_2 : \mathcal{M} \times \mathsf{SaltSp} \to \mathcal{M}'$ are hash functions, $\mathsf{SaltSp} = \{0,1\}^\ell$, and ν_{id} is the upper bound on the number of devices that can be registered under an identity, id.

5   Extension

Consider the realistic scenario where an entity with identity id possesses several devices say, laptop, desktop, smartphone/i-pad etc., any one of which can be used to sign messages under id. Or, the scenario of smart-home where multiple devices, registered against the id of the owner, can generate signature on the data they transmit. One natural strategy appears to be the use of two-level hierarchical IBS [GS02,Rüc10]. However, that typically comes with increased cost in terms of signing/verification time as well as signature length. In this section, we propose an interesting extension of the Pan-Wagner generic constructions [PW21] which realizes the above functionality with hardly any increase in terms of computation/communication cost.

In Fig. 8, we detail the construction in the random oracle model which supports registering more than one signing devices under a single identity. See Remark 3, where we briefly discuss about a similar extension for the chameleon hash based construction. The construction in Fig. 8 differs from the original hash-based construction [PW21] mainly in the key generation. Let ν_{id} denote the number of signing devices registered under the identity id. The corresponding user is issued ν_{id} keys by running the key generation algorithm of the non-adaptive IBS scheme IBS′ on each transformed identity $id'_i = \mathcal{H}(id, i, r)$ for $i \in [\nu_{id}]$, where r is a fixed random value for id. As a result, it may appear that the key size increases by up to ν_{id} times compared to the original. However, the key components are distributed amongst the signing devices under the possession of id and are stored locally in that device. Note that we retain the same salt r across all the signing devices registered under a single identity. However, the construction as well as the security reduction has the flexibility to allow unique salt for each signing device.

Modeling the security of the above mentioned scenario requires suitable modification in the security game of IBS. In particular, the attacker should be allowed to adaptively corrupt any signing devices of its choice by specifying (i, id) through the key oracle, where $i \in [\nu_{id}]$. We start with a model where the attacker is barred from asking any sign queries on the devices that it may corrupt, as it can generate those signatures itself. On the other hand, attacker is allowed to even corrupt all but one of the devices under the *target identity* to produce a forgery with respect to the uncorrupted device. Let us denote this model by WModel-II, which is formalized as follows:

WModel-II: For the sake of completeness, we first extend the definition of IBS scheme that allows registration of multiple devices under the same identity.

Definition 7. *It consists of four PPT algorithms -* Setup, KeyGen, Sign *and* Ver.

- Setup: *It takes as input a security parameter* κ, *and outputs public parameters and master secret key pair* (pp, msk).
- KeyGen: *It takes as input public parameters* pp, *master secret key* msk *and an identity* id $\in \mathcal{ID}$, *where* $\mathcal{ID}$ *is the identity space, and outputs a signing key* $sk_{id} = (\{sk_{id_i}\}_{i \in [\nu_{id}]})$, *where* ν_{id} *is the number of signing devices registered under the identity* id.[2]
- Sign: *It takes as input public parameters* pp, *a message* $m \in \mathcal{M}$, *where* $\mathcal{M}$ *is the message space, and the* k-th *component key* sk_{id_k} *for some* $\in [\nu_{id}]$, *and outputs a device index and signature pair* (k, σ).
- Ver: *It takes as input public parameters* pp, *an identity* id, *a device index* k, *a message* m *and a signature* σ. *It outputs a value 1, if* σ *is a valid signature for* (id, k, m), *else it outputs 0.*

---

[2] The key components sk_{id_i} may share some common information, e.g., a random salt.

Correctness: For all $(\mathsf{pp}, \mathsf{msk}) \leftarrow \mathsf{Setup}(1^\kappa)$, for all $\mathsf{id} \in \mathcal{ID}$, $\{\mathsf{sk}_{\mathsf{id}_i}\}_{i \in [\nu_{\mathsf{id}}]} \leftarrow \mathsf{KeyGen}(\mathsf{pp}, \mathsf{msk}, \mathsf{id})$, for all $k \in [\nu_{\mathsf{id}}]$ and for all $m \in \mathcal{M}$, it is required that

$$\mathsf{Ver}(\mathsf{pp}, \mathsf{id}, k, m, \mathsf{IBS.Sign}(\mathsf{pp}, m, \mathsf{sk}_{\mathsf{id}_k})) = 1.$$

$\mathsf{Exp}_{\mathcal{A}}^{\mathsf{WModel\text{-}II}}(\kappa)$:	$\mathcal{O}_{\mathsf{key}}(\mathsf{id}, k)$:	$\mathcal{O}_{\mathsf{sign}}(\mathsf{id}, k, m)$:
1: $\mathcal{Q}_{\mathsf{key}} := \emptyset, \mathcal{Q}_{\mathsf{sign}} := \emptyset, \mathcal{L}_{\mathsf{sk}} := \emptyset, \mathcal{L}_{\mathsf{sync}} := \emptyset$	1: **if** $\exists (\mathsf{id}, k, *) \in \mathcal{Q}_{\mathsf{sign}}$ **then**	1: **if** $(\mathsf{id}, k) \in \mathcal{Q}_{\mathsf{key}}$ **then**
2: $(\mathsf{pp}, \mathsf{msk}) \leftarrow \mathsf{Setup}(1^\kappa)$	2: abort ▷ Restriction	2: abort ▷ Restriction
3: $(\mathsf{id}^*, k^*, m^*, \sigma^*) \leftarrow \mathcal{A}^{\{\mathcal{O}_{\mathsf{sign}}, \mathcal{O}_{\mathsf{key}}\}}(1^\kappa, \mathsf{pp})$	3: **end if**	3: **end if**
4: **if** $(\mathsf{id}^*, k^*) \in \mathcal{Q}_{\mathsf{key}}$ or $(\mathsf{id}^*, k^*, m^*) \in \mathcal{Q}_{\mathsf{sign}}$ **then**	4: **if** $(\mathsf{id}, *) \notin \mathcal{L}_{\mathsf{sync}}$ **then**	4: **if** $(\mathsf{id}, *) \notin \mathcal{L}_{\mathsf{sync}}$ **then**
5: **return** 0	5: $\mathsf{sk}_{\mathsf{id}} \leftarrow \mathsf{KeyGen}(\mathsf{pp}, \mathsf{msk}, \mathsf{id})$	5: $\mathsf{sk}_{\mathsf{id}} \leftarrow \mathsf{KeyGen}(\mathsf{pp}, \mathsf{msk}, \mathsf{id})$
6: **end if**	6: $\mathcal{L}_{\mathsf{sync}} := \mathcal{L}_{\mathsf{sync}} \cup \{(\mathsf{id}, \mathsf{sk}_{\mathsf{id}})\}$	6: $\mathcal{L}_{\mathsf{sync}} := \mathcal{L}_{\mathsf{sync}} \cup \{(\mathsf{id}, \mathsf{sk}_{\mathsf{id}})\}$
7: **return** $\mathsf{Ver}(\mathsf{pp}, \mathsf{id}^*, k^*, m^*, \sigma^*)$	7: **end if**	7: **end if**
	8: **if** $(\mathsf{id}, k) \notin \mathcal{Q}_{\mathsf{key}}$ **then**	8: **if** $(\mathsf{id}, k, m) \notin \mathcal{Q}_{\mathsf{sign}}$ **then**
	9: $\mathcal{Q}_{\mathsf{key}} := \mathcal{Q}_{\mathsf{key}} \cup \{(\mathsf{id}, k)\}$	9: $\mathcal{Q}_{\mathsf{sign}} := \mathcal{Q}_{\mathsf{sign}} \cup \{(\mathsf{id}, k, m)\}$
	10: $\mathcal{L}_{\mathsf{sk}} := \mathcal{L}_{\mathsf{sk}} \cup \{(\mathsf{id}, k, \mathsf{sk}_{\mathsf{id}_k})\}$ ▷ Using $\mathcal{L}_{\mathsf{sync}}$	10: $\mathcal{L}_{\mathsf{sk}} := \mathcal{L}_{\mathsf{sk}} \cup \{(\mathsf{id}, k, \mathsf{sk}_{\mathsf{id}_k})\}$ ▷ Using $\mathcal{L}_{\mathsf{sync}}$
	11: **end if**	11: **end if**
	12: **return** $\mathsf{sk}_{\mathsf{id}_k}$ ▷ $\mathsf{sk}_{\mathsf{id}_k}$ available in $\mathcal{L}_{\mathsf{sk}}$	12: **return** $\mathsf{Sign}(\mathsf{pp}, m, \mathsf{sk}_{\mathsf{id}_k})$ ▷ $\mathsf{sk}_{\mathsf{id}_k}$ available in $\mathcal{L}_{\mathsf{sk}}$

Fig. 9. Experiment for EUF-ID-CMA security of an identity-based signature scheme supporting multiple devices registered under the same identity, with the restriction that $\mathcal{Q}_{\mathsf{key}}$ and $\mathcal{Q}_{\mathsf{sign}}$ do not share any common pair $(\mathsf{id}, k) \in \mathcal{ID} \times [\nu_{\mathsf{id}}]$. The first three lines in both $\mathcal{O}_{\mathsf{key}}$ and $\mathcal{O}_{\mathsf{sign}}$ capture this restriction. The list $\mathcal{L}_{\mathsf{sync}}$ maintains a unique key $\mathsf{sk}_{\mathsf{id}}$ for each identity id, ensuring that common information shared among key components for different devices is preserved

Formally, the security model, WModel-II, is defined below.

Definition 8 (WModel-II). *An IBS scheme is said to be secure with respect to the model* WModel-II, *if for all PPT algorithms* $\mathcal{A}$, *the advantage*

$$\mathsf{Adv}_{\mathcal{A}}^{\mathsf{WModel\text{-}II}}(\kappa) := \Pr\left[\mathsf{Exp}_{\mathcal{A}}^{\mathsf{WModel\text{-}II}}(\kappa) = 1\right]$$

in $\mathsf{Exp}_{\mathcal{A}}^{\mathsf{WModel\text{-}II}}(\kappa)$ *defined in the left column of Fig. 9 is a negligible function in* κ, *where* $\mathcal{A}$ *is provided access to signature oracle* $\mathcal{O}_{\mathsf{sign}}$ *(shown in the right column) and key-gen oracle* $\mathcal{O}_{\mathsf{key}}$ *(shown in the middle column) at most polynomial number of times, and* $\mathcal{Q}_{\mathsf{key}}$ *is the set of identity-device index pairs on which key-gen queries were made and* $\mathcal{Q}_{\mathsf{sign}}$ *is the set of identity-index-message triplets on which signature queries were made.*

Remark 2. One can remove the restrictions imposed on $\mathcal{A}$ in WModel-II, as depicted in lines 1–3 of $\mathcal{O}_{\mathsf{sign}}$ and $\mathcal{O}_{\mathsf{key}}$ in Fig. 9. This will result in an extension of the standard EUF-ID-CMA security model of IBS that allows registering multiple signing devices under a single identity.

The model attempts to capture the following real-world scenario: A couple of signing devices are compromised, so the attacker can generate a signature on any message using those devices. Also, the attacker learns some signatures for the uncompromised devices. The goal of the attacker is to forge on behalf of one of the uncompromised devices. It is immediate that this is essentially a generalization of the model, WModel-I, from single device to multiple devices under a single identity. That is, when $\nu_{id} = 1$ for all $id \in \mathcal{ID}$, we get WModel-I. The following theorem shows that the modified scheme is tightly secure in the WModel-II security model.

Theorem 3. *Suppose there exists an adversary $\mathcal{A}$ who can break the EUF-ID-CMA security with respect to WModel-II (c.f., Definition 8) of the target identity-based signature scheme IBS (c.f., Fig. 8) in the random oracle model. Then, using $\mathcal{A}$ as a subroutine, we can create PPT algorithm S for breaking the EUF-naCMA security (c.f., Definition 6) of the primitive scheme IBS' with the following relation of the respective advantages:*

$$\mathsf{Adv}_{\mathcal{A}}^{\mathsf{WModel\text{-}II}}(\kappa) \leq \mathsf{Adv}_{\mathsf{S}}^{\mathsf{EUF\text{-}naCMA}}(\kappa) + (q_k + 2q_s) \cdot \frac{q_h}{2^\ell}$$

where q_k, q_s, and q_h are the numbers of key queries, sign queries and hash queries respectively.

Proof. The proof needs to handle multiple registrations under a single identity. Other than that, it closely resembles the reduction in Theorem 1. Similar to the proof of Theorem 1, we consider here hybrid arguments through two different games: G_0 and G_1. If $\mathcal{A}$ is an attacker in game G_i, then its advantage in winning game G_i is denoted by $\mathsf{Adv}_{\mathcal{A}}^{\mathsf{G}_i}(\kappa)$. We treat $\mathcal{H}_1$ and $\mathcal{H}_2$ as random oracles.

Let G_0 be defined as the security model WModel-II. Let G_1 be defined similarly to G_0, except that whenever the random oracle values, such as $\mathcal{H}_1(id, i, r)$ and $\mathcal{H}_2(m, s)$, are already defined while answering key or sign queries, the game is aborted. Following a similar argument as considered in the proof of Theorem 1, the gap of the advantages of an attacker $\mathcal{A}$ in these games is given by:

$$\left| \mathsf{Adv}_{\mathcal{A}}^{\mathsf{G}_0}(\kappa) - \mathsf{Adv}_{\mathcal{A}}^{\mathsf{G}_1}(\kappa) \right| \leq (q_k + 2q_s) \cdot q_h / 2^\ell. \tag{7}$$

Let $\mathcal{A}$ be a WModel-II attacker of the target scheme IBS. We present a reduction of the EUF-naCMA security of IBS' from the WModel-II security of IBS in the context of environment G_1. To achieve this, we construct an algorithm (simulator), denoted by S, which breaks the EUF-naCMA security of the primitive scheme IBS' using $\mathcal{A}$ as a black-box.

The simulator S prepares the lists $\mathcal{Q}'_{\mathsf{key}}$ and $\mathcal{Q}'_{\mathsf{sign}}$, along with additional lists $\mathcal{L}'_{\mathsf{key}}$ and $\mathcal{L}'_{\mathsf{sign}}$, in a manner similar to the proof of Theorem 1, except that the salts r_i and s_j are stored in a separate list $\mathcal{L}_r$. Specifically, after updating $\mathcal{L}'_{\mathsf{key}}$ and $\mathcal{L}'_{\mathsf{sign}}$ with the replied keys and signatures from the EUF-naCMA challenger $\mathcal{C}$, the entries in $\mathcal{L}'_{\mathsf{key}}$ become $(i, id'_i, \mathsf{sk}_{id'_i})$ and those in $\mathcal{L}'_{\mathsf{sign}}$ become $(i, j, id'_i, m'_j, \sigma'_{ij})$.

To follow the rules of the WModel-II game, we consider two lists, $\mathcal{Q}_{\mathsf{key}}$ and $\mathcal{L}_{id}$, which store entries of the form (id, k) and (i, id, k), respectively, where k denotes the device index registered under the identity id, and i denotes the index mapping to the entry of $\mathcal{L}'_{\mathsf{sign}}$. Note that $\mathcal{Q}_{\mathsf{key}}$ and $\mathcal{L}_{id}$ are updated while answering key and sign queries of $\mathcal{A}$, respectively. Additionally, to maintain a unique random salt for all keys and signatures associated with each identity, a list $\mathcal{L}_r$ is used, which is updated in either the key oracle or the sign oracle. Essentially, when an identity id appears for the first time in a key query or a sign query, a random salt r is sampled, and $\mathcal{L}_r$ is updated with the entry (id, r). Similar to the proof of Theorem 1, two counters, ctr and $\overline{\mathsf{ctr}}$, are used to accurately map the key and sign queries to $\mathcal{L}'_{\mathsf{key}}$ and $\mathcal{L}'_{\mathsf{sign}}$, respectively.

The simulator answers all the queries using the only resources $\mathcal{L}'_{\mathsf{key}}$ and $\mathcal{L}'_{\mathsf{sign}}$, as shown in Fig. 10. When $\mathcal{A}$ produces a forgery $(\mathrm{id}^*, k^*, m^*, \sigma^*)$ with

---

$\mathcal{O}_{\mathsf{key}}(\mathrm{id}, k)$:
1: **if** $\exists(*, \mathrm{id}, k) \in \mathcal{L}_{\mathrm{id}}$ **then**
2:     $\boxed{\mathsf{abort}}$ ▷ Due to the restriction on security
3: **end if**
4: **if** $(\mathrm{id}, k) \in \mathcal{Q}_{\mathsf{key}}$ **then**
5:     **return** $\mathsf{sk}_{\mathrm{id}_k}$ ▷ Available in the list $\mathcal{L}_{\mathsf{sk}}$
6: **end if**
7: **if** $\nexists(\mathrm{id}, *) \in \mathcal{L}_r$ **then**   ▷ id appearing for the first time in $\mathcal{Q}_{\mathsf{key}} \cup \mathcal{Q}_{\mathrm{id}}$
8:     $r \xleftarrow{\mathsf{U}} \mathsf{SaltSp}$
9:     $\mathcal{L}_r := \mathcal{L}_r \cup \{(\mathrm{id}, r)\}$
10: **end if**
11: $\mathcal{Q}_{\mathsf{key}} := \mathcal{Q}_{\mathsf{key}} \cup \{(\mathrm{id}, k)\}$
12: $\overline{\mathsf{ctr}} := \overline{\mathsf{ctr}} + 1$
13: set $i := \overline{\mathsf{ctr}}$
14: set $x_i := (\mathrm{id}, k, r, \mathrm{id}'_i)$ ▷ id'_i is available at the i-th entry of $\mathcal{L}'_{\mathsf{key}}$
15: **if** $\exists(*, *, *, \mathrm{id}'_i) \in \mathcal{L}_{\mathcal{H}_1}$ **then**
16:     $\boxed{\mathsf{abort}}$ ▷ As defined in G_1
17: **end if**
18: $\mathcal{L}_{\mathcal{H}_1} := \mathcal{L}_{\mathcal{H}_1} \cup \{x_i\}$
19: set $\mathsf{sk}_{\mathrm{id}_k} := (\mathsf{sk}_{\mathrm{id}'_i}, k, r)$ ▷ $\mathsf{sk}_{\mathrm{id}'_i}$ is available at the i-th entry of $\mathcal{L}'_{\mathsf{key}}$
20: $\mathcal{L}_{\mathsf{sk}} := \mathcal{L}_{\mathsf{sk}} \cup \{(\mathrm{id}, k, \mathsf{sk}_{\mathrm{id}_k})\}$
21: **return** $\mathsf{sk}_{\mathrm{id}_k}$ ▷ k-th component of $\mathsf{sk}_{\mathrm{id}}$

$\mathcal{O}_{\mathcal{H}_1}(\mathrm{id}, k, r)$
1: **if** $(k, \mathrm{id}, r, \mathcal{H}_1(\mathrm{id}, k, r)) \notin \mathcal{L}_{\mathcal{H}_1}$ **then**
2:     $\mathrm{id}' \xleftarrow{\mathsf{U}} \mathcal{ID}'$
3:     set $\mathcal{H}_1(\mathrm{id}, k, r) := \mathrm{id}'$
4:     $\mathcal{L}_{\mathcal{H}_1} := \mathcal{L}_{\mathcal{H}_1} \cup \{(k, \mathrm{id}, r, \mathcal{H}_1(\mathrm{id}, r)\}$
5: **end if**
6: **return** $\mathcal{H}_1(k, \mathrm{id}, r)$

$\mathcal{O}_{\mathcal{H}_2}(m, s)$
1: **if** $(m, s, \mathcal{H}_2(m, s)) \notin \mathcal{L}_{\mathcal{H}_2}$ **then**
2:     $m' \xleftarrow{\mathsf{U}} \mathcal{M}'$
3:     set $\mathcal{H}_2(m, s) := m'$
4:     $\mathcal{L}_{\mathcal{H}_2} := \mathcal{L}_{\mathcal{H}_2} \cup \{(m, s, \mathcal{H}_2(m, s)\}$
5: **end if**
6: **return** $\mathcal{H}_2(m, s)$

$\mathcal{O}_{\mathsf{sign}}(\mathrm{id}, k, m)$:
1: **if** $(\mathrm{id}, k) \in \mathcal{Q}_{\mathsf{key}}$ **then**
2:     $\boxed{\mathsf{abort}}$ ▷ Due to the restriction on security
3: **end if**
4: **if** $\nexists(\mathrm{id}, *) \in \mathcal{L}_r$ **then**   ▷ id appearing for the first time in $\mathcal{Q}_{\mathsf{key}} \cup \mathcal{Q}_{\mathrm{id}}$
5:     $r \xleftarrow{\mathsf{U}} \mathsf{SaltSp}$
6:     $\mathcal{L}_r := \mathcal{L}_r \cup \{(\mathrm{id}, r)\}$
7: **end if**
8: **if** $\nexists(*, \mathrm{id}, k) \in \mathcal{L}_{\mathrm{id}}$ **then** ▷ Sign query on id and k appearing for the first time
9:     $\mathsf{ctr} := \mathsf{ctr} + 1$
10:     set $i := \mathsf{ctr}$
11:     set $x_i := (\mathrm{id}, k, r, \mathrm{id}'_i)$ ▷ id'_i is available at the i-th entry of $\mathcal{L}'_{\mathsf{key}}$
12:     **if** $\exists(*, *, *, \mathrm{id}'_i) \in \mathcal{L}_{\mathcal{H}_1}$ **then**
13:        $\boxed{\mathsf{abort}}$ ▷ As defined in G_1
14:     **end if**
15:     $\mathcal{L}_{\mathcal{H}_1} := \mathcal{L}_{\mathcal{H}_1} \cup \{x_i\}$
16:     $\mathcal{L}_{\mathrm{id}} := \mathcal{L}_{\mathrm{id}} \cup \{(i, \mathrm{id}, k)\}$
17: **end if**
18: $\exists i \in [q_s]$ such that $(i, \mathrm{id}, k) \in \mathcal{L}_{\mathrm{id}}$
19: $\mathsf{ctr}_i := \mathsf{ctr}_i + 1$
20: set $j := \mathsf{ctr}_i$
21: $s \xleftarrow{\mathsf{U}} \mathsf{SaltSp}$
22: set $y_j := (m, s, m'_j)$ ▷ m'_j is available at the (i, j)-th entry of $\mathcal{L}'_{\mathsf{sign}}$
23: **if** $\exists(*, *, m'_j) \in \mathcal{L}_{\mathcal{H}_2}$ **then**
24:     $\boxed{\mathsf{abort}}$ ▷ As defined in G_1
25: **end if**
26: $\mathcal{L}_{\mathcal{H}_2} := \mathcal{L}_{\mathcal{H}_2} \cup \{y_j\}$
27: **return** $(k, \sigma_{ij} := (\sigma'_{ij}, r, s))$ ▷ σ'_{ij} is available at the (i, j)-th entry $\mathcal{L}'_{\mathsf{sign}}$

Fig. 10. Illustration of handling key queries, sign queries, and random oracle queries

$\sigma^* = (\sigma'^*, r^*, s^*)$, S computes $\mathrm{id}'^* \leftarrow \mathcal{H}_1(\mathrm{id}^*, k^*, r^*)$ and $m'^* \leftarrow \mathcal{H}_2(m^*, s^*)$ and sends $(\mathrm{id}'^*, m'^*, \sigma'^*)$ as a forgery to $\mathcal{C}$. The remaining part of the proof is very straightforward and is thus omitted.

Remark 3. It is easy to see that a chameleon hash-based extension of [PW21] can be obtained with similar modifications as in Fig. 8. Furthermore, the security reduction of the chameleon hash-based construction in WModel-II can be shown in the standard model using an approach analogous to that used for the above construction in the ROM.

Removing the restrictions imposed on the sign and key oracles in WModel-II will lead to an extension of the standard EUF-ID-CMA model (see Remark 2). Similar to the proof of Theorem 2, an IBS with WModel-II security can be shown to be secure in that extended model with a degradation linear in q_k. In fact, the proof is more or less the same as that of Theorem 2, except that the partitioning now happens over the extended ID-space; that is, $\{(\mathrm{id}, i) : \mathrm{id} \in \mathcal{ID} \text{ and } i \in [\nu_{\mathrm{id}}]\}$ needs to be partitioned, instead of $\mathcal{ID}$.

6   Concluding Remark

Pan and Wagner [PW21] proposed generic conversion from non-adaptive to adaptively secure IBS that they claimed achieve tight security. In this work, we have identified certain gaps in their approach and proposed new reductions that address these gaps while arguing why the technique as proposed in [PW21] is unlikely to yield tight reduction. We also proposed a functional extension of the Pan-Wagner technique that allows registering multiple devices under the same identity. Augmenting the Pan-Wagner generic technique to realize tight security would be an interesting research problem arising out of this work.

Acknowledgement. We would like to thank the anonymous reviewers of Indocrypt 2024 for their comments and suggestions, which helped us in polishing the technical and editorial content of this paper.

References

[BNN04] Bellare, M., Namprempre, C., Neven, G.: Security proofs for identity-based identification and signature schemes. In: Cachin, C., Camenisch, J.L. (eds.) EUROCRYPT 2004. LNCS, vol. 3027, pp. 268–286. Springer, Heidelberg (2004). https://doi.org/10.1007/978-3-540-24676-3_17

[BNN09] Bellare, M., Namprempre, C., Neven, G.: Security proofs for identity-based identification and signature schemes. J. Cryptol. **22**(1), 1–61 (2009)

[CC03] Choon, J.C., Hee Cheon, J.: An identity-based signature from gap Diffie-Hellman groups. In: Desmedt, Y.G. (ed.) PKC 2003. LNCS, vol. 2567, pp. 18–30. Springer, Heidelberg (2003). https://doi.org/10.1007/3-540-36288-6_2

[CDP21] Chatterjee, S., Dimri, A., Pandit, T.: Identity-based signature and extended forking algorithm in the multivariate quadratic setting. In: Adhikari, A., Küsters, R., Preneel, B. (eds.) INDOCRYPT 2021. LNCS, vol. 13143, pp. 387–412. Springer, Cham (2021). https://doi.org/10.1007/978-3-030-92518-5_18

[CKMS16] Chatterjee, S., Koblitz, N., Menezes, A., Sarkar, P.: Another look at tightness II: practical issues in cryptography. In: Phan, R.C.-W., Yung, M. (eds.) Mycrypt 2016. LNCS, vol. 10311, pp. 21–55. Springer, Cham (2017). https://doi.org/10.1007/978-3-319-61273-7_3

[CLND19] Chen, J., Ling, J., Ning, J., Ding, J.: Identity-based signature schemes for multivariate public key cryptosystems. Comput. J. 62(8), 1132–1147 (2019)

[CMS11] Chatterjee, S., Menezes, A., Sarkar, P.: Another look at tightness. In: Miri, A., Vaudenay, S. (eds.) SAC 2011. LNCS, vol. 7118, pp. 293–319. Springer, Heidelberg (2012). https://doi.org/10.1007/978-3-642-28496-0_18

[Cor00] Coron, J.-S.: On the exact security of full domain hash. In: Bellare, M. (ed.) CRYPTO 2000. LNCS, vol. 1880, pp. 229–235. Springer, Heidelberg (2000). https://doi.org/10.1007/3-540-44598-6_14

[DH76] Diffie, W., Hellman, M.: New directions in cryptography. IEEE Trans. Inf. Theory 22(6), 644–654 (1976)

[ElG84] ElGamal, T.: A public key cryptosystem and a signature scheme based on discrete logarithms. In: Blakley, G.R., Chaum, D. (eds.) CRYPTO 1984. LNCS, vol. 196, pp. 10–18. Springer, Heidelberg (1985). https://doi.org/10.1007/3-540-39568-7_2

[FS86] Fiat, A., Shamir, A.: How to prove yourself: practical solutions to identification and signature problems. In: Odlyzko, A.M. (ed.) CRYPTO 1986. LNCS, vol. 263, pp. 186–194. Springer, Heidelberg (1987). https://doi.org/10.1007/3-540-47721-7_12

[GS02] Gentry, C., Silverberg, A.: Hierarchical ID-based cryptography. In: Zheng, Y. (ed.) ASIACRYPT 2002. LNCS, vol. 2501, pp. 548–566. Springer, Heidelberg (2002). https://doi.org/10.1007/3-540-36178-2_34

[Hes02] Hess, F.: Efficient identity based signature schemes based on pairings. In: Nyberg, K., Heys, H. (eds.) SAC 2002. LNCS, vol. 2595, pp. 310–324. Springer, Heidelberg (2003). https://doi.org/10.1007/3-540-36492-7_20

[LPLL20] Lee, Y., Park, J.H., Lee, K., Lee, D.H.: Tight security for the generic construction of identity-based signature (in the multi-instance setting). Theor. Comput. Sci. 847, 122–133 (2020)

[Luy19] Van Luyen, L.: An improved identity-based multivariate signature scheme based on rainbow. Cryptography 3(1) (2019)

[Pat02] Paterson, K.G.: ID-based signatures from pairings on elliptic curves. Electron. Lett. 38(18), 1025–1026 (2002)

[PW21] Pan, J., Wagner, B.: Short identity-based signatures with tight security from lattices. In: Cheon, J.H., Tillich, J.-P. (eds.) PQCrypto 2021 2021. LNCS, vol. 12841, pp. 360–379. Springer, Cham (2021). https://doi.org/10.1007/978-3-030-81293-5_19

[Rüc10] Rückert, M.: Strongly unforgeable signatures and hierarchical identity-based signatures from lattices without random oracles. In: Sendrier, N. (ed.) PQCrypto 2010. LNCS, vol. 6061, pp. 182–200. Springer, Heidelberg (2010). https://doi.org/10.1007/978-3-642-12929-2_14

[Sha84] Shamir, A.: Identity-based cryptosystems and signature schemes. In: Blakley, G.R., Chaum, D. (eds.) CRYPTO 1984. LNCS, vol. 196, pp. 47–53. Springer, Heidelberg (1985). https://doi.org/10.1007/3-540-39568-7_5

[SPMC23] Sageloli, É., Pébereau, P., Méaux, P., Chevalier, C.: Shorter and faster identity-based signatures with tight security in the (Q)ROM from lattices. In: Tibouchi, M., Wang, X. (eds.) ACNS 2023. LNCS, vol. 13905, pp. 634–663. Springer, Cham (2023)

[ZLGL19] Zhang, X., Liu, S., Dawu, G., Liu, J.K.: A generic construction of tightly secure signatures in the multi-user setting. Theoret. Comput. Sci. **775**, 32–52 (2019)

Efficient Revocable Linkable Ring Signatures

R. Kabaleeshwaran$^{(\boxtimes)}$ and Bachala Khandava Kishan

Department of Computer Science and Engineering, Indian Institute of Information
Technology Design and Manufacturing Kurnool, Kurnool, Andhra Pradesh, India
kabaleesh@iiitk.ac.in

Abstract. Linkable ring signature allows a user to create an anonymous
signature on behalf of a group, which can be traced for disclosing multiple signatures from the same user. It improves privacy and anonymity
in applications like privacy-preserving e-voting and e-cash. On the other
hand, the advanced anonymity might compromise decentralized applications like cryptocurrencies and the possibility of money laundering or
terrorist financing may emerge. Although Linkable Ring Signatures may
help detect policy violations like double-spending attacks in e-voting,
and multiple transaction detection attacks in e-cash protocols without
compromising anonymity, they lack mechanisms that would allow revoking malicious users who fail to comply with these policies.

In this paper, we introduce an efficient revocable and linkable ring
signature scheme that features mandatory revocability and linkability,
and we prove its security in the random oracle model. Our construction
builds on Liu et al.'s Schnorr-based linkable ring signatures. Compared
to Zhang et al.'s revocable linkable ring signatures, our scheme significantly improves efficiency, reducing the time complexity of the signature
and verification algorithms by a factor of four while also cutting the size
of the signature in half.

Keywords: Ring Signatures · Linkable Ring Signatures · Revocable
Ring Signatures · Schnorr-based linkable ring signature

1 Introduction

1.1 Ring Signatures

A ring signature allows group members to sign a message on behalf of the group
without revealing the signer's identity. The main security aspect of this cryptographic primitive is that the prover can't determine which one of the members
signed the message. Ring signatures were introduced in the pioneering work [20],
but many constructions and variants have been proposed in the literature like
[1,7].

A ring signature differs from an ordinary group signature scheme in that
it allows for spontaneous grouping. There is no central authority or manager

© The Author(s), under exclusive license to Springer Nature Switzerland AG 2025
S. Mukhopadhyay and P. Stănică (Eds.): INDOCRYPT 2024, LNCS 15495, pp. 305–325, 2025.
https://doi.org/10.1007/978-3-031-80308-6_14

controlling group formation or the revocation of a signer's identity. Hence, any person can form a group by collecting the public keys of potential members without their knowledge or consent. Once the ring signature is generated, it becomes infeasible to determine whether two signatures, generated using different ring members, originate from the same member. This provides strong anonymity.

However, this unconditional anonymity introduces a problem. In some applications, such as e-voting or e-cash, one would want multiple actions to be linked to the same person's identity without revealing who that person is. This leads to the case of *linkable ring signatures*.

1.2   Linkable Ring Signatures

To fulfill the need for linkability in certain applications, linkable ring signatures (LRS) were proposed by Liu et al. in 2004 [15]. In this variant, the anonymity of the signer is maintained, but if the same signer provides more than one signature, these signatures are linkable. This property makes LRS particularly suitable for applications where detecting double voting or multiple transactions is necessary without compromising anonymity.

Since the inception of linkable ring signatures, several improvements have been proposed, including schemes with constant-size signatures [2,21], unconditional anonymity [14], and traceability [9,10]. Further improvements include enhanced security features [16], certificate-based LRS [4], and LRS with separability [22], which provide greater flexibility depending on the application.

While traditional ring signatures guarantee complete unlinkability between signatures, linkable ring signatures introduce a middle ground: anonymity is preserved, but a verifier can distinguish whether multiple signatures come from the same user.

1.3   Revocable and Traceable Ring Signatures

In addition to the standard ring signature scheme, traceable (or revocable) ring signatures have emerged as a variant that reduces the anonymity provided by conventional ring signatures, while still preserving the spontaneous nature of group formation [3,5,17]. Unlike group signatures, which require a central manager to oversee membership and revoke anonymity, traceable and revocable ring signatures provide mechanisms to control anonymity without the need for a central authority. These schemes are particularly useful in applications requiring both privacy and accountability, such as e-voting or privacy-preserving transactions.

In revocable ring signatures, a set of pre-defined revocation authorities can reveal the signer's identity at any time, a property known as mandatory revocability [13]. This ensures that anonymity can be revoked when necessary, without compromising the decentralized nature of the system. On the other hand, traceable ring signatures [10] provide conditional traceability, where the signer's identity is only revealed if they submit multiple signatures using the same private key, making it ideal for applications that need to detect repeat actions like double voting. Additionally, convertible or verifiable ring signatures allow the signer to voluntarily reveal their identity when necessary, providing flexibility in situations where proof of authorship may be required, while still maintaining anonymity when desired [12,17].

Revocable or traceable ring signatures are essential in real-world scenarios that require both anonymity and accountability. They have been applied in confidential transactions [18], where financial transparency might necessitate the revelation of the signer's identity, and privacy-preserving payment systems [6], where anonymity is preserved unless explicitly needed.

1.4   Related Work

Liu et al. [15] introduced the first spontaneous anonymous group signature in which one can generate the ring spontaneously without any requirement of a group manager or centralized setup allowing ad hoc group. Their construction have mandatory linkability but lack revocability. It is proven secure against CPA attacks and supports efficient e-voting without the registration phase. Also Liu et al. [14] introduces linkability with unconditional anonymity. According to [11, Theorem 1] LRS cannot have unconditional anonymity. But [14] proved their theorem is not always true and proposed an LRS with unconditional anonymity, our scheme construction is inspired by [14] in addition we also include the revocability property with better efficiency. The Dual Ring [23] introduces a novel approach for generic ring construction which can be constructed using several canonical identification schemes like the Schnorr identification scheme. Liu et al. [13] proposed bilinear pairing-based revocable ring signatures. But their scheme lacks linkability property. Mandatory linkability and revocability can be very useful in real-world applications like e-voting [21] (i.e. prevent double voting), e-cash [6,18] (helps in preventing double spending in decentralized systems).

Zhang et al. [24] introduced ring signature with mandatory linkability and mandatory revocability where the revocation authority can revoke the real signer under any circumstances. It inherits the properties of group signatures (spontaneous group generation and anonymity revocation).

1.5   Our Contribution

Recognizing the impact of revocable and linkable ring signature (RLRS) schemes, an efficient implementation can greatly enhance performance for high-volume usage. We propose a more efficient revocable ring signature scheme inspired by Liu et al.'s [15] linkable ring signatures, incorporating suitable aggregated commitment components during the generation of ring signatures. In contrast, Zhang et al.'s [24] construction requires the computation of similar commitment components for each user in the signer ring.

Our RLRS scheme introduces two key elements on top of Liu et al.'s LRS scheme to improve efficiency. The first element pertains to the aggregated commitment components, while the second addressesthe randomness associated with the revocation mechanism. As a result, our approach halves the signature size and reduces the time complexity of the signing and verification algorithms by a factor of four.

2   Preliminaries

2.1   Notation

We denote $\lambda > 0$ as the security parameter. Y is the ring consisting of n public keys. Y is a subset of the set of all possible public keys, For a positive integer n, we denote the set $\{1, ..., n\}$ as $[1, n]$. Let A be a randomized algorithm. We denote the operation of this algorithm on input ip and its output op as $op \leftarrow A(ip)$. For a non-empty set S, $x \in S$ denotes the element x belongs to the set S.

2.2   Computational Assumptions

First, we define the cyclic group generator.

Definition 1 (Cyclic Group Generator). *Given the security parameter λ, the cyclic group generator algorithm $\mathcal{G}$ returns (p, G, g), where G is the cyclic group of prime order p and g is the generator of the group G.*

The following definitions of Discrete Logarithms and Decisional Diffie-Hellman assumptions are taken from [24].

Definition 2 (Discrete Log Assumption). *The Discrete Logarithm problem in a cyclic group generator $\mathcal{G}$ is defined as follows: Given (p, G, g) from the output of $\mathcal{G}(\lambda)$ and an element $y \in G$, find an integer x such that $y = g^x$. We say that the (t, ϵ)-DL assumption holds in $\mathcal{G}$ if there does not exist any probabilistic polynomial time (PPT) algorithm running in time t has an advantage of at least ϵ in solving the discrete logarithm problem in $\mathcal{G}$.*

In the cyclic group setting, the discrete logarithm (DL) relation of a group G consists of $(g, y) \in G$ with the existence of $x \in \mathbb{Z}_p$ such that $y = g^x$, for any $g \in G$. From the above definition, it is clear that computing a witness for the statement with DL relation is computationally hard.

Definition 3 (Decisional Diffie-Hellman(DDH) Assumption). *The Decisional Diffie-Hellman (DDH) problem in a cyclic group generator $\mathcal{G}$ is defined as follows: Given (p, G, g) and a triple $(g^a, g^b, g^c) \in G^3$, decide whether $c = ab$ or not. We say that the (t, ϵ)-DDH assumption holds in $\mathcal{G}$ if there does not exist any PPT algorithm running in time t has an advantage of at least ϵ over random guessing in solving the DDH problem in $\mathcal{G}$.*

2.3   El Gamal Public Key Encryption Scheme

In our scheme we use the El Gamal encryption scheme [8] consisting of the following four PPT algorithms which are defined as follows:

- $param \leftarrow$ Setup(λ): On input security parameter λ algorithm returns $param = \{p, G, g\}$, where G is the group with prime order p and g is the generator of that group.
- $(sk, pk) \leftarrow$ KeyGen($param$): On input $param = \{p, G, g\}$ outputs a secret and public key pair (sk, pk) where sk is chosen at random from $\mathbb{Z}_p$ and $pk = g^{sk}$.
- $C \leftarrow$ Encrypt(M, pk): On input message M and receivers public key pk the sender randomly picks a number $r \in \mathbb{Z}_p$ and computes the ciphertext $C_1 = g^r$ mod p and $C_2 = M \cdot pk^r$ mod p. The final output is $C = \{C_1, C_2\}$.
- $M \leftarrow$ Decrypt(sk, C): On input $C = \{C_1, C_2\}$, and receiver's secret key sk, recovers the message by computing $M = C_2/C_1^{sk}$.

We state the security of the El Gamal encryption scheme, for the complete proof, refer to [8].

Theorem 1. *If the DDH assumption holds in $\mathcal{G}$, then the El Gamal encryption scheme is IND-CPA secure.*

3   Revocable Linkable Ring Signature Definition

The following definition of revocable linkable ring signature (RLRS) scheme is taken from [24] and it is defined using six probabilistic polynomial time (PPT) algorithms (SETUP, KEYGEN, SIGN, VERIFY, LINK and REVOKE) which are defined below.

★ **param** ← **Setup**(λ). On input a security parameter λ, outputs the public parameters *param*.

★ **(sk, pk)** ← **KeyGen(param)**. On input a public parameter *param*, outputs a secret key and public key pair (sk, pk). Here we denote $\mathcal{SK}$ as the secret key space and $\mathcal{PK}$ as the public key space.

★ σ ← **Sign(event, n, Y, sk, pk$_R$, M)**. Given an event identifier *event*, group size n, a set Y of n public keys $\{pk_1, \ldots, pk_n\}$ such that $pk_i \in \mathcal{PK}$ for $i \in [1, n]$, a secret key $sk \in \mathcal{SK}$ whose corresponding public key is contained in Y, a revocation authority's public key $pk_R \in \mathcal{PK}$ and a message M, it returns a ring signature σ.

★ **0/1** ← **Verify(event, n, Y, pk$_R$, M, σ)**. Given an event identifier *event*, a group size n, a set $Y = \{pk_1, \ldots, pk_n\}$ containing n public keys $pk_i \in \mathcal{PK}$ for $i \in [1, n]$, a revocation authority's public key $pk_R \in \mathcal{PK}$ and a message-signature pair (M, σ), it returns 1 if the message-signature pair is valid; otherwise, it returns 0.

⋆ **Linked/Unlinked** ← **Link(event, n_1, n_2, Y_1, Y_2, M_1, M_2, σ_1, σ_2).**
Given an event identifier *event*, group sizes n_1 and n_2, two sets Y_1 and Y_2
containing n_1 and n_2 public keys respectively, where all the public keys are
from $\mathcal{PK}$ and two valid message-signature pairs (M_1, σ_1) and (M_2, σ_2), it
returns "linked" if the message-signature pairs are signed by the same user;
otherwise, it returns "unlinked".

⋆ **pk** ← **Revoke(n, Y, σ, sk_R).** Given a group size n, set of public keys $Y =$
$\{pk_1, \ldots, pk_n\}$ such that $pk_i \in \mathcal{PK}$ for $i \in [1, n]$, a valid ring signature σ
and revoking authority secret key $sk_R \in \mathcal{SK}$ corresponding to $pk_R \in \mathcal{PK}$, it
returns a pk in Y.

Correctness. *RLRS schemes must satisfy:*

- **Verification Correctness.** Signatures generated according to the specification must be accepted during verification.
- **Linking Correctness.** If two signatures are generated for the same event according to the specification, then they should be linked if and only if the two signatures share a common signer.
- **Revoking Correctness:** If a signature is generated by any signer who passed verification can be revoked by the revoking authority.

3.1 Security Definitions of RLRS:

In the formal security definition, adversary $\mathcal{A}$ use the following oracles:

- **pk** ← **JO($\perp$)** : The joining oracle, on request, adds a new user to the system. It returns the new user's public key $pk \in \mathcal{PK}$.
- **sk_i** ← **CO(pk_i)** : The corruption oracle, on input a public key $pk_i \in \mathcal{PK}$, which is a query output of JO, returns the corresponding secret key $sk_i \in \mathcal{SK}$.
- σ ← **SO(event, n, Y, pk_π, pk_R, M)** : The signing oracle, on input an event identifier *event*, a group size n, a set $Y = \{pk_1, \ldots, pk_n\}$ of n public keys from $\mathcal{PK}$, the signers public key $pk_\pi \in Y$, a revocation authority's public key $pk_R \in \mathcal{PK}$ and a message M, returns a valid ring signature denoted as σ ← SIGN($event, n, Y, sk_\pi, pk_R, M$).

Unforgeability: It is defined using a game between a challenger $\mathcal{B}$ and an
adversary $\mathcal{A}$ which is defined as:

a. $\mathcal{B}$ runs the SETUP procedure with a security parameter λ and receives *param* as output.
b. $\mathcal{A}$ can adaptively query the oracles *JO*, *CO*, and *SO*.
c. $\mathcal{A}$ provides $\mathcal{B}$ with an event identifier *event*, a group size n, a set $Y^* = \{pk_1, \ldots, pk_n\}$ consisting of n public keys (where each $pk_i \in Y^*$ for $i \in [1, n]$), a revocation authority's public key $pk_R \in \mathcal{PK}$, a forgery message and ring signature pair (M^*, σ^*).

$\mathcal{A}$ wins the game if the following conditions are satisfied:

1. VERIFY$(event, n, Y^*, pk_R, M^*, \sigma^*) = 1$;
2. All public keys in Y^* have been generated by querying JO;
3. None of the public keys in Y^* were submitted as inputs to CO; and
4. The forgery signature σ^* was not previously generated by SO.

The success probability of $\mathcal{A}$ in winning the unforgeability is expressed as:

$$\text{Adv}_{\mathcal{A}}^{\text{Unf}}(\lambda) = \Pr[\mathcal{A} \text{ wins the game}].$$

Definition 4 (Unforgeability). *A RLRS scheme is existentially unforgeable against adaptive chosen message and chosen public key attacks if for all PPT adversaries $\mathcal{A}$, the $Adv_{\mathcal{A}}^{Unf}(\lambda)$ is negligible.*

Anonymity: The anonymity is defined as a game between a challenger $\mathcal{B}$ and an adversary $\mathcal{A}$ which is defined as:

1. $\mathcal{B}$ runs the SETUP algorithm on a security parameter λ and outputs *param*.
2. $\mathcal{A}$ can make queries to *Join Oracle (JO)* adaptively.
3. $\mathcal{A}$ provides $\mathcal{B}$ with an event identifier *event*, a group size n, a set Y of n public keys, where all public keys in Y are generated by JO, the revocation authority's public key $pk_R \in \mathcal{PK}$ and a message M. $\mathcal{B}$ parses Y as $\{pk_1, \ldots, pk_n\}$, chooses π uniformly at random from $\{1, \ldots, n\}$, and computes a *Challenge Signature* σ_π generated using SIGN with the corresponding secret key sk_π. The signature σ_π is given to $\mathcal{A}$.
4. Finally, $\mathcal{A}$ guesses $\pi' \in \{1, \ldots, n\}$.

$\mathcal{A}$ wins the anonymity game if $\pi' = \pi$. The success probability of $\mathcal{A}$ in winning the Anonymity is expressed as:

$$\text{Adv}_{\mathcal{A}}^{\text{Anon}}(\lambda) = \left| \Pr[\pi' = \pi] - \frac{1}{n} \right|.$$

Definition 5 (Anonymity). *A RLRS scheme is computationally anonymous if for all PPT adversaries $\mathcal{A}$, the $Adv_{\mathcal{A}}^{Anon}(\lambda)$ is negligible.*

Linkability: The likability is defined as a game between a challenger $\mathcal{B}$ and an adversary $\mathcal{A}$, who is given access to JO, CO, and SO, is defined as follows:

a. $\mathcal{B}$ runs the SETUP algorithm on a security parameter λ and outputs *param*.
b. $\mathcal{A}$ can query JO, CO, and SO adaptively.
c. $\mathcal{A}$ provides $\mathcal{B}$ with an event identifier *event*, two group sizes n_1 and n_2 with the assumption $n_1 \leq n_2$ without loss of generality, two sets Y_1 and Y_2 with n_1 and n_2 public keys respectively, two message-signature pairs (M_1, σ_1) and (M_2, σ_2), and a revocation authority's public key $pk_R \in \mathcal{PK}$.

$\mathcal{A}$ wins the game if:

1. All the public keys in $Y_1 \cup Y_2$ are outputs of JO;
2. VERIFY$(event, n_i, Y_i, pk_R, M_i, \sigma_i) = 1$ and σ_i is not the output of SO, for $i = 1, 2$;
3. CO has been queried less than 2 times, i.e., $\mathcal{A}$ can have at most one user secret key
4. LINK$(event, \sigma_1, \sigma_2) =$ Unlinked.

The success probability of $\mathcal{A}$ in winning the Linkability game is expressed as:

$$\text{Adv}_{\mathcal{A}}^{\text{Link}}(\lambda) = \Pr[\mathcal{A} \text{ wins the game}].$$

Definition 6 (Linkability). *A RLRS scheme is linkable if for all PPT adversaries $\mathcal{A}$, $Adv_{\mathcal{A}}^{Link}(\lambda)$ is negligible.*

Revocability: The revocability property in an RLRS scheme is mandatory, meaning that the probability of a signer generating a signature without their identity being revealed by the revocation authority should be negligible. The revocability is defined as a game between a challenger $\mathcal{B}$ and an adversary $\mathcal{A}$, who is given access to JO, CO, and SO, is defined as follows:

a. $\mathcal{B}$ runs the SETUP algorithm on a security parameter λ and outputs $param$.
b. $\mathcal{A}$ can query JO, CO, and SO adaptively.
c. $\mathcal{A}$ can obtain at most one private key of a ring member from CO.
d. $\mathcal{A}$ provides $\mathcal{B}$ with an event identifier $event$, a group size n, a set Y containing n public keys, a message M, a revocation authority's public key pk_R, and a ring signature σ.

$\mathcal{A}$ wins the game if:

1. VERIFY$(event, n, Y, pk_R, M, \sigma) = 1$,
2. All public keys in Y are outputs of JO;
3. σ is not an output of SO;
4. CO has been queried less than two times i.e., $\mathcal{A}$ can obtain at most one private key, denoted as x_π and
5. $pk_j =$ REVOKE(n, Y, σ, sk_R), where $j \neq \pi$.

The success probability of adversary $\mathcal{A}$ in winning the Revocability game is expressed as:

$$\text{Adv}_{\mathcal{A}}^{\text{Revoke}}(\lambda) = \Pr[\mathcal{A} \text{ wins the game}].$$

Definition 7 (Revocability). *A RLRS scheme is revocable if for any PPT adversary $\mathcal{A}$, $Adv_{\mathcal{A}}^{Revoke}(\lambda)$ is negligible.*

Non-slanderability: Informally, the non-slanderability ensures that the attacker should not be able to accuse an honest user of generating a signature that is determined to be linked with a malicious signature generated by the attacker. The non-slanderability is defined as a game between a challenger $\mathcal{B}$ and an adversary $\mathcal{A}$, who is given access to JO, CO, and SO, as follows:

a. $\mathcal{B}$ runs the SETUP algorithm on a security parameter λ and outputs *param*.
b. $\mathcal{A}$ can query JO, CO, and SO adaptively.
c. $\mathcal{A}$ provides $\mathcal{B}$ with an event identifier, a group size n, a set Y of n public keys, a message M, a revocation authority's public key pk_R, and a public key of an insider $pk_\pi \in Y$ such that pk_π has not been queried to CO or included as the insider public key in any query to SO. $\mathcal{B}$ uses the corresponding secret key sk_π to run SIGN($event, n, Y, sk_\pi, pk_R, M$) and produces σ for $\mathcal{A}$.
d. $\mathcal{A}$ queries the oracles with arbitrary interleaving. Specifically, $\mathcal{A}$ can query CO for any public key except pk_π.
e. $\mathcal{A}$ provides a group size n^*, Y^* with n^* public keys, a message M^*, a revocation authority's public key pk_R, and a signature $\sigma^* \neq \sigma$.

$\mathcal{A}$ wins the game if:

1. VERIFY($event, n^*, Y^*, pk_R, M^*, \sigma^*$) = 1.
2. σ^* is not an output of SO;
3. All public keys in Y^* and Y are outputs of JO;
4. pk_π has not been queried to CO
5. LINK(σ^*, σ) = Linked.

The success probability of adversary $\mathcal{A}$ in winning the Non-Slanderability game is expressed as:

$$\mathrm{Adv}_{\mathcal{A}}^{\mathrm{NS}}(\lambda) = \Pr[\mathcal{A} \text{ wins the game}].$$

Definition 8 (Non-Slanderability). *A RLRS scheme is non-slanderable if for all PPT adversaries $\mathcal{A}$, $Adv_{\mathcal{A}}^{NS}$ is negligible.*

4   Efficient Revocable Linkable Ring Signatures

In this section, we present our construction of revocable linkable ring signatures and analyze their security in the random oracle model. Our approach builds upon the Schnorr-based Linkable Ring Signature (LRS) scheme introduced by Liu et al. [14]. We describe how we extend their LRS scheme to incorporate revocability.

Liu et al.'s LRS scheme utilizes two hash functions: $H : \{0,1\}^* \to G$ and $H' : \{0,1\}^* \to \mathbb{Z}_p$. The function H generates the linkable tag $t = H(event)^{sk_\pi}$, while H' is used to create the aggregated challenge term $\sum_{i=1}^{n} c_i$ for the Schnorr-based ring signature on behalf of all signers $Y = \{pk_1, \ldots, pk_n\}$. This includes the ring-commitment term K, which relates to the discrete logarithm (DL) relation of the secret-public key pairs, and the link-commitment term K', which captures the DL relation between $e = H(event)$ and the linkable tag $t = e^{sk}$. Notably, a similar conceptualization has been extended in the DualRing scheme [23].

In 2019, Zhang et al. [24] introduced revocable linkable ring signatures within the discrete logarithm framework. Their construction utilizes four commitment terms, $\{a_{j,i}, \bar{a}_{j,i}\}$, for each user $i \in [1, n]$ and $j \in [1, 2]$, in contrast to the aggregated term approach of Liu et al.'s LRS. The signer ring is defined as $Y = \{pk_1, \ldots, pk_n\}$. The terms $a_{1,i}$ and $\bar{a}_{1,i}$ encapsulate the ring signature features alongside the associated challenge terms S'_{i+1} and S''_{i+1}. In contrast, $a_{2,i}$ facilitates revocability through an El Gamal encryption of the signer's public key pk_π, while $\bar{a}_{2,i}$ ensures linkability.

We enhance Zhang et al.'s construction by using aggregated commitments instead of non-aggregated ones, which can lead to increased computational efficiency and compactness in the ring signature. Drawing inspiration from Liu et al.'s [15] linkable ring signature that employs aggregated commitments, we integrate the revocability feature from Zhang et al.'s approach while leveraging aggregated commitments during signature generation.

In summary, our construction builds on Liu et al.'s LRS framework by introducing a Schnorr-type signature term $\tilde{z}$ and an aggregated commitment term K'' as additional inputs to the hash function H'. This is combined with the El Gamal encryption of the signer's public key pk_π using randomness u. The term $\tilde{z} = r_z - c_\pi u$ conceals the signature knowledge on the El Gamal encryption randomness u, while K'' corresponds to the aggregated commitment term for the DL relation of the pair $\{g, C_1 = g^u\}$.

4.1   Our RLRS Scheme

In this section, we provide a detailed construction of our Revocable Linkable Ring Signature (RLRS) scheme. Our RLRS Scheme is defined using six PPT algorithms which are defined as:

Setup(λ): Run $\mathcal{G}(\lambda)$ which returns (p, G, g), where G is a cyclic group of prime order p such that the discrete logarithm problem is intractable and g is a generator of G. Consider two cryptographic hash functions $H : \{0, 1\}^* \rightarrow G$ and $H' : \{0, 1\}^* \rightarrow \mathbb{Z}_p$. It returns the public parameter $param$ as (p, G, g, H, H').

KeyGen$(param)$: A user randomly chooses x from $\mathbb{Z}_p$ and computes $Z = g^x$. It returns the secret key $sk = x$ and the corresponding public key $pk = Z$.

Sign$(event, n, Y, sk_\pi, pk_R, M)$: On input the event identifier $event$, a number of users n included in the ring of signers $Y = \{pk_1, \ldots, pk_n\} = \{Z_1, \ldots, Z_n\}$, sk_π is the secret key corresponding to the public key pk_π such that $pk_\pi \in Y$ (i.e., $\pi \in [1, n]$), pk_R is the revoking authority's public key, M is the message to be signed. The signer, using the knowledge of $sk_\pi = x_\pi$, computes the following:

1. Compute $e = H(event)$ and $t = e^{x_\pi}$.

2. Choose u uniformly at random from $\mathbb{Z}_p$ and compute the ciphertext $C = \{C_1, C_2\}$ using El Gamal encryption as

$$C_1 \leftarrow g^u, \qquad C_2 \leftarrow Z_\pi(pk_R)^u. \tag{1}$$

3. Choose $r_x, r_z, c_1, \ldots, c_{\pi-1}, c_{\pi+1}, \ldots, c_n$ uniformly at random from $\mathbb{Z}_p$ and compute

$$K = g^{r_x} \prod_{\substack{i=1 \\ i \neq \pi}}^{n} Z_i^{c_i} \quad \text{(Ring signature component)}, \tag{2}$$

$$K' = e^{r_x} t^{\sum_{\substack{i=1 \\ i \neq \pi}}^{n} c_i} \quad \text{(Linkable tag component)}, \tag{3}$$

$$K'' = pk_R^{r_z} \prod_{\substack{i=1 \\ i \neq \pi}} \left(\frac{C_2}{Z_i}\right)^{c_i} \quad \text{(Revocable component)}. \tag{4}$$

4. Find c_π such that

$$c_1 + \cdots + c_n \mod p = H'(Y \parallel event \parallel t \parallel M \parallel K \parallel K' \parallel K''). \tag{5}$$

5. Compute

$$\tilde{x} = r_x - c_\pi x_\pi \mod p \text{ and } \tilde{z} = r_z - c_\pi u \mod p. \tag{6}$$

6. Output the ring signature $\sigma = \{t, \tilde{x}, \tilde{z}, c_1, \ldots, c_n, C = \{C_1, C_2\}\}$.

Verify$(event, n, Y, M, pk_R, \sigma)$: Given the input, first compute $e = H(event)$ then parse the ciphertext $C = \{C_1, C_2\}$. Then compute

$$c_0 = H'\left(Y \parallel event \parallel t \parallel M \parallel g^{\tilde{x}} \prod_{i=1}^{n} Z_i^{c_i} \parallel e^{\tilde{x}} t^{\sum_{i=1}^{n} c_i} \parallel pk_R^{\tilde{z}} \prod_{i=1}^{n} \left(\frac{C_2}{Z_i}\right)^{c_i}\right) \tag{7}$$

Then check if $c_0 = \sum_{i=1}^{n} c_i \mod p$ output 1, otherwise output 0.

Link$(event, n_1, n_2, Y_1, Y_2, M_1, M_2, \sigma_1, \sigma_2)$: On input the event identifier $event$, two group of ring of signers Y_1, Y_2 with sizes n_1 and n_2 respectively, and two message-signature pairs (M_1, σ_1) and (M_2, σ_2), where $\sigma_1 = (t_1, \cdot)$ and $\sigma_2 = (t_2, \cdot)$, output *Linked* if $t_1 = t_2$, otherwise output *Unlinked*.

Revoke(n, Y, σ, sk_R): On input ring of signer Y of size n, a ring signature σ, and the secret key $sk_R = x_R$ corresponding to revocation authority pk_R computes the following:

1. First checks whether the signature is valid. If yes, continue. Otherwise, abort.
2. To revoke the anonymity of the real signer, the revocation authority computes as follows:
   (a) Parse the signature σ as $\{t, \tilde{x}, \tilde{z}, c_1, c_2, \ldots, c_n, C = \{C_1, C_2\}\}$.
   (b) Compute $Z_\pi = C_2/C_1^{x_R}$.
3. Output Z_π as a public key of the real signer for a given ring signature σ.

4.2   Correctness of Our RLRS Scheme

– **Verification Correctness**: From Eq. 7 of our revocable linkable ring signature construction, it is sufficient to prove that

$$K = g^{\tilde{x}} \prod_{i=1}^{n} Z_i^{c_i}, \quad K' = e^{\tilde{x}} t^{\sum_{i=1}^{n} c_i}, \quad K'' = pk_R^{\tilde{z}} \prod_{i=1}^{n} \left(\frac{C_2}{Z_i}\right)^{c_i},$$

where elements $r_x, r_z, \{c_i\}, u$ are chosen uniformly at random from $\mathbb{Z}_p$ and the components $\tilde{x}, \tilde{z}, K, K', K'', C = \{C_1, C_2\}$ are computed as in SIGN algorithm. Let $\pi \in [1, n]$ denote the actual signer's index. First, consider the following derivation for the component K,

$$g^{\tilde{x}} \prod_{i=1}^{n} Z_i^{c_i} = g^{r_x - c_\pi x_\pi} (g^{x_\pi})^{c_\pi} \prod_{\substack{i=1 \\ i \neq \pi}}^{n} Z_i^{c_i} = g^{r_x} \prod_{\substack{i=1 \\ i \neq \pi}}^{n} Z_i^{c_i} = K,$$

where the first equality follows from Eq. 6 and the definition of public key $Z_\pi = g^{x_\pi}$ and the last equality follows from Eq. 2. Similarly, consider,

$$e^{\tilde{x}} t^{\sum_{i=1}^{n} c_i} = e^{r_x - c_\pi x_\pi} t^{c_\pi} t^{\sum_{\substack{i=1 \\ i \neq \pi}} c_i} = e^{r_x} t^{\sum_{\substack{i=1 \\ i \neq \pi}} c_i} = K',$$

where the first equality follows from Eq. 6 and the second equality from the identity $t = e^{x_\pi}$ and the last equality from Eq. 3.

Finally, consider the following derivation for K'',

$$pk_R^{\tilde{z}} \prod_{i=1}^{n} \left(\frac{C_2}{Z_i}\right)^{c_i} = pk_R^{r_z - c_\pi u} (C_2/Z_\pi)^{c_\pi} \prod_{\substack{i=1 \\ i \neq \pi}}^{n} \left(\frac{C_2}{Z_i}\right)^{c_i} = pk_R^{r_z} \prod_{\substack{i=1 \\ i \neq \pi}}^{n} \left(\frac{C_2}{Z_i}\right)^{c_i} = K''.$$

The first equality follows from Eq. 6 and the last equality follows from Eq. 4. The second equality follows from the identity $C_2/Z_\pi = pk_R^u$, which is obtained from the El Gamal ciphertext components $C_1 = g^u$ and $C_2 = Z_\pi (pk_R)^u$.

– **Linking Correctness.** The linking correctness is ensured as the signer calculates the linking tag as:

$$t = e^{x_\pi}, \text{ where } e = H(event).$$

Thus, the user can only generate the linking tag once, for a given event.

– **Revoking Correctness.** When the signer adheres to the protocol, the revocation authority can successfully retrieve the signer's public key by decrypting the ciphertext as follows:

$$Z_\pi = \frac{C_2}{C_1^{x_R}},$$

where x_R is the private key of the revocation authority, and Z_π represents the actual signer's public key.

4.3   Security Analysis of Our RLRS Scheme

Theorem 2 (Unforgeablility). *Our revocable and linkable ring signature scheme is unforgeable in the random oracle model if the Discrete Logarithm (DL) assumption holds in $\mathcal{G}$.*

Proof. We prove this by a contradiction assumption. Assume that the DL problem is hard in the cyclic group generator $\mathcal{G}$. Suppose that our RLRS Scheme is not unforgeable, i.e., there exists a PPT adversary $\mathcal{A}$ who breaks the unforgeability of our RLRS Scheme with some non-negligible probability in the random oracle model. Now we construct a simulator $\mathcal{B}$ that breaks the DL assumption in $\mathcal{G}$. Consider the DL instance (p, G, g), $y \in G$, $\mathcal{B}$'s goal is to compute x from $\mathbb{Z}_p$ such that $y = g^x$.

It is easy to see that DL assumption is reducible to DL with Multiple instances. The reduction transform the DL instance y into multiple instances $\{y_i\}_{i=1}^n$, where $y_i = g^{\alpha_i} y^{\beta_i}$ and $\alpha_i, \beta_i \in \mathbb{Z}_p^*$, for $i \in [1, n]$. Once we could solve at least one of the instances (say $x_i = \alpha_i + x\beta_i$ such that $y_i = g^{x_i}$), the reduction can compute the given DL solution by computing $x = (x_i - \alpha_i)/\beta_i$. So, we provide the reduction-based multiple instance DL assumption.

Setup: Given n discrete logarithm problem instances $(X_1, \ldots, X_n)$ and a generator $g \in G$, $\mathcal{B}$'s goal is to find at least one $x_i \in \mathbb{Z}_p$ such that $X_i = g^{x_i}$ for $i \in [1, n]$.
$\mathcal{B}$ simulates the oracles as follows:

Random Oracle H: $\mathcal{B}$ randomly selects $\theta \in \mathbb{Z}_p$ and returns g^θ.
Random Oracle H': $\mathcal{B}$ randomly selects $\alpha \in \mathbb{Z}_p$ and returns it, provided that the value has not been assigned already.

Joining Oracle JO: Assume that $\mathcal{A}$ can query JO at most n' times, where $n' \geq n$. $\mathcal{B}$ randomly chooses a subset I_n of size n from these queries. $\mathcal{B}$ assigns Z_i, for $i \in [1, n]$, to these n indexes. Without loss of generality, we assume $1, \ldots, n$ denote these indexes (where $\mathcal{B}$ does not know the corresponding secret key) and $n + 1, \ldots, n'$ denote the other indexes. For the remaining $n' - n$ indexes, $\mathcal{B}$ generates the secret-public key pairs according to the KEYGEN algorithm. Upon the j-th query, $\mathcal{B}$ returns the corresponding public key, for $j \in [1, n']$.
Corruption Oracle CO: On input a public key pk that is an output from JO, $\mathcal{B}$ checks whether it corresponds to the subset I_n. If it does, $\mathcal{B}$ aborts. Otherwise, $\mathcal{B}$ outputs the corresponding secret key sk_j, for $j \in [n + 1, n']$.
Signing Oracle SO: On input a signing query for an event identifier *event*, a set of public keys $Y = \{Z_1, \ldots, Z_n\}$, the signer's public key Z_π where $\pi \in [1, n]$, and a message M, $\mathcal{B}$ simulates as follows:

- If $H(event)$ has not been queried, perform the H-query of *event* as described above. Set $e = H(event)$. Note that $\mathcal{B}$ knows the discrete logarithm θ of e to the base g, i.e., $e = g^\theta$.
- If Z_π does not correspond to any element in the set I_n, $\mathcal{B}$ knows the secret key and computes the signature according to the algorithm.

– Otherwise, assume Z_π is the π-th index from JO such that $\pi \in [1, n]$. Now $\mathcal{B}$ sets $t = X_\pi^\theta$ and chooses u uniformly at random from $\mathbb{Z}_p$ and computes C as described in SIGN algorithm using El Gamal encryption scheme.
  • $\mathcal{B}$ chooses $\tilde{x}, \tilde{z}, c_i$ uniformly at random from $\mathbb{Z}_p$, for all $i \in [1, n]$ and sets the value $\sum_{i=1}^n c_i \mod p$ as

$$H' \left(Y \parallel event \parallel t \parallel M \parallel g^{\tilde{x}} \prod_{i=1}^n Z_i^{c_i} \parallel e^{\tilde{x}} t^{\sum_{i=1}^n c_i} \parallel pk_R^{\tilde{z}} \prod_{i=1}^n \left(\frac{C_2}{Z_i} \right)^{c_i} \right).$$

If a collision occurs, i.e., the value $\sum_{i=1}^n c_i \mod p$ has already been assigned to some H' query, repeat this step.

Finally, $\mathcal{B}$ returns the signature $\sigma = (t, \tilde{x}, \tilde{z}, c_1, \ldots, c_n, C)$. It is easy to see that the simulated signature components are identically distributed as in our RLRS construction.

Output: For one successful simulation, suppose the forgery of $\mathcal{A}$ is $\sigma_1 = \{t_1, \tilde{x}_1, \tilde{z}_1, c_1^1, \ldots, c_{n''}^1, C\}$ on an event $event$ and a set of public keys Y'' such that it is a subset of those public keys corresponding to indexes in I_n. Without loss of generality, let $n'' = n$. By the assumption of the random oracle model, $\mathcal{A}$ has a query $H(event)$, denoted by e, and a query $H'(Y'' \parallel event \parallel t \parallel M \parallel K \parallel K' \parallel K'')$, where

$$K = g^{\tilde{x}_1} \prod_{i=1}^n Z_i^{c_i^1}, \quad K' = e^{\tilde{x}_1} t^{\sum_{i=1}^n c_i^1}, \quad K'' = pk_R^{\tilde{z}_1} \prod_{i=1}^n \left(\frac{C_2}{Z_i} \right)^{c_i^1}.$$

Let $K = g^\delta$ for some $\delta \in \mathbb{Z}_p$. Suppose this occurs at the ℓ-th query of H', and $\mathcal{B}$ returns c_0^1. Since $c_0^1 = c_1^1 + \ldots + c_n^1 \mod p$ and by the assumption of the random oracle model, at least one c_i^1 (where $1 \le i \le n$) is determined after c_0^1 is returned by $\mathcal{B}$. In the best case, if there is only one c_i^1 not determined at the ℓ-th query, one rewind allows $\mathcal{B}$ to discover the secret x_i. By assumption of random oracle $\mathcal{A}$ queries $H'(Y' \parallel event \parallel t \parallel M \parallel \bar{K} \parallel \bar{K}' \parallel \bar{K}'')$ where

$$K = g^\delta, \quad \bar{K} = g^{\bar{\delta}}$$

where

$$\delta = \tilde{x} + \sum_{i=1}^n x_i c_i = \tilde{x}' + \sum_{i=1}^n x_i c_i' = \bar{\delta}.$$

By the assumption of the random oracle model, at least one c_i' (where $1 \le i \le n$) is determined after c_0' is returned by $\mathcal{B}$ i.e. there is one such index $j \in [1, n]$ such that $c_i \ne c_j$ for some $i = j$. From the above equation $\mathcal{A}$ could compute the secret x_j as

$$\tilde{x} + x_j c_j = \tilde{x}' + x_j c_j' \mod p \implies x_j = \frac{\tilde{x}' - \tilde{x}}{c_j - c_j'} \mod p.$$

In the worst case, where all n values c_i^1 (for $1 \leq i \leq n$) are not determined at the ℓ-th H'-query, $\mathcal{B}$ rewinds with the same input tape for $\mathcal{A}$ and answers all queries consistently until the ℓ-th H'-query. For the i-th rewind at the ℓ-th H'-query, c_0^i is randomly picked in $\mathbb{Z}_p$ and returned. $\mathcal{B}$ repeats this process until n simulations of $\mathcal{A}$ have completed.

Suppose all n simulations are successful. $\mathcal{A}$ produces a total of n forgeries $\{t_i, \tilde{x}_i, \tilde{z}_i, c_1^i, \ldots, c_n^i, C_i\}$ for all $i \in [1, n]$. Since $c_0^i \neq c_0^j$ for any $i \neq j$, $i, j \in \{1, \ldots, n\}$, there are n distinct linear equations:

$$\delta = \tilde{x}_i + \sum_{j=1}^{n} c_j^i x_j \quad \text{for } i = 1, \ldots, n.$$

Thus, up to n x_i values can be obtained. According to the Forking Lemma [19], the probability of a successful rewind simulation is at least $\frac{\epsilon}{4}$, where ϵ is the probability that adversary $\mathcal{A}$ successfully forges a signature. Therefore, the probability that $\mathcal{B}$ successfully breaks the Discrete Logarithm (DL) Problem is at least $\frac{\epsilon}{4}$. $\square$

Theorem 3 (Anonymity). *Our RLRS scheme is anonymous in the random oracle model if the DDH assumption holds in $\mathcal{G}$.*

Proof. We will prove this by contradiction. Assume that the DDH assumption is hard in the cyclic group generator $\mathcal{G}$. Suppose, for the sake of contradiction, that our RLRS Scheme is not anonymous; that is, there exists a PPT adversary $\mathcal{A}$ capable of breaking the anonymity of our RLRS Scheme with non-negligible probability in the random oracle model. We can then construct a simulator $\mathcal{B}$ that can break the DDH assumption in $\mathcal{G}$.

Consider the DDH instance $g, \alpha = g^a, \beta = g^b, \gamma = g^c$, $\mathcal{B}$'s goal is decide whether $c = ab$ or not. First, $\mathcal{B}$ chooses π uniformly at random from $[1, n]$, where n is the number of signers in the ring.

Now $\mathcal{A}$ has access to JO, which s/he can query at most n' times where $n' \geq n$, returns a DL instance $pk = g^{sk}$ where sk is chosen uniformly at random from $\mathbb{Z}_p$.

The challenge signature σ_π is created using a randomly selected public key from Y. For computing the signature $\mathcal{B}$ assigns $e = \beta$ and $t = \gamma$ sets $Z_\pi = \alpha$. $\mathcal{B}$ randomly picks $u \in \mathbb{Z}_p$, and computes the ciphertext $C = \{C_1, C_2\}$ using El Gamal encryption as

$$C_1 \leftarrow g^u \qquad C_2 \leftarrow Z_\pi (pk_R)^u.$$

Then $\mathcal{B}$ randomly chooses $\tilde{x}, \tilde{z}, c_1, \ldots, c_n$ from $\mathbb{Z}_p$ and computes

$$K = g^{\tilde{x}} \prod_{i=1}^{n} Z_i^{c_i}, \quad K' = e^{\tilde{x}} t^{\sum_{i=1}^{n} c_i} \quad K'' = pk_R^{\tilde{z}} \prod_{i=1}^{n} \left(\frac{C_2}{Z_i} \right)^{c_i}$$

and sets hash oracle outcome

$$\sum_{i=1}^{n} c_i = H'\left(Y \parallel event \parallel t \parallel M \parallel K \parallel K' \parallel K''\right),$$

$$\sigma_\pi = \{t, \tilde{x}, \tilde{z}, c_1, \ldots, c_n, C = \{C_1, C_2\}\}.$$

and sends to $\mathcal{A}$. Suppose $\mathcal{A}$ guesses the signer's index as $j \in [1, n]$ and returns j to $\mathcal{B}$. By convention, $\mathcal{A}$ returns 0 if it cannot identify the signer. $\mathcal{B}$ returns 1 if $j = \pi$; returns 0 if $j = 0$; and returns 1 or 0 with equal probability otherwise. Then

$$\begin{aligned}
\Pr[\mathcal{B}(\alpha,\beta,\gamma) &= b \mid b = 1] \\
&= \Pr[\mathcal{B}(\alpha,\beta,\gamma) = b \mid b = 1, \mathcal{A}(\sigma_\pi) = \pi]\Pr[\mathcal{A}(\sigma_\pi) = \pi] \\
&\quad + \Pr[\mathcal{B}(\alpha,\beta,\gamma) = b \mid b = 1, \mathcal{A}(\sigma_\pi) \neq \pi, j \neq 0]\Pr[\mathcal{A}(\sigma_\pi) \neq \pi, j \neq 0] \\
&\geq 1\left(\frac{1}{n} + \frac{1}{Q(\lambda)}\right) + \frac{1}{2}\left(1 - \frac{1}{n} - \frac{1}{Q(\lambda)}\right) \\
&\geq \frac{1}{2} + \frac{1}{2n} + \frac{1}{2Q(\lambda)}.
\end{aligned}$$

If $b = 0$ then all the signers have equal probability to sign the signature from $\mathcal{A}$ point of view. Thus $\mathcal{A}$ can do no better than random guessing.

$$\begin{aligned}
\Pr[\mathcal{B}(\alpha,\beta,\gamma) &= b \mid b = 0] \\
&= \Pr[\mathcal{B}(\alpha,\beta,\gamma) = b \mid b = 0, \mathcal{A}(\sigma_\pi) = \pi]\Pr[\mathcal{A}(\sigma_\pi) = \pi] \\
&\quad + \Pr[\mathcal{B}(\alpha,\beta,\gamma) = b \mid b = 0, \mathcal{A}(\sigma_\pi) \neq \pi]\Pr[\mathcal{A}(\sigma_\pi) \neq \pi] \\
&\geq 0\left(\frac{1}{n}\right) + \frac{1}{2}\left(1 - \frac{1}{n}\right).
\end{aligned}$$

Combining both the probabilities we have

$$\begin{aligned}
\Pr[\mathcal{B}(\alpha,\beta,\gamma) = b] &= \Pr[\mathcal{B}(\alpha,\beta,\gamma) = b \mid b = 1]\Pr[b = 1] \\
&\quad + \Pr[\mathcal{B}(\alpha,\beta,\gamma) = b \mid b = 0]\Pr[b = 0] \\
&\geq \frac{1}{4}\left(1 + \frac{1}{n} + \frac{1}{Q(\lambda)}\right) + \frac{1}{4}\left(1 - \frac{1}{n}\right) \\
&\geq \frac{1}{2} + \frac{1}{4Q(\lambda)}.
\end{aligned}$$

Therefore, $\mathcal{A}$ solves the DDH problem with probability nonnegligibly more than $\frac{1}{2}$. □

Theorem 4 (Linkability). *Our RLRS scheme is linkable in the random oracle model if the DL assumption holds in $\mathcal{G}$.*

Proof. The proof is similar to the (Unforgeability) Theorem 2, except that $\mathcal{A}$ is allowed to have at most one secret key, sk_π which s/he gets from CO. From the contradiction, suppose $\mathcal{A}$ produces two valid signatures

$$\sigma = \{t, \tilde{x}, \tilde{z}, c_1, \ldots, c_n, C\} \text{ and } \bar{\sigma} = \{\bar{t}, \bar{\tilde{x}}, \bar{\tilde{z}}, \bar{c}_1, \ldots, \bar{c}_n, \bar{C}\},$$

where $t = e^{x_\pi}$ and $\bar{t} = e^{x'_\pi}$ denotes the linkable tags of the above signatures respectively. Now if the *event* is fixed for both the runs, if we rewind to get back a different value for H' we can get back another signature

$$\sigma' = \{t', \tilde{x}', \tilde{z}', c'_1, \ldots, c'_n, C'\}.$$

As in Theorem 2, from the signature pairs σ and σ', $\mathcal{B}$ derives the secret key x_π as

$$x_\pi = \frac{\tilde{x}' - \tilde{x}}{c_j - c'_j} \mod p,$$

where $t = e^{x_\pi} = t'$, as *event* is fixed and $pk_\pi = g^{sk_\pi}$.

Similarly, rewind simulation for the second signature $\bar{\sigma}$, $\mathcal{B}$ obtains with some non-negligible probability of $\bar{\sigma}'$. Now $\mathcal{B}$ derives the secret key x'_π as

$$x'_\pi = \frac{\bar{\tilde{x}}' - \bar{\tilde{x}}}{\bar{c}_j - \bar{c}'_j} \mod p.$$

In the linkability, $\mathcal{A}$ can have at most one secret key and hence $x'_\pi = x_\pi$. This implies that $t = \bar{t}$ and hence the signatures σ and $\bar{\sigma}$ are 'Linked'. $\mathcal{A}$ can break the DL problem if the rewind simulation is successful. Hence $\mathcal{B}$ can break the DL assumption if the rewind is successful.

Theorem 5 (Non-Slanderability). *Our RLRS scheme is non-slanderable in the random oracle model if the DL assumption holds in $\mathcal{G}$.*

Proof $\mathcal{A}$ gives simulator $\mathcal{B}$ a public key Z_π an event identifier *event*, list of public keys of the ring as Y, revocable user public key pk_R and a message M. $\mathcal{B}$ generates a signature $\sigma_\pi = \{t, \cdot\}$ where t is the linking tag computed using $sk_\pi = x_\pi$ and sends it to $\mathcal{A}$. $\mathcal{A}$ has access to all the oracles and can query them any time except that $\mathcal{A}$ is not allowed to query Z_π the signers public key to CO. $\mathcal{A}$ produces a forgery signature $\sigma^* = \{t^*, \cdot\}$ such that both the signatures are linked $t = t^* = e^{x_\pi} = e^{x^*_\pi}$ where $e = H(event)$ as both the signatures are linked $x_\pi = x^*_\pi$ which implies that $\mathcal{A}$ knows the corresponding x_π value corresponding to the public key Z_π. This contradicts that $\mathcal{A}$ is not allowed to query CO for Z_π. □

Theorem 6 (Revocability). *Our RLRS scheme is revocable in the random oracle model if the construction of our RLRS scheme is unforgeable.*

Proof. We use the same setting as in Unforgeablility Theorem 2. The only difference is that the adversary $\mathcal{A}$ can get one private key $sk_\pi = x_\pi$ corresponding to some $pk_\pi = Z_\pi$ in Y from the corruption oracle CO. Since all the public keys are discrete logarithm instances generated independently, $\mathcal{A}$ will not be able to find the corresponding secret key under our assumption. For contradiction if suppose $\mathcal{A}$ successfully generates a valid signature $\sigma = \{t, \tilde{x}, \tilde{z}, c_1, \ldots, c_n, C = \{C_1, C_2\}\}$ where $C_1 = g^u$, $C_2 = Z_i pk_R^u$ for some randomly chosen $u \in \mathbb{Z}_p$, and pk_R is the revoking authority's public key. Since our scheme is unforgeable, a valid signature is strictly generated by $sk_\pi = x_\pi$. There are two cases to break the revokability of our scheme:

1. **Case 1**: $\mathcal{A}$ randomly selects $r_x, r_z, c_1, ..., c_{\pi-1}, c_{\pi+1}, ..., c_n \in \mathbb{Z}_p$ and $\mathcal{A}$ computes:

$$K = g^{r_x} \prod_{\substack{i=1 \\ i \neq \pi}}^{n} Z_i^{c_i}, \quad K' = e^{r_x} t^{\sum_{\substack{i=1 \\ i \neq \pi}}^{n} c_i}, \quad K'' = pk_R^{r_z} \prod_{\substack{i=1 \\ i \neq \pi}}^{n} \left(\frac{C_2}{Z_i}\right)^{c_i}.$$

Now to close the ring $\mathcal{A}$ needs to know the secret key $sk_j \neq sk_\pi$ which contradicts our assumption that $\mathcal{A}$ only has one secret key.

2. **Case 2**: $\mathcal{A}$ randomly selects $r_x, r_z, c_1, ..., c_{\pi-1}, c_{\pi+1}, ..., c_n \in \mathbb{Z}_p$ and $\mathcal{A}$ computes:

$$K = g^{r_x} \prod_{\substack{i=1 \\ i \neq \pi}}^{n} Z_i^{c_i}, \quad K' = e^{r_x} t^{\sum_{\substack{i=1 \\ i \neq \pi}}^{n} c_i}, \quad K'' = pk_R^{r_z} \prod_{\substack{i=1 \\ i \neq \pi}}^{n} \left(\frac{C_2}{Z_i}\right)^{c_i}.$$

Now in order to close the ring $\mathcal{A}$ uses the secret key sk_π and computes $\tilde{x} = r_x - c_\pi x$, $\tilde{z} = r_z - c_\pi u \mod p$. However, this construction will fail to pass the verification as any honest verifier will follow the protocol and compute

$$c_0 = H'\left(Y \parallel event \parallel t \parallel M \parallel g^{\tilde{x}} \prod_{i=1}^{n} Z_i^{c_i} \parallel e^{\tilde{x}} t^{\sum_{i=1}^{n} c_i} \parallel pk_R^{\tilde{z}} \prod_{i=1}^{n} \left(\frac{C_2}{Z_i}\right)^{c_i}\right)$$

and if $j \neq \pi$, then the value of c_0 will differ from the sum of all c_i values for $i \in [1, n]$. This contradicts the requirement for the generated signature to be valid.

$\square$

5   Comparison

In this section, we compare our RLRS scheme with Zhang et al. [24] in Table 1. We start by providing some computational notations as follows:

- T_E : The time for one modular exponentiation computation
- T_M : The time for one modular multiplication computation
- T_H : The time for one hash function computation
- T_A : The time for one modular addition computation
- n : The number of public keys in the ring
- λ : The length of the elements in $\mathbb{Z}_p$

Our RLRS scheme is based on the Schnorr construction, allowing us to compare its performance with existing Schnorr-based schemes. To our knowledge, Zhang et al. [24] are the only authors to present a scheme that combines both revocability and linkability in the discrete logarithm setting. In contrast, Liu et al. [13] introduced a revocable ring signature scheme in a pairing-based context

Table 1. Comparison of revocable linkable ring signatures

Scheme	Signing time	Verify time	Revoke time	|Sign|
RLRS [24]	$(8n-1)T_E + (5n+1)T_M + (2n+2)T_H + 2T_A$	$8nT_E + 5nT_M + 2nT_H$	$T_M + T_E$	$(2n+5)\lambda$
Our RLRS	$(2n+5)T_E + 2nT_M + 2T_H + (n+2)T_A$	$(2n+5)T_E + (2n+2)T_M + 2T_H + nT_A$	$T_M + T_E$	$(n+4)\lambda$

that supports mandatory revocability but does not provide linkability. Additionally, Liu et al. [15] developed a scheme that ensures linkability but lacks revocability.

As detailed in Sect. 4, our RLRS construction for ring signature generation utilizes aggregated commitment terms, while Zhang et al. [24] compute aggregated commitments for each signer within the ring. This allows us to reduce the number of hash computations from $\mathcal{O}(n)$ to $\mathcal{O}(1)$ in both the SIGN and VERIFY algorithms. In addition to optimizing hash computations, we also achieved improvements in exponentiation and multiplication operations. Compared to Zhang et al.'s RLRS scheme, our approach reduces the time complexity of both the SIGN and VERIFY algorithms by approximately a factor of four.

6   Conclusion

We present a ring signature scheme that incorporates mandatory revocability and linkability, offering significantly improved efficiency compared to the previous RLRS scheme [24]. Additionally, for future work, exploring generic constructions for RLRS schemes inspired by DualRing [23] and achieving logarithmic signature sizes are intriguing open challenges.

References

1. Abe, M., Ohkubo, M., Suzuki, K.: 1-out-of-n signatures from a variety of keys. In: Zheng, Y. (ed.) ASIACRYPT 2002. LNCS, vol. 2501, pp. 415–432. Springer, Heidelberg (2002). https://doi.org/10.1007/3-540-36178-2_26
2. Au, M.H., Chow, S.S.M., Susilo, W., Tsang, P.P.: Short linkable ring signatures revisited. In: Atzeni, A.S., Lioy, A. (eds.) EuroPKI 2006. LNCS, vol. 4043, pp. 101–115. Springer, Heidelberg (2006). https://doi.org/10.1007/11774716_9
3. Au, M.H., Liu, J.K., Susilo, W., Yuen, T.H.: Constant-size ID-based linkable and revocable-iff-linked ring signature. In: Barua, R., Lange, T. (eds.) INDOCRYPT 2006. LNCS, vol. 4329, pp. 364–378. Springer, Heidelberg (2006). https://doi.org/10.1007/11941378_26
4. Au, M.H., Liu, J.K., Susilo, W., Yuen, T.H.: Certificate based (linkable) ring signature. In: Dawson, E., Wong, D.S. (eds.) ISPEC 2007. LNCS, vol. 4464, pp. 79–92. Springer, Heidelberg (2007). https://doi.org/10.1007/978-3-540-72163-5_8
5. Au, M.H., Liu, J.K., Susilo, W., Yuen, T.-H.: Secure id-based linkable and revocable-iff-linked ring signature with constant-size construction. Theor. Comput. Sci. **469**, 1–14 (2013)

6. Chen, Y., He, D., Bao, Z., Luo, M., Choo, K.-K.R.: A post-quantum privacy-preserving payment protocol in vehicle to grid networks. IEEE Trans. Intell. Veh. (2024)

7. Dodis, Y., Kiayias, A., Nicolosi, A., Shoup, V.: Anonymous identification in *Ad Hoc* groups. In: Cachin, C., Camenisch, J.L. (eds.) EUROCRYPT 2004. LNCS, vol. 3027, pp. 609–626. Springer, Heidelberg (2004). https://doi.org/10.1007/978-3-540-24676-3_36

8. ElGamal, T.: A public-key cryptosystem and a signature scheme based on discrete logarithms. IEEE Trans. Inf. Theory **31**(4), 469–472 (1985). (Conference version appeared in CRYPTO 1984, pp. 10–18)

9. Fujisaki, E.: Sub-linear size traceable ring signatures without random oracles. In: Kiayias, A. (ed.) CT-RSA 2011. LNCS, vol. 6558, pp. 393–415. Springer, Heidelberg (2011). https://doi.org/10.1007/978-3-642-19074-2_25

10. Fujisaki, E., Suzuki, K.: Traceable ring signature. In: Okamoto, T., Wang, X. (eds.) PKC 2007. LNCS, vol. 4450, pp. 181–200. Springer, Heidelberg (2007). https://doi.org/10.1007/978-3-540-71677-8_13

11. Jeong, I.R., Kwon, J.O., Lee, D.H.: Ring signature with weak linkability and its applications. IEEE Trans. Knowl. Data Eng. **20**(8), 1145–1148 (2008)

12. Lee, K.C., Wen, H.A., Hwang, T.: Convertible ring signature. IEE Proc. Commun. **152**(4), 411–414 (2005)

13. Liu, D.Y., Liu, J.K., Mu, Y., Susilo, W., Wong, D.S.: Revocable ring signature. J. Comput. Sci. Technol. **22**(6), 785–794 (2007)

14. Liu, J.K., Au, M.H., Susilo, W., Zhou, J.: Linkable ring signature with unconditional anonymity. IEEE Trans. Knowl. Data Eng. **26**(1), 157–165 (2013)

15. Liu, J.K., Wei, V.K., Wong, D.S.: Linkable spontaneous anonymous group signature for ad hoc groups. In: Wang, H., Pieprzyk, J., Varadharajan, V. (eds.) ACISP 2004. LNCS, vol. 3108, pp. 325–335. Springer, Heidelberg (2004). https://doi.org/10.1007/978-3-540-27800-9_28

16. Liu, J.K., Wong, D.S.: Linkable ring signatures: security models and new schemes. In: Gervasi, O., et al. (eds.) ICCSA 2005. LNCS, vol. 3481, pp. 614–623. Springer, Heidelberg (2005). https://doi.org/10.1007/11424826_65

17. Lv, J., Wang, X.: Verifiable ring signature. In: Proceedings of DMS 2003 - The 9th International Conference on Distributed Multimedia Systems, pp. 663–667 (2003)

18. Noether, S., Mackenzie, A.: Ring confidential transactions. Ledger **1**, 1–18 (2016)

19. Pointcheval, D., Stern, J.: Security proofs for signature schemes. In: Maurer, U. (ed.) EUROCRYPT 1996. LNCS, vol. 1070, pp. 387–398. Springer, Heidelberg (1996). https://doi.org/10.1007/3-540-68339-9_33

20. Rivest, R.L., Shamir, A., Tauman, Y.: How to leak a secret. In: Boyd, C. (ed.) ASIACRYPT 2001. LNCS, vol. 2248, pp. 552–565. Springer, Heidelberg (2001). https://doi.org/10.1007/3-540-45682-1_32

21. Tsang, P.P., Wei, V.K.: Short linkable ring signatures for E-voting, E-cash and attestation. In: Deng, R.H., Bao, F., Pang, H.H., Zhou, J. (eds.) ISPEC 2005. LNCS, vol. 3439, pp. 48–60. Springer, Heidelberg (2005). https://doi.org/10.1007/978-3-540-31979-5_5

22. Tsang, P.P., Wei, V.K., Chan, T.K., Au, M.H., Liu, J.K., Wong, D.S.: Separable linkable threshold ring signatures. In: Canteaut, A., Viswanathan, K. (eds.) INDOCRYPT 2004. LNCS, vol. 3348, pp. 384–398. Springer, Heidelberg (2004). https://doi.org/10.1007/978-3-540-30556-9_30
23. Yuen, T.H., Esgin, M.F., Liu, J.K., Au, M.H., Ding, Z.: *DualRing*: generic construction of ring signatures with efficient instantiations. In: Malkin, T., Peikert, C. (eds.) CRYPTO 2021. LNCS, vol. 12825, pp. 251–281. Springer, Cham (2021). https://doi.org/10.1007/978-3-030-84242-0_10
24. Zhang, X., Liu, J.K., Steinfeld, R., Kuchta, V., Yu, J.: Revocable and linkable ring signature. In: Liu, Z., Yung, M. (eds.) Inscrypt 2019. LNCS, vol. 12020, pp. 3–27. Springer, Cham (2020). https://doi.org/10.1007/978-3-030-42921-8_1

Quantum Cryptography

Quantum Cryptanalysis of ZUC and Related Resource Estimation

Suman Dutta[1,2]([✉]), Anirban Ghatak[2], Anupam Chattopadhyay[1], and Subhamoy Maitra[2]

[1] CCDS, Nanyang Technological University, Singapore, Singapore
sumand.iiserb@gmail.com, anupam@ntu.edu.sg
[2] Applied Statistics Unit, Indian Statistical Institute, Kolkata, India
subho@isical.ac.in

Abstract. The ZUC stream cipher is integral to modern mobile communication standards like 4G and 5G, playing a key role in securing data transmission across global networks. Although multiple attempts have been made to perform classical cryptanalysis of ZUC, it is surprising that no concrete efforts have been made towards its quantum cryptanalysis and related resource estimation. This paper presents a comprehensive quantum resource estimation of ZUC in the context of Grover-based quantum key-recovery attacks. We introduce novel circuit optimization techniques, such as modular quantum doubling and successive quantum modular addition, and implement reversible quantum logic circuit synthesis for both linear and non-linear functions to reduce the quantum resources required for implementing ZUC. We obtain that a full-scale Grover's search on ZUC-128 requires approximately $1.5 \cdot 2^{87}$ Clifford + T gates, with a T-depth of $1.42 \cdot 2^{83}$, and an overall circuit depth of $1.42 \cdot 2^{84}$, whereas for ZUC-256, these numbers are approximately $1.67 \cdot 2^{151}$, $1.57 \cdot 2^{147}$, and $1.57 \cdot 2^{148}$, respectively. Additionally, further evaluation of ZUC against NIST's MAXDEPTH criterion confirms its security, albeit with a narrow margin, suggesting potential areas for future optimization and further analysis of ZUC's resilience against quantum attacks.

Keywords: Grover's search algorithm · Quantum circuit synthesis · Quantum resource estimation · MAXDEPTH criterion · ZUC cipher

1 Introduction

Quantum computing has emerged as an exciting applied discipline of quantum physics, offering exponentially faster computational capabilities compared to classical systems. Since the early nineties, several landmark quantum algorithms have been developed, including the Deutsch-Jozsa algorithm [9], Simon's algorithm [23], Shor's factoring algorithm [22], and Grover's search algorithm [12]. While Shor's algorithm poses a direct threat to public key cryptosystems,

© The Author(s), under exclusive license to Springer Nature Switzerland AG 2025
S. Mukhopadhyay and P. Stănică (Eds.): INDOCRYPT 2024, LNCS 15495, pp. 329–355, 2025.
https://doi.org/10.1007/978-3-031-80308-6_15

Grover's search, the Simon and the Deutsch-Jozsa algorithms present a significant challenge to symmetric key cryptography. Initially, it was believed that doubling the key size would protect symmetric ciphers from quantum attacks. However, recent studies have shown that this approach is inadequate.

Despite the absence of fully error-corrected, large-scale quantum computers, advancements in quantum technology have intensified the urgency of evaluating the security of cryptographic schemes against potential quantum attacks. Governments and corporations have made significant investments in quantum technology, including the development of qubits, hardware, and software solutions. Consequently, in the last couple of decades, quantum resource estimation has become a critical area of study. The seminal work on Grover's attack against AES [11,30] marked the beginning of research into the security of symmetric ciphers against quantum adversaries, with several cryptosystems being assessed using NIST's standardization criterion for quantum attack resilience [4,5,13]. The estimation of quantum resources - such as the number of required qubits and quantum gates - provides essential insight into the security of various ciphers and offers a realistic timeline for preparing against quantum threats.

The ZUC stream cipher, developed by the Data Assurance and Communication Security Research Center of the Chinese Academy of Sciences, plays a crucial role in securing modern mobile communication standards, such as 4G LTE and 5G networks. Since the release of its updated version, ZUC 1.6, in June 2011, the cipher has undergone multiple rounds of classical cryptanalysis [14,28,29], yet, surprisingly, no concrete effort has been made to evaluate its resilience against quantum attacks.

In this paper, we present the first comprehensive quantum cryptanalysis of the ZUC stream cipher using Grover's search algorithm. Our work includes detailed quantum resource estimation for mounting a Grover-based key-recovery attack on ZUC. We propose novel circuit optimization techniques, such as modular quantum doubling and successive quantum modular addition, aimed at reducing the quantum resource overhead. Additionally, we implement quantum logic synthesis for both linear and non-linear components of ZUC, contributing to a more efficient quantum implementation of the cipher. To the best of our knowledge, our optimized implementations represent the very first instance of a complete quantum circuit implementation of both ZUC-128 and ZUC-256. Our findings thus provide a foundational benchmark for future research into the quantum security of ZUC and similar cryptographic schemes, such as SNOW [10].

1.1   Organization and Contribution

Here we describe the structure of the paper and highlight the contributions. Note that Sect. 3 and Sect. 4 are the main contributory sections.

– Section 2 presents the preliminaries, including a brief description of ZUC-128 and ZUC-256, along with an outline of Grover's search algorithm and NIST's MAXDEPTH criterion in the context of stream cipher cryptanalysis.

- In Sect. 3, we define novel circuit optimization techniques for implementing quantum doubling and quantum addition modulo $2^n - 1$, with the least possible resource requirements compared to the state-of-the-art results. Further, we present the optimal implementation of linear functions (L_1, L_2) following [19, Algorithm 1]. We conclude our contributions in this section with a technique for quantum implementation of an *arbitrary* S-box following [15] and present the related resource estimations.

- In Sect. 4, we provide a step-by-step quantum resource estimation for implementing ZUC. First, we calculate the resource requirements for one round of ZUC, as shown in Table 9. Next, we estimate the resources required for complete encryption, presented in Table 10. Since Grover's oracle requires both encryption and decryption in each cycle, we double the encryption resources and add the resources needed for keystream comparison (cf. Table 12). Finally, by multiplying the resources from Table 12 by $\pi/4\sqrt{2^k}$, where $k = 128, 256$, we estimate the total quantum resources required to perform a full-scale Grover's search on ZUC. The results are summarized in Table 13.

  We find that a full-scale Grover's search on ZUC-128 requires approximately $1.5 \cdot 2^{87}$ Clifford + T gates, with a T-depth of $1.42 \cdot 2^{83}$ and a circuit depth of $1.42 \cdot 2^{84}$. Similarly, an exhaustive Grover's search on ZUC-256 requires around $1.67 \cdot 2^{151}$ Clifford + T gates, with a T-depth of $1.57 \cdot 2^{147}$ and a circuit depth of $1.57 \cdot 2^{148}$. Alternatively, aiming for a reduction in the T-depth, an exhaustive Grover's search requires $1.24 \cdot 2^{88}$ and $1.38 \cdot 2^{152}$ Clifford + T gates for ZUC-128 and ZUC-256, respectively, reducing the T-depth to $1.42 \cdot 2^{81}$ and $1.57 \cdot 2^{145}$. Finally, multiplying the total gate count by the circuit depth, we obtain $1.07 \cdot 2^{172}$ for ZUC-128 and $1.31 \cdot 2^{300}$ for ZUC-256, which slightly exceeds the MAXDEPTH criterion of NIST, confirming ZUC's resilience against exhaustive Grover's search. The results are summarized in Table 14. We note that, to our knowledge, this is the first reported instance of a rigorous and comprehensive analysis of the quantum security of ZUC using the aforementioned NIST guidelines.

- Section 5 concludes the paper with a summary of the results of this paper and briefly outlines the application and scope of our results in related research problems.

Additionally, we also design and implement a toy-version of ZUC in qiskit, with appropriate modifications of the parameters, and demonstrate the keystream generation. A schematic diagram of the toy-ZUC (Fig. 7) and the observed histogram (Fig. 8) have been provided in Appendix A. For the corresponding qiskit circuit, one can refer to our GitHub repository [32].

2   Preliminaries

In this section, we present a brief description of two variants of the ZUC cipher along with a few necessary definitions and results, which will be used in the development of the main results of the paper.

2.1 Design of ZUC-128

There are two phases in the operation of the ZUC cipher: the initialization phase and the keystream generation phase. The schematic diagrams of the initialization phase and the key-stream generation phase of ZUC are shown in Fig. 1 and Fig. 2, respectively.

The 128-bit key (seed) and the 128-bit initialization vector (IV) are first loaded and processed for 32 rounds during the initialization phase; the key-stream generation commences afterward. We next describe the structure of ZUC and briefly discuss its operation in a step-by-step manner, adhering to the material in [2,16].

There are three interconnected operational layers in ZUC:

1. The Linear Feedback Shift Register (LFSR),
2. The Bit-Reorganization Layer (BR),
3. The Non-Linear Function Layer (NLF).

Brief operational descriptions of the above layers are provided below.

Linear Feedback Shift Register (LFSR). The stream cipher ZUC is based on a single LFSR in the *Fibonacci configuration*, i.e., there is no feedback input at any state other than the state with the maximum index. It contains sixteen states, each a 31-bit register, with the taps defined by the following primitive polynomial, $f(x) = x^{16} - (2^{15}x^{15} + 2^{17}x^{13} + 2^{21}x^{10} + 2^{20}x^4 + (1 + 2^8))$ over the field $\mathbb{F}_p$, where $p = 2^{31} - 1$. Thus, the update function of the LFSR is given by

$$\mathsf{feedback}_f(s_0, \ldots, s_{15}) = 2^{15}s_{15} + 2^{17}s_{13} + 2^{21}s_{10} + 2^{20}s_4 + (1 + 2^8)s_0.$$

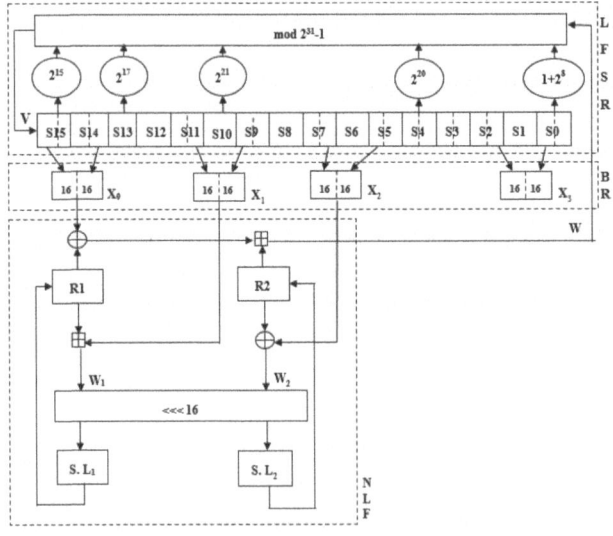

Fig. 1. ZUC circuit diagram for initialization phase

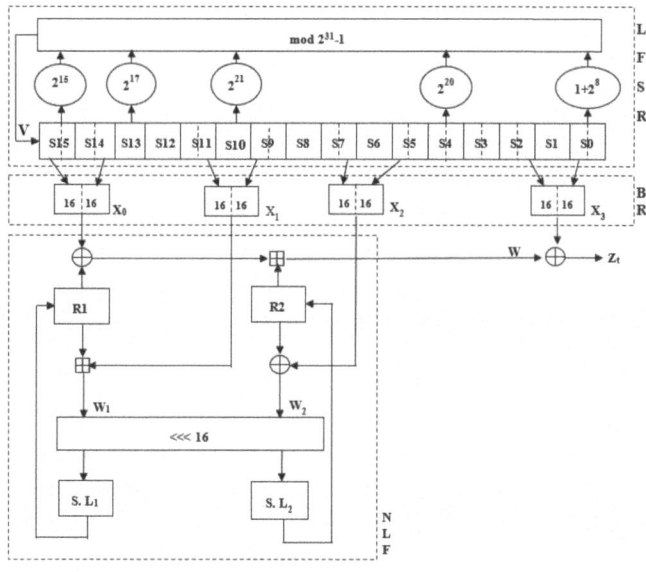

Fig. 2. ZUC circuit diagram for key-stream generation phase

Here, the choice of the primitive polynomial and the size of the underlying field ensures the maximal period of $p^{16} - 1 \approx 2^{496}$.

Bit-Reorganisation (BR). In the bit-reorganization layer, the bits from the LFSR $s_0, s_2, s_5, s_7, s_9, s_{11}, s_{14}, s_{15}$ are chosen and re-organized following Table 1. Here s_{iH} implies the 16 high bits of s_i, and s_{iL} refers to the 16 lower bits of s_i.

Table 1. Bit-Reorganisation (BR) algorithm

Input: $(s_0, \ldots, s_{15})$
Output: (X_0, X_1, X_2, X_3)
1. Define 32-bit word $X_0 = s_{15H} \| s_{14L}$;
2. Define 32-bit word $X_1 = s_{11L} \| s_{9H}$;
3. Define 32-bit word $X_2 = s_{7L} \| s_{5H}$;
4. Define 32-bit word $X_3 = s_{2L} \| s_{0H}$;
5. Return 128-bit string (X_0, X_1, X_2, X_3);

Non-linear Function (NLF). In every round, the non-linear function layer takes (X_0, X_1, X_2) as input, and depending on the internal states of the 32-bit registers R_1 and R_2, it generates the output W while updating the registers R_1 and R_2 as described in Table 2.

The S-box (substitution box) S used inside the non-linear function is defined using the two-component S-boxes S_0 and S_1, as provided in [16, Table 3.1,

Table 2. Non-linear function of ZUC

Input: (X_0, X_1, X_2) and R_1, R_2
Output: W
1. Define 32-bit word $W = (X_0 \oplus R_1) \boxplus R_2$;
2. Define 32-bit word $W_1 = R_1 \boxplus X_1$;
3. Define 32-bit word $W_2 = R_2 \oplus X_2$;
4. Update 32-bit word $R_1 = S(L_1(W_{1L} \| W_{2H}))$;
5. Update 32-bit word $R_2 = S(L_2(W_{2L} \| W_{1H}))$;
6. Return 32-bit word W;

Table 3.2]. Note that, $S_0, S_1 : \{0,1\}^8 \rightarrow \{0,1\}^8$, while $S : \{0,1\}^{32} \rightarrow \{0,1\}^{32}$. Thus, for any tuple (x_1, x_2, x_3, x_4), where $x_i \in \{0,1\}^8$, S is applied as: $S(x_1 \| x_2 \| x_3 \| x_4) := S_0(x_1) \| S_1(x_2) \| S_0(x_3) \| S_1(x_4)$. The linear maps L_1 and L_2 used in the non-linear function are given by:

$$L_1(X) = X \oplus (X \lll 2) \oplus (X \lll 10) \oplus (X \lll 18) \oplus (X \lll 24),$$
$$L_2(X) = X \oplus (X \lll 8) \oplus (X \lll 14) \oplus (X \lll 22) \oplus (X \lll 30).$$

Initialization and Keystream Generation. First, the key and the initialization vector (IV) are loaded into the LFSR and the registers R_1, R_2 are initialized with 0^{32} - this phase is termed as Round 0. If the key is denoted as $k_{15} \| \ldots \| k_0$ and the IV is given by $iv_{15} \| \ldots \| iv_0$, then the i-th state $(0 \leq i \leq 15)$ of the LFSR is loaded as: $s_i = k_i \| d_i \| iv_i$, where $d_i's$ are 15-bit constants, defined as

$$d_0 = \text{0x44d7}, \quad d_1 = \text{0x26bc}, \quad d_2 = \text{0x626b}, \quad d_3 = \text{0x135e}$$
$$d_4 = \text{0x5789}, \quad d_5 = \text{0x35e2}, \quad d_6 = \text{0x7135}, \quad d_7 = \text{0x09af}$$
$$d_8 = \text{0x4d78}, \quad d_9 = \text{0x2f13}, \quad d_{10} = \text{0x6bc4}, \quad d_{11} = \text{0x1af1}$$
$$d_{12} = \text{0x5e26}, \quad d_{13} = \text{0x3c4d}, \quad d_{14} = \text{0x789a}, \quad d_{15} = \text{0x47ac}.$$

After the loading process, the initialization phase continues for another 32 rounds. Based on the current states of the LFSR and the registers R_1, R_2, the states in the next round are updated as follows.

- Linear Feedback: $v := 2^{15} s_{15} + 2^{17} s_{13} + 2^{21} s_{10} + 2^{20} s_4 + (1 + 2^8) s_0$,
- Non-linear Feedback: $w := NLF(X_0, X_1, X_2)$,
- Feedback to the LFSR: $s_{16} := v + w \mod (2^{31} - 1)$.

2.2 Design of ZUC-256

The initial proposal for the ZUC cipher featured a 128-bit key and a 128-bit IV. Later, an upgraded version of ZUC with a 256-bit key and an 184-bit IV was proposed in [1], differing from the 128-bit ZUC only in the loading mechanism. In ZUC-256, the 256-bit key is divided into 32 key bytes, $K_0, \ldots, K_{31}$, and the

184-bit IV is divided into 17 IV bytes, $IV_0, \ldots, IV_{16}$, and eight 6-bit strings, $IV_{17}, \ldots, IV_{24}$. Each of the constants $d_0, \ldots, d_{15}$ has a length of 7 bits.

The ZUC Design Team from the Chinese Academy of Sciences later published an addendum defining a variant of the ZUC-256 design with a new initialization scheme that supports a 128-bit IV. Compared to the first ZUC-256 proposal, this new key/IV setup algorithm avoids the division of the whole key/IV byte. The key/IV loading scheme of the modified version is described as follows.

– The 128-bit IV is divided as: $IV \leftarrow IV_{15} \| \ldots \| IV_0$ (each part is 8 bits).
– The 256-bit key is distributed as: $K \leftarrow K_{31} \| \ldots \| K_0$ (each part is 8 bits).
– 16 loading constants are used, based on the binary expansion of π, given by 11.00100100001111110110101010001000100001011010001100000…, where each of the constants $d_0, \ldots, d_{15}$ is 7 bits long, described as:

$$d_0 = 1100100, \quad d_1 = 1000011, \quad d_2 = 1111011, \quad d_3 = 0101010,$$
$$d_4 = 0010001, \quad d_5 = 0000101, \quad d_6 = 1010001, \quad d_7 = 1000010,$$
$$d_8 = 0011010, \quad d_9 = 0110001, \quad d_{10} = 0011000, \quad d_{11} = 1100110,$$
$$d_{12} = 0010100, \quad d_{13} = 0101110, \quad d_{14} = 0000001, \quad d_{15} = 1011100.$$

In ZUC-256, the key/IV loading mechanism is as follows.

$$s_0 = K_0 \| d_0 \| K_{16} \| K_{24}, \quad s_1 = K_1 \| d_1 \| K_{17} \| K_{25}, \quad s_2 = K_2 \| d_2 \| K_{18} \| K_{26},$$
$$s_3 = K_3 \| d_3 \| K_{19} \| K_{27}, \quad s_4 = K_4 \| d_4 \| K_{20} \| K_{28}, \quad s_5 = K_5 \| d_5 \| K_{21} \| K_{29},$$
$$s_6 = K_6 \| d_6 \| K_{22} \| K_{30}, \quad s_7 = K_7 \| d_7 \| IV_0 \| IV_8, \quad s_8 = K_8 \| d_8 \| IV_1 \| IV_9,$$
$$s_9 = K_9 \| d_9 \| IV_2 \| IV_{10}, \quad s_{10} = K_{10} \| d_{10} \| IV_3 \| IV_{11}, \quad s_{11} = K_{11} \| d_{11} \| IV_4 \| IV_{12},$$
$$s_{12} = K_{12} \| d_{12} \| IV_5 \| IV_{13}, \quad s_{13} = K_{13} \| d_{13} \| IV_6 \| IV_{14}, \quad s_{14} = K_{14} \| d_{14} \| IV_7 \| IV_{15},$$
$$s_{15} = K_{15} \| d_{15} \| K_{23} \| K_{31}.$$

2.3 Grover's Search on Stream Ciphers

Grover's search algorithm is a quantum algorithm that provides a quadratic speed-up for searching unsorted databases. The problem can also be mapped to finding the support of a Boolean function, $f : \{0,1\}^n \rightarrow \{0,1\}$ in $\mathcal{O}\left(\sqrt{2^n}\right)$ many queries, whereas any known classical algorithm would require $\mathcal{O}(2^n)$ many queries. The technique used in Grover's search is called amplitude amplification, where it identifies the concerned set of points and increases their amplitude while decreasing the amplitudes of the other data points.

Given any key, $K \in \{0,1\}^k$ and an initial value, $IV \in \{0,1\}^m$, let $ZUC_{K,IV} = ks$ denote the key-stream generated from ZUC. For any given key stream ks of length ρ, we can apply Grover's search algorithm for key recovery as follows.

1. Construct the oracle of a Boolean function f which takes K, IV as input and satisfies $f(K, IV) = 1$ if $ZUC_{K,IV} = ks$, and 0 otherwise.

2. Begin with k many qubits all initialized to $|0\rangle$, followed by k Hadamard gates to obtain an equal superposition of all possible keys.

$$|0\rangle^{\otimes k} \xrightarrow{H^{\otimes k}} 2^{-k/2} \sum_{j=0}^{2^k-1} |K_j\rangle = |\mathcal{K}\rangle.$$

3. Pass $|\mathcal{K}\rangle$ through the quantum oracle U_f, and using the phase kickback, we obtain $\frac{1}{\sqrt{2^k}} \sum_{j=0}^{2^k-1} (-1)^{f(K_j)}|K_j\rangle$. Therefore, the amplitude becomes negative only when $f(K_j) = 1$, i.e., when $ZUC_{K_j,IV} = ks$, and remains unchanged otherwise.
4. Apply the Grover's iterate $(2\langle\mathcal{K}||\mathcal{K}\rangle - I)$ to reflect the superposition with respect to $|\mathcal{K}\rangle$.
5. Repeat Step 3 and Step 4 for about $\mathcal{O}\left(\sqrt{2^k}\right)$ times.
6. Finally, measure all k qubits, and the observed bit-pattern is the desired key, $K_0 \in \{0,1\}^k$ with a probability close to 1.

Note that, since most of the time, the IV is public information, the first step can be modified so that f takes only K as input and satisfies $f(K) = 1$ if ZUK_K generates the key-stream ks.

Here, one must consider the scenario where two different keys, K_i and K_j, generate the same key-stream. To increase the success probability of the attack, we need longer key-stream bits, i.e., more plaintext-ciphertext pairs. Given the key size k, we can design the function $C_\rho(K) : 0, 1^k \rightarrow 0, 1^\rho$, which takes the key K as input and produces a key-stream of length ρ. Then, the collision probability is given by $\Pr\left(C_\rho(K_i) = C_\rho(K_j)\right) \approx \frac{2^k-1}{2^\rho}$. Now, if we set $\rho = k + c$ for some constant c, the collision probability becomes approximately $\mathcal{O}\left(2^{-c}\right)$. Therefore, the collision probability is significantly reduced even for a small value of c. While designing Grover's oracle, we construct the circuit (denoted as $\mathcal{ENC}$) of ZUC that generates a key stream of length $k + c$. The key stream is then matched with the given key stream. If they match, the target qubit is flipped; if they don't match, it remains unchanged. The schematic diagram of such an oracle is shown in Fig. 3.

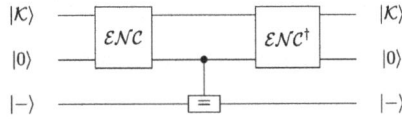

Fig. 3. Grover's oracle for the stream cipher ZUC [4]

2.4   Resource Estimation and MAXDEPTH Criterion

In our present paper, we construct quantum circuits for ZUC and provide resource estimates in terms of the number of qubits, Toffoli gates, CNOT gates,

NOT gates, and the depth of the circuit. Generally, the size and depth correspond to computation time, while the number of qubits corresponds to memory size. We have assumed full parallelism while constructing the circuits, meaning any number of gates can be applied simultaneously, provided the gates act on a disjoint set of qubits. The reasons for using Toffoli gates in the circuit design are as follows.

1. Toffoli gate networks can be directly simulated using classical reversible simulators.
2. For implementations on actual quantum hardware, Toffoli gate circuits can be efficiently debugged, and faults can be localized through binary search.

In this paper, we first present a quantum resource estimation in terms of Toffoli gates. Next, by leveraging the fact that a Toffoli gate can be decomposed into 7 T gates and 8 Clifford gates, with a T-depth of 4 and a total depth of 8 [3], we extend the resource estimation to the Clifford+T gate set. It is important to note that logical T gates are usually considered significantly more resource-intensive than Clifford gates when implemented on a surface code. In the process of quantum resource estimation, T-depth also plays a crucial role. Therefore, we also incorporate the Toffoli decomposition from [20], which utilizes 7 T gates, 9 Clifford gates, and 4 additional ancilla qubits, reducing the T-depth to 1. We provide a comparative summary of the resource estimates for both decompositions, presenting the results in tabular form to highlight the optimal configuration.

We assume that an adversary is bounded by a constraint on the circuit depth that they can use for Grover's search. In this regard, NIST proposed a parameter 'MAXDEPTH' with plausible values ranging from 2^{40} to 2^{96}. Here, we impose the bound of 2^{40}. The idea is that a cipher with key size of 128 bits, which can be attacked with a gate count of 2^{130} for MAXDEPTH= 2^{40}, will be considered VULNERABLE, is influenced by the following observations from [18]:

– On page 16, it is stated that any attack that breaks the relevant security definition must require computational resources comparable to or greater than the requirements for key search on a block cipher with a 128-bit key, such as AES128.
– On page 18, NIST also provided the estimates for optimal key recovery for AES128 as 2^{170}/MAXDEPTH.

Therefore, for MAXDEPTH 2^{40}, a cipher with key size of 128 bits can be considered VULNERABLE if there exists an attack with a gate count less than 2^{130}.

3 New Combinatorial Tools for Circuit Optimization

In this section, we introduce novel circuit optimization techniques essential for the quantum implementation of ZUC. These techniques include modular doubling, consecutive modular addition, linear transformations, and S-box implementation.

3.1   Doubling Modulo $2^n - 1$

Given an n-length bit-pattern $\mathbf{x} = x_{n-1}, x_{n-2}, \ldots, x_0$, we know that the doubling operation produces an $n + 1$-length bit-pattern $2\mathbf{x} = x_{n-1}, x_{n-2}, \ldots, x_0, 0$. If $x_{n-1} = 1$, to perform modulo $2^n - 1$, we first compute the modulo 2^n and then add 1 to the least significant bit (LSB). Since the LSB is 0 after doubling, the new LSB becomes the same as x_{n-1}. Moreover, if $x_{n-1} = 0$, then $2\mathbf{x}$ is less than 2^n, so the modulo operation is not performed, and again, the new LSB remains the same as x_{n-1}. The entire process of doubling modulo $2^n - 1$ can be described as: $|x_{n-1}, x_{n-2}, \ldots, x_0\rangle \rightarrow |x_{n-2}, \ldots, x_0, x_{n-1}\rangle$. Figure 4a describes the schematic diagram of the quantum circuit for doubling modulo $2^n - 1$.

Proposition 1. *The quantum implementation of doubling modulo $2^n - 1$ can be achieved through rewiring of the qubits, requiring no additional resources.*

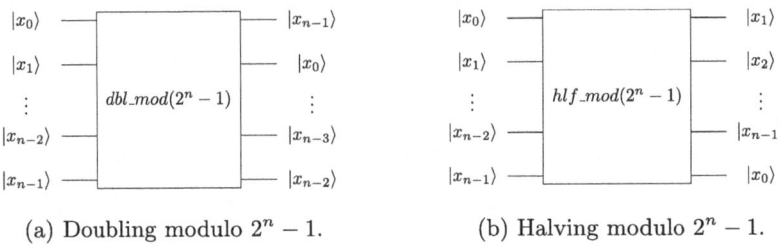

(a) Doubling modulo $2^n - 1$.   (b) Halving modulo $2^n - 1$.

Fig. 4. Quantum circuit for doubling and halving modulo $2^n - 1$

Remark 1. Note that the doubling modulo operation described above does not transform $2^n - 1$ to 0. To achieve this, an additional ancilla qubit initialized to $|0\rangle$ can be introduced, targeted by an n-controlled Toffoli gate, which sets the ancilla to $|1\rangle$ if and only if all the qubits are in the state $|1\rangle$. Then, using n many CNOT gates controlled by the ancilla, each qubit can be reset to $|0\rangle$. However, during the implementation of ZUC, the result of the doubling operation undergoes addition modulo $2^n - 1$, so retaining $2^n - 1$ does not present any issues.

Remark 2. An interesting, albeit elementary, observation relates the above construction of doubling an n-bit pattern modulo $2^n - 1$ to cyclic error-correcting codes of length n as follows. Recall that a linear code $\mathcal{C}$ is termed a cyclic code if every cyclic shift of a codeword $\mathbf{c} \in \mathcal{C}$ is another codeword $\mathbf{c}' \in \mathcal{C}$. Vectors of the form $\mathbf{c} = (c_0, c_1, \cdots, c_{n-1})$ over a field $\mathbb{F}_q$ correspond to polynomials $\sum_{i=0}^{n-1} c_i X^i \in \mathbb{F}_q[X]$ of degree $\leq n - 1$. The polynomial representation of the codewords of a cyclic code involves interpreting the operation of the cyclic shift

$$\mathbf{c} = (c_0, c_1, \cdots, c_{n-1}) \longrightarrow \mathbf{c}' = (c_{n-1}, c_0, c_1, \cdots, c_{n-2})$$

as multiplication of the corresponding codeword polynomial by X *modulo* $X^n - 1$:

$$X \sum_{i=0}^{n-1} c_i X^i = \sum_{i=0}^{n-2} c_i X^{i+1} + c_{n-1} X^n \equiv c_{n-1} + c_0 X + \ldots + c_{n-2} X^{n-1} \quad \mod (X^n - 1).$$

The analogy with the doubling of an n-bit pattern modulo $2^n - 1$ is evident, replacing the indeterminate X with 2 for a '2-adic' expansion.

Additionally, one can obtain the quantum circuit for halving modulo $2^n - 1$ by reversing the doubling modulo $2^n - 1$ circuit, as in Fig. 4b, which is again essentially a rewiring of the qubits: $|x_{n-1}, \ldots, x_1, x_0\rangle \rightarrow |x_0, x_{n-1}, \ldots, x_1\rangle$.

While implementing the LFSR update function in the quantum ZUC, we first obtain the appropriate 2's power of the designated $s_i \in \{0, 1\}^n$ in-place by repetitively using the doubling circuit and then get back the original s_i (in-place) once it has been added to the overall sum, without using any additional quantum resources. Without our optimization, each modular doubling (halving) would require an additional $4n$ Toffoli and $10n$ CNOT gates, following [7,25].

$$s_i \xrightarrow{\;dbl\_mod(2^n-1)\;} 2s_i \xrightarrow{\;dbl\_mod(2^n-1)\;} 2^2 s_i \ldots 2^{l-1} s_i \xrightarrow{\;dbl\_mod(2^n-1)\;} 2^l s_i$$

$$\text{and, } 2^l s_i \xrightarrow{\;hlf\_mod(2^n-1)\;} 2^{l-1} s_i \xrightarrow{\;hlf\_mod(2^n-1)\;} 2^{l-2} s_i \ldots 2s_i \xrightarrow{\;hlf\_mod(2^n-1)\;} s_i.$$

3.2 Consecutive Addition Modulo $2^{31} - 1$ $(add\_mod(2^{31} - 1))$

The LFSR update function in ZUC requires the implementation of addition modulo $2^{31} - 1$ for several 31-bit integers sequentially. Here, we follow the integer addition circuit with an incoming carry, as proposed by Cuccaro et al. [7]. We use an ancilla qubit, initialized to $|0\rangle$, to store the incoming carry bit. The fundamental idea is that the final carry bit (the high bit) from one addition will be used as the incoming carry for the subsequent addition. Storing the carry bit also requires an additional ancilla qubit, initialized to $|0\rangle$. If the 32-nd bit from an intermediate addition is 1, i.e., the sum is greater than $2^{31} - 1$, ignoring the MSB effectively implements the addition modulo 2^{31}. Then, we add 1 to the sum by using the most significant bit (MSB) of the addition as an incoming carry for the next addition. Note that, after every addition, the incoming carry bit becomes $|0\rangle$ and, therefore, can be used for storing the MSB of the following addition.

Moreover, the first addition operation always has an incoming carry 0. Therefore, we can implement this using integer addition without an incoming carry, which requires $2(31 - 2) = 58$ NOT gates, $(5 \cdot 31 - 3) = 152$ CNOT gates, and $(2 \cdot 31 - 1) = 61$ Toffoli gates, with a circuit depth of $(2 \cdot 31 + 4) = 66$ (refer to ([7], Fig. 6)). Similarly, the final addition operation can have a final carry of 1, which needs to be added to the overall addition using (a modified version of) addition modulo 2^{31} circuit (due to [7]). That, in turn, requires an additional $2(31 - 2) = 58$ NOT gates, $(3 \cdot 31 - 7) = 86$ CNOT gates, and $(2 \cdot 31 - 3) = 59$ Toffoli gates, with a circuit depth of $(2 \cdot 31 + 4) = 66$. Hence, we have the following proposition.

Proposition 2. *Following [7], the in-place implementation of a k-length series of addition modulo $2^{31} - 1$ requires $58(k+1)$ NOT gates, $\left[152 + (k-1)156 + 86\right] = (156k + 82)$ CNOT gates, and $(61k + 59)$ Toffoli gates, with a circuit depth of $\left[66 + (k-1)68 + 66\right] = (68k + 64)$.*

In ZUC, each round of LFSR update requires 5-consecutive additions. Thus, from Proposition 2, one round of LFSR update requires a maximum of 348 NOT gates, 862 CNOT gates, and 364 Toffoli gates, with a circuit depth of 336. Without this optimization, five modular additions would require approximately $5 * 2 * (5 * 31 - 3) = 1520$ CNOT and $5 * 2 * (2 * 31 - 1) = 610$ Toffoli gates following [7, Table 1].

3.3   Quantum Implementation of Linear Transformations

Here, we present the optimal quantum implementation of the linear transformations L_1 and L_2, following [19, Algorithm 1]. In [19], Patel et al. proposed a CNOT-based quantum circuit synthesis algorithm for any given linear transformation by partitioning the matrix into m different parts, and the required number of CNOT gates depends upon the choice of m. Here, we implement the algorithm in qiskit and present the optimal depth quantum circuit with corresponding gate counts. From Sect. 2.1, the linear transformation L_1 is defined as $L_1(X) = X \oplus (X \lll 2) \oplus (X \lll 10) \oplus (X \lll 18) \oplus (X \lll 24)$ i.e., $L_1(x_i) = x_i \oplus x_{i-2 \bmod 32} \oplus x_{i-10 \bmod 32} \oplus x_{i-18 \bmod 32} \oplus x_{i-24 \bmod 32}$, for all $0 \le i \le 31$. For easy reading, we have provided the matrix representation of L_1 in the GitHub [32].

Here, we implement the linear reversible circuit synthesis algorithm due to [19] using qiskit and observe that for $m = 7$, the number of required CNOT gates is minimum, which is 201, with a corresponding circuit depth of 55. On the other hand, when $m = 10$, although the CNOT count increases by 1, the circuit depth reduces to 46. Table 3 presents a comparative analysis of the different values of m with the required number of CNOT gates and circuit depth for each case.

Table 3. CNOT counts and CNOT depth for the quantum circuit implementation of L_1 for different values m

m	2	3	4	5	6	7	8	9	10	11	12	13	14	15	16
CNOT count	485	341	295	235	232	201	210	202	202	206	212	214	216	224	232
CNOT depth	221	125	97	66	67	55	50	50	46	46	46	46	46	46	45

Remark 3. In ([19], Sect. 5), m was arbitrarily chosen to be round($\log_2 n/2$). According to their assumption, for $n = 32$, m should be 3. However, from Table 3, we observe that for $m = 3$, both the CNOT count and CNOT depth are much higher than the optimal CNOT count of 201 (for $m = 7$) and the optimal CNOT

depth of 45 (for $m = 16$). In this paper, we chose $m = 10$ (which is $2 \log_2 n$) for simultaneously optimizing both the CNOT count and CNOT depth, and obtained a CNOT count of 202, matching the suggested theoretical bound of $\mathcal{O}\left(n^2 / \log_2 n\right)$ from [19].

Similarly, the linear transformation L_2 is defined as $L_2(X) = X \oplus (X \lll 8) \oplus (X \lll 14) \oplus (X \lll 22) \oplus (X \lll 30)$ i.e., $L_2(x_i) = x_i \oplus x_{i-8 \bmod 32} \oplus x_{i-14 \bmod 32} \oplus x_{i-22 \bmod 32} \oplus x_{i-30 \bmod 32}$, for all $0 \leq i \leq 31$. From the matrix representation of L_2 (provided in [32]), one can observe that $L_2 = L_1^\dagger$, i.e., L_2 is the unitary inverse of L_1. Therefore, the quantum circuit of L_2 would be the exact reverse of L_1, and the corresponding quantum circuit requires the exact same number of CNOT gates for different values of m. The quantum circuit for L_1 and L_2 with $m = 10$ have been implemented in qiskit, and the corresponding circuit diagrams are provided in [32]. The optimal quantum implementation of L_1 and L_2 and the CNOT requirements with the corresponding circuit depths have been summarized in Proposition 3.

Proposition 3. *In ZUC, the quantum circuit of L_1 and L_2 can be implemented using a maximum of 202 CNOT gates, with a CNOT depth of 46 each.*

Since L_1 and L_2 act on different sets of qubits, they can be applied simultaneously, causing the overall CNOT depth to be 46.

3.4   Quantum Implementation of an S-Box

In this section, we present the quantum implementation of S-boxes following [15]. While optimized quantum implementations of 3-bit and 4-bit S-boxes exist [6,8], arbitrary 8-bit S-boxes are less studied. Strategies to reduce the resource requirements for the quantum implementations of AES S-box [11,30] are also not directly applicable to arbitrary 8-bit S-boxes.

In this regard, we present the quantum circuit implementation of any S-box, employing the MMD-based reversible logic synthesis method as described in [15], along with some tricks and tweaks. As this is an in-place implementation, it does not necessitate additional working qubits, except for the implementation of multi-controlled Toffoli gates, which may require ancilla qubits initialized and returned to $|0\rangle$. The associated quantum resource requirements are summarized in Table 4.

Table 4. Resource requirements for the quantum implementation of S_0 and S_1 following [15]

	NOT	CNOT	Toffoli	Toffoli3	Toffoli4	Toffoli5	Toffoli6	Toffoli7	depth
S_0	5	102	199	236	202	136	53	10	936
S_1	4	110	211	221	189	132	62	9	930

Further decomposition of the n-controlled Toffoli gates following Fig. 6 requires 5 additional ancilla qubits (specifically for decomposing Toffoli-7 gates, which can be used for other MCT gates as well) and the total number of required Toffoli gates for each of these S-boxes is given by

$$S_0 : 199 + (236 \times 3) + (202 \times 5) + (136 \times 7) + (53 \times 9) + (10 \times 11) = 3456,$$
$$S_1 : 211 + (221 \times 3) + (189 \times 5) + (132 \times 7) + (62 \times 9) + (9 \times 11) = 3400.$$

Similarly, the circuit depth of each of these S-boxes becomes:

$$S_0 : 936 - 637 + (236 \times 3) + (202 \times 3) + (136 \times 5) + (53 \times 5) + (10 \times 7) = 2628,$$
$$S_1 : 930 - 613 + (221 \times 3) + (189 \times 3) + (132 \times 5) + (62 \times 5) + (9 \times 7) = 2580.$$

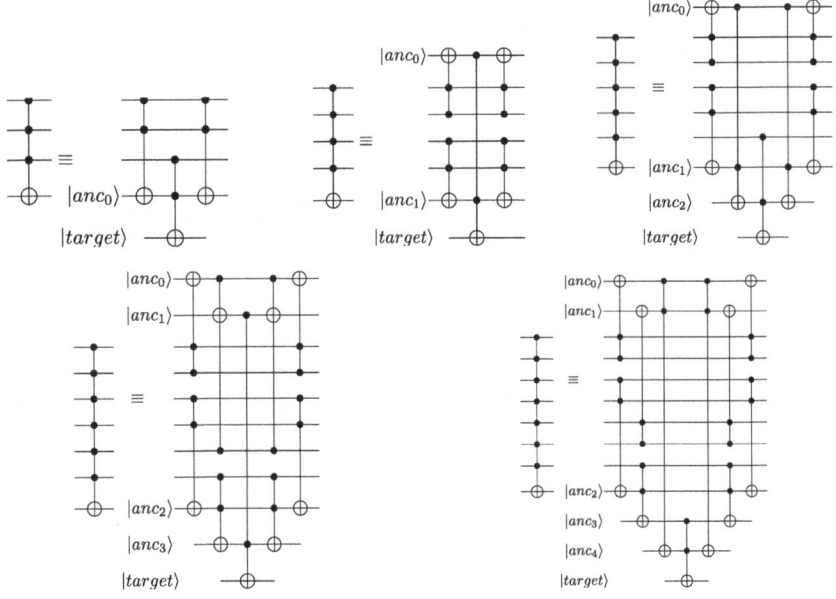

Fig. 5. Toffoli decomposition of multi-controlled Toffoli gates

As discussed in Sect. 2, for any given 32-bit tuple (x_1, x_2, x_3, x_4), where $x_i \in \{0, 1\}^8$, the S-boxes S_0 and S_1 are applied following $S(x_1 \| x_2 \| x_3 \| x_4) := S_0(x_1) \| S_1(x_2) \| S_0(x_3) \| S_1(x_4)$, i.e., each of them is applied twice. However, since each of these S-boxes is applied to a different set of qubits, the overall circuit depth of S will be $\max\{2628, 2580\} = 2628$. The complete resource requirements to implement the S-box S, along with the individual S-boxes, are summarized in Table 5.

Table 5. Resource requirements for S_0 and S_1 following [15]

	#ancilla	#NOT	#CNOT	#Toffoli	depth
S_0	5	5	102	3456	2628
S_1	5	4	110	3400	2580
S	5	18	424	13712	2628

4   Quantum Resource Estimation of ZUC

In this section we present the quantum circuit implementation of ZUC-128 and ZUC-256 and provide the resource requirements to mount the Grover-based attack on ZUC to recover the secret key.

Here we present the quantum circuit implementation of ZUC-128 and ZUC-256, along with the resource requirements to mount a Grover-based key recovery attack on ZUC. From Sect. 2, we know that there are three major components in the design of ZUC: the update of the Linear Feedback Shift Register (LFSR), Bit Reorganization (BR), and the update of the Non-linear Function (NLF). Besides these, there is key/IV loading at the beginning of the process and the generation of W at the end. During the initialization phase, W is not XOR-ed with X_3. Instead, the rightmost bit is removed, and the remaining 31 bits are added to the highest component of the LFSR. On the other hand, during the keystream generation phase, W is XOR-ed with X_3, and produces the desired keystream. We now estimate the maximum quantum resource requirements for the implementation of each of these individual steps.

4.1   Resource Requirements for Key/IV Loading

Storing the 16 registers of the LFSR, each of size 31-bit requires a total of $31 \times 16 = 496$ qubits.

For ZUC-128, copying a 128-bit key to the LFSR requires a maximum of 128 CNOT gates. Assuming IVs are known, 128-bit IV can be incorporated in the LFSR using a maximum of 128 NOT gates. Similarly, d_i can also be loaded in the LFSR using a maximum of 240 NOT gates. Therefore, the key/IV (LFSR) loading in ZUC-128 requires a maximum of 128 CNOT gates and 368 NOT gates, depending upon the Hamming weight of the initial values (IVs) and the d_i's.

Similarly, in ZUC-256, copying a 256-bit key to the LFSR requires a maximum of 256 CNOT gates. Moreover, assuming IVs are known, the loading of IV and d_i requires a maximum of 240 NOT gates, depending on the Hamming weight of these bit-patterns. Overall, we can conclude that the key/IV loading in ZUC-128 and ZUC-256 requires a maximum of 512 Clifford (CNOT+NOT) gates, which can be applied simultaneously, having a circuit depth 1.

4.2 Resource Requirements for LFSR Update

To perform the LFSR operation, we consider s_{16}, a 31-qubit memory location, initialized to $|0\rangle^{\otimes 31}$ where we copy s_0 using 31 CNOT gates, simultaneously. From Sect. 3.1, we know that doubling does not require any additional quantum resources, we first obtain $2^8 s_0, 2^{20} s_4, 2^{21} s_{10}, 2^{17} s_{13}, 2^{15} s_{15}$ in-place, and then sequentially add them in-place to s_{16} modulo $(2^n - 1)$. In each round of LFSR update, there are 5-consecutive addition modulo $2^{31} - 1$ adders, which requires a maximum of 348 NOT gates, 862 CNOT gates, and 364 Toffoli gates, with a circuit depth of 336 (refer to Proposition 2). Finally, we get back $s_0, s_4, s_{10}, s_{13}, s_{15}$ by using the modular halving circuit as shown in Fig. 4b.

Proposition 4. *The quantum implementation of the one-round LFSR update in ZUC requires 31 working qubits, 2 additional ancilla qubits for addition operations, a maximum of 348 NOT gates, 893 CNOT gates, and 364 Toffoli gates, with a circuit depth of 337.*

Additionally, after each clock cycle, we rename s_i from the previous round as s_{i-1} for the next round, for all $i : 1 \leq i \leq 15$. As mentioned earlier, since we perform the LFSR update operation in place of s_{16}, the s_{16} from the previous round is renamed as s_{15} for the following round.

Table 6. Resource requirements for one round of LFSR update

	#Qubits	#NOT	#CNOT	#Toffoli	depth
ZUC-128, ZUC-256	31+2	348	893	364	337

4.3 Resource Requirements for Bit Reorganization

Since the bit-reorganization step of ZUC is simply rearranging the high-bits and low-bits of designated registers (s_i), it is essentially a re-wiring of the qubits. Therefore, this step does not require any additional quantum resources.

4.4 Resource Requirements for Updating the Non-linear Functions

Since R_1 and R_2 both are of size 32 bits a total of 64 qubits are required to store R_1 and R_2. First, we perform in-place addition modulo 2^{32} to R_1 following [7] with the 16 low bits of s_{11} and 16 high bits of s_9. This requires $2(31 - 2) = 58$ NOT gates, $5(31+1) - 7 = 153$ CNOT gates, and $2(31+1) - 3 = 61$ Toffoli gates, with a circuit depth of $2(31+1) + 2 = 66$. Simultaneously, we obtain the in-place XOR at R_2 with the 16 low bits of s_7 and 16 high bits of s_5 using 32 CNOT gates at once. The 16-bit left rotation on the updated values (W_1 and W_2) can be achieved by re-wiring the qubits, requiring no additional resources. Moreover, from Sect. 3.3, we know that the implementation of the linear transformations L_1 and L_2 requires a maximum of 202 CNOT gates each, adding to another 404

CNOT gates. Further, from Sect. 3.3, we know that the S-box implementation S_0 and S_1 requires a total of 18 NOT gates, 424 CNOT gates, 13712 Toffoli gates, and 5-additional ancilla qubits.

The output bits from SL_1 and SL_2 are the updated R_1 and R_2, respectively.

Proposition 5. *The update function of R_1 and R_2 requires a total of $(58 + 18) = 76$ NOT gates, $(153 + 32 + 404 + 424) = 1013$ CNOT gates, and $(61 + 13712) = 13773$ Toffoli gates, with a circuit depth of $(66 + 1 + 46 + 2628) = 2741$.*

Table 7. Resource requirements for one round of NLF update

	#Qubits	#NOT	#CNOT	#Toffoli	depth
ZUC-128, ZUC-256	64+5	76	1013	13773	2741

4.5 Resource Requirements for Key Stream Generation

Storing W requires 32 ancilla qubits initialized to $|0\rangle^{32}$, and the information of R_1 can be copied to W using 32 CNOT gates simultaneously. Then, W is XORed with the 16 high bits of s_{15} and 16 low bits of s_{14}, using 32 CNOT gates at once. After that, we perform addition modulo 2^{32} with R_2 in-place to W, using one additional ancilla qubit, $2(31 - 2) = 58$ NOT gates, $5(32) - 7 = 153$ CNOT gates, and $2(32) - 3 = 61$ Toffoli gates, with a circuit depth of $2(32) + 2 = 66$ following [7, Table 1].

Before the keystream generation, ZUC is initialized for 32 rounds where the rightmost bit of W is discarded, and the rest is added to the LFSR using 2-ancilla qubits, $2(31 - 2) = 58$ NOT gates, $5(31) + 1 = 156$ CNOT gates, and $2(31) - 1 = 61$ Toffoli gates, where the circuit depth is $2(31) + 6 = 68$. After initialization, during the keystream generation phase, W is XOR-ed with the 16 low bits of s_2 and 16 high bits of s_0 using 32 CNOT gates at once. The output is the desired bitstream Z.

Proposition 6. *In the initialization phase of ZUC, generating W and adding them to LFSR requires $(58 + 58) = 116$ NOT gates, $(32 + 32 + 153 + 156) = 373$ CNOT gates, and $(61 + 61) = 122$ Toffoli gates, with a circuit depth of $(1 + 1 + 66 + 68) = 136$.*

Similarly, the keystream generation phase of ZUC requires 58 NOT gates, $(32 + 32 + 153 + 32) = 249$ CNOT gates, and 61 Toffoli gates, where the circuit depth becomes $(1 + 1 + 66 + 1) = 69$.

The total amount of quantum resources required for one round of initialization/ keystream generation has been summarized in Proposition 7 and Table 9.

Table 8. Resource requirements for generation of W

		#Qubits	#NOT	#CNOT	#Toffoli	depth
Initialization	ZUC-128, 256	32+2	116	373	122	136
KS generation	ZUC-128, 256	32+2	58	249	61	69

Proposition 7.

- *(Initialization).* One complete round of initialization in *ZUC-128/256* requires a total of $(496 + 31 + 64 + 32) = 623$ *working qubits,* $\max\{2, 5, 2\} = 5$ *ancilla qubits initialized and returned as* $|0\rangle$, $(368/240 + 348 + 76 + 116) = 908/780$ *NOT gates,* $(128/256 + 893 + 1013 + 373) = 2407/2535$ *CNOT gates, and* $(364 + 13773 + 122) = 14259$ *Toffoli gates, with a maximum circuit depth of* $(1 + 337 + 2741 + 136) = 3215$.
- *(Keystream Generation).* One complete round keystream generation in *ZUC-128/256* requires a maximum of $(348 + 76 + 58) = 482$ *NOT gates,* $(893 + 1013 + 249) = 2155$ *CNOT gates, and* $(364 + 13773 + 61) = 14198$ *Toffoli gates, where the circuit depth is* $(337 + 2741 + 69) = 3147$.

Table 9. Resource requirements for one complete round of initialization and keystream generation in ZUC-128/256

		#Qubits	#NOT	#CNOT	#Toffoli	depth
Initialization	ZUC-128	623+5	908	2407	14259	3215
	ZUC-256	623+5	780	2535	14259	3215
KS generation	ZUC-128/256	–	482	2155	14198	3147

4.6 Resource Requirements for Grover's Search and MAXDEPTH Criteria

In ZUC, before the keystream generation begins, the cipher is initialized for 32 rounds. From Table 9, 32-round initialization of ZUC-128/256 requires a maximum of 29056/24960 NOT gates, 77024/81120 CNOT gates, and 456288 Toffoli gates. Additionally, from Sect. 2.3, finding the correct key using Grover's search requires a keystream longer than the key size. Since each round of the keystream generation phase produces a keystream of length 32, for ZUC-128 and ZUC-256, respectively, 5 and 9 rounds of keystream generation would be sufficient. This requires an additional 2410/4338 NOT gates, 10775/19395 CNOT gates, and 70990/127782 Toffoli gates for ZUC-128 and ZUC-256, respectively. Starting from the key/IV loading to the complete keystream generation before the oracle search, denoted by $\mathcal{ENC}$, the complete quantum resource requirements have been summarized as follows.

Table 10. Quantum resource requirements for ZUC $\mathcal{ENC}$

	#NOT	#CNOT	#Toffoli	depth
ZUC-128	31466	87799	527278	118615
ZUC-256	29298	100515	584070	131203

From [17], it is known that a Toffoli gate can be decomposed in terms of Clifford + T gates, using 7 T gates and 9 Clifford gates, where the T-depth is 6. Using a meet-in-the-middle algorithm, Amy et al. [3] derived the optimal T-depth quantum circuit without using additional ancilla qubits. The circuit contains 7 T-gates and 9 Clifford gates, achieving a T-depth of 3. Additionally, they also provided a T-depth 4 circuit decomposition of Toffoli that requires 7T gates and 8 Clifford gates. The circuit is shown in Fig. 6a.

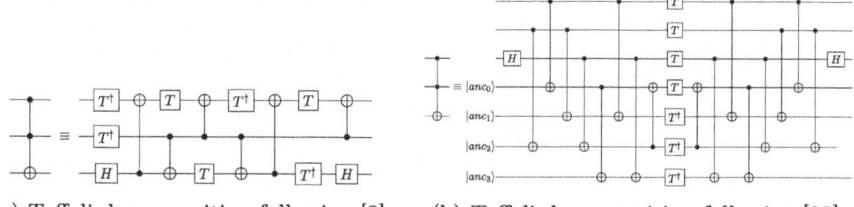

(a) Toffoli decomposition following [3].   (b) Toffoli decomposition following [20].

Fig. 6. Toffoli decomposition using T gates

In [20], Selinger further decomposed Toffoli gates into a circuit with T-depth 1 using 7 T gates and 18 Clifford (16 CNOT, 2 Hadamard) gates. To achieve the T-depth one, the circuit requires the use of 4-additional ancilla qubits (initialized and returned as $|0\rangle$). The corresponding circuit is presented in Fig. 6b.

Here, for the quantum resource estimation of ZUC, we follow the Toffoli decomposition circuits due to [3] (7T + 9 Clifford, T-depth 3) and [20] (7T + 18 Clifford, T-depth 1). Since parallel Toffoli gates contribute to the circuit depth only once, we assume that the number of Toffoli gates contributing to the circuit depth equals the total circuit depth. Therefore, when estimating T-depth, we multiply the T-depth of a single Toffoli gate by the circuit's depth. The corresponding resource estimation has been summarized in Table 11.

To estimate the resource requirements for Grover's oracle, we first encrypt the cipher using all possible keys and decrypt it for subsequent rounds. Between these processes, the generated keystream is compared to the target keystream using a multi-controlled Toffoli gate, where the number of controlling qubits matches the length of the generated keystream. Given that an n-controlled Toffoli gate can be simplified using $(32n - 84)$ T gates [26], the number of T gates required for executing one round of Grover's search (amplitude amplification) can be

Table 11. Quantum resource requirements for ZUC $\mathcal{ENC}$ in terms of Clifford+T gates. The cells with the lowest Clifford gates, lowest circuit depth, and the least T-depth are highlighted in green

		#Clifford	#T	T-depth	depth
Figure 6a	ZUC-128	4337489	3690946	474460	948920
	ZUC-256	4802373	4088490	524812	1049624
Figure 6b	ZUC-128	9610269	3690946	118615	830305
	ZUC-256	10643073	4088490	131203	918421

estimated as $(32n - 84) + 2 \cdot \#T$ gates in $\mathcal{ENC}$. Similarly, the number of Clifford gates can be estimated as $2 \cdot \#$Clifford gates in $\mathcal{ENC}$, and the T-depth can be estimated using $2 \cdot \#$T-depth in $\mathcal{ENC}$.

As discussed earlier, an n-controlled Toffoli gate can also be decomposed into $2n - 3$ CNOT (Clifford) gates, requiring $n - 2$ ancilla qubits, which are initialized and returned to the $|0\rangle$ state. This approach reduces the number of T gates at the expense of additional ancilla qubits. For ZUC-128 and ZUC-256, where the keystream lengths are 160 and 288, respectively, the n-controlled Toffoli implementation requires 317 CNOT gates and 158 ancilla qubits for ZUC-128, and 573 CNOT gates with 286 ancilla qubits for ZUC-256. The complete resource requirements for one round of Grover's search, considering different trade-offs, are summarized as follows (Table 12).

Table 12. Quantum resource requirements for one round of Grover's search. The best results of the columns for ZUC-128/256 are highlighted in green

MCT var.	Toffoli		#anc	#Clifford	#T	T-depth	depth
	Figure 6a	ZUC-128	5	8674978	7386928	948920	1897840
From [26]		ZUC-256	5	9604746	8186112	1049624	2099248
$32n - 84$ T	Figure 6b	ZUC-128	5	19220538	7386928	237230	1660610
		ZUC-256	5	21286146	8186112	262406	1836842
	Figure 6a	ZUC-128	158	8675295	7381892	948920	1897840
Figure 6		ZUC-256	286	9605319	8176980	1049624	2099248
$2n - 3$ CNOT	Figure 6b	ZUC-128	158	19220855	7381892	237230	1660610
		ZUC-256	286	21286719	8176980	262406	1836842

Successful key-recovery using Grover's search requires the oracle and Grover's iterate to run for approximately $\frac{\pi}{4}\sqrt{2^k}$ times. Thus, for ZUC-128 and ZUC-256, we multiply the single round gate count by 2^{64} and 2^{128}, respectively, with an additional factor of $\pi/4$. The overall gate count for the complete Grover's search has been summarized in Table 13.

Table 13. Quantum resource requirements for a complete Grover's search. We highlight the lowest gate counts and the least T-depth of the column in green

MCT	Toffoli		anc	#Clifford	#T	Clifford+T	T-depth	depth
[26]	Figure 6a	ZUC-128	5	$1.62 \cdot 2^{86}$	$1.38 \cdot 2^{86}$	$1.5 \cdot 2^{87}$	$1.42 \cdot 2^{83}$	$1.42 \cdot 2^{84}$
		ZUC-256	5	$1.8 \cdot 2^{150}$	$1.53 \cdot 2^{150}$	$1.67 \cdot 2^{151}$	$1.57 \cdot 2^{147}$	$1.57 \cdot 2^{148}$
	Figure 6b	ZUC-128	5	$1.8 \cdot 2^{87}$	$1.38 \cdot 2^{86}$	$1.24 \cdot 2^{88}$	$1.42 \cdot 2^{81}$	$1.24 \cdot 2^{84}$
		ZUC-256	5	$1.99 \cdot 2^{151}$	$1.53 \cdot 2^{150}$	$1.38 \cdot 2^{152}$	$1.57 \cdot 2^{145}$	$1.37 \cdot 2^{148}$
Figure 6	Figure 6a	ZUC-128	158	$1.62 \cdot 2^{86}$	$1.38 \cdot 2^{86}$	$1.5 \cdot 2^{87}$	$1.42 \cdot 2^{83}$	$1.42 \cdot 2^{84}$
		ZUC-256	286	$1.8 \cdot 2^{150}$	$1.53 \cdot 2^{150}$	$1.67 \cdot 2^{151}$	$1.57 \cdot 2^{147}$	$1.57 \cdot 2^{148}$
	Figure 6b	ZUC-128	158	$1.8 \cdot 2^{87}$	$1.38 \cdot 2^{86}$	$1.24 \cdot 2^{88}$	$1.42 \cdot 2^{81}$	$1.24 \cdot 2^{84}$
		ZUC-256	286	$1.99 \cdot 2^{151}$	$1.53 \cdot 2^{150}$	$1.38 \cdot 2^{152}$	$1.57 \cdot 2^{145}$	$1.37 \cdot 2^{148}$

Table 14. Cost of Grover's search on ZUC under MAXDEPTH. The final column provides the NIST's standardization criterion

Toffoli		Clifford+T(G)	depth (D)	$G \cdot D$	NIST [18]
Figure 6a	ZUC-128	$1.5 \cdot 2^{87}$	$1.42 \cdot 2^{84}$	$1.07 \cdot 2^{172}$	2^{170}
	ZUC-256	$1.67 \cdot 2^{151}$	$1.57 \cdot 2^{148}$	$1.31 \cdot 2^{300}$	2^{298}
Figure 6b	ZUC-128	$1.24 \cdot 2^{88}$	$1.24 \cdot 2^{84}$	$1.54 \cdot 2^{172}$	2^{170}
	ZUC-256	$1.38 \cdot 2^{152}$	$1.37 \cdot 2^{148}$	$1.89 \cdot 2^{300}$	2^{298}

From Table 13, the Clifford + T gate counts, T-depth, and overall depth are the same when using the MCT decomposition from [26] or Fig. 6. Thus, either row can be used for estimating the MAXDEPTH.

5 Conclusion

In this paper, we have presented a comprehensive quantum resource estimation for ZUC in the context of Grover-based quantum key-recovery attacks. By introducing novel circuit optimization techniques, such as modular quantum doubling and successive quantum modular addition, and constructing quantum logic circuits for the linear transformations L_1, L_2, as well as non-linear transformations like S-boxes, we provided detailed quantum resource estimates for implementing ZUC and performing its quantum cryptanalysis.

We found that a full-scale Grover's search on ZUC-128 requires approximately $1.5 \cdot 2^{87}$ Clifford + T gates, with a T-depth of $1.42 \cdot 2^{83}$, and overall circuit depth of $1.42 \cdot 2^{84}$. Similarly, a key-recovery attack on ZUC-256 deploys around $1.67 \cdot 2^{151}$ Clifford + T gates, with a T-depth of $1.57 \cdot 2^{147}$, and overall circuit depth of $1.57 \cdot 2^{148}$. Alternatively, if we focus on reducing the T-depth, an exhaustive Grover's search requires $1.24 \cdot 2^{88}$ and $1.38 \cdot 2^{152}$ Clifford + T gates for ZUC-128 and ZUC-256, respectively, while reducing the T-depth to $1.42 \cdot 2^{81}$ and $1.57 \cdot 2^{145}$. Finally, by multiplying the total gate count with the circuit depth, we obtain $1.07 \cdot 2^{172}$ for ZUC-128 and $1.31 \cdot 2^{300}$ for ZUC-256. Since NIST's standardization criterion is bounded by 2^{170} for ZUC-128 and 2^{298} for ZUC-256, ZUC remains secure against exhaustive Grover's search under the MAXDEPTH criterion. Our work, thus, breaks new ground in the following conjoined directions. First, as far as we are aware, we present the first complete and optimized quantum circuit implementation of ZUC-128 and ZUC-256. Secondly, a rigorous analysis of Grover's search on our implementation leads to the first ever quantum security results pertaining to ZUC, in the framework of NIST criteria.

In addition to the above, this study makes several significant additions to the state-of-the-art for related problems as follows.

1. Our generalization of circuit optimization techniques, such as modular quantum doubling and successive quantum modular addition, for any $n \in \mathbb{N}$, is applicable to the broad area of quantum circuit optimization.
2. Optimization of the quantum circuit synthesis of the S-box using advanced quantum circuit synthesis tools like RevKit [24] and the Munich Quantum Toolkit (MQT) [27] could be used to significantly reduce both gate count and circuit depth.
3. The novel quantum cryptanalysis techniques in our comprehensive framework for ZUC can be readily extended to similar stream ciphers like SNOW to provide quantum resource estimations.

Acknowledgements. The authors would like to thank the anonymous reviewers for their detailed comments that improved the technical content as well as the presentation of the paper. The last author (Subhamoy Maitra) acknowledges the support of MeitY, Government of India, related to the initiative "Cluster - Cryptography, Information Security Education and Awareness (ISEA) Project Phase - III". The first author (Suman Dutta) and the third author (Anupam Chattopadhyay) acknowledge the support of 'MoE AcRF Tier 1 award RT10/23'.

Disclosure of Interests. The authors have no competing interests to declare that are relevant to the content of this article.

A   Implementing Toy-ZUC and Related Resource Estimation

In the NISQ era, with limited resources available, it is almost impossible to mount a full-scale attack on any existing cryptosystem using quantum algorithms such as Grover's or Shor's. However, it is standard practice to scale down the size of the original cipher by designing a toy version that retains the fundamental structure. This scaled-down version is then implemented on existing quantum platforms to verify whether the required resources align with theoretical estimations. Additionally, this approach allows for the estimation of overall resource requirements for such attacks on the original cipher and provides an opportunity to analyze their security according to existing standards. In this context, we present the toy-ZUC model and implement it in the Qiskit simulator (qasm_simulator) as well as on the real quantum devices provided by IBM through the cloud.

A.1 Design of Toy-ZUC

In this section, we introduce a simplified design that resembles the original ZUC in terms of functional components specifically tailored for implementation in Qiskit. Given the constraints of the limited number of accessible qubits in the *qasm_simulator*, we assume that the Toy-ZUC operates with an 8-bit key, $K = k_7 k_6 \ldots k_0$, and an 8-bit initialization vector (IV), $IV = iv_7 iv_6 \ldots iv_0$.

Considering the original ZUC-128/256, the Toy-ZUC is also based on a Fibonacci LFSR with eight states, each represented by a 2-bit register, initialized as $s_i = (k_i | iv_i)$, i.e., iv_i as the LSB of s_i and k_i as the MSB s_i. The taps are defined by the primitive polynomial $f(x) = x^8 - (x^7 + x^5 + 2x + 1)$ over $\mathbb{F}_3$. Therefore, the update function of the LFSR is given by $\mathsf{feedback}_f(s_0, \ldots, s_7) = s_7 + s_5 + 2s_1 + s_0$. Here, the choice of a primitive polynomial of degree 8 over the finite field $\mathbb{F}_3$ ensures the maximal period of $3^8 - 1$.

In the bit-reorganization layer, the bits from all the LFSR states are chosen and re-organized as: $X_0 = s_{7H} \| s_{6L}$, $X_1 = s_{5L} \| s_{4H}$, $X_2 = s_{3L} \| s_{2H}$, $X_3 = s_{1L} \| s_{0H}$, where s_{iH} implies the MSB of s_i, and s_{iL} refers to the LSB of s_i.

In every round, the non-linear function layer takes (X_0, X_1, X_2) as input, and depending on the internal states of the 2-bit registers R_1 and R_2, it generates the output W while updating the registers R_1 and R_2 as described in Table 15.

Table 15. Toy-ZUC Non-linear function

Input: (X_0, X_1, X_2) and R_1, R_2
Output: W
1. Define 2-bit word $W = (X_0 \oplus R_1) \boxplus R_2$;
2. Define 2-bit word $W_1 = R_1 \boxplus X_1$;
3. Define 2-bit word $W_2 = R_2 \oplus X_2$;
4. Update 2-bit word $R_1 = S(\mathcal{L}(W_{1L} \| W_{2H}))$;
5. Update 2-bit word $R_2 = S(\mathcal{L}(W_{2L} \| W_{1H}))$;
6. Return 2-bit word W

Here $\boxplus$ denotes addition modulo 2^2.

The linear transformation $\mathcal{L}$ used inside the non-linear function of Toy-ZUC is given by $\mathcal{L}(X) = X \lll_2 1$. Whereas the 2-input 2-output S-box, S, is defined by the permutation 1 3 2 0. The outputs from the S-box define the updated values for the registers R_1 and R_2. A schematic diagram illustrating different layers of the Toy-ZUC is shown in Fig. 7.

A.2 Qiskit Implementation of Toy-ZUC

Here, we implement a specific example of the Toy-ZUC defined above using qiskit, with the key, $K = 11100101$, and the $IV = 01101001$. The corresponding

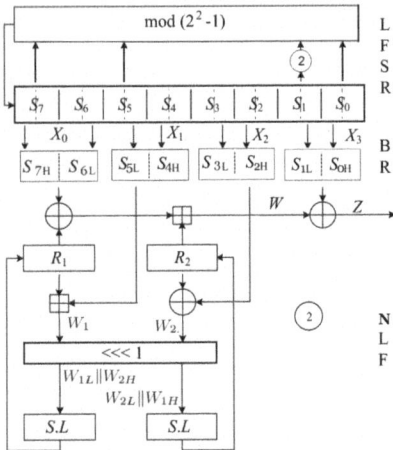

Fig. 7. Toy-ZUC circuit diagram for key-stream generation

qiskit code and the circuit diagram producing the keystream are provided in the GitHub [32].

1. The LFSR is initially loaded with the key and IV as: $S_7\|S_6\|\dots\|S_0 = 10\|11\|11\|00\|01\|10\|00\|11$. Assuming $R_1 = 00$, $R_2 = 00$, the keystream generated after the first round:

$$W = [(S_{7H}S_{6L} \oplus 00) \boxplus 00] \oplus S_{1L}S_{0H} = [(11 \oplus 00) \boxplus 00] \oplus 01 = 10.$$

2. In the second round, the LFSR is updated as $S_8 = S_0 + 2S_1 + S_5 + S_7$ mod (3), which is $11 + 00 + 11 + 10 \mod (3) = 10$. Thus, the updated LFSR can be described as: $S_7\|S_6\|\dots\|S_0 = 10\|10\|11\|11\|00\|01\|10\|00$. Simultaneously, W_1, W_2 are updated as $W_1 = R_1 \boxplus S_{5L}S_{4H} = 00 \boxplus 10 = 10$ and $W_2 = R_2 \oplus S_{3L}S_{2H} = 00 \oplus 11 = 11$. Hence, the updated values of R_1 and R_2 are given by $R_1 = S\left(L(W_{1L}W_{2H})\right) = S\left(L(01)\right) = S(10) = 10$, and $R_2 = S\left(L(W_{2L}W_{1H})\right) = S\left(L(11)\right) = S(11) = 00$. Therefore, the keystream generated after the second round is given by:

$$W = [(S_{7H}S_{6L} \oplus 10) \boxplus 00] \oplus S_{1L}S_{0H} = [(10 \oplus 10) \boxplus 00] \oplus 00 = 00.$$

3. In the third round, the LFSR is updated as $S_8 = S_0 + 2S_1 + S_5 + S_7$ mod (3), which is $00 + 01 + 11 + 10 \mod (3) = 11$. Thus, the updated LFSR can be described as: $S_7\|S_6\|\dots\|S_0 = 11\|10\|10\|11\|11\|00\|01\|10$. Simultaneously, W_1 and W_2 are updated as $W_1 = R_1 \boxplus S_{5L}S_{4H} = 10 \boxplus 11 = 01$ and $W_2 = R_2 \oplus S_{3L}S_{2H} = 00 \oplus 00 = 00$. Hence, the updated values of R_1 and R_2 are given by $R_1 = S\left(L(W_{1L}W_{2H})\right) = S\left(L(10)\right) = S(01) = 11$, and $R_2 = S\left(L(W_{2L}W_{1H})\right) = S\left(L(00)\right) = S(00) = 01$. Therefore, the keystream generated after the third round is given by:

$$W = [(S_{7H}S_{6L} \oplus 11) \boxplus 01] \oplus S_{1L}S_{0H} = [(10 \oplus 11) \boxplus 01] \oplus 11 = 01.$$

4. Similarly, running the Toy-ZUC for another round, the LFSR is updated as $S_8 = S_0 + 2S_1 + S_5 + S_7 \mod (3)$, which is $10 + 10 + 10 + 11 \mod (3) = 00$. The updated LFSR can be described as: $S_7 \| \ldots \| S_0 = 00 \| 11 \| 10 \| 10 \| 11 \| 11 \| 00 \| 01$. Simultaneously, W_1 and W_2 are updated as $W_1 = R_1 \boxplus S_{5L} S_{4H} = 11 \boxplus 01 = 00$ and $W_2 = R_2 \oplus S_{3L} S_{2H} = 01 \oplus 10 = 11$. Hence, the updated values of R_1 and R_2 are given by $R_1 = S(L(W_{1L}W_{2H})) = S(L(01)) = S(10) = 10$, and $R_2 = S(L(W_{2L}W_{1H})) = S(L(10)) = S(01) = 11$. Therefore, the keystream generated after the third round is given by:

$$W = [(S_{7H} S_{6L} \oplus 10) \boxplus 11] \oplus S_{1L} S_{0H} = [(01 \oplus 10) \boxplus 11] \oplus 00 = 01.$$

Hence, after running the Toy-ZUC four rounds, we obtain the keystream 01010010. The Toy-ZUC was implemented in Qiskit using 30 qubits, $(9*7+50) = 113$ CNOT gates, and $(9*3+7) = 34$ Toffoli gates. The quantum circuit has been executed using the *qasm_simulator*, and the observed keystream matched the theoretical result. The corresponding histogram is presented in Fig. 8.

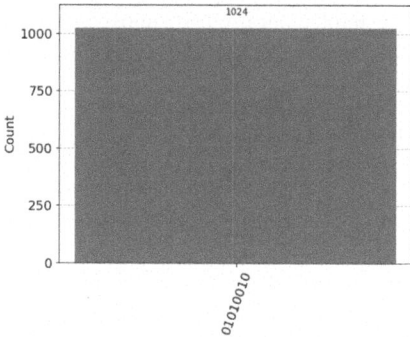

Fig. 8. Keystream obtained from running four rounds of Toy-ZUC using the *qasm_simulator*

Note that while running Grover's search on Toy-ZUC, one can start with an equal superposition of all possible keys by using Hadamard gates on the qubits initialized to $|0\rangle^8$. After running the cipher for four rounds, the Boolean function $f : \{0,1\}^8 \to \{0,1\}$, defined as $f(01010010) = 1$ and 0 otherwise, can be used as Grover's oracle to recover the desired key.

References

1. The ZUC-256 Stream Cipher (2018). http://www.is.cas.cn/ztzl2016/zouchongzhi/201801/W020180416526664982687.pdf
2. ETSI/SAGE Technical Report, Version 2.0. Specification of the 3GPP confidentiality and integrity algorithms 128-EEA3 and 128-EIA3 - Document 4: design and evaluation report (2011). https://www.gsma.com/about-us/wp-content/uploads/2014/12/EEA3_EIA3_Design_Evaluation_v2_0.pdf

3. Amy, M., Maslov, D., Mosca, M., Roetteler, M.: A meet-in-the-middle algorithm for fast synthesis of depth-optimal quantum circuits. IEEE Trans. Comput.-Aided Design Integrat. Circuits Syst. **32**(6), 818–830 (2013). https://doi.org/10.1109/TCAD.2013.2244643

4. Anand, R., Maitra, A., Maitra, S., Mukherjee, C.S., Mukhopadhyay, S.: Quantum resource estimation for FSR based symmetric ciphers and related Grover's attacks. In: Adhikari, A., Küsters, R., Preneel, B. (eds.) INDOCRYPT 2021. LNCS, vol. 13143, pp. 179–198, Springer, Cham (2021). https://doi.org/10.1007/978-3-030-92518-5_9

5. Bathe, B., Anand, R., Dutta, S.: Evaluation of Grover's algorithm toward quantum cryptanalysis on ChaCha. Quant. Inf. Process. **20**(12), 1–19 (2021). https://doi.org/10.1007/s11128-021-03322-7

6. Chun, M., Baksi, A., Chattopadhyay, A.: DORCIS: depth optimized quantum implementation of substitution boxes (2023). ePrint: https://eprint.iacr.org/2023/286

7. Cuccaro, S.A., Draper, T.G., Kutin, S.A., Moulton, D.P.: A new quantum ripple-carry addition circuit. arXiv preprint arXiv:quant-ph/0410184 (2004)

8. Dasu, V.A., Baksi, A., Sarkar, S., Chattopadhyay, A.: LIGHTER-R: optimized reversible circuit implementation for sboxes. In: The 32nd IEEE International System-on-Chip Conference, pp. 260–265 (2019). https://doi.org/10.1109/SOCC46988.2019.1570548320

9. Deutsch, D., Jozsa, R.: Rapid solution of problems by quantum computation. Proc. Roy. Soc. Lond. **439**(1907), 553–558 (1992). https://doi.org/10.1098/rspa.1992.0167

10. Ekdahl, P., Johansson, T.: SNOW - a new stream cipher. In: Proceedings of first NESSIE Workshop, Belgium (2000). https://api.semanticscholar.org/CorpusID:8172481

11. Grassl, M., Langenberg, B., Roetteler, M., Steinwandt, R.: Applying Grover's algorithm to AES: quantum resource estimates. In: Takagi, T. (eds.) Post-quantum Cryptography, PQCrypto 2016. LNCS, vol. 9606, pp. 29–43, Springer, Cham (2016). https://doi.org/10.1007/978-3-319-29360-8_3

12. Grover, L.K.: A fast quantum mechanical algorithm for database search. In: Proceeding of the 28th ACM Symposium on Theory of Computing, STOC 1996, pp. 212–219 (1996). https://doi.org/10.1145/237814.237866

13. Jang, K., Baksi, A., Breier, J., Seo, H., Chattopadhyay, A.: Quantum implementation and analysis of Default. Cryptogr. Commun. (2023). https://doi.org/10.1007/s12095-023-00666-y

14. Liu, F., Meier, W., Sarkar, S., Wang, G., Ito, R., Isobe, T.: New cryptanalysis of ZUC-256 initialization using modular differences. IACR Trans. Symmet. Cryptol. **2022**(3), 152–190 (2022). https://doi.org/10.46586/tosc.v2022.i3.152-190

15. Miller, D.M., Maslov, D., Dueck, G.W.: A transformation based algorithm for reversible logic synthesis. In: Proceedings of the 40th annual Design Automation Conference, DAC 2003, pp. 318–323 (2003). https://doi.org/10.1145/775832.775915

16. Design and Cryptanalysis of ZUC. SCSSN. Springer, Singapore (2021). https://doi.org/10.1007/978-981-33-4882-0

17. Nielsen, M.A., Chuang, I.L.: Quantum computation and quantum information. In: 10th Anniversary Edition, Cambridge University Press (2011). https://doi.org/10.1017/CBO9780511976667

18. NIST. Submission requirements and evaluation criteria for the postquantum cryptography standardization process (2016). http://www.nist.gov/pqcrypto

19. Patel, K.N., Markov, I.L., Hayes, J.P.: Optimal synthesis of linear reversible circuit. Quant. Inf. Comput. **8**(3), 282–294 (2008). https://doi.org/10.26421/QIC8.3-4-4

20. Selinger, P.: Quantum circuits of T-depth one. Phys. Rev. A **87**(4) (2013). https://doi.org/10.1103/PhysRevA.87.042302

21. Selinger, P.: Efficient Clifford+T approximation of single-qubit operators. Quant. Inf. Computat. **15**(1–2), 159–180 (2015). https://doi.org/10.26421/QIC15.1-2-10

22. Shor, P.W.: Polynomial-time algorithms for prime factorization and discrete logarithms on a quantum computer. SIAM J. Comput. **26**(5) (1997). https://doi.org/10.1137/S0097539795293172

23. Simon, D.R.: On the power of quantum computation. SIAM J. Comput. **26**(5), 1474–1483 (1997). https://doi.org/10.1137/S0097539796298637

24. Soeken, M., Frehse, S., Wille, R., Drechsler, R. RevKit: an open source toolkit for the design of reversible circuits. In: De Vos, A., Wille, R. (eds.) RC 2011. LNCS, vol. 7165, pp. 64–76. Springer, Heidelberg (2012). https://doi.org/10.1007/978-3-642-29517-1_6

25. Takahashi, Y., Tani, S., Kunihiro, N.: Quantum addition circuits and unbounded fan-out. Quant. Inf. Comput. **10**(9), 872–890 (2010). arXiv:0910.2530

26. Wiebe, N., Roetteler, M.: Quantum arithmetic and numerical analysis using repeat-until-success circuits. Quant. Inf. Computat. **16**(1–2), 134–178 (2016). arxiv:1406.2040

27. Wille, R., et al.: The MQT handbook: a summary of design automation tools and software for quantum computing (2024). arXiv:2405.17543

28. Wu, H., Huang, T., Nguyen, P.H., Wang, H., Ling, S.: Differential attacks against stream cipher ZUC. In: Wang, X., Sako, K. (eds.) ASIACRYPT 2012, LNCS, vol. 7658, pp. 262–277. Springer, Heidelberg (2012). https://doi.org/10.1007/978-3-642-34961-4_17

29. Yang, J., Johansson, T., Maximov. A.: Spectral analysis of ZUC-256. IACR Trans. Symmet. Cryptol. **2020**(1), 266–288 (2020). https://doi.org/10.13154/tosc.v2020.i1.266-288

30. Zou, J., Wei, Z., Sun, S., Liu, X., Wu, W.: Quantum circuit implementations of AES with fewer qubits. In: Moriai, S., Wang, H. (eds.) ASIACRYPT 2020, LNCS, vol. 12492, pp. 697–726. Springer, Cham (2020). https://doi.org/10.1007/978-3-030-64834-3_24

31. ZUC Design Team. An addendum to the ZUC-256 stream cipher (2021). ePrint: https://eprint.iacr.org/2021/1439

32. GitHub. https://github.com/sumand96/ZUC-Quantum-cryptanalysis

A Parametric Class of Mutually Unbiased Bases Using Resolvable Block Designs

Ajeet Kumar, Rakesh Kumar, and Subhamoy Maitra[✉]

Applied Statistics Unit, Indian Statistical Institute,
203 B T Road, Kolkata 700 108, India
subho@isical.ac.in

Abstract. Mutually Unbiased Bases (MUBs) have important applications in several domains, particularly in Quantum Cryptography where the bases are used in designing the protocols related to Quantum Key Distribution (QKD). In this paper we consider parameterization of MUBs so that one can explore several classes of them for various applications. For dimension $d = s^2$, we present the construction of affine-parametric classes with $MOLS(s) + 2$ many MUBs, where $MOLS(s)$ is the number of Mutually Orthogonal Latin Squares of dimension s. If s is a power of prime, then $MOLS(s) = s - 1$, and the number of MUBs will be $s + 1$. Considering the first one to be the identity matrix, in our construction, each of the rest $MOLS(s) + 1$ MUBs will have at least $s(s - 1)$ free parameters, that cannot be absorbed by a global unitary operation. In comparison to Goyeneche et al.'s paper (2015), our result produces larger number of MUBs as well as free parameters in most of the cases. This can help in exploring various choices of MUBs in the protocols for higher dimensional QKDs and other applications of MUBs related to quantum information.

Keywords: Quantum Cryptography · Quantum Key Distribution · Combinatorial Structures · Mutually Unbiased Bases · Parallel class · Resolvable block design

1 Introduction

The problem of determining the maximum number of mutually unbiased bases (MUBs) for any given dimension d, has been a problem of interest in the domain of quantum information theory for decades. Introduced by Schwinger [13] in the year 1960, MUBs are now widely used in different domains, including quantum state discrimination, quantum error correction, quantum key distribution, dense coding, entanglement swapping, covariant cloning, etc. In particular, we need to consider the applications of MUBs in Quantum Key Distribution (QKD) protocols [1,3], which are the most important part of Quantum Cryptography. To present the problem, let us explain the scenario as follows.

© The Author(s), under exclusive license to Springer Nature Switzerland AG 2025
S. Mukhopadhyay and P. Stănică (Eds.): INDOCRYPT 2024, LNCS 15495, pp. 356–373, 2025.
https://doi.org/10.1007/978-3-031-80308-6_16

One can relate 'ket' with a unit vector, that can be written as $|v\rangle$. Such a vector belongs to some complex vector space, say V and physically can be understood as a quantum bit or qubit. One can define the inner product between two d-dimensional complex vectors $|u\rangle = (u_1, \ldots, u_d), |v\rangle = (v_1, \ldots, v_d)$ as $\langle u|v\rangle = u_1 v_1^* + \ldots + u_d v_d^*$, where v_i^* is complex conjugate transpose of v_i and this is a complex number. The modulus of a complex number $z = x + iy$ is $\sqrt{x^2 + y^2}$. The angle between two vectors $|u\rangle$ and $|v\rangle$ is related to absolute value of their inner product $|\langle u|v\rangle|$. With this as background, let us present the definition of MUBs.

Definition 1. Two orthonormal bases in the d-dimensional complex Hilbert space $\mathbb{C}^d$, $\{|e_1\rangle, \ldots, |e_d\rangle\}$ and $\{|f_1\rangle, \ldots, |f_d\rangle\}$ are called Mutually Unbiased if

$$|\langle e_i|f_j\rangle| = \frac{1}{\sqrt{d}}, \ \forall i, j \in \{1, 2, \ldots, d\}.$$

Similarly, some r orthonormal bases are called Mutually Unbiased Bases (MUBs) if they are pairwise Mutually Unbiased.

An example will be useful here for $d = 2$. Consider the bases

$$M_0 = \{|0\rangle, |1\rangle\},$$
$$M_1 = \left\{ \frac{|0\rangle + |1\rangle}{\sqrt{2}}, \frac{|0\rangle - |1\rangle}{\sqrt{2}} \right\},$$
$$M_2 = \left\{ \frac{|0\rangle + i|1\rangle}{\sqrt{2}}, \frac{|0\rangle - i|1\rangle}{\sqrt{2}} \right\}$$

if you consider two vectors $|u\rangle, |v\rangle$ from a specific basis M_i, then naturally $|\langle u|v\rangle| = 0$. On the other hand, if $|u\rangle$ is from M_i, and $|v\rangle$ is from M_j, with $i \neq j$, then $|\langle u|v\rangle| = \frac{1}{\sqrt{2}}$. Note that, the bases M_0, M_1 are used in the BB84 QKD protocol [1], involving four quantum states, whereas M_0, M_1, M_2 are used when the protocol is extended with the six states [3].

Since its inception, it has been shown that finding MUBs for any given dimension d has a direct connection with discrete mathematics, especially with combinatorics [9,15]. A major result in this direction by Wootters et al. [15] is that, in a complex Hilbert space of dimensions d, the maximum possible number of mutually unbiased bases is $(d+1)$, provided they exist. For Hilbert space $\mathbb{C}^d$, the existence of $d+1$ many MUBs makes the set complete. Whenever $d = p^k$ where p is some prime and $k \in \mathbb{N}$, the construction of the complete set of MUBs is known from the work of Wootters et al. [15] using the properties of Galois fields. However, constructing complete sets of MUBs for Hilbert space, whose dimensions are not prime power, remains open for decades. The only known generic lower bound on the number of MUBs is $p^r + 1$, where p^r is the smallest prime power in the prime decomposition of the dimension (this lower bound is shown using tensor products of the Wootters-Fields construction). While in certain dimensions this lower bound has been improved [16], there exists no composite dimension (except single prime power) in which the exact number of MUBs is known. For the smallest composite dimension, six, the number of MUBs is known to be no more than seven and no less than three, from the general bounds. However, which of the numbers in between is the exact number of MUBs in dimension six is unknown (apart from the fact that it cannot be six, following from a general result by Weiner [17]). It was first conjectured by Zauner in 1999 that there are no more than three MUBs in dimension six [18], and this conjecture has not been resolved to date, despite substantial efforts.

With this background and context, let us now explain the problem that we are considering here. In [6], the following two points have been shown.

- Any set of m real MUBs existing in dimension $N > 2$ can admit the introduction of $\frac{(m-1)N}{2}$ free parameters that cannot be absorbed by a global unitary operation.
- As consequence, there are $m = k+1$ mutually unbiased bases in every dimension $N = k^2$ with $\frac{k^3}{2}$ free parameters, where k is even.

In the second case, the statement needs to be corrected as for dimension $N = k^2$, it is not guaranteed to have $k + 1$ MUBs. This is only possible when k is a prime power [16, Sect. 5]. For example, taking $N = 6^2 = 36$, we can only construct 5 MUBs which is the lower bound, and no other better result is known to the best of our knowledge. Thus, here obtaining $m = k+1 = 6+1 = 7$ MUBs are not possible. We believe that the authors of [6] actually meant the situation when k is a prime power of 2. If k is a power of an odd prime, then MUBs will not be real. In fact, we will not get even a pair of MUBs. But if k is a power of even prime, say $k = 2^e$ then $k+1$ MUBs will be real through MOLS construction [16], because there exists a real Hadamard matrix of order $k = 2^e$.

Now, related to the Mutually Orthogonal Latin Squares (MOLS), let us outline the constructions for MUBs. We take the dimension $d = s^2$, and consider the Resolvable Block Design (RBD) having s^2 entities. To construct RBD, we start with Mutually Orthogonal Latin Square (MOLS) of order s. For each MOLS of order s, we have $MOLS(s) + 2$ parallel classes in the RBD of order s^2. We convert each of the parallel classes into orthonormal bases using the Hadamard matrix of order s. If there exists a real Hadamard matrix of order s then we will get only real MUBs. However, if we do not have the real ones, then we need to consider a complex Hadamard matrix of order s, which is guaranteed in every dimension. Using complex Hadamard matrices of order s, we will obtain complex MUBs.

It is well known that $N_{MOLS} \leq s - 1$ for all s, where N_{MOLS} is the number of MOLS of order s [16]. If this bound is attained we say that there is a *complete* set of mutually orthogonal Latin squares of order s. A construction of complete sets of MOLS of order s is known if s is a prime power [8, Theorem 8.3]. For any other value, no construction for a complete set of Mutually Orthogonal Latin Squares is known so far. In fact, if s is not a prime power, the largest value of N_{MOLS} of order s is usually considerably smaller than $s - 1$, and for several values it is known that $s - 1$ MOLS of order s cannot exist. The minimum value of N_{MOLS} is known to be 2 for all s except for $s = 2$ or 6 [16]. The smallest case for which the exact number of $MOLS(s)$ is not known is for $s = 10$ [16].

One should note that when $s = p^e$, a power of prime, then $d = p^{2e}$, and $d+1$ MUBs can be constructed [15]. On the other hand, the number of MUBs we are considering here is much lesser, almost square-root of that. However, here and also in [6], the motivation is to introduce free parameters in the MUBs, and thus we explore lesser number of MUBs, but with larger number of free parameters.

1.1   Organization and Contribution

In this paper, we consider introduction of free parameters in the sets of MUBs. Note that, in our case, we consider the dimension to be d, with $d = s^2$. Thus, written in this manner, the contribution of [6] tells that any set of m real MUBs existing in dimension $d > 2$ can admit the introduction of $\frac{(m-1)d}{2} = \frac{(m-1)s^2}{2}$ free parameters that cannot be absorbed by a global unitary operation. In our construction, we obtain $MOLS(s) + 2$ MUBs, which is independent of the existence of m real MUBs. We consider the first MUB in our construction as the identity matrix. Each of the rest $MOLS(s) + 1$ MUBs has $s(s-1)$, i.e., in total $(MOLS(s)+1)s(s-1)$ free parameters. In case, when $MOLS(s) = s-1$, the total number of free parameters is $s^2(s-1)$. These results are presented in Sect. 2. Exact comparisons with the existing works and that of [6] in particular are presented in Subsect. 2.3. Section 3 concludes the paper. Before proceeding further, let us now present some existing results in this direction.

1.2   Notation, Definition and Background

The concepts of combinatorial block design, in particular Resolvable Block Design(RBD) and its basic characteristics can be found concisely at Sect. 2.1 of [9] and Sect. 2.1 of [10].

A d-dimensional complex-square matrix U is called Unitary if U satisfies $U^\dagger U = UU^\dagger = I$, where $U^\dagger$ denotes the conjugate-transpose of U and I denotes Identity matrix. When Unitary matrix is a diagonal matrix then we denote them by D where non zero diagonal entries are $(D)_{ii} = \frac{\exp(\iota\theta_i)}{\sqrt{d}}$ for some θ_i. A special class of Unitary matrix is Permutation matrix, which we denote by P, which has entries from set $\{0,1\}$, such that every row and column contain exactly one non zero entry. I is an example of permutation matrix. For permutation matrix we have $P^T P = I$. Two unitary matrices U_1 and U_2 of equal dimension, are said to be equivalent ($U_1 \approx U_2$) if there exist diagonal unitary matrices D_1, D_2 and permutation matrices P_1, P_2 such that $U_1 = D_1 P_1 U_2 P_2 D_2$. Note that this is an *equivalence relation*.

There is a specific class of unitary matrices, known as the Hadamard matrix, which we will denote by H. An d-dimensional unitary matrix $H = (h_{ij}) : 1 \leq i,j \leq d$ is called Hadamard if $|h_{ij}| = \frac{1}{\sqrt{d}}$ for all $i,j \in \{0,1,\ldots,d\}$. Since multiplication of arbitrary phase factor to all the elements of the row or column does not change the Hadamard property of the matrix, hence any Hadamard matrix is equivalent to a Hadamard matrix, having all its entries of first row and first column as 1. Such Hadamard matrix is called a dephased Hadamard matrix.

If there are independent parameters in a Hadamard matrices, then there exist continuous of inequivalent Hadamard matrix. Since absolute value of all the entries of Hadamard matrix is one, hence the independent parameter occurs in the phase of the entries of Hadamard matrix. All the Hadamard matrix lying on this continuum is called family of Hadamard matrix , stemming from a starting point, which is by convention taken as the dephased Hadamard matrix i.e. all the entries of first row and first column is 1. The family is Affine if there exist a set of $H(\mathcal{R})$, stemming from a dephased Hadamard matrix, associated with the subspace $\mathcal{R}$ of the real space of $d \times d$ matrix, with zero in the first row and column such that $H(\mathcal{R}) = \{H \circ \exp(\iota R) : R \in \mathcal{R})$. Here $\circ$ denotes Hadamard product and $R \in \mathbb{R}^{d^2}$. We say that Hadamard matrix $H(\mathcal{R})$ an $m-$ parameters affine family if R contains m free parameters and generate an $m-$dimensional subspace of $\mathcal{R}$ i.e. $dim(R) = m$. Thus, this implies that the phases are the linear combination of the variables. And if m is the dimension of the subspace of R generated by the linear combination of the variables, then H is called $m-$ parameters affine family. Note that m is essentially number of independent linear combinations of the variables occurring in the phases of the entries of H. For more on Affine families of Hadamard matrices one may refer Section II of [5, 12]

A pair of Hadamard matrix (H_i, H_j) is called mutually unbiased if

$$\frac{1}{\sqrt{d}} H_i^\dagger H_j = H_k$$

where H_k is another Hadamard matrix. A set $\{H_1, H_2, \ldots H_r\}$ of r Hadamard matrix is Mutually Unbiased Hadamard Matrices (MUHM) if every pair is unbiased. A MUB matrix will be denoted by M, which is a matrix whose column would consist of basis vectors of a MUB. Hence M is also Unitary matrix. Through this paper we will denote MUBs by their MUB matrices. The diagonal unitary matrix having affine parametric phase factors, would be denoted by $D(\theta)$, where θ is added to indicate that phase factors are variables.

The Basic Idea from [9]. Our construction follows the method provided in [9] using *Resolvable Block Designs* (RBD). For more details about RBD see [9] and the references therein. To proceed, we need to refer [9, Construction 1] first. We follow the same notation as in [9] and present the necessary background. In this direction, the construction of an orthonormal basis using a parallel class from an RBD (X, A), is as follows.

Construction 1.

1. In a design (X, A), choose the elements of X as any set of orthonormal basis vectors of $\mathbb{C}^d$. That is, if $|X| = d$, then $X = \{|\psi_1\rangle, |\psi_2\rangle, \ldots, |\psi_d\rangle\}$, such that $\langle \psi_i | \psi_j \rangle = \delta_{ij}$. Hence A, which contains blocks made out of the elements from X, would now consist of blocks with the elements from the set of chosen orthonormal basis vectors.

2. Let $B = \{b_1, b_2, \ldots, b_s\}$ be one of the parallel class of the design (X, A), where b_i's are disjoint blocks containing elements from X. Since B is a parallel class, this implies $X = b_1 \cup b_2 \cup \ldots \cup b_s$, and $b_i \cap b_j = \phi$ for all $1 \le i \ne j \le s$.

3. Consider one of the blocks $b_r = \{|\psi_{r_1}\rangle, |\psi_{r_2}\rangle, \ldots, |\psi_{r_{n_r}}\rangle\} \in B$ and let $|b_r| = n_r$. Corresponding to this block, choose any $n_r \times n_r$ unitary matrix whose elements are say u_{ij}^r, $i, j = 1, 2, \ldots, n_r$.

4. Next construct n_r many vectors in the following manner, using b_r and u_{ij}^r.

$$|\phi_i^r\rangle = u_{i1}^r|\psi_{r_1}\rangle + u_{i2}^r|\psi_{r_2}\rangle + \ldots + u_{in_r}^r|\psi_{r_{n_r}}\rangle = \sum_{k=1}^{n_r} u_{ik}^r|\psi_{r_k}\rangle : i = 1, 2, \ldots, n_r.$$

5. In a similar manner, corresponding to each block $b_j \in B$, construct n_j many vectors where $|b_j| = n_j$, using any $n_j \times n_j$ unitary matrix. Since $\sum_{j=1}^{s} n_j = d$, we will get exactly d many vectors.

Let us construct the matrix M_B of size $d \times d$ having column vectors as $|\phi_i^r\rangle$. Therefore, $M_B = (|\phi_1^1\rangle, \ldots, |\phi_{n_1}^1\rangle, |\phi_1^2\rangle, \ldots, |\phi_{n_2}^2\rangle, \ldots, |\phi_1^s\rangle, \ldots, |\phi_{n_s}^s\rangle)$. From [9, Lemma 1], the $|\phi_i^r\rangle$'s corresponding to a parallel class B of X form an orthonormal set of basis vectors. Hence M_B is a unitary matrix. In this regard, we have the following lemma as per Construction 1. Later Example 1 provides a more detailed view on this.

Lemma 1. *Refer to Construction 1 and the unitary matrix M_B above. If X consists of the computational basis vectors, then $M_B = P_B H$, where P_B is a permutation matrix of size $d \times d$ and H is a block diagonal matrix consisting of unitary matrices of size $n_j \times n_j$ $(j = 1, 2, \ldots, s)$ as block matrices.*

Proof. Using Construction 1, we obtain the set of orthonormal basis vectors $|\phi_i^l\rangle$'s over $d-$ dimensional vector space, by choosing elements of X as computational basis vectors. That is, $X = \{|i\rangle : i = 1, 2, \ldots, d\}$, where $|i\rangle$ is a column vector of size $d \times 1$ which has all the entries as zero except the ith one, which is 1.

Let us refer to [9, Construction 1]. From each block B_l of size n_l, one gets of n_l many basis vectors. These basis vectors can be arranged as columns of $d \times n_l$ size matrix i.e., $[|\phi_1^l\rangle \ldots |\phi_{n_l}^l\rangle]$. From construction this matrix can be written as

$$\left[|\phi_1^l\rangle \ldots |\phi_{n_l}^l\rangle\right] = \left[|\psi_{l_1}\rangle |\psi_{l_2}\rangle \ldots |\psi_{l_{n_l}}\rangle\right] H^{l\dagger},$$

where $H^{l\dagger}$ is the Hadamard matrix of size $n_l \times n_l$ and $|\psi_{l_i}\rangle \in X$. Now concatenating all the column vectors constructed from different blocks, which are s in numbers, we have the unitary matrix M_B, where

$$M_B = \left[|\phi_1^1\rangle \ldots |\phi_{n_1}^1\rangle |\phi_1^2\rangle \ldots |\phi_{n_2}^2\rangle \ldots |\phi_1^s\rangle \ldots |\phi_{n_s}^s\rangle\right] =$$

$$\left[|\psi_{1_1}\rangle \ldots |\psi_{1_{n_1}}\rangle |\psi_{2_1}\rangle \ldots |\psi_{2_{n_2}}\rangle \ldots |\psi_{s_1}\rangle \ldots |\psi_{s_{n_s}}\rangle\right] \begin{pmatrix} H^{1\dagger} & 0 & \ldots & 0 & 0 \\ 0 & H^{2\dagger} & \ldots & 0 & 0 \\ \vdots & \vdots & \ddots & \vdots & \vdots \\ 0 & 0 & \ldots & H^{(s-1)\dagger} & 0 \\ 0 & 0 & \ldots & 0 & H^{s\dagger} \end{pmatrix}.$$

Given that $|\phi_i\rangle$'s are the orthonormal basis vectors, we have the unitary matrix $M_B = \left[|\phi_1^1\rangle \ldots |\phi_{n_1}^1\rangle |\phi_1^2\rangle \ldots |\phi_{n_s}^s\rangle\right]$. This is also evident from above as it is a product of two unitary matrices. Note that $|\psi_{l_i}\rangle \in X$, which are the columns of computational basis vectors. Since the blocks of the parallel class forms the partition of X, hence each computational basis vector appears exactly once in any one of the blocks. Thus $\left[|\psi_{1_1}\rangle \ldots |\psi_{1_{n_1}}\rangle |\psi_{1_1}\rangle \ldots |\psi_{1_{n_1}}\rangle \ldots |\psi_{1_1}\rangle \ldots |\psi_{1_{n_1}}\rangle\right]$ is a permutation matrix and let us name it as P_B, which is decided by the parallel class $\mathcal{P}$. Since $H^{j^\dagger}$ are unitary matrix, we have $M_B = P_B H$, where H is a block diagonal Hadamard matrix where each block consists of a Hadamard matrix whose size is equal to the size of the blocks of the parallel class and P is a permutation matrix, which is decided by the elements in the blocks of the parallel class. $\qquad\square$

Thus we see that the number of unitary matrices, constructed using the RBDs, is dependent on number of parallel classes. In this regard, we have the existing theorem from [10], which in turn follows from the result of [16] which we present here.

Theorem 1. *Consider an RBD* (X, A) *such that* $|X| = s^2$, *then one can construct* $MOLS(s) + 2$ *many parallel classes, each having* s *many blocks of size* s *and any two blocks from different parallel classes will have exactly one point in common.*

We also need the following technical results.

Lemma 2. *Let* M_1 *and* M_2 *be a pair of MUBs in* d-*dimensional vector space. Then* $M_1^\dagger M_2$ *is a Hadamard matrix.*

Proof. Let $M_1 = [v_1\, v_2\, \ldots v_d]$ and $M_2 = [w_1\, w_2\, \ldots w_d]$. Then, from Definition 1, $(M_1^\dagger M_2)_{i,j} = \overline{v_i}.w_j = \frac{\exp(\iota\phi_{ij})}{\sqrt{d}}$ for some ϕ_{ij} called phase factor. Further since M_1 and M_2 being unitary matrix, therefore $M_1^\dagger M_2$ is also a unitary. Hence $M_1^\dagger M_2$ is a Hadamard matrix. $\qquad\square$

Note that, if H is a Hadamard matrix, then it is also an MUB with respect to the Identity matrix. Since we can generate a pair of MUBs in parametric form using Construction 1, one can use this idea to construct a Hadamard matrix in parametric form for any composite dimension. We now state the following technical results that will be used later for parameterization.

Lemma 3. *If* D_1 *is a diagonal matrix and* P_1 *is a permutation matrix then* $D_1 P_1 = P_1 D_2$ *for some diagonal matrix* D_2, *having the same diagonal entries as that of* D_1.

Proof. For a permutation matrix, we have $P_1^{-1} = P_1^T$. Hence $D_2 = P_1^T D_1 P_1$. Thus if P_1 is modified by changing the lth column to the kth column then P_1^T will also be modified by the movement of the lth row to the kth row. Thus $(D_2)_{kk} = (D_1)_{ll}$, and hence D_2 will be a diagonal matrix having the same diagonal entries as in D_1. In fact, if P_1 is represented by σ, then $(D_2)_{\sigma(l)\sigma(l)} = (D_1)_{ll}$. $\qquad\square$

Lemma 4. *For any square matrix M if $MD_1 = D_2M$, where D_1 and D_2 are diagonal matrices, then $D_1 = D_2 = \alpha I$, where α is some constant.*

Proof. Since D_1 and D_2 are diagonal matrices, they can be expressed as

$$D_1 = \mathrm{diag}(d_{11}, d_{12}, \ldots, d_{1n}), \quad D_2 = \mathrm{diag}(d_{21}, d_{22}, \ldots, d_{2n}),$$

Since we have, $MD_1 = D_2M$, consider the $(ij)^{th}$ elements for both sides. We will have,

$$(MD_1)_{ij} = M_{ij}d_{1j} = d_{2i}M_{ij} = (D_2M)_{ij}.$$

This equation implies that either $M_{ij} = 0$ or $d_{1j} = d_{2i}$. Since M is any square matrix, so we have $d_{1j} = d_{2i}$ for all ij's. Hence all the diagonal elements are constant, say α. So, $D_1 = D_2 = \alpha I$. □

From the above, we immediately have the following result..

Corollary 1. *For a general block diagonal matrix M_B if $M_B D_1 = D_2 M_B$, where D_1 and D_2 is a diagonal matrix then $D_1 = D_2 = D_I$ where D_I is a block diagonal matrix, such that non zero block matrices are of the form $\alpha_r I_r$. The size of the blocks is equal to the size of the blocks in the matrix M_B and α_r is a constant for each block.*

2 Construction of Affine Parametric Form of MUBs for Square Dimension, $d = s^2$

If $|X| = d = s^2$, then we can construct RBD (X, A) such that it has $MOLS(s)+2$ many parallel classes, each having s many blocks of size s such that blocks from different parallel classes will have exactly one point in common. Suppose, the $MOLS(s) + 2$ many parallel classes are denoted by $C_1, C_2, \ldots, C_w$.

Then, the above construction method would produce $MOLS(s)+2$ number of MUBs. Moreover, when $s = p$, a prime number, $MOLS(p) = p - 1$ and in such a scenario, one can construct a $(p+1)$ many MUBs of dimension $p^2 = d$. Toward the end of this Section, we estimate the maximum permissible free parameters in the newly constructed set of MUBs. Let us now describe the following construction in line with Theorem 1 (see [10] for more details).

Construction 2. Let (X, A) be an RBD such that $|X| = d = s^2$ for some $s \in \mathbb{N}$. Suppose, the set X is represented as $X = \{1, 2, \ldots, s^2\}$ and the $MOLS(s) + 2$ many parallel classes are denoted by $A = \{C_0, C_1, \ldots, C_w\}$, where each C_i represents a partition of X. Assuming each element $k \in X$ as a d dimensional vector, having only non-zero entry as 1 at the kth position, each block of a parallel class turns out to be a $d \times s$ dimensional matrix. Concatenation of s many such matrices corresponding to a parallel class provides a $d \times d$ dimensional matrix for the same parallel class C_i. Following the construction method, discussed before, we can construct the set of $w + 1$ many sparse MUBs [9]. Denoting the set as $\{M_0, M_1, \ldots, M_w\}$, where $M_i = P_i \mathbb{H}_i$. Here, P_i is the d dimensional permutation matrix having 1's at the $(m, n)^{th}$ position if and only if m is the nth entry in the parallel class C_i, with $1 \leq n \leq s^2$.

Here $w = \text{MOLS}(s) + 1$. The orthonormal bases constructed form MUBs, because any pair of blocks from two different parallel classes have exactly one element in common, which is called the intersection number (μ). The intersection number plays a critical role in construction of the approximate MUBs using RBDs [9,11]. Note that P_i is determined by the parallel classes C_i. Since the first entry of the first block of every parallel class is 1, we have $(P_i)_{11} = 1, \forall i$. Further, $\mathbb{H}_i$ is a block diagonal matrix, where each block matrix consists of a Hadamard matrix of order s.

Now using the set of $w + 1$ MUBs, we obtain a set of w many Mutually Unbiased Hadamard matrices $\{H_1, \ldots, H_w\}$ where $H_i = M_0^\dagger M_i = \mathbb{H}_0^\dagger P_0^T P_i \mathbb{H}_i$, where $P_0^\dagger = P_0^T$. This follows from Lemma 2. Note that starting with I_0 the identity matrix, it forms a set of $w + 1$ many MUBs.

Note that when all the Hadamard matrices in $\mathbb{H}$ are identical, say H, then $\mathbb{H} = \mathbb{I}_s \otimes H$ and hence we have

$$M_i = P_i \left(\mathbb{I}_s \otimes H \right),$$

where $\mathbb{I}_s$ denotes the identity matrix of order s. In this situation we have $H_i = \left(\mathbb{I}_s \otimes H^\dagger \right) P_0^T P_i \left(\mathbb{I}_s \otimes H \right)$.

Example 1. Let $|X| = 2^2$ such that $X = \{1, 2, 3, 4\}$. The underlying parallel classes can be represented as follows.

$$C_0 : \{1, 2\}, \{3, 4\}, C_1 : \{1, 3\}, \{2, 4)\}, C_2 : \{1, 4\}, \{2, 3\}\}.$$

Consider the two dimensional Hadamard matrix as $H = \frac{1}{\sqrt{2}} \begin{smallmatrix} 1 & 1 \\ 1 & -1 \end{smallmatrix}$. Therefore the sparse MUBs are denoted by

$$M_0 = P_0 \cdot M_0 = \begin{pmatrix} 1 & 0 & 0 & 0 \\ 0 & 1 & 0 & 0 \\ 0 & 0 & 1 & 0 \\ 0 & 0 & 0 & 1 \end{pmatrix} \frac{1}{\sqrt{2}} \begin{pmatrix} 1 & 1 & 0 & 0 \\ 1 & -1 & 0 & 0 \\ 0 & 0 & 1 & 1 \\ 0 & 0 & 1 & -1 \end{pmatrix} = \frac{1}{\sqrt{2}} \begin{pmatrix} 1 & 1 & 0 & 0 \\ 1 & -1 & 0 & 0 \\ 0 & 0 & 1 & 1 \\ 0 & 0 & 1 & -1 \end{pmatrix} = P_0 \left(\mathbb{I} \otimes H \right),$$

$$M_1 = P_1 \cdot M_0 = \begin{pmatrix} 1 & 0 & 0 & 0 \\ 0 & 0 & 1 & 0 \\ 0 & 1 & 0 & 0 \\ 0 & 0 & 0 & 1 \end{pmatrix} \frac{1}{\sqrt{2}} \begin{pmatrix} 1 & 1 & 0 & 0 \\ 1 & -1 & 0 & 0 \\ 0 & 0 & 1 & 1 \\ 0 & 0 & 1 & -1 \end{pmatrix} = \frac{1}{\sqrt{2}} \begin{pmatrix} 1 & 1 & 0 & 0 \\ 0 & 0 & 1 & 1 \\ 1 & -1 & 0 & 0 \\ 0 & 0 & 1 & -1 \end{pmatrix} = P_1 \left(\mathbb{I} \otimes H \right),$$

$$M_2 = P_2 \cdot M_0 = \begin{pmatrix} 1 & 0 & 0 & 0 \\ 0 & 0 & 1 & 0 \\ 0 & 0 & 0 & 1 \\ 0 & 1 & 0 & 0 \end{pmatrix} \frac{1}{\sqrt{2}} \begin{pmatrix} 1 & 1 & 0 & 0 \\ 1 & -1 & 0 & 0 \\ 0 & 0 & 1 & 1 \\ 0 & 0 & 1 & -1 \end{pmatrix} = \frac{1}{\sqrt{2}} \begin{pmatrix} 1 & 1 & 0 & 0 \\ 0 & 0 & 1 & 1 \\ 0 & 0 & 1 & -1 \\ 1 & -1 & 0 & 0 \end{pmatrix} = P_2 \left(\mathbb{I} \otimes H \right).$$

Using the above matrices, we obtain the following set of two Mutually Unbiased Hadamard matrices:

$$H_1 = M_0^\dagger \cdot M_1 = \frac{1}{\sqrt{2}} \begin{pmatrix} 1 & 1 & 0 & 0 \\ 1 & -1 & 0 & 0 \\ 0 & 0 & 1 & 1 \\ 0 & 0 & 1 & -1 \end{pmatrix}^\dagger \frac{1}{\sqrt{2}} \begin{pmatrix} 1 & 1 & 0 & 0 \\ 0 & 0 & 1 & 1 \\ 1 & -1 & 0 & 0 \\ 0 & 0 & 1 & -1 \end{pmatrix} = \frac{1}{2} \begin{pmatrix} 1 & 1 & 1 & 1 \\ 1 & 1 & -1 & -1 \\ 1 & -1 & 1 & -1 \\ 1 & -1 & -1 & 1 \end{pmatrix},$$

$$H_2 = M_0^\dagger \cdot M_2 = \frac{1}{\sqrt{2}} \begin{pmatrix} 1 & 1 & 0 & 0 \\ 1 & -1 & 0 & 0 \\ 0 & 0 & 1 & 1 \\ 0 & 0 & 1 & -1 \end{pmatrix}^\dagger \frac{1}{\sqrt{2}} \begin{pmatrix} 1 & 1 & 0 & 0 \\ 0 & 0 & 1 & 1 \\ 0 & 0 & 1 & -1 \\ 1 & -1 & 0 & 0 \end{pmatrix} = \frac{1}{2} \begin{pmatrix} 1 & 1 & 1 & 1 \\ 1 & 1 & -1 & -1 \\ 1 & -1 & 1 & -1 \\ -1 & 1 & 1 & -1 \end{pmatrix}.$$

Along with Identity Matrix I_4, we obtain a set of 3 MUBs.

2.1   Introducing the Parameters

Next, we show that using the above method we can introduce affine parameters in a set of w many mutually unbiased Hadamard matrices, so constructed thereby demonstrating the existence of a parametric form of mutually unbiased bases, in every dimension of the form $d = s^2$. We do this by exploiting the fact that $\mathbb{H}$ is block diagonal matrix where each block is a Hadamard matrix of order s and noting that $D_i(\theta)\mathbb{H}$ is also blocked Hadamard matrix, where $D_i(\theta)$ is a diagonal unitary matrix, with diagonal entries of the form $\exp(\iota\theta_i)$, where θ_i is a independent parameter.

Thus using above we have $\mathbb{M}_i = P_i D_i(\theta)\mathbb{H}$. Therefore,

$$\mathbb{M}_i^\dagger \mathbb{M}_j = \mathbb{H}_0^\dagger D_i^\dagger(\theta) P_i^T P_j D_j(\theta)\mathbb{H}.$$

Note that $P^\dagger = P^T$ for any permutation matrix P. Note that the product of two permutations is another permutation and the product of two diagonal matrices is again a diagonal matrix. Thus using the Lemma 3 above, the term $D_i(\theta)P_i^T P_j D_j(\theta)$ can be written as $\tilde{P}_j \tilde{D}_j(\theta)$, where $\tilde{D}_j(\theta)$ is a unitary diagonal matrix having diagonal entries of the form $\exp(\iota\theta_i)$, for some θ_i where θ_i's are independent parameters and $\tilde{P}_j$ is some permutation matrix. Hence total independent parameters are equal to the dimension of the matrix which is s^2. Thus, the set of Mutually Unbiased Hadamard matrices are

$$\{H_1 = \mathbb{M}_0^\dagger \mathbb{M}_1, H_2 = \mathbb{M}_0^\dagger \mathbb{M}_2, \ldots, H_w = \mathbb{M}_0^\dagger \mathbb{M}_w\},$$

where

$$H_i = \mathbb{M}_0^\dagger \mathbb{M}_i = \mathbb{H}^\dagger \tilde{P}_j \tilde{D}_j(\theta)\mathbb{H}.$$

Now referring to Corollary 1, if D is a block diagonal matrix, such that each block matrix is $\alpha_r I_s$, where I_s is the identity matrix of order s then D commutes with block diagonal matrix H, which contain block matrix of size s. Hence

$$\tilde{D}_j(\theta) = \tilde{D}_{j1}(\theta)\tilde{D}_{j2}(\theta),$$

where, $\tilde{D}_{j2}(\theta)$ is a block diagonal matrix where each block is of form $\exp(\iota\theta_j)I_s$ and $\tilde{D}_{j1}(\theta)$ is block diagonal matrix having diagonal entries of the form $\exp(\iota\theta_j)$. Hence from Corollary 1, $\tilde{D}_{j2}(\theta)$ will commute with $\mathbb{H}$. Thus

$$H_i = \mathbb{H}^\dagger \tilde{P}_j \tilde{D}_j(\theta)\mathbb{H} = \mathbb{H}^\dagger \tilde{P}_j \tilde{D}_{j1}(\theta)\tilde{D}_{j2}(\theta)\mathbb{H} = \mathbb{H}^\dagger \tilde{P}_j \tilde{D}_{j1}(\theta)\mathbb{H}\tilde{D}_{j2}(\theta).$$

Since all the θ_j's are free parameters, we can absorb s many of them in $\tilde{D}_{j2}(\theta)$. Further, multiplying the Unitary Diagonal Matrix from left to an MUB matrix does not affect the equivalence of the MUBs, as it corresponds to multiplying an MUB vector with some arbitrary phase. Thus

$$H_i \equiv \mathbb{H}^\dagger \tilde{P}_j \tilde{D}_{j1}(\theta) \mathbb{H},$$

where the number of independent parameters become

$$s^2 - s = s(s-1).$$

Furthermore, note that such set $\{I_0, H_1, \ldots, H_w\}$ also forms a class of MUBs for $\mathbb{C}^d$. Thus, in this process, we also construct a class of $MOLS(s) + 2$ many affine-parametric MUBs for dimension $d = s^2$ (We say affine parameters because all the parameters (variables) are linear). One should note that, while determining the number of free parameters, we can pull out the redundant parameters only from columns, and not from the rows, otherwise the MUB-structures will get destroyed. Therefore, each such basis matrix, except Identity (I_0) contains at least $s(s-1)$ many free parameters.

Further note that while constructing block diagonal Hadamard matrices $\mathbb{H}$ in Construction 2, we had the liberty to choose s many Hadamard matrices of size s. Now if each of these Hadamard matrices contains r_i^k many free parameters, $i = 1, 2, \ldots, s$, then the total number of free parameters in each of these bases, $\mathbf{H}_k$ would further be increased by $\left(\sum_{i=1}^{s} r_i^k\right)$. Moreover, the $\mathbf{H}_k \equiv \mathbb{H}_0^\dagger \tilde{P}_k \tilde{D}_{k1}(\theta) \mathbb{H}_k$. The free parameter in $\mathbb{H}_0^\dagger$, by virtue of the construction, would be common to all the $\mathbf{H}_k$'s. Thus each $\mathbf{H}_k$ would have the number of independent parameters given by

$$\left(\sum_{i=1}^{s} r_i^0 + \sum_{i=1}^{s} r_i^k\right) + s(s-1),$$

where the $\sum_{i=1}^{s} r_i^0$ many parameters would be common to all the MUHM. Thus our main contribution in this regard can be summarized in the following theorem.

Theorem 2. *For dimension $d = s^2$ let $w = MOLS(s) + 1$, there exists a set of MUBs $\{I, H_1, H_2, \ldots, H_w\}$ consisting of the identity matrix and the MUHMs, such that each Hadamard matrix H_i have at least $s(s-1)$ many independent affine parameters, that cannot be absorbed by a global unitary operation.*

Proof. The proof follows from discussion above, The fact that the parameters cannot be absorbed by any global unitary operation is clear when one note that each of the affine parameters in H_l is independent from any other affine parameters of say H_k, coupled with the fact that I Identity matrix is also present in the set, hence these parameters cannot be absorbed by any global unitary operations. $\square$

Now let us illustrate this with an example for $d = 4$. The matrices $\mathbb{M}_0^\dagger \mathbb{M}_1$ and $\mathbb{M}_0^\dagger \mathbb{M}_2$ can be made affine parametric MUBs, each having $2(2-1) = 2$ free parameters, by pulling out parameters only from the columns and not from the rows, in the following way.

$$\mathbb{M}_0^\dagger \mathbb{M}_1 = \frac{1}{2} \begin{pmatrix} e^{i\theta_1} & e^{i\theta_1} & e^{i\theta_2} & e^{i\theta_2} \\ e^{i\theta_1} & e^{i\theta_1} & -e^{i\theta_2} & -e^{i\theta_2} \\ e^{i\theta_3} & -e^{i\theta_3} & e^{i\theta_4} & -e^{i\theta_4} \\ e^{i\theta_3} & -e^{i\theta_3} & -e^{i\theta_4} & e^{i\theta_4} \end{pmatrix} = \frac{1}{2} \begin{pmatrix} 1 & 1 & 1 & 1 \\ 1 & 1 & -1 & -1 \\ e^{i\alpha} & -e^{i\alpha} & e^{i\beta} & -e^{i\beta} \\ e^{i\alpha} & -e^{i\alpha} & -e^{i\beta} & e^{i\beta} \end{pmatrix},$$

$$\mathbb{M}_0^\dagger \mathbb{M}_2 = \frac{1}{2} \begin{pmatrix} e^{i\phi_1} & e^{i\phi_1} & e^{i\phi_2} & e^{i\phi_2} \\ e^{i\phi_1} & e^{i\phi_1} & -e^{i\phi_2} & -e^{i\phi_2} \\ e^{i\phi_4} & -e^{i\phi_4} & e^{i\phi_3} & -e^{i\phi_3} \\ -e^{i\phi_4} & e^{i\phi_4} & e^{i\phi_3} & -e^{i\phi_3} \end{pmatrix} = \frac{1}{2} \begin{pmatrix} 1 & 1 & 1 & 1 \\ 1 & 1 & -1 & -1 \\ e^{i\gamma} & -e^{i\gamma} & e^{i\delta} & -e^{i\delta} \\ -e^{i\gamma} & e^{i\gamma} & e^{i\delta} & -e^{i\delta} \end{pmatrix}.$$

The matrices $\{I_d, \mathbb{M}_0^\dagger \mathbb{M}_1(\alpha, \beta), \mathbb{M}_0^\dagger \mathbb{M}_2(\gamma, \delta)\}$ form a class of three affine parametric MUBs for dimension 4. Other than the identity matrix, there are 2 free parameters in each of them. Note that here $4 = 2^2$, for which we have $2 + 1 = 3$ affine parametric MUBs. Thus when s is some prime power, we have following corollary.

Corollary 2. *Consider $d = q^2$, where q is some prime power. Then there exists a set of q many MUHMs each having atleast $q(q-1)$ independent affine parameters that cannot be absorbed by a global unitary operation.*

2.2   Parametric Class of Hadamard Matrices

As noted above, the set $\{H_1 = \mathbb{M}_0^\dagger \mathbb{M}_1, H_2 = \mathbb{M}_0^\dagger \mathbb{M}_2, \dots, H_w = \mathbb{M}_0^\dagger \mathbb{M}_w\}$ forms Mutually Unbiased Hadamard matrices, i.e., each H_i is a Hadamard matrix. Each of these matrices has $s(s-1)$ many free parameters. Since the Hadamard property of a matrix remains unaffected even when the rows are multiplied by some arbitrary phase, it can be used to reduce further the free parameter, by multiplying suitable diagonal unitary matrix from left. Consider any H_i as given above, i.e., $H_i \equiv \mathbb{H}^\dagger \tilde{P}_j \tilde{D}_{j1}(\theta)\mathbb{H}$. Now $\tilde{P}_j \tilde{D}_{j1}(\theta) = \bar{D}_{j1}(\theta)\tilde{P}_j$, where $\bar{D}_{j1}(\theta)$ is a diagonal unitary matrix having same entries as $\tilde{D}_{j1}(\theta)$. However, the first entry of $\bar{D}_{j1}(\theta)\tilde{P}_j$ is also 1. This is because the $\tilde{P}_j$ is a permutation matrix with the first entry as 1. Now as done in the case of MUB, for further reducing the parameters, we expressed the $\tilde{D}_j$ and product of two diagonal matrix viz $\tilde{D}_{j1}\tilde{D}_{j2}$ such that $\tilde{D}_{j2}$ commute with $\mathbb{H}$. Hence we again express $\tilde{D}_{j1} = \tilde{D}_{j3}\tilde{D}_{j4}$ such that $\tilde{D}_{j3}$ commutes with $\mathbb{H}_0^\dagger$. Thus, by choosing suitable $\tilde{D}_{j3}$, we can reduce the number of affine parameters remaining in $\tilde{D}_{j4}$. However, the first entry of $\tilde{D}_{j3}\tilde{P}_j$ is also 1. This is because the $\tilde{P}_j$ is a permutation matrix with the first entry $(\tilde{P}_j)_{11} = 1$. Hence only $s-1$ independent parameters can be chosen, using which one can further reduce $s-1$ parameters in $\tilde{D}_{j4}$. Hence number of independent affine parameters are $s(s-1) - (s-1) = (s-1)^2$.

Continuing with our example of $|X| = 2^2$, we have the following matrices:

$$\mathbb{M}_0^\dagger \mathbb{M}_0 = M_0^\dagger D(\theta) P_0 M_0 = M_0^\dagger D(\theta) M_0 = \mathbb{I}_d,$$

$$M_0^\dagger M_1 = M_0^\dagger D(\theta) P_1 M_0 = M_0^\dagger D(\theta) M_1.$$

Assuming the parameters in $D(\theta)$ as $\theta_1, \theta_2, \theta_3$ and θ_4, we obtain

$$M_0^\dagger M_1 = \frac{1}{2} \begin{pmatrix} e^{i\theta_1} & e^{i\theta_1} & e^{i\theta_2} & e^{i\theta_2} \\ e^{i\theta_1} & e^{i\theta_1} & -e^{i\theta_2} & -e^{i\theta_2} \\ e^{i\theta_3} & -e^{i\theta_3} & e^{i\theta_4} & -e^{i\theta_4} \\ e^{i\theta_3} & -e^{i\theta_3} & -e^{i\theta_4} & e^{i\theta_4} \end{pmatrix} = \frac{1}{2} \begin{pmatrix} 1 & 1 & 1 & 1 \\ 1 & 1 & -1 & -1 \\ 1 & -1 & e^{i\theta} & -e^{i\theta} \\ 1 & -1 & -e^{i\theta} & e^{i\theta} \end{pmatrix},$$

and $M_0^\dagger M_2 = M_0^\dagger D(\phi) P_2 M_0 = M_0^\dagger D(\phi) M_2.$

Again, assuming the parameters in $D(\phi)$ as ϕ_1, ϕ_2, ϕ_3 and ϕ_4, we obtain

$$M_0^\dagger M_2 = \frac{1}{2} \begin{pmatrix} e^{i\phi_1} & e^{i\phi_1} & e^{i\phi_2} & e^{i\phi_2} \\ e^{i\phi_1} & e^{i\phi_1} & -e^{i\phi_2} & -e^{i\phi_2} \\ e^{i\phi_4} & -e^{i\phi_4} & e^{i\phi_3} & -e^{i\phi_3} \\ -e^{i\phi_4} & e^{i\phi_4} & e^{i\phi_3} & -e^{i\phi_3} \end{pmatrix} = \frac{1}{2} \begin{pmatrix} 1 & 1 & 1 & 1 \\ 1 & 1 & -1 & -1 \\ 1 & -1 & e^{i\phi} & -e^{i\phi} \\ -1 & 1 & e^{i\phi} & -e^{i\phi} \end{pmatrix}.$$

One can now check that the matrices $M_0^\dagger M_1$ and $M_0^\dagger M_2$ form a class of affine parametric Hadamard matrices, each having $(2-1)(2-1) = 1$ free parameter. However, they might not be mutually unbiased, since the inner product between the ith column of $M_0^\dagger M_1$ and the jth column of $M_0^\dagger M_2$ may not be equal to $\frac{1}{2}$ in certain cases. Thus, here we need additional efforts.

It is to be noted that the method described in Construction 2 enables us to construct the following interesting class of MUBs in square dimension.

Corollary 3. *In square dimension $d = s^2$, we can have a set of $MOLS(s) + 2$ many affine-parametric MUBs having the same eigenvalues and components of eigen-vectors are permutations of one of the MUBs.*

Proof. Following Construction 2, we begin with an RBD (X, C) with $|X| = s^2$ and $C = \{C_0, C_1, \dots, C_w\}$ representing the parallel classes. Corresponding to each parallel class, we can construct a class of sparse MUBs $\{M_0, M_1, \dots, M_w\}$, such that $M_i = P_i (\mathbb{I}_s \otimes H)$, where P_i represents the permutation matrix and H is a Hadamard matrix as defined earlier. Observe that, if M_i, M_j are MUBs, so are $M_i P_i^T$ and $M_j P_j^T$, since

$$\left(M_i P_i^T \right)^\dagger M_j P_j^T = \left(P_i^T \right)^\dagger M_i^\dagger M_j P_j^T.$$

Now, consider the set of MUBs $\{M_0 P_0^T, M_1 P_1^T, \dots, M_w P_w^T\}$. Note that the eigenvalues of $(\mathbb{I}_s \otimes H)$ are the same as the eigenvalues (each eigenvalue has multiplicity as order of H) of H and multiplication by a permutation matrix does not change the eigenvalues of the resultant matrix, it only changes the eigenvectors.

Suppose, λ denotes the eigenvalues of $(\mathbb{I}_s \otimes H)$ with an eigenvector $\mathbf{Y}$, i.e.

$$(\mathbb{I}_s \otimes H) \mathbf{Y} = \lambda \mathbf{Y}.$$

Now $M_i P_i^T = P_i (\mathbb{I}_s \otimes H) P_i^T$ will also have the same eigenvalues, irrespective of the choice of permutation matrices. Suppose the corresponding eigenvector is given by $\mathbf{Z}$, i.e.,

$$P_i (\mathbb{I}_s \otimes H) P_i^T \mathbf{Z} = \lambda \mathbf{Z} \Rightarrow P_i^T \left[P_i (\mathbb{I}_s \otimes H) P_i^T \mathbf{Z} \right] = \lambda P_i^T \mathbf{Z}$$
$$\Rightarrow (\mathbb{I}_s \otimes H) \left(P_i^T \mathbf{Z} \right) = \lambda \left(P_i^T \mathbf{Z} \right).$$

Thus, $\mathbf{Y} = \left(P_i^T \mathbf{Z} \right) \Rightarrow \mathbf{Z} = P_i \mathbf{Y}$. Hence, the class of $MOLS(s) + 2$ MUBs $\{ M_0 P_0^T, M_1 P_1^T, \ldots, M_w P_w^T \}$ have the same eigenvalues and the corresponding eigenvectors are given by $P_i \mathbf{Y}$, where $\mathbf{Y}$ is the eigenvector of M_0 corresponding to the same eigenvalue. □

Example 2. Let us continue with the earlier example of $d = 2^2$. Here,

$$M_0 = \frac{1}{\sqrt{2}} \begin{pmatrix} 1 & 1 & 0 & 0 \\ 1 & -1 & 0 & 0 \\ 0 & 0 & 1 & 1 \\ 0 & 0 & 1 & -1 \end{pmatrix}$$

and the set of eigenvalues of M_0 is given by ± 1. One of the eigen-vectors corresponding to the eigenvalue 1 is given by $\mathbf{Y} = \left(0, 0, 1 + \sqrt{2}, 1 \right)^T$. The matrix $M_1 P_1^T$ is given by

$$\begin{pmatrix} 1 & 0 & 1 & 0 \\ 0 & 1 & 0 & 1 \\ 1 & 0 & -1 & 0 \\ 0 & 1 & 0 & -1 \end{pmatrix}$$

and the eigen-vector of $M_1 P_1^T$ corresponding to the eigenvalue 1 is given by $\mathbf{Z} = \left(0, 1 + \sqrt{2}, 0, 1 \right)^T = P_1 \mathbf{Y}$. This follows from the previous corollary.

2.3   Comparison with [6]

In this initiative, we construct a class of $MOLS(s) + 2$ many affine-parametric MUBs for dimension $d = s^2$, each having $s(s - 1)$ many free parameters other than the identity matrix. In an example above corresponding to Construction 2 (the RBD construction), we presented 3 real MUBs in dimension 4. After applying the appropriate unitary transformation, we obtained one identity matrix and two real Hadamard matrices. Each Hadamard matrix contains at least 2 parameters, giving a total of 4 counting both matrices. We could also prepare examples with exactly four parameters in total. This is similar to the result of Goyeneche et al. [6], which also introduced 4 free parameters in dimension 4.

For dimension 16, our approach introduces at least 48 parameters compared to 64 as in [6]. Here the work of [6] is having an advantage, because there are nine real MUBs (maximum possible real MUBs in dimension 16), and our RBD construction provides only 5. We could introduce at least 12 parameters in each MUB, whereas it is only 8 in each MUB of dimension 16 for [6]. In any dimension other than a power of four, i.e., 4^k, our method introduces higher number of parameters than [6], because our construction of introducing parameters is not based on the existence of real ones, and most dimensions generally admit at most two real MUBs [2].

In dimension 9, we have 4 MUBs through our RBD construction and we can introduce at least 6 parameters in each MUB making a total of at least 18 parameters in three, considering the first one as identity. So, we have introduced parameters in every dimension irrespective of whether we have real or complex MUBs through MOLS construction. In this dimension, other than identity, no real MUB is available, and hence the construction of [6] cannot be applied at all. Thus, our result of Theorem 2 provides a broader class with more parameters in general and this is an improvement over [6].

2.4 Towards Generalizing the Idea Towards Construction of Affine Parametric Class of Hadamard Matrices for $d = k \times s$

The Resolvable Block Design (RBD) approach enables us to construct at least MOLS(s) + 2 many MUBs in $d = s^2$, because for this situation we have RBD(X, A) which has MOLS(s) + 2 parallel classes, where each parallel class consists of constant block size (s) and any pair of blocks from different parallel classes have exactly one point in common, i.e., $\mu = 1$. Further $\beta = 1$, which is the bound on absolute value of the dot product between any pair of vectors from different bases. Such properties cannot be achieved when dimension is not square, i.e., $d = k \times s$ and $k \neq s$.

In such a situation one can always construct RBD(X, A) which has at least two parallel classes having $\mu = 1$ and $\beta = 1$. For this we provide the following technique for constructing RBD(X, A) with $|X| = d = k \times s$, having two parallel class namely $\mathcal{P}_k$ and $\mathcal{P}_s$, where $\mathcal{P}_k$ consists of s blocks each of constant size k and $\mathcal{P}_s$ consists of k blocks each of constant size s such that any pair of block from different parallel classes, has exactly one element in common.

Construction 3. Consider $d = k \times s$, where k and s are positive integers.

1. Let the elements of $X = \{1, 2, \ldots, s, s+1, \ldots, 2s, 2s+1 \ldots, ks\}$
2. Define $\mathcal{P}_s = \{\mathcal{B}_1^s = \{1, 2, \ldots, s\}, \mathcal{B}_2^s = \{s+1, s+2, \ldots, 2s\}, \ldots, \mathcal{B}_k^s = \{(k-1)s+1, (k-1)s+2, \ldots, ks\}$. Hence, $\mathcal{P}_s$ is a parallel class having k many blocks $\{\mathcal{B}_i^s\}_{i=1,2,\ldots,k}$, where each block has exactly s elements.
3. Now construct $\mathcal{P}_k$ having s many blocks $\{\mathcal{B}_i^k\}_{i=1,2,\ldots,s}$, such that every block has exactly one element from each $\mathcal{B}_i^s$, hence each $\{\mathcal{B}_i^k\}_{i=1,2\ldots s}$ is of size k. Hence $\mathcal{P}_k$ can be constructed in large number of ways. One simple way would be $\mathcal{P}_k = \{\mathcal{B}_1^k = \{1, s+1, \ldots, (k-1)s+1\}, \mathcal{B}_2^k = \{2, s+2, \ldots, (k-1)s+2\}, \ldots, \mathcal{B}_s^k = \{s, 2s, \ldots, ks\}\}$.

Using these two parallel classes, we can obtain at least a pair of MUBs. Note that once P_s is fixed, there are $\sum_j^s j^k$ many different P_k's possible, which in turn may be exploited to get obtain an affine parametric Hadamard matrix. However, examining their equivalence appears to be very challenging task. To characterize, the constructed affine parametric Hadamard matrix, using $\text{RBD}(X, \{\mathcal{P}_s, \mathcal{P}_k\})$, we prove following lemma, which essentially reproduces the results of [4], in terms of number of affine parameters. However, further analysis is required to see if Hadamard matrices are equivalent to the one provided in [4].

Lemma 5. *For dimension* $d = k \times s$, *there exists a Hadamard matrix* H_i *having at least* $(k-1)(s-1)$ *many affine parameters, such that first row and first column consist of 1.*

Proof. Using Construction 2, with RBD having only two parallel classes, and applying the diagonal unitary matrix $D(\theta)$ as in Theorem 2, we obtain the set

$$\{\mathbb{I}, H_1 = \mathbb{M}_0^\dagger \mathbb{M}_1 \equiv \mathbb{H}_0^\dagger \widetilde{P}_j \widetilde{D}_{j1}(\theta) \mathbb{H}_1\}.$$

Here H_i is a Hadamard matrix with at least $d - k = k(s - 1)$ many free parameters. Since the Hadamard property of a matrix remains unaffected even when the rows are multiplied by some arbitrary phase, we use this to further reduce the number of free parameters, by multiplying suitable diagonal unitary matrix from left.

Consider H_1 as given above, i.e., $H_1 \equiv \mathbb{H}_0^\dagger \widetilde{P}_j \widetilde{D}_{j1}(\theta) \mathbb{H}_1$. Now $\widetilde{P}_j \widetilde{D}_{j1}(\theta) = \bar{D}_{j1}(\theta) \widetilde{P}_j$, where $\bar{D}_{j1}(\theta)$ is a diagonal unitary matrix having same entries as $\widetilde{D}_{j1}(\theta)$. Since we can also multiply the phase factor to the rows of the H_1, we use this to further reduce the number of affine parameters, that are introduced because of $\widetilde{D}_{j1}(\theta)$. Let us write $\bar{D}_{j1}(\theta) = \bar{D}_{j1}^1(\theta) \bar{D}_{j1}^2(\theta)$, such that $\bar{D}_{j1}^1(\theta)$ is a block diagonal unitary matrix, where block diagonals are of the form $\exp(\iota \alpha_r) I_k$, where $r = 1, 2, \ldots, s$. Hence the matrix $\bar{D}_{j1}^1(\theta)$ will commute with $\mathbb{H}_0^\dagger$. Thus, by choosing suitable α_r we can reduce the number of affine parameters remaining in $\bar{D}_{j1}^2(\theta)$. However, the first entry of $\bar{D}_{j1}(\theta) \widetilde{P}_j$ is also 1. This is because the $\widetilde{P}_j$ is a permutation matrix with the first entry $(\widetilde{P}_j)_{11} = 1$. Thus $\alpha_1 = 0$. Hence only $s - 1$ independent α_r's can be chosen, using which one can further reduce $s - 1$ parameters. Hence number of independent affine parameters are $k(s - 1) - (s - 1) = (k-1)(s-1)$. That the first row and the first column of $H_1 \equiv \mathbb{H}_0^\dagger \bar{D}_{j1}^2(\theta) \widetilde{P}_j \mathbb{H}_1$ follows from fact that each block of $\mathbb{H}_0$ and $\mathbb{H}_1$ have first row and first column having 1, along with the fact that $(\widetilde{P}_j)_{11} = 1$. $\qquad\square$

As done previously, while choosing the Hadamard matrices for $\mathbb{H}_0$, if they have $\{k_i\}_{i=1,\ldots,s}$ many independent parameters, then the constructed Hadamard matrix of order d will have extra parameters. Similarly for $\mathbb{H}_1$ if the Hadamard matrices have $\{r_i\}_{i=1,\ldots s}$ many independent parameters, then the maximum number of free parameters in each H_1 would be given by $\left(\sum_{i=1}^k k_i + \sum_{i=1}^s r_i\right) + (k - 1)(s - 1)$. One should also refer to [12, Sect. 2.2], [14, Sect. 4.5], [5] and [7, Proposition 2.9] in this direction.

For example, using two parallel classes for $d = 2 \times 3 = 6$, from above analysis we get $(k - 1)(s - 1) = 2$ parameters corresponding to the Hadamard matrix.

For the RBD having two parallel classes constructed for $d = 6$, we obtain the resultant parametric Hadamard matrix as

$$
\mathrm{M}_0^\dagger \mathrm{M}_1 (\alpha, \beta) = \frac{1}{\sqrt{6}}
\begin{pmatrix}
1 & 1 & 1 & 1 & 1 & 1 \\
1 & 1 & 1 & -1 & -1 & -1 \\
1 & \omega & \omega^2 & e^{i\alpha} & \omega e^{i\alpha} & \omega^2 e^{i\alpha} \\
1 & \omega & \omega^2 & -e^{i\alpha} & -\omega e^{i\alpha} & -\omega^2 e^{i\alpha} \\
1 & \omega^2 & \omega & e^{i\beta} & \omega^2 e^{i\beta} & \omega e^{i\beta} \\
1 & \omega^2 & \omega & -e^{i\beta} & -\omega^2 e^{i\beta} & -\omega e^{i\beta}
\end{pmatrix} .
$$

Thus $\{\mathrm{I}, \mathrm{M}_0^\dagger \mathrm{M}_1 (\alpha, \beta)\}$ form a pair of MUBs in $d = 6$ with 2 parameters.

3   Conclusion

In this paper, we consider the introduction of parameters in MUBs. The MUBs have applications in several areas of quantum information and cryptology. Thus, such parameterization may provide better flexibility in choosing form a larger class of options. Our results show improvements over the work of [6] in terms of the number of MUBs as well as parameters for most of the dimensions $d = s^2$. Further, we also present directions to generalize this for dimensions of the form $d = k \times s$. As we are discussing about free parameters, that cannot be absorbed by a global unitary operation, we need to analyze this more critically. In this regard, we require a closer look in terms of equivalence between two sets of MUBs.

Acknowledgments. The authors would like to thank the anonymous reviewers for their detailed comments that improved the editorial as well as technical presentation of the paper. The last author (Subhamoy Maitra) acknowledges the support of MeitY, Government of India, related to the initiative "Cluster - Cryptography, Information Security Education and Awareness (ISEA) Project Phase - III".

References

1. Bennett, C.H., Brassard, G.: Quantum cryptography: public key distribution and coin tossing. In: Proceedings of IEEE International Conference on Computers, Systems and Signal Processing, pp. 175–179 (1984)
2. Boykin, P.O., Sitharam, M., Tarifi, M., Wocjan, P.: Real mutually unbiased bases. arXiv preprint arXiv:quant-ph/0502024 (2005)
3. Bruß, D.: Optimal eavesdropping in quantum cryptography with six states. Phys. Rev. Lett. **81**(14), 3018–3021 (1998)
4. Dita, P.: Some results on the parametrization of complex Hadamard matrices. J. Phys. A: Math. Gen. **37**(20), 5355 (2004)
5. Goyeneche, D.: A new method to construct families of complex Hadamard matrices in even dimensions. J. Math. Phys. **54**, 032201 (2013). https://doi.org/10.1063/1.4794068
6. Goyeneche, D., Gomez, S.: Mutually unbiased bases with free parameters. Phys. Rev. A **92**(6), 23–25 (2015). https://doi.org/10.1103/PhysRevA.92.062325

7. Haagerup, U.: Orthogonal maximal Abelian-subalgebras of the $n \times n$ matrices and cyclic n-roots. Odense Universitet (1996)
8. Hedayat, A.S., Sloane, N.J.A., Stufken, J.: Orthogonal Arrays: Theory and Applications. Springer (2012)
9. Kumar, A., Maitra, S.: Resolvable block designs in construction of approximate real MUBs that are sparse. Cryptogr. Commun. (2021)
10. Kumar, A., Maitra, S., Roy, S.: Almost Perfect mutually unbiased bases that are sparse. arXiv preprint arXiv:2402.03964 (2024)
11. Kumar, A., Maitra, S.: Further constructions of AMUBs for non-prime power composite dimensions. arXiv preprint arXiv:2402.04231 (2024)
12. McNulty, D., Weigert, S.: Isolated Hadamard matrices from mutually unbiased product bases. J. Math. Phys. **53**(12) (2012)
13. Schwinger, J.: Unitary operator bases. Proc. Natl. Acad. Sci. U.S.A. **46**(4), 570–579 (1960). https://doi.org/10.1073/pnas.46.4.570
14. Tadej, W., Zyczkowski, K.: A concise guide to complex Hadamard matrices. Open Syst. Inf. Dyn. **13**(2), 133–177 (2006)
15. Wootters, W.K., Fields, B.D.: Optimal state-determination by mutually unbiased measurements. Ann. Phys. **191**, 363–381 (1989). https://doi.org/10.1016/0003-4916(89)90322-9
16. Wocjan, P., Beth, T.: New construction of mutually unbiased bases in square dimensions. arXiv preprint arXiv:quant-ph/0407081 (2004)
17. Weiner, M.: A gap for the maximum number of mutually unbiased bases. Proc. Am. Math. Soc. **141**(6), 1963–1969 (2013)
18. Zauner, G.: Quantum designs: foundations of a non-commutative design theory. Int. J. Quant. Inf. **9**(01), 445–507 (2011)

Author Index

© The Editor(s) (if applicable) and The Author(s), under exclusive license
to Springer Nature Switzerland AG 2025
S. Mukhopadhyay and P. Stănică (Eds.): INDOCRYPT 2024, LNCS 15495, pp. 375–376, 2025.
https://doi.org/10.1007/978-3-031-80308-6

GPSR Compliance

The European Union's (EU) General Product Safety Regulation (GPSR) is a set of rules that requires consumer products to be safe and our obligations to ensure this.

If you have any concerns about our products, you can contact us on ProductSafety@springernature.com

In case Publisher is established outside the EU, the EU authorized representative is:

Springer Nature Customer Service Center GmbH
Europaplatz 3
69115 Heidelberg, Germany

The manufacturer's authorised representative in the EU is Springer
Nature Customer Service Centre GmbH, Europaplatz 3, 69115 Heidelberg,
Germany. If you have any concerns regarding our products, please
contact ProductSafety@springernature.com

Printed and bound by CPI Group (UK) Ltd, Croydon, CR0 4YY
29/04/2026
02099537-0013